10/24/87 ✓

D1118611

MECHANICS OF
MATERIALS
BORES.

PRINCIPLES OF DYNAMICS

Prentice-Hall International Series in Dynamics

Y. C. FUNG, *Editor*

PRENTICE-HALL, INC.
PRENTICE-HALL INTERNATIONAL, INC., UNITED KINGDOM AND EIRE
PRENTICE-HALL OF CANADA, LTD., CANADA

PRINCIPLES OF DYNAMICS

by

Donald T. Greenwood

Professor, Department of Aeronautical
and Astronautical Engineering
University of Michigan

PRENTICE-HALL, INC.
Englewood Cliffs, New Jersey

PRENTICE-HALL INTERNATIONAL, INC., *London*
PRENTICE-HALL OF AUSTRALIA, PTY., LTD., *Sydney*
PRENTICE-HALL OF CANADA, LTD., *Toronto*
PRENTICE-HALL OF INDIA (PRIVATE) LTD., *New Delhi*
PRENTICE-HALL OF JAPAN, INC., *Tokyo*

© 1965 by Prentice-Hall, Inc., Englewood Cliffs, N.J.
All rights reserved. No part of this book may be repro-
duced in any form, by mimeograph or any other means,
without permission in writing from the publisher. Library
of Congress Catalog Card Number 64–19700.

Current printing (last digit):
18 17 16 15 14 13 12

Printed in the United States of America
70897—C

PREFACE

Although there has been a steady improvement in the depth of most introductory courses in dynamics during recent years, the place of an additional
course at the intermediate level remains an important one. A course at this
level is normally taken by first-year graduate students or by undergraduate
seniors. The greater maturity of these students permits the presentation of
the subject from a more advanced viewpoint, with additional mathematical
knowledge assumed, and with the inclusion of illustrative examples having
more than the usual complexity. Through the study of the general theory and
its application in these examples, and by the solution of a variety of problems
of comparable difficulty, most students can attain a real competence in
dynamics.

This textbook has evolved from a set of notes which accompany the first
of three courses forming a sequence in the general area of flight mechanics.
The nature of this sequence explains the presence of several rocket and
satellite problems among the illustrative examples. Nevertheless, an effort
has been made to treat the subject of dynamics in a rather general context
with the liberal use of idealizations such as particles, massless rods, uniform
disks, and so on, without requiring that the configurations approximate
practical designs in any particular area of present-day technology.

The introductory chapter reviews some of the basic concepts of Newtonian mechanics and gives a short discussion of units and their definitions.
There is also a review of those topics in vector analysis which are most commonly used in dynamics. This is in accord with the general policy of giving
brief explanations or summaries of new mathematical topics as they arise.

It has been my observation that one of the principal sources of difficulty
for students of the vectorial approach to dynamics is one of kinematics.
Consequently, the kinematical foundations of particle motion are discussed
rather thoroughly in Chapter 2. Motion in a plane and also general three-
dimensional motion are included. Particular attention is given to rotating
reference frames and to vector derivatives relative to these frames.

With this background in kinematics, a general vectorial development of
the dynamics of a single particle and of systems of particles is given in the
next two chapters. Thus we find that most of the basic principles of dynamics
are developed and applied to a general set of particles before the introduction
of systems with distributed mass.

Chapter 5 is concerned with orbital motion. The discussion is almost
entirely limited to motion in an inverse-square gravitational field. In addition

v

to the derivation of the orbital trajectories, some attention is given to the time of flight, the determination of orbits, and to elementary perturbation theory.

Beginning with Chapter 6, extensive usage is made of the Lagrangian formulation of the equations of motion. The principle of virtual work is often associated with the study of statics. But it is included here because the ideas of virtual displacement and virtual work are fundamental in the derivation of Lagrange's equations and in obtaining a clear understanding of generalized forces. The introduction of Lagrange's equations of motion at this point goes very smoothly in the classroom because the students now have a sufficient fundamental background. Furthermore, they have been motivated to find ways of easing the kinematical difficulties in problem formulation.

Chapters 7 and 8 present the kinematics and dynamics of rigid bodies with particular emphasis upon rotational motion in three dimensions. In addition to the general analysis of free and forced motions of rigid bodies, special attention is given to the forced motion of axially symmetric bodies using the complex notation method. Matrix notation is introduced and is extensively used in these chapters.

Matrix notation is continued in the final chapter which is devoted to vibration theory. The finding of eigenvalues and the diagonalization of matrices, which were previously associated with the problem of obtaining principal axes of inertia, are now extended to the solution for the natural modes of vibration of systems with many degrees of freedom. Other topics such as Rayleigh's principle, the use of symmetry, and the free and forced vibrations of damped systems are also included.

The material contained in this text can be covered in about four semester hours at the senior or fifth-year levels. It is my opinion, however, that the discussion of homework problems and illustrative examples should be given an important place in the overall allotment of class time; hence the course could easily be extended to six semester hours. On the other hand, a course of three semester hours could be arranged by omitting portions of Chapters 5 and 9, and by spending less time on problems.

The major portion of this book was written during a sabbatical leave from the University of Michigan. I am particularly appreciative of the aid of Professor H. D. Christensen and others of the Aerospace Engineering faculty at the University of Arizona for providing a place to write and for their helpful discussions. Also, the comments and suggestions of Professor Y. C. Fung of Caltech were of great value during the final preparation of the manuscript. Finally, I wish to thank my wife who typed the manuscript, helped with the proofreading, and provided encouragement throughout this period.

DONALD T. GREENWOOD

Ann Arbor, Michigan

CONTENTS

1

INTRODUCTORY CONCEPTS

The science of *mechanics* is concerned with the study of the interactions of material bodies. *Dynamics* is that branch of mechanics which consists of the study of the *motions* of interacting bodies and the description of these motions in terms of postulated laws.

In this book we shall concentrate on the dynamical aspects of *Newtonian* or *classical nonrelativistic* mechanics. By omitting quantum mechanics, we eliminate the study of the interactions of elementary particles on the atomic or nuclear scale. Further, by omitting relativistic effects, we eliminate from consideration those interactions involving relative speeds approaching the velocity of light, whether they occur on an atomic or on a cosmical scale. Nor shall we consider the very large systems studied by astronomers and cosmologists, involving questions of long-range gravitation and the curvature of space.

Nevertheless, over a broad range of system dimensions and velocities, Newtonian mechanics is found to be in excellent agreement with observation. It is remarkable that nearly three centuries ago, Newton, aided by the discoveries of Galileo and other predecessors, was able to state these basic laws of motion and the law of gravitation in essentially the same form as they are used at present. Upon this basis, but using the mathematical and physical discoveries and notational improvements of later investigators, we shall present a modern version of classical dynamics.

1–1. ELEMENTS OF VECTOR ANALYSIS

Scalars, Vectors, and Tensors. Newtonian mechanics is, to a considerable extent, vectorial in nature. Its basic equation relates the applied force and the acceleration (both vector quantities) in terms of a scalar constant of proportionality called *the mass*. In contrast to Newton's *vectorial* approach, Euler, Lagrange, and Hamilton later emphasized the *analytical* or algebraic approach in which the differential equations of motion are obtained by performing certain operations on a scalar function, thereby simplifying the analysis in certain respects. Our approach to the subject will be vectorial

1

for the most part, although some of the insights and procedures of analytical mechanics will also be used.

Because vector operations are so important in the solution of dynamical problems, we shall review briefly a few of the basic vector operations. First, however, let us distinguish among scalars, vectors, and other tensors of higher order.

A *scalar* quantity is expressible as a single, real number. Common examples of scalar quantities are mass, energy, temperature, and time.

A quantity having direction as well as magnitude is called a *vector*.[1] Common vector quantities are force, moment, velocity, and acceleration. If one thinks of a vector quantity existing in a three-dimensional space, the essential characteristics can be expressed geometrically by an arrow or a directed line segment of proper magnitude and direction in that space. But the vector can be expressed equally well by a group of three real numbers corresponding to the components of the vector with respect to some frame of reference; for example, a set of cartesian axes. If one writes the numbers in a systematic fashion, such as in a column, then one can develop certain conventions which relate the position in the column to a given component of the vector. This concept can be extended readily to mathematical spaces with more than three dimensions. Thus, one can represent a vector in an *n*-dimensional space by a column of *n* numbers.

So far, we have seen that a scalar can be expressed as a single number and that a vector can be expressed as a column of numbers, that is, as a one-dimensional array of numbers. Scalars and vectors are each special cases of *tensors*. Scalars are classed as zero-order tensors, whereas vectors are first-order tensors. In a similar fashion, a second-order tensor is expressible as a two-dimensional array of numbers; a third-order tensor is expressible as a three-dimensional array of numbers, and so on. Note, however, that an array must also have certain transformation properties to be called a tensor. An example of a second-order tensor is the inertia tensor which expresses the essential features of the distribution of mass in a rigid body, as it affects the rotational motion.

We shall have no occasion to use tensors of order higher than two; hence no more than a two-dimensional array of numbers will be needed to express the quantities encountered. This circumstance enables us to use matrix notation, where convenient, rather than the more general but less familiar tensor notation. (Matrix notation will be introduced in Chapter 7 in the study of the rotational motion of rigid bodies.)

For the most part, we shall be considering motions which can be described mathematically using a space of no more than three dimensions; that is,

---

[1] In addition, vectors must have certain transformation properties. For example, equal vectors must remain equal after a rotation of axes. See Sec. 7–6 for a discussion of these rotation equations.

each matrix or array will have no more than three rows or columns and each vector will have no more than three components. An exception will be found in the study of vibration theory in Chapter 9 where we shall consider eigen-vectors in a multidimensional space.

Types of Vectors. Considering the geometrical interpretation of a vector as a directed line segment, it is important to recall that its essential features include *magnitude* and *direction*, but *not location*. This is not to imply that the location of a vector quantity, such as a force, is irrelevant in a physical sense. The location or point of application can be very important, and this will be reflected in the details of the mathematical formulation; for example, in the evaluation of the coefficients in the equations of motion. Nevertheless, the rules for the mathematical manipulation of vectors do not involve loca-tion; therefore, from the mathematical point of view, the only quantities of interest are magnitude and direction.

But from the *physical* point of view, vector quantities can be classified into three types, namely, *free vectors*, *sliding vectors*, and *bound vectors*. A vector quantity having the previously discussed characteristics of magnitude and direction, but no specified location or point of application, is known as a *free vector*. An example of a free vector is the translational velocity of a nonrotating body, this vector specifying the velocity of any point in the body. Another example is a force vector when considering its effect upon translational motion.

On the other hand, when one considers the effect of a force on the rotational motion of a rigid body, not only the magnitude and direction of the force, but also its line of action is important. In this case, the moment acting on the body depends upon the line of action of the force, but is independent of the precise point of application along that line. A vector of this sort is known as a *sliding vector*.

The third type of vector is the *bound vector*. In this case, the magnitude, direction, and point of application are specified. An example of a bound vector is a force acting on an elastic body, the elastic deformation being dependent upon the exact location of the force along its line of action.

Note again that all mathematical operations with vectors involve only their free vector properties of magnitude and direction.

Equality of Vectors. We shall use boldface type to indicate a vector quantity. For example, $\mathbf{A}$ is a vector of magnitude A, where A is a scalar.

Two vectors $\mathbf{A}$ and $\mathbf{B}$ are equal if $\mathbf{A}$ and $\mathbf{B}$ have the same magnitude and direction, that is, if they are represented by parallel line segments of equal length which are directed in the same sense. It can be seen that the transla-tion of either $\mathbf{A}$ or $\mathbf{B}$, or both, does not alter the equality since they are con-sidered as free vectors.

Unit Vectors. If a positive scalar and a vector are multiplied together (in either order), the result is another vector having the same direction, but whose magnitude is multiplied by the scalar factor. Conversely, if a vector is multiplied by a negative scalar, the direction of the resulting vector is reversed, but the magnitude is again multiplied by a factor equal to the magnitude of the scalar. Thus one can always think of a given vector as the product of a scalar magnitude and a vector of unit length which designates its direction. We can write

$$\mathbf{A} = A\mathbf{e}_A \qquad (1\text{-}1)$$

where the scalar factor A specifies the magnitude of $\mathbf{A}$ and the *unit vector* $\mathbf{e}_A$ shows its direction (Fig. 1-1).

Addition of Vectors. The vectors $\mathbf{A}$ and $\mathbf{B}$ can be added as shown in Fig. 1-2 to give the *resultant vector* $\mathbf{C}$. To add $\mathbf{B}$ to $\mathbf{A}$, translate $\mathbf{B}$ until its origin coincides with the terminus or arrow of $\mathbf{A}$. The vector sum is indicated by the line directed from the origin of $\mathbf{A}$ to the arrow of $\mathbf{B}$. It can be seen that

$$\mathbf{C} = \mathbf{A} + \mathbf{B} = \mathbf{B} + \mathbf{A} \qquad (1\text{-}2)$$

since, for either order of addition, the vector $\mathbf{C}$ is the same diagonal of the parallelogram formed by using $\mathbf{A}$ and $\mathbf{B}$ as sides. This is the *parallelogram*

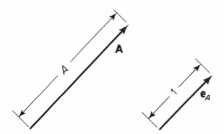

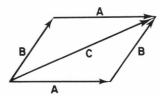

Fig. 1-1. A vector and its corresponding unit vector.

Fig. 1-2. The parallelogram rule of vector addition.

rule of vector addition. Since the order of the addition of two vectors is unimportant, vector addition is said to be *commutative.*

This procedure can be extended to find the sum of more than two vectors. For example, a third vector $\mathbf{D}$ can be added to the vector $\mathbf{C}$ obtained previously, giving the resultant vector $\mathbf{E}$. From Fig. 1-3, we see that

$$\mathbf{E} = \mathbf{C} + \mathbf{D} = (\mathbf{A} + \mathbf{B}) + \mathbf{D} \qquad (1\text{-}3)$$

But we need not have grouped the vectors in this way. Referring again to Fig. 1-3, we see that

$$\mathbf{E} = \mathbf{A} + (\mathbf{B} + \mathbf{D}) \qquad (1\text{-}4)$$

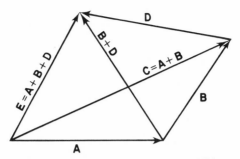

Fig. 1-3. The polygon rule of vector addition.

From Eqs. (1-3) and (1-4) we obtain

$$(\mathbf{A} + \mathbf{B}) + \mathbf{D} = \mathbf{A} + (\mathbf{B} + \mathbf{D}) = \mathbf{A} + \mathbf{B} + \mathbf{D} \qquad (1\text{--}5)$$

illustrating that vector addition is *associative*.

Because of the commutative and associative properties of vector addition, we can dispense with the parentheses in a series of additions and perform the additions in any order. Furthermore, using the graphical procedure of Fig. 1–3, we see that the resultant vector $\mathbf{E}$ is drawn from the origin of the first vector $\mathbf{A}$ to the terminus of the last vector $\mathbf{D}$, thus closing the polygon. This generalization of the parallelogram rule is termed the *polygon rule of vector addition*. A similar procedure applies for the case where all vectors do not lie in the same plane.

It is important to note that certain physical quantities that are apparently vectorial in nature do not qualify as true vectors in the sense that the usual rules for vector operations do not apply to them. For example, a finite rotational displacement of a rigid body is not a true vector quantity because the order of successive rotations is important, and therefore it does not follow the commutative property of vector addition. Further discussion of this topic will be found in Chapter 7.

Subtraction of Vectors. The negative of the vector $\mathbf{A}$ is the vector

$$-\mathbf{A} = -A\mathbf{e}_A \qquad (1\text{--}6)$$

which has the same magnitude as $\mathbf{A}$ but is opposite in direction. In other words, the vector $-\mathbf{A}$ is equal to the product of the vector $\mathbf{A}$ and the scalar -1.

Subtracting a vector $\mathbf{B}$ from another vector $\mathbf{A}$ is equivalent to adding its negative.

$$\mathbf{A} - \mathbf{B} = \mathbf{A} + (-\mathbf{B}) \qquad (1\text{--}7)$$

In particular, for the case where $\mathbf{A} = \mathbf{B}$, we may use the *distributive law* for multiplication of a vector and a scalar to obtain

$$A - A = (1 - 1)A = 0 \tag{1-8}$$

In general, the distributive law applies to either the scalar or the vector. Thus,

$$(n + m)A = nA + mA \tag{1-9}$$

and

$$n(A + B) = nA + nB \tag{1-10}$$

Components of a Vector. If a given vector **A** is equal to the sum of several vectors with differing directions, these vectors can be considered as *component vectors* of **A**. Since component vectors defined in this way are not unique, it is the usual practice in the case of a three-dimensional space to specify three directions along which the component vectors must lie. These directions are indicated by three linearly independent unit vectors, that is, a set of unit vectors such that none can be expressed as a linear combination of the others.

Suppose we choose the unit vectors e_1, e_2, and e_3 with which to express the given vector **A**. Then we can write

$$A = A_1 e_1 + A_2 e_2 + A_3 e_3 \tag{1-11}$$

where the scalar coefficients A_1, A_2, and A_3 are now determined uniquely. A_1, A_2, and A_3 are known as the *scalar components*, or simply the *components*, of the vector **A** in the given directions.

If another vector **B** is expressed in terms of the same set of unit vectors, for example,

$$B = B_1 e_1 + B_2 e_2 + B_3 e_3 \tag{1-12}$$

then the components of the vector sum of **A** and **B** are just the sums of the corresponding components.

$$\begin{aligned} A + B = (A_1 + B_1)e_1 \\ + (A_2 + B_2)e_2 \\ + (A_3 + B_3)e_3 \end{aligned} \tag{1-13}$$

This result applies, whether or not e_1, e_2, and e_3 form an orthogonal triad of unit vectors.

Now consider a case where the unit vectors are mutually orthogonal, as in the cartesian coordinate system of Fig. 1–4. The vector **A** can be expressed in terms of the scalar components A_x, A_y, and A_z, that is, it can be *resolved* as follows:

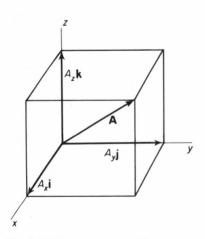

Fig. 1-4. The components of a vector in a cartesian coordinate system.

$$A = A_x\mathbf{i} + A_y\mathbf{j} + A_z\mathbf{k} \tag{1-14}$$

where $\mathbf{i}$, $\mathbf{j}$, and $\mathbf{k}$ are unit vectors in the directions of the positive x, y, and z axes, respectively.

From Fig. 1–4, it can be seen that the component vectors $A_x\mathbf{i}$, $A_y\mathbf{j}$, and $A_z\mathbf{k}$ form the edges of a rectangular parallelepiped whose diagonal is the vector $\mathbf{A}$. A similar situation occurs for the case of nonorthogonal or *skewed* unit vectors, except that the parallelepiped is no longer rectangular. Nevertheless, a vector along a diagonal of the parallelepiped has its components represented by edge lengths. In this geometrical construction, we are dealing with free vectors, and it is customary to place the origins of the vector $\mathbf{A}$ and the unit vectors at the origin of the coordinate system.

It is important to note that, for an *orthogonal* coordinate system, the components of a vector are identical with the orthogonal projections of the given vector onto the coordinate axes. For the case of a skewed coordinate system, however, the scalar components are *not* equal, in general, to the corresponding orthogonal projections. This distinction will be important in the discussions of Chapter 8 concerning the analysis of rigid body rotation by means of Eulerian angles; for, in this case, a skewed system of unit vectors is used.

Scalar Product. Consider the two vectors $\mathbf{A}$ and $\mathbf{B}$ shown in Fig. 1–5. The *scalar product* or *dot product* is

$$\mathbf{A} \cdot \mathbf{B} = AB \cos \theta \tag{1-15}$$

Since the cosine function is an even function, it can be seen that

$$\mathbf{A} \cdot \mathbf{B} = \mathbf{B} \cdot \mathbf{A} \tag{1-16}$$

implying that the scalar multiplication of vectors is commutative.

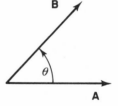

Fig. 1-5. Multiplication of two vectors.

The scalar product can also be considered as the product of the magnitude of one vector and the orthogonal projection of the second vector upon it. Now, it can be seen from Fig. 1–6 that the sum of the projections of vectors $\mathbf{A}$ and $\mathbf{B}$ onto a third vector $\mathbf{C}$ is equal to the projection of $\mathbf{A} + \mathbf{B}$ onto $\mathbf{C}$. Therefore, noting that the multiplication of scalars is distributive, we obtain

$$(\mathbf{A} + \mathbf{B}) \cdot \mathbf{C} = \mathbf{A} \cdot \mathbf{C} + \mathbf{B} \cdot \mathbf{C} \tag{1-17}$$

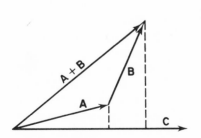

Fig. 1-6. The distributive law for the dot product.

Thus, the *distributive* property applies to the scalar product of vectors.

Now consider the dot product of two vectors **A** and **B**, each of which is expressed in terms of a given set of unit vectors $\mathbf{e}_1$, $\mathbf{e}_2$, and $\mathbf{e}_3$. From Eqs. (1–11) and (1–12), we obtain

$$\begin{aligned} \mathbf{A} \cdot \mathbf{B} = A_1 B_1 + A_2 B_2 + A_3 B_3 + (A_1 B_2 + A_2 B_1)\mathbf{e}_1 \cdot \mathbf{e}_2 \\ + (A_1 B_3 + A_3 B_1)\mathbf{e}_1 \cdot \mathbf{e}_3 + (A_2 B_3 + A_3 B_2)\mathbf{e}_2 \cdot \mathbf{e}_3 \end{aligned} \quad (1\text{–}18)$$

For the common case where the unit vectors form an orthogonal triad, the terms involving dot products of different unit vectors are all zero. For this case, we see from Eq. (1–18) that

$$\mathbf{A} \cdot \mathbf{B} = A_1 B_1 + A_2 B_2 + A_3 B_3 \quad (1\text{–}19)$$

Vector Product. Referring again to Fig. 1–5, we define the *vector product* or *cross product* as follows:

$$\mathbf{A} \times \mathbf{B} = AB \sin \theta \, \mathbf{k} \quad (1\text{–}20)$$

where **k** is a unit vector perpendicular to, and out of, the page. In general, the direction of **k** is found by the right-hand rule, that is, it is perpendicular to the plane of **A** and **B** and positive in the direction of advance of a right-hand screw as it rotates in the sense that carries the first vector **A** into the second vector **B**. The angle of this rotation is θ. It is customary, but not necessary, to limit θ to the range $0 \le \theta \le \pi$.

Using the right-hand rule, it can be seen that

$$\mathbf{A} \times \mathbf{B} = -\mathbf{B} \times \mathbf{A} \quad (1\text{–}21)$$

indicating that the vector product is *not commutative*.

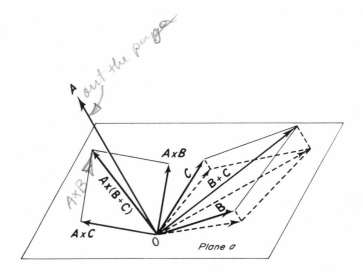

Fig. 1-7. The distributive law for the vector product.

Another way of visualizing the vector product $\mathbf{A} \times \mathbf{B}$ is to note again that it must be perpendicular to both $\mathbf{A}$ and $\mathbf{B}$. Thus if planes a and b pass through a common origin O and are normal to $\mathbf{A}$ and $\mathbf{B}$, respectively, then $\mathbf{A} \times \mathbf{B}$ must be directed along the line of intersection of these planes with the sense again being determined by the right-hand rule. The magnitude of $\mathbf{A} \times \mathbf{B}$ is equal to A times the projection of $\mathbf{B}$ onto plane a, or conversely, it is equal to B times the projection of $\mathbf{A}$ onto plane b.

Now let us use this approach to evaluate the vector product $\mathbf{A} \times (\mathbf{B} + \mathbf{C})$ (Fig. 1–7). First, we note that the vectors $\mathbf{A} \times \mathbf{B}$, $\mathbf{A} \times \mathbf{C}$, and $\mathbf{A} \times (\mathbf{B} + \mathbf{C})$ must all lie in plane a that is normal to $\mathbf{A}$. Furthermore, it is apparent that the projection of $\mathbf{B} + \mathbf{C}$ onto plane a is equal to the vector sum of the projections of $\mathbf{B}$ and $\mathbf{C}$ onto the same plane. Thus, the parallelogram formed by the vectors $\mathbf{A} \times \mathbf{B}$ and $\mathbf{A} \times \mathbf{C}$ is similar to that formed by the projections of $\mathbf{B}$ and $\mathbf{C}$ onto a, there being a rotation through 90 degrees in plane a and a multiplication by a scalar factor A. Therefore, we obtain that

$$\mathbf{A} \times (\mathbf{B} + \mathbf{C}) = \mathbf{A} \times \mathbf{B} + \mathbf{A} \times \mathbf{C} \tag{1-22}$$

showing that the *distributive law* applies to vector products.

Using the distributive law, we can evaluate the vector product $\mathbf{A} \times \mathbf{B}$ in terms of the cartesian components of each. Thus,

$$\mathbf{A} \times \mathbf{B} = (A_x \mathbf{i} + A_y \mathbf{j} + A_z \mathbf{k}) \times (B_x \mathbf{i} + B_y \mathbf{j} + B_z \mathbf{k}) \tag{1-23}$$

But

$$\mathbf{i} \times \mathbf{i} = \mathbf{j} \times \mathbf{j} = \mathbf{k} \times \mathbf{k} = 0$$
$$\mathbf{i} \times \mathbf{j} = -\mathbf{j} \times \mathbf{i} = \mathbf{k}$$
$$\mathbf{j} \times \mathbf{k} = -\mathbf{k} \times \mathbf{j} = \mathbf{i} \tag{1-24}$$
$$\mathbf{k} \times \mathbf{i} = -\mathbf{i} \times \mathbf{k} = \mathbf{j}$$

and therefore,

$$\mathbf{A} \times \mathbf{B} = (A_y B_z - A_z B_y)\mathbf{i} + (A_z B_x - A_x B_z)\mathbf{j} + (A_x B_y - A_y B_x)\mathbf{k} \tag{1-25}$$

This result can be expressed more concisely as the following determinant:

$$\mathbf{A} \times \mathbf{B} = \begin{vmatrix} \mathbf{i} & \mathbf{j} & \mathbf{k} \\ A_x & A_y & A_z \\ B_x & B_y & B_z \end{vmatrix} \tag{1-26}$$

In general, if the sequence $\mathbf{e}_1$, $\mathbf{e}_2$, and $\mathbf{e}_3$ forms a right-handed set of mutually orthogonal unit vectors, then the vector product can be expressed in terms of the corresponding components as follows:

$$\mathbf{A} \times \mathbf{B} = \begin{vmatrix} \mathbf{e}_1 & \mathbf{e}_2 & \mathbf{e}_3 \\ A_1 & A_2 & A_3 \\ B_1 & B_2 & B_3 \end{vmatrix} \tag{1-27}$$

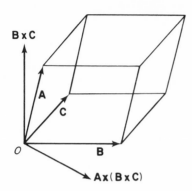

Fig. 1-8. Triple products of vectors.

Scalar Triple Product. The product $\mathbf{A} \cdot (\mathbf{B} \times \mathbf{C})$ is known as the *scalar triple product.* Looking at Fig. 1–8, we see that $\mathbf{B} \times \mathbf{C}$ is a vector whose magnitude is equal to the area of a parallelogram having $\mathbf{B}$ and $\mathbf{C}$ as sides and whose direction is perpendicular to the plane of that parallelogram. Considering this plane to be horizontal for the moment, we note that $\mathbf{A} \cdot (\mathbf{B} \times \mathbf{C})$ is just the area of the base multiplied by the projection of $\mathbf{A}$ onto the vertical; that is, its magnitude is the *volume* of the parallelepiped having $\mathbf{A}$, $\mathbf{B}$, and $\mathbf{C}$ as edges. The sign of $\mathbf{A} \cdot (\mathbf{B} \times \mathbf{C})$ is positive or negative, depending upon whether or not $\mathbf{A}$ and $\mathbf{B} \times \mathbf{C}$ lie on the same side of the plane of $\mathbf{B}$ and $\mathbf{C}$. Of course, if $\mathbf{A}$, $\mathbf{B}$, and $\mathbf{C}$ lie in the same plane, the product is zero.

We can find the scalar triple product in terms of cartesian components by using equations in the form of Eqs. (1–14) and (1–26) to obtain

$$\mathbf{A} \cdot (\mathbf{B} \times \mathbf{C}) = \mathbf{A} \cdot \begin{vmatrix} \mathbf{i} & \mathbf{j} & \mathbf{k} \\ B_x & B_y & B_z \\ C_x & C_y & C_z \end{vmatrix} = \begin{vmatrix} A_x & A_y & A_z \\ B_x & B_y & B_z \\ C_x & C_y & C_z \end{vmatrix} \qquad (1\text{–}28)$$

From the rules for determinants, we note that the interchange of any two vectors (that is, any two rows) results in a change in the sign of the product. Furthermore, an even number of such interchanges results in a cyclic permutation of the vectors, but no change in the sign of the product. Therefore, since the dot product is commutative, the dot and the cross in a scalar triple product may be interchanged, or a cyclic permutation of the vectors may occur, without affecting the result.

$$\mathbf{A} \cdot (\mathbf{B} \times \mathbf{C}) = (\mathbf{A} \times \mathbf{B}) \cdot \mathbf{C} \qquad (1\text{–}29)$$

Also,

$$\mathbf{A} \cdot (\mathbf{B} \times \mathbf{C}) = \mathbf{B} \cdot (\mathbf{C} \times \mathbf{A}) = \mathbf{C} \cdot (\mathbf{A} \times \mathbf{B}) \qquad (1\text{–}30)$$

These results can be interpreted geometrically by noting that the parallelepiped whose edges are formed by the vectors $\mathbf{A}$, $\mathbf{B}$, and $\mathbf{C}$, has a volume that is independent of the order of these vectors. The order, as we have seen, influences the sign of the result but not its magnitude.

Vector Triple Product. It can be seen from Fig. 1–8 that the *vector triple product* $\mathbf{A} \times (\mathbf{B} \times \mathbf{C})$ lies in the plane of $\mathbf{B}$ and $\mathbf{C}$. On the other hand, $(\mathbf{A} \times \mathbf{B}) \times \mathbf{C}$ lies in the plane of $\mathbf{A}$ and $\mathbf{B}$ and is not, in general, equal to

A × (B × C). Therefore the vector product does not obey the associative law, and parentheses are needed to show the required groupings.

One can evaluate **A × (B × C)** in terms of cartesian components by twice using the determinant form for the cross product as given by Eq. (1–26). Collecting terms and simplifying, the result can be written

$$\mathbf{A} \times (\mathbf{B} \times \mathbf{C}) = (\mathbf{A} \cdot \mathbf{C})\mathbf{B} - (\mathbf{A} \cdot \mathbf{B})\mathbf{C} \tag{1–31}$$

Note that the cartesian system was used as a matter of convenience. The vector form of the final result indicates its general validity, independent of the choice of a coordinate system.

Derivative of a Vector. In the vectorial treatment of mechanics, one often encounters the concept of a *vector function* of a scalar variable, that is, a quantity whose magnitude and direction are dependent upon the value of a scalar variable. For example, consider the vector function **A**(u). Suppose that the vector **A** of Fig. 1–9 corresponds to a certain value u of the scalar variable. Similarly, the vector **A + ΔA** corresponds to $u + \Delta u$. In other words, the change **ΔA** in the vector corresponds to a change Δu in the scalar. Then the *derivative* of the vector **A** with respect to the scalar u is defined by the limit

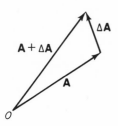

Fig. 1-9. Derivative of a vector.

$$\frac{d\mathbf{A}}{du} = \lim_{\Delta u \to 0} \frac{\Delta \mathbf{A}}{\Delta u} \tag{1–32}$$

assuming that the limit exists.

For the case where the vector is differentiated with respect to time, geometrical significance can be attached to the derivative. It is simply the velocity of the point of the vector when its origin is fixed.

Using a similar procedure, one can evaluate the derivative of the vector sum **A + B**. Noting that

$$\Delta(\mathbf{A} + \mathbf{B}) = \Delta \mathbf{A} + \Delta \mathbf{B}$$

we find that the differentiation of vectors is *distributive*.

$$\frac{d(\mathbf{A} + \mathbf{B})}{du} = \frac{d\mathbf{A}}{du} + \frac{d\mathbf{B}}{du} \tag{1–33}$$

Also, one can show that

$$\frac{d}{du}(g\mathbf{A}) = \frac{dg}{du}\mathbf{A} + g\frac{d\mathbf{A}}{du} \tag{1–34}$$

$$\frac{d}{du}(\mathbf{A} \cdot \mathbf{B}) = \frac{d\mathbf{A}}{du} \cdot \mathbf{B} + \mathbf{A} \cdot \frac{d\mathbf{B}}{du} \tag{1–35}$$

$$\frac{d}{du}(\mathbf{A} \times \mathbf{B}) = \frac{d\mathbf{A}}{du} \times \mathbf{B} + \mathbf{A} \times \frac{d\mathbf{B}}{du} \tag{1-36}$$

As an example of differentiation, consider the vector $\mathbf{A}$ expressed in terms of its cartesian components:

$$\mathbf{A} = A_x \mathbf{i} + A_y \mathbf{j} + A_z \mathbf{k}$$

For the common case where differentiation is with respect to time, and the unit vectors have a fixed orientation in space, we obtain, using Eqs. (1–14), (1–33), and (1–34), that

$$\frac{d\mathbf{A}}{dt} = \frac{dA_x}{dt}\mathbf{i} + \frac{dA_y}{dt}\mathbf{j} + \frac{dA_z}{dt}\mathbf{k} \tag{1-37}$$

In the more general case where $\mathbf{A}$ is expressed in terms of the unit vectors $\mathbf{e}_1$, $\mathbf{e}_2$, and $\mathbf{e}_3$ which may change their orientation in space, we can write

$$\mathbf{A} = A_1 \mathbf{e}_1 + A_2 \mathbf{e}_2 + A_3 \mathbf{e}_3$$

Then, we obtain

$$\frac{d\mathbf{A}}{dt} = \frac{dA_1}{dt}\mathbf{e}_1 + \frac{dA_2}{dt}\mathbf{e}_2 + \frac{dA_3}{dt}\mathbf{e}_3 + A_1\frac{d\mathbf{e}_1}{dt} + A_2\frac{d\mathbf{e}_2}{dt} + A_3\frac{d\mathbf{e}_3}{dt}$$

or, expressing the result using a dot over a symbol to indicate its time derivative,

$$\frac{d\mathbf{A}}{dt} = \dot{A}_1\mathbf{e}_1 + \dot{A}_2\mathbf{e}_2 + \dot{A}_3\mathbf{e}_3 + A_1\dot{\mathbf{e}}_1 + A_2\dot{\mathbf{e}}_2 + A_3\dot{\mathbf{e}}_3 \tag{1-38}$$

where we note that the time rate of change of a unit vector is always perpendicular to that unit vector. Equations of this sort will be developed more extensively in the study of kinematics in Chapter 2.

1–2 NEWTON'S LAWS OF MOTION

In his *Principia*, published in 1687, Sir Isaac Newton stated the laws upon which classical mechanics is based. Using modern terminology, these laws can be stated as follows:

I.   *Every body continues in its state of rest, or of uniform motion in a straight line, unless compelled to change that state by forces acting upon it.*

II. *The time rate of change of linear momentum of a body is proportional to the force acting upon it and occurs in the direction in which the force acts.*

III. *To every action there is an equal and opposite reaction; that is, the mutual forces of two bodies acting upon each other are equal in magnitude and opposite in direction.*

The Laws of Motion for a Particle. An understanding of Newton's laws of motion is most easily achieved by applying them to the study of the motion of *particles*, where a *particle* is defined as a mass concentrated at a point. Later, when we consider the case of bodies with a continuous distribution of mass, the generalization of the dynamical methods from the discrete to the continuous case will be seen to be quite straightforward.

Now let us state three basic laws applying to the motion of a particle. The first is the *law of motion* which summarizes Newton's first two laws. It can be expressed by the equation

$$\mathbf{F} = k \frac{d}{dt}(m\mathbf{v}) = km\mathbf{a} \tag{1-39}$$

where the product $m\mathbf{v}$ is known as the *linear momentum* **p**, that is,

$$\mathbf{p} = m\mathbf{v} \tag{1-40}$$

and where m is the mass of the particle, $\mathbf{a}$ is its acceleration, $\mathbf{F}$ is the applied external force, and k is a positive constant whose value depends upon the choice of units. The mass m is considered to be constant, since we are not concerned with relativistic effects and any variable-mass systems are treated as collections of particles.

Because of the fundamental nature of Eq. (1–39), the units are chosen such that $k = 1$. So, with a proper choice of units, Eq. (1–39) simplifies to

$$\mathbf{F} = m\mathbf{a} \tag{1-41}$$

which is the usual statement of the law of motion.

The second basic law is the *law of action and reaction:*

When two particles exert forces on each other, these interaction forces are equal in magnitude, opposite in sense, and directed along the straight line joining the particles.

This is essentially a statement of Newton's third law as it applies to two particles, but the *collinearity* of the interaction forces has been mentioned specifically. The added requirement of collinearity will be found to be essential for the conservation of angular momentum of an isolated mechanical system and applies to all mechanical or gravitational interaction forces. It *does not* apply, however, to certain forces between moving, charged particles, a situation which will not concern us in this book.

The third basic law is the *law of addition of forces:*

Two forces **P** *and* **Q** *acting simultaneously on a particle are together equivalent to a single force* $\mathbf{F} = \mathbf{P} + \mathbf{Q}$.

By similar reasoning, we can conclude that the simultaneous action of *more than two* forces on a particle produces the same motion as a single force equal to their vector sum.

Newton stated the law of addition of forces as a corollary to his laws of motion. Note that it also implies that a single force may be replaced by its component forces in a dynamical calculation. Thus, if we consider a particle of mass m to be moving with respect to a fixed cartesian system under the action of a force $\mathbf{F}$, we can write

$$\mathbf{F} = F_x \mathbf{i} + F_y \mathbf{j} + F_z \mathbf{k} \tag{1-42}$$

and the resulting acceleration components are found from

$$F_x = m\ddot{x}$$
$$F_y = m\ddot{y} \tag{1-43}$$
$$F_z = m\ddot{z}$$

where the total acceleration is

$$\mathbf{a} = \ddot{x}\mathbf{i} + \ddot{y}\mathbf{j} + \ddot{z}\mathbf{k} \tag{1-44}$$

In other words, the single vector equation given in Eq. (1–41) is equivalent to the three scalar equations of Eq. (1–43).

That these equations can be verified experimentally indicates the actual independence of the component accelerations and also that the mass m is a single scalar quantity.[2]

Frames of Reference. In our previous discussion we have not concerned ourselves with the question of what is a proper reference frame from which to measure the accelerations to be expected in accordance with the laws of motion. The approach which we shall take is to define an *inertial* or *Newtonian* reference frame to be any rigid set of coordinate axes such that particle motion relative to these axes is described by Newton's laws of motion.

In an attempt to find an example of an inertial frame in the physical world, one might consider first a system fixed in the earth. Such a choice would be adequate for most cases where the distances traveled are short relative to the earth's radius and where the velocities are small compared to the velocity of escape from the earth. If one were to analyze long-range missiles or satellites, however, the earth would be a completely inadequate approximation to an inertial frame.

A much better approximation would be a system whose origin is at the center of the earth and whose axes are not rotating with respect to the "fixed" stars. Even here, however, measurements of satellites traveling far from the earth would easily show deviations from Newton's laws due to the neglected gravitational forces exerted by the sun, moon, and planets.

Finally, one could choose as a Newtonian reference frame a system at the center of the sun (or, more exactly, at the center of mass of the solar

---

[2] This, of course, presupposes that the particle velocity is small compared to the velocity of light and therefore that relativistic effects can be neglected.

system) and nonrotating with respect to the so-called fixed stars. This would appear to be adequate for the foreseeable future.

So let us assume that we have found an inertial reference frame, and therefore that Newton's laws apply for motions relative to this frame. It can be shown that any other reference frame that is not rotating but is translating with a uniform velocity relative to an inertial frame is itself an inertial frame. For example, if system B is translating at a constant velocity $\mathbf{v}_{rel}$ with respect to an inertial system A, then, denoting the velocity of a particle as viewed by observers on A and B by $\mathbf{v}_A$ and $\mathbf{v}_B$, respectively, we see that

$$\mathbf{v}_A = \mathbf{v}_B + \mathbf{v}_{rel} \qquad (1\text{--}45)$$

Differentiating with respect to time and noting that the derivative of $\mathbf{v}_{rel}$ is zero, we obtain

$$\mathbf{a}_A = \mathbf{a}_B \qquad (1\text{--}46)$$

where $\mathbf{a}_A$ and $\mathbf{a}_B$ are the accelerations of the particle as viewed from systems A and B, respectively. Now the total force applied to the particle is independent of the motion of the observer. So from the Newtonian point of view, observers on systems A and B see identical forces, masses, and accelerations, and therefore, Eq. (1–41) is equally valid for each observer.

One can summarize by saying that the existence of an inertial frame implies the existence of an infinite number of other inertial frames, all having no rotation rate relative to the fixed stars but translating with constant velocities relative to each other. Thus, even Newtonian mechanics has no single, preferred frame of reference.

1-3. UNITS

When one attempts to apply the law of motion as given by Eq. (1–41), one is immediately faced with the problem of choosing a proper set of units with which to express the quantities of interest. As we have seen previously, when we set the proportionality constant k equal to unity in Eq. (1–39), the sizes of the units used to specify force, mass, and acceleration are no longer arbitrary. Furthermore, as we shall see, the selection of units for any two of these quantities fixes the units to be used in measuring the third quantity since the proportionality constant is assumed to be dimensionless, that is, it is a pure number and has no units associated with it.

In considering the problem of units let us first discuss the so-called *dimensions* associated with each unit.

Dimensions. It can be seen that the units which are used in the measurement of physical quantities may differ *quantitatively* as well as *qualitatively*. For example, the foot and the inch differ in magnitude but are qualitatively the same in that both are units of length. On the other hand, the foot and the second are qualitatively different. Now, the qualitative aspects of a given unit are characterized by its *dimensions*. Thus, the foot and the inch both have the dimension of *length*. Similarly, the hour and the second have the dimension of *time*.

It turns out that all units used in the study of mechanics can be expressed in terms of only three dimensions. By common agreement, the dimensions of length and of time are considered to be fundamental. Thus, a physical quantity such as velocity has the dimensions of length per unit time, written $[LT^{-1}]$, regardless of whether the units chosen are mi/hr, ft/sec, or even knots. Similarly, acceleration has the dimensions $[LT^{-2}]$.

A characteristic of the equations of physics is that they must exhibit *dimensional homogeneity*. By this, we mean that any terms which are added or subtracted must have the same dimensions and it also implies, of course, that the expressions on each side of an equality must have the same dimensions. Furthermore, any argument of a transcendental function, such as the trigonometric function, exponential function, Bessel function, and so on, must be dimensionless; that is, all exponents associated with the fundamental dimensions of the argument must be zero. One will sometimes find apparent exceptions to the requirement of dimensional homogeneity, but in all cases some of the coefficients will prove to have unsuspected dimensions or else the equation is an empirical approximation not based on physical law. In checking for dimensional homogeneity, one should note that the unit of angular displacement, the radian, is dimensionless.

If we note that the equations of mechanics are principally of the form of Eq. (1–41), or its integrals with respect to space or time (or the corresponding moments), then the requirement of dimensional homogeneity implies that the fundamental dimensions corresponding to mass and force cannot be chosen independently. If one chooses mass to be the third fundamental dimension, then the dimensions of force are determined, and vice versa. Thus, we see that there are two obvious possibilities for choices of fundamental dimensions: (1) the *absolute* system in which mass, length, and time are the fundamental dimensions; (2) the *gravitational* system in which force, length, and time are the fundamental dimensions. Both types of systems are in wide use in this country; the former being used more extensively by physicists and other scientists, the latter by engineers.

Systems of Units. Within either the absolute system or the gravitational system, many sets of fundamental units can be chosen. Absolute systems of units in common use include the cgs or centimeter-gram-second system and the mks or meter-kilogram-second system where, of course, the centimeter (or meter) is the fundamental unit of length, the gram (or kilogram) is the fundamental unit of mass, and the second is the fundamental unit of time.

On the other hand, the *English gravitational system*, which is the system that we shall use, employs the foot as the fundamental unit of length, the pound as the fundamental unit of *force*, and the second as the fundamental unit of time. In this case, the unit of mass, the *slug*, is a *derived unit* rather than a fundamental unit. A slug is that mass which is given an acceleration of 1 ft/sec by an applied force of 1 lb. In terms of fundamental units, we see that

$$1 \text{ slug} = 1 \text{ lb sec}^2/\text{ft}$$

since, from Eq. (1–41) and the principle of dimensional homogeneity, the unit of mass must equal the unit of force divided by the unit of acceleration.

Note particularly that we always use the term *pound* as a unit of force and never as a unit of mass. A possible source of confusion arises from the fact that the legal standard of mass is based on the absolute system and, in that system, a standard unit of mass is the pound, also called the pound-mass. To avoid confusion, however, we shall use the gravitational system exclusively.

Some of the quantities that are most commonly used in dynamics are listed in Table 1–1. The letters F, L, and T refer to the dimensions of force, length, and time, respectively.

TABLE 1–1. GRAVITATIONAL UNITS AND DIMENSIONS

Quantity	Gravitational Units	Dimensions
Length	ft	$[L]$
Time	sec	$[T]$
Force	lb	$[F]$
Mass	lb sec²/ft (slug)	$[FT^2L^{-1}]$
Velocity	ft/sec	$[LT^{-1}]$
Acceleration	ft/sec²	$[LT^{-2}]$
Energy (work)	ft lb	$[FL]$
Angular Velocity	rad/sec	$[T^{-1}]$
Moment	lb ft	$[FL]$
Moment of Inertia	lb ft sec²	$[FLT^2]$
Linear Momentum	lb sec	$[FT]$
Angular Momentum	lb ft sec	$[FLT]$
Linear Impulse	lb sec	$[FT]$
Angular Impulse	lb ft sec	$[FLT]$

Conversion of Units. When making checks of dimensional homogeneity, and even when performing numerical computations, it is advisable to carry along the units, treating them as algebraic quantities. This algebraic manipulation of units often requires the conversion from one set of units to another set having the same dimensions. Normally, one converts to three basic units, such as pounds, feet, and seconds, rather than carrying along derived units, such as slugs, or perhaps several different units of length. In this conversion process, one does not change the magnitudes of any of the physical quantities, but only their form of expression.

A convenient method of changing units is to multiply by one or more fractions whose magnitude is unity, but in which the numerator and the denominator are expressed in different units. Suppose, for example, that one wishes to convert knots into inches per second. One finds that

$$1 \text{ knot} = 1 \text{ nautical mi/hr} \qquad 1 \text{ ft} = 12 \text{ in.}$$
$$1 \text{ nautical mi} = 6080 \text{ ft} \qquad 1 \text{ hr} = 3600 \text{ sec}$$

Therefore,

$$1 \text{ knot} = \left(\frac{1 \text{ naut. mi}}{1 \text{ hr}}\right)\left(\frac{6080 \text{ ft}}{1 \text{ naut. mi}}\right)\left(\frac{12 \text{ in.}}{1 \text{ ft}}\right)\left(\frac{1 \text{ hr}}{3600 \text{ sec}}\right)$$
$$= 20.27 \text{ in./sec}$$

Weight and Mass. We have seen that, although we shall use the pound as a unit of *force*, one of the units of mass in the absolute system is also known as the *pound*. This emphasizes the need for a clear distinction between *weight* and *mass* since the weight of an object is expressed in units of force.

Briefly, the *weight* of a body is the force with which that body is attracted toward the earth. Its *mass* is the quantity of matter in the body, irrespective of its location in space. For example, if a given body is moved from a valley to the top of an adjacent mountain, its mass is constant; but its weight is less on the mountain top because the magnitude of the gravitational attraction decreases with increasing elevation above sea level.

Another approach is provided by the equation

$$w = mg \tag{1–47}$$

where w is the weight, m is the mass, and g is the local acceleration of gravity, that is, g is the acceleration that would result if the body were released from rest in a vacuum at that location.

Although the interpretation of Eq. (1–47) appears to be straightforward, certain complications arise in the determination of the acceleration of gravity. First, the acceleration of gravity is measured with respect to a reference frame fixed in the earth. This is not an inertial frame in the strict sense because the earth is rotating in space. Therefore, as we shall show more clearly in Chapter 2, the so-called inertial forces must be included in a calculation of the acceleration of gravity. Thus it turns out that centrifugal as well as gravitational forces enter the problem. The effect of the centrifugal force is to cause a slight change in the magnitude and direction of the acceleration of gravity, the amount of the change depending upon the latitude.

A second reason for the slight variation in the acceleration of gravity with latitude is the oblateness of the earth, the polar radius being about 0.3 per cent smaller than the equatorial radius. The two factors previously mentioned result in a variation of the acceleration of gravity at the earth's surface which can be approximated by the expression

$$g = 32.26 - 0.17 \cos^2 \theta \text{ ft/sec}^2$$

where g is the local acceleration of gravity and θ is the latitude of the point in question.

For our purposes, it will generally be sufficient to use the value $g = 32.2$ ft/sec for the acceleration of gravity at the surface of the earth. Furthermore, in any discussion of orbits about the earth, we shall use the symbol g_0 which includes gravitational effects only and refers to a spherical earth.

1-4. THE BASIS OF NEWTONIAN MECHANICS

Now that we have presented the basic laws of motion and have discussed the question of units briefly, let us consider in greater detail the fundamental assumptions of Newtonian mechanics. In particular, let us consider the concepts of space, time, mass, and force from the viewpoint of Newtonian mechanics and indicate experimental procedures for defining the corresponding fundamental units. In the latter process, we shall not be concerned so much with the practical utility of the operational definitions as with their theoretical validity.

Space. Newton conceived of space as being infinite, homogeneous, isotropic, and absolute. The infinite nature of space follows from the implicit assumption that ordinary, Euclidean geometry applies to it. By *homogeneous* and *isotropic*, it is meant that the local properties of space are independent of location or direction.

In claiming that space is absolute, Newton assumed the existence of a primary inertial frame. In his view, it was nonrotating relative to the fixed stars and also was fixed relative to the center of the universe which was interpreted as being located at the center of mass of the solar system. Of course, Newton realized that any other coordinate system that is translating uniformly with respect to his primary frame would serve equally well as an inertial frame. On philosophical grounds, however, he chose to give special preference to the inertial frame that is fixed at the "center of the universe."

We, too, saw in the discussion of Eq. (1–46) that the law of motion is not changed by the transformation to another system that is translating uniformly with respect to a given inertial system. But any *rotational* motion of a reference frame relative to a given inertial frame will produce apparent acceleration terms which change the form of the basic equation of motion. So we can conclude that Newtonian mechanics requires an absolute reference for rotational motion, but translational motion is relative in that an arbitrary uniform translation may be superimposed.

The spatial relationships associated with Newtonian mechanics are measured in units of length. As a standard unit of length, one can choose the distance between two marks on a bar at a standard temperature, or perhaps one could define the standard unit of length in terms of the wavelength of the light corresponding to a given spectral line measured under standard conditions. In either event, the unit is obtained in a rather direct fashion and, in this theory, does not depend upon the motion of either the standard or the observer.

Time. Newton conceived of time as an absolute quantity, that is, the same for all observers and, furthermore, independent of all objects of the physical world. He considered that a definition of time in terms of natural phenomena such as the rotation of the earth was, at best, an approximation to the uniform flow of "true" time.

Later, Mach[3] contended that time is merely an abstraction arrived at by changes in the physical world. For example, we say that object *A* moves uniformly if equal changes in the displacement of *A* correspond to equal changes in the position of another system such as the hands of a clock. But one cannot show that the motion of the clock is uniform *in itself;* a comparison must be made with another system. Thus, time is not an absolute entity, but is assumed in order to express the interrelationships of the motions of the physical world.

In deciding which physical process to use as a basis for the definition of a standard unit of time, we *assume* the validity of certain physical laws and choose a process that, according to the physical theory, proceeds at a uniform rate, or at a constant frequency, and is relatively easy to measure. For example, according to Newtonian mechanics, the rotation rate of the earth relative to the fixed stars is constant except for a very small retarding effect due to tidal friction. So, for most purposes, the definition of a standard unit of time based upon the rotation rate of the earth, as measured by the interval between successive transits of a given star, would seem to be satisfactory.

Also, we shall adopt the Newtonian assumption that time is the same for all observers. Note that this allows for a finite propagation velocity of light if compensation is made for the time of transmission of clock synchronizing signals from one place to another. No relativistic corrections are required if the relative velocities of all bodies are much smaller than the velocity of light.

---

[3]E. Mach, *The Science of Mechanics* (La Salle, Ill.: The Open Court Publishing Co., 1942). The first edition was published in 1893.

Mass. We have seen that reasonably straightforward operational definitions can be given for the fundamental units of length and of time. The
third fundamental unit can be chosen as a unit of mass or of force, depending
upon whether the absolute or the gravitational system of dimensions is considered to be most appropriate. Let us consider first the choice of mass as
a fundamental dimension.

Newton defined mass as "quantity of matter" which was calculated as
the product of volume and density. Although this definition may give one
an intuitive concept of mass, it is not a satisfactory operational definition
but merely replaces one undefined term by another.

In an attempt to find a more satisfactory definition of mass, we might
take a given body A as a standard. Another body B is assumed to have a
mass equal to that of A if bodies A and B balance when tested on a beam
balance. In other words, their masses are equal if their *weights* are equal.
If one assumes further that mass is an additive property, that is, that the
mass of several bodies acting together is equal to the sum of their individual
masses, then one can establish a set of standard masses which are equal to
various multiples or fractional parts of the original primary standard. Using
the standard masses, one can in theory determine the mass of a given body
to an arbitrary precision.

This relatively convenient procedure of comparing the masses of two
bodies by comparing their weights has certain deficiencies when used as a
fundamental definition of mass. One difficulty is that it assumes the equivalence of *inertial mass* and *gravitational mass*. The mass that we are interested
in determining is the inertial mass m in the equation of motion $\mathbf{F} = m\mathbf{a}$.
On the other hand, the weighing procedure compares the magnitudes of the
forces of gravity acting on the bodies. As we shall see in Sec. 5–1, the force
of gravity acting on a body is proportional to its mass, but there is no *a
priori* reason for the equality of gravitational mass and inertial mass. This
equality must be experimentally determined.[4]

In comparing the masses of two bodies by comparing their weights, the
assumption is made that, during the balancing procedure, the bodies are
motionless in an inertial frame. For this case, the upward force exerted on
each body by the balance is equal to the downward force of gravity on that
body. This results in a net force of zero which, in accordance with the law
of motion, corresponds to zero acceleration in an inertial frame, in agreement
with the initial assumption.

Using this method, if one wishes to determine the mass of a moving body,
an additional assumption must be made; namely, that *the mass of a body is
independent of its motion.* In other words, its mass while in motion is assumed

---

[4]It is interesting to note that Newton tacitly assumed the equivalence of inertial and
gravitational mass. In more recent times, this *principle of equivalence has* been one of
the postulates of Einstein's general theory of relativity.

to be the same as the mass which was measured while at rest. This assumption of Newtonian mechanics is in agreement with experimental results for cases where the velocity is very small compared with the velocity of light and relativistic effects are negligible.

Now suppose that the mass of a given body is to be determined by weighing, using a calibrated spring scale. The calibration can be accomplished by using a set of standard masses which are obtained in a manner similar to that used previously except that a spring scale is used in place of a beam balance. (Note that in obtaining standard masses, no calibration of the scale is required since weights are merely matched.)

The use of a spring scale to measure the mass of a body relative to the mass of a primary standard will be observed to have all the shortcomings of the beam balance. In addition, the calibration of the scale is accurate only if the weighings occur at a location with an acceleration of gravity equal to the calibration value. This follows from the definition of weight given by Eq. (1–47) and the fact that the spring scale measures weight or force directly rather than matching weights as in the case of the beam balance.

In order to avoid the theoretical difficulties involved in determining mass by means of weight measurements, Mach proposed a method of measuring *inertial mass* ratios by measuring accelerations. A dynamically isolated system is assumed, consisting of two mutually interacting masses. Then, making the experimental proposition that the accelerations are opposite in direction and lie along the line connecting their centers, he defines the mass ratio to be the inverse ratio of their acceleration magnitudes. Designating the bodies as A and B, the mass ratio is

$$\frac{m_A}{m_B} = \frac{a_B}{a_A} \tag{1–48}$$

where a_A and a_B are the magnitudes of the corresponding accelerations. Then, if mass A is a standard unit, the mass of body B can, in theory, be determined.

Mach does not specify the nature of the interaction in his definition. It might be gravitational or perhaps a direct connection as by a rigid rod or elastic rope. Another possibility is a momentary interaction such as occurs in an impact.

Besides being a *dynamic* method of measuring mass, Mach's procedure has the virtue of not specifically assuming the concept of force. Given the definitions of mass, length, and time, the unit of force can then be defined in accordance with the law of motion as the effort required to produce a unit acceleration of a unit mass. Furthermore, the law of action and reaction follows from the law of motion and Mach's experimental proposition in the definition of a mass ratio.

Nevertheless, Mach's procedure has its shortcomings. First is the practical difficulty of finding a system of two interacting masses that is dynamically isolated. The earth's gravitational attraction would be an extraneous influence for experiments performed near the earth, although the effect could be minimized by confining the motion to a horizontal plane by constraints with very small friction. Another possibility is to cause the masses to hit each other. As will be seen more clearly in Sec. 4–7, the impact will cause very large accelerations of short duration, and during this interval, the effects of other influences will be relatively minor. But these procedures do not appear to be practical for mass determinations of ordinary objects.

A second objection to Mach's procedure is its tacit assumption that the accelerations are measured relative to an inertial frame of reference. Since the criterion for an inertial frame is that Newton's law of motion applies in this frame, we see that, again, the definition of mass assumes the validity of a law which presupposes a definition of mass.

A dynamic method of mass measurement such as Mach proposes would seem to be the least objectionable on theoretical grounds. But, as a practical matter, the measurement of mass by a weighing procedure will undoubtedly continue to be widely used.

Another important point in the discussion of mass is that Newtonian mechanics assumes the *principle of conservation of mass*. Thus, while a certain quantity of water may be converted to the vapor phase, its mass is not changed in the process. Similarly, if a rocket is fired, the total mass of the rocket body and ejected gases remains constant. This principle enables one to consider a wide variety of physical systems as a group of particles, constant in number, with the mass of each particle remaining the same in spite of changes in its motion or its physical state.

Force. We have seen that if standard units of length, time, and mass are chosen, then the corresponding unit of force is determined in accordance with Newton's law of motion. Thus, a unit force applied to a unit mass results in a unit acceleration.

Now, suppose that *force*, length, and time are chosen as fundamental dimensions. What is a suitable operational procedure for defining a standard unit of force?

We shall define a unit force in terms of the effort required to produce a certain extension in a given standard spring under standard conditions. Then, noting that springs connected in series transmit equal forces, we can mark the extensions produced by the standard force in each of a set of similar springs, thereby producing a group of standard springs. Next, we observe that if n standard springs are connected in parallel and if each is stretched the standard amount, then the force transmitted by the n springs is n standard force units. So, by connecting a given spring in series with various numbers of parallel standard springs, the given spring may be calibrated in standard force units and subsequently used for force measurements.

This definition of force, which is based upon an elastic deformation, has the advantage of not requiring the measurements to be made in an inertial system if the spring is assumed to have negligible mass. In a practical case, the so-called inertial forces associated with the mass of the spring must be small compared to the force transmitted by the spring.

Nevertheless, there are objections to the choice of force as a fundamental dimension. But before we discuss these objections, let us first note that the forces that concern us in mechanics are of two general types, namely, *contact* forces and *field* forces. Contact forces refer to a direct mechanical push or pull which is transmitted by material means. For example, a rod or a rope, or even air or water pressure could be used to apply contact forces to a body at its surface. On the other hand, field forces are associated with action at a distance, such as in the cases of gravitational and electrical forces. Also, field forces are often applied throughout a body rather than at the surface.

Now, difficulties arise when one attempts to apply to field forces the concept of a force as a push or pull that is measured by an elastic deflection. For example, if a body is falling freely in a uniform gravitational field, the gravitational force does not produce any elastic deflection and therefore this method cannot be used to measure it. Actually, the gravitational force is often found by counteracting it with a contact force such that the acceleration is zero and then measuring this contact force. This is the process of weighing the body. The assumption is made, of course, that the gravitational force is not a function of velocity or acceleration.

More generally, field forces are calculated by observing accelerations and obtaining the corresponding forces from Newton's law of motion. In other words, accelerations are observed and then forces are *inferred* in accordance with the law of motion.

The concept of force as a fundamental quantity in the study of mechanics has been criticized by various scientists and philosophers of science from shortly after Newton's enunciation of the laws of motion until the present time. Briefly, the idea of a force, and a field force in particular, was considered to be an intellectual construction which has no real existence. It is merely another name for the product of mass and acceleration which occurs in the mathematics of solving a problem. Furthermore, the idea of force as a *cause* of motion should be discarded since the assumed cause and effect relationships cannot be proved.

We shall adopt the viewpoint that the existence of contact forces can be detected and measured by springs or by other means of measuring elastic deformations. Field forces will be calculated from observed accelerations, using Newton's laws of motions, or else from rules governing the force that have been established from such observations. For the most part, we shall avoid questions of cause and effect. For example, suppose a stone is tied to the end of a string and is whirled in a circular path. Does the tension in the string cause the stone to follow a circular path, or does the motion in a circular path cause the tension in the string? Conceivably, either viewpoint could be taken. But generally it is preferable to note that forces and accelerations occur simultaneously and neither is specifically cause nor effect.

We have seen that the definitions of the fundamental units of mechanics depend to some extent upon Newton's laws of motion and these are the laws to be demonstrated. Thus, the logic has a certain circularity. Nevertheless, the validity of the laws of motion has been established for mechanical systems with a wide range of velocities and spatial dimensions. So long as the distances involved are larger than atomic dimensions and the velocities are much smaller than the velocity of light, the fundamental laws of Newtonian mechanics apply remarkably well.

1-5. D'ALEMBERT'S PRINCIPLE

Newton's law of motion for a particle is given by Eq. (1–41). Suppose we write it in the form

$$\mathbf{F} - m\mathbf{a} = 0 \tag{1–49}$$

where $\mathbf{F}$ is the sum of the external forces acting on the particle, m is its mass, and $\mathbf{a}$ is the acceleration of the particle relative to an inertial reference frame. Now, if we consider the term $-m\mathbf{a}$ to represent another force, known as an *inertial force* or *reversed effective force*, then Eq. (1–49) states that the vector sum of all forces, external and inertial, is zero. But this is just the form of the force summation equation of statics and the methods of analysis for statics problems apply to it, including more advanced methods, such as virtual work (Sec. 6–4). Thus, in a sense, the dynamics problem has been reduced to a statics problem. Briefly, then, d'Alembert's principle states that the laws of static equilibrium apply to a dynamical system if the inertial forces, as well as the actual external forces, are considered as applied forces acting on the system. Of course, one still must solve the differential equations of dynamics rather than the algebraic equations of statics, but the point of view in setting up the equations is similar to that of statics.

Considerable care must be used in setting up the equations of motion by means of d'Alembert's principle. In particular, the inertial forces should not be confused with the external forces comprising the total force $\mathbf{F}$ that is applied to the particle. In order to keep the distinction clear on force diagrams, we will designate inertial forces by a dashed arrow and external forces by a solid arrow, as shown in Fig. 1–10. By *external forces*, we mean contact forces and gravitational or other field forces applied to the particle.

Another approach is to think of the inertial force as a reaction force exerted by the particle on its surroundings in accordance with the law of action and reaction. This viewpoint is convenient when the motion of a particle is given and one desires to calculate the force it exerts on the remainder of the system. For example, if a particle is whirled in a circular path by means of a string attached to a fixed point, then the inertial force of the particle is the so-called centrifugal force which is equal to the tensile force in the string. On the other hand, the external force on the particle is the centripetal force of the string acting radially inward toward the fixed point at the center of the circular path. By the law of action and reaction, the centripetal force is equal in magnitude to the centrifugal force but is opposite in direction.

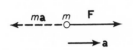

Fig. 1-10. Inertial force for an accelerating particle.

Example 1-1. A particle of mass m is supported by a massless wire of length l that is attached to a point O of a box having a constant acceleration **a** to the right (Fig. 1–11). Find the angle θ corresponding to a condition of equilibrium, that is, such that the particle will remain at constant

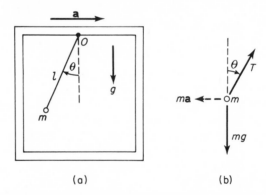

(a)   (b)

Fig. 1-11. Simple pendulum in an accelerating box.

θ if released from this position. What is the tensile force T in the wire in this case? [The direction of the acceleration due to gravity is shown by the vertical arrow in Fig. 1–11(a).]

The external forces acting on the particle are the weight mg and the force T due to the wire, as shown in Fig. 1–11(b). In addition, one notes that the particle has an acceleration **a** to the right for constant θ, and therefore the inertial force is of magnitude ma and is directed to the left.

Now, according to d'Alembert's principle, the external and inertial forces must add vectorially to zero. So, taking horizontal components, we obtain

$$T \sin \theta - ma = 0$$

Similarly, taking vertical components,

$$T \cos \theta - mg = 0$$

Solving these two equations, we obtain

$$\theta = \tan^{-1}\left(\frac{a}{g}\right)$$

and

$$T = m\sqrt{a^2 + g^2}$$

REFERENCES

1. Housner, G. W., and D. E. Hudson, *Applied Mechanics—Dynamics*, 2nd ed. Princeton, N.J.: D. Van Nostrand Co., Inc., 1959.

2. Jammer, M., *Concepts of Space*. Cambridge, Mass.: Harvard University Press, 1954.

3. ———, *Concepts of Force*. Cambridge, Mass.: Harvard University Press, 1957.

4. ———, *Concepts of Mass*. Cambridge, Mass.: Harvard University Press, 1961.

5. Lass, H., *Vector and Tensor Analysis*. New York: McGraw-Hill, Inc., 1950.

6. Lindsay, R. B., and H. Margenau, *Foundations of Physics*, New York: Dover Publications, Inc., 1957.

7. Mach, E., *The Science of Mechanics*, 6th ed. LaSalle, Ill.: Open Court Publishing Company, 1960.

8. Phillips, H. B., *Vector Analysis*. New York: John Wiley & Sons, Inc., 1933.

PROBLEMS

1–1. Forces P and Q are applied to a particle of mass m. The magnitudes of these forces are constant but the angle θ between their lines of action can be varied, thereby changing the acceleration of the particle. If the maximum possible acceleration is three times as large as the minimum acceleration, what value of θ is required to attain an acceleration which is the mean of these two values?

1–2. In terms of a fixed cartesian system, the position of automobile A is $\mathbf{r}_a = 100t\mathbf{i}$ ft and the position of airplane P is $\mathbf{r}_p = 1000\mathbf{i} + 300t\mathbf{j} + 2000\mathbf{k}$ ft where the time t is measured in seconds. Find: (a) the velocity of P relative to A and (b) the minimum separation between P and A.

1–3. Given the following triad of unit vectors:

$$e_1 = l_1 i + l_2 j + l_3 k$$
$$e_2 = m_1 i + m_2 j + m_3 k$$
$$e_3 = n_1 i + n_2 j + n_3 k$$

where i, j, k are the usual cartesian unit vectors. (a) Write 3 equations involving the l's, m's, and n's which apply for any unit triad. (b) What additional equation applies if e_1, e_2, and e_3 are coplanar? (c) What equations apply if e_1, e_2, and e_3 are mutually orthogonal?

1–4. Consider the vector $A = A_x i + A_y j + A_z k$. Find the components A_1, A_2, A_3 of the vector A in a skewed coordinate system whose axes have directions specified by the following unit vector triad:

$$e_1 = i, \qquad e_2 = (i + j)/\sqrt{2}, \qquad e_3 = (i + k)/\sqrt{2}.$$

1–5. If units of length, mass, and force were chosen as the fundamental units, what would the dimensions of time and acceleration be?

1–6. Suppose that units of force, length, and time are chosen such that the density of water and the acceleration of gravity are both of unit magnitude. If the pound is taken as the unit of force, what are the sizes of the units of length and time?

1–7. At a certain moment during reentry, a satellite is moving horizontally in the earth's atmosphere. The satellite carries an accelerometer which measures the contact force acting on a small mass moving with the satellite. If the accelerometer reading is 64 ft/sec² in a direction opposite to the velocity of the satellite, and if the acceleration of gravity at that point is 32 ft/sec², what is the absolute acceleration of the satellite? If the speed of the satellite is 10,000 ft/sec, what is the angular rate of rotation of its velocity vector?

1–8. The particles at B and C are of equal mass m and are connected by strings to each other and to points A and D as shown, all points remaining in the same horizontal plane. If points A and D move with the same acceleration a along parallel paths, solve for the tensile force in each of the strings. Assume that all points retain their initial relative positions.

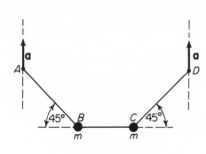

Fig. P1-8

2

KINEMATICS OF A PARTICLE

Kinematics is the study of the motions of particles and rigid bodies, disregarding the forces associated with these motions. It is purely mathematical in nature and does not involve any physical laws such as Newton's laws.

In this chapter we are concerned primarily with the kinematics of a particle, that is, with the motion of a point. Depending upon the circumstances, we may at times choose to consider the point as being attached to a rigid body and at other times as being the location of an individual particle. In any event, we shall be interested in calculating such quantities as the position, velocity, and acceleration vectors of the point.

In discussing the motion of a point, one must specify a frame of reference, since the motion will be different, in general, when viewed from different reference frames. We saw in Sec. 1–2 that Newton's laws of motion apply in a particular set of reference frames known as *inertial frames*. These frames, nonrotating but translating uniformly relative to one another, thus constitute a special or preferred set in writing the dynamical equations of a particle. From the viewpoint of kinematics, however, there are no preferred frames of reference since physical laws are not involved in the derivation of kinematical equations. Thus, in the strict sense, all motion is considered to be relative and no reference frame is more fundamental or absolute than another. Nevertheless, we shall at times refer to absolute motions because of the dynamical background, or for ease in exposition.

2–1. POSITION, VELOCITY, AND ACCELERATION OF A POINT

The position of a point P relative to the XYZ reference frame (Fig. 2–1) is given by the vector $\mathbf{r}$ drawn from the origin O to P. If the point P moves on a curve C, then the velocity $\mathbf{v}$ is in the direction of the tangent to the curve at that point and has a magnitude equal to the speed[1] with which it moves along the curve.

---

[1]*Speed* is a scalar quantity and is equal to the magnitude of the velocity. Sometimes the term *velocity* is used loosely in the same sense, the meaning generally being clear from the context.

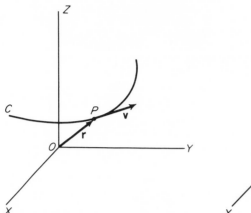

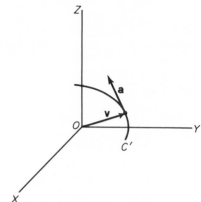

Fig. 2-1. Position and velocity vectors of a particle P as it moves along curve C.

Fig. 2-2. The hodograph showing velocity and acceleration vectors.

Considering the velocity **v** as a free vector, we can imagine that it is drawn with its origin at the origin of the coordinate system at each instant of time, as shown in Fig. 2–2. The path C' of the tip of the velocity vector **v** is then known as the *hodograph*. Now, in general, the velocity of the tip of a vector drawn from the origin in a given coordinate system is simply the time derivative of the vector in that system. So the velocity of the tip of the vector **v** along the hodograph is the *acceleration* **a** of point P relative to the XYZ system. Hence we summarize by stating that the velocity of point P is

$$\mathbf{v} = \frac{d\mathbf{r}}{dt} \tag{2-1}$$

and the corresponding acceleration is

$$\mathbf{a} = \frac{d\mathbf{v}}{dt} = \frac{d^2\mathbf{r}}{dt^2} \tag{2-2}$$

Note that a knowledge of the path C does not determine the motion unless the speed along the path is known at all times. Similarly, a knowledge of C' does not determine the motion unless one knows the starting point in space as well as the rate of travel along the hodograph at all times. If both C_i and C' are known and are compatible, then the motion is nearly always determined, an exception being the case of rectilinear motion at a varying speed. We always assume that C is continuous, corresponding to a finite velocity of the point.

2–2. ANGULAR VELOCITY

In studying the kinematics of a particle, we are primarily concerned with the translation of a point relative to a given reference frame, that is, with the position vector of the point and with its derivatives with respect to time. On the other hand, the general motion of a rigid body involves changes of *orientation* as well as changes of location. It is the rate of change of orientation that is expressed by means of the angular velocity vector.

Consider the motion of the rigid body shown in Fig. 2–3 during the infinitesimal interval Δt. Anticipating the results of Chasles' theorem (Sec. 7–8), we note that the infinitesimal displacement during this interval can be considered as a translational displacement Δs of all points in the body plus a rotational displacement $\Delta \theta$ about an axis through the *base point* A' fixed in the body. The order of performing the translation and rotation is immaterial, but Fig. 2–3 shows the case where the translation occurs first. Thus a typical point P moves to P' and the base point A moves to A', each undergoing the same displacement Δs. Then the infinitesimal rotation $\Delta \theta$ occurs, moving P' to P'' while A' does not move since it is on the axis of rotation.

The infinitesimal angular displacement $\Delta \theta$ is a vector whose magnitude

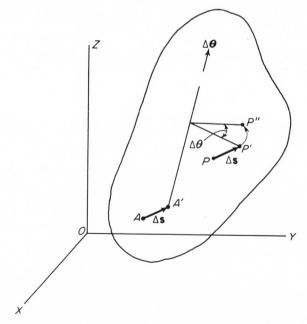

Fig. 2-3. Infinitesimal displacement of a rigid body.

is equal to the angle of rotation and whose direction is along an axis determined by those points not displaced by the infinitesimal rotation. The sense is in accordance with the right-hand rule. We define the angular velocity $\boldsymbol{\omega}$ as follows:

$$\boldsymbol{\omega} = \lim_{\Delta t \to 0} \frac{\Delta \boldsymbol{\theta}}{\Delta t} \tag{2-3}$$

Now, if we should analyze the same infinitesimal displacement of the body but choose a different base point, only the translational part of the motion would be changed; the rotational displacement would be identical. Thus we can see that the angular velocity is a property of the body as a whole and is not dependent upon the choice of a base point. Therefore the angular velocity $\boldsymbol{\omega}$ is a free vector.

Note that measurements of the angular velocity of a body as viewed from different reference frames will, in general, produce different results. Therefore, when giving the angular velocity, the reference frame should be stated or clearly implied. An inertial frame is usually assumed if no statement is made to the contrary. Also, recall again that angular velocity refers to the motion of a rigid body, or, in essence, to the motion of three rigidly connected points that are not collinear. The term has no unique meaning for the motion of a point or a vector (or a straight line) in three-dimensional space. Finally, the angular velocity vector will usually change both its magnitude and direction continuously with time. Only in the simpler cases is the motion confined to rotation about an axis fixed in space or in the body.

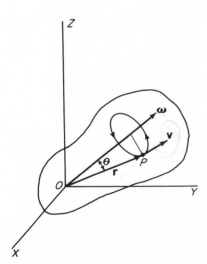

Fig. 2-4. Rigid body rotation about a fixed point.

2–3. RIGID BODY MOTION ABOUT A FIXED POINT

As we have seen, the rotational motion relative to a given system does not depend upon the choice of the base point. So, in order to concentrate upon the rotational aspects of the motion, assume that the base point of the rigid body is fixed at the origin O of the cartesian system XYZ, as shown in Fig. 2–4. Let us calculate the velocity $\mathbf{v}$ relative to the XYZ system of a point P that is fixed in the body. If the XYZ system is fixed in inertial space, then $\mathbf{v}$ is the absolute velocity of point P and $\boldsymbol{\omega}$ is the absolute angular velocity of the body.

In general, the rotation at any instant is taking place about an axis passing through the fixed base point. This axis is known as the *instantaneous axis of rotation*. Consider first the case where the rotation takes place about an axis that is fixed in the XYZ system. Then $\boldsymbol{\omega}$ has a fixed direction and the path of point P is a circle of radius $r \sin \theta$, as can be seen from Fig. 2–4. The speed with which point P moves along the circle is

$$\dot{s} = \omega r \sin \theta \qquad (2\text{–}4)$$

where s is the displacement of P along its path, ω is the magnitude of $\boldsymbol{\omega}$, r is the constant length of the position vector drawn from O to P, and θ is the angle between $\boldsymbol{\omega}$ and $\mathbf{r}$. The velocity of P is of magnitude $\dot{s}$ and is directed along the tangent to the path. Thus we can write

$$\mathbf{v} = \boldsymbol{\omega} \times \mathbf{r} \qquad (2\text{–}5)$$

In the general case where ω and θ may vary with time, the displacement Δs of point P during the infinitesimal interval Δt is

$$\Delta s = \omega r \, \Delta t \sin \theta \qquad (2\text{–}6)$$

where higher-order terms have been neglected. It is assumed that $\Delta \omega$ and $\Delta \theta$ approach zero as the interval Δt approaches zero, implying that $\dot{\boldsymbol{\omega}}$ is finite. Of course, Δr is zero, since point P and the base point O are fixed in the body.

Dividing Eq. (2–6) by Δt and letting Δs and Δt approach zero, we again obtain Eq. (2–4). Furthermore, the direction of $\mathbf{v}$ is again normal to $\boldsymbol{\omega}$ and $\mathbf{r}$. Hence we see that Eq. (2–5) is valid for the general case of rigid-body rotation about a fixed point.

Differentiating Eq. (2–5) with respect to time, we obtain an expression for the acceleration $\mathbf{a}$ of point P.

$$\mathbf{a} = \boldsymbol{\omega} \times \dot{\mathbf{r}} + \dot{\boldsymbol{\omega}} \times \mathbf{r}$$

or, noting that $\dot{\mathbf{r}}$ is the velocity $\mathbf{v}$ and using Eq. (2–5),

$$\mathbf{a} = \boldsymbol{\omega} \times (\boldsymbol{\omega} \times \mathbf{r}) + \dot{\boldsymbol{\omega}} \times \mathbf{r} \qquad (2\text{–}7)$$

where, of course, all vectors are measured relative to the XYZ system.

It can be seen that the term $\boldsymbol{\omega} \times (\boldsymbol{\omega} \times \mathbf{r})$ in Eq. (2–7) is a vector that is directed radially inward from point P toward the instantaneous axis of rotation and perpendicular to it. It is called the *centripetal acceleration*. The term $\dot{\boldsymbol{\omega}} \times \mathbf{r}$ is often called the *tangential acceleration*. Note, however, that the so-called tangential acceleration is in a direction tangent to the path of P only if $\dot{\boldsymbol{\omega}}$ is parallel to the plane of $\boldsymbol{\omega}$ and $\mathbf{r}$. This would occur, for example, if $\boldsymbol{\omega}$ retains a constant direction and changes in magnitude only.

2-4. TIME DERIVATIVE OF A UNIT VECTOR

We have obtained the derivative with respect to time of the position vector **r** of a point P in a rigid body that is rotating about a fixed point O (Fig. 2-4). A similar situation exists in the calculation of the rates of change of unit vectors. As was the case with the position vector to point P, the unit vectors are each of constant length. Furthermore, we have seen that the time derivative of a vector can be interpreted as the velocity of the tip of the vector when the other end is fixed. So let us calculate the velocities of the unit vectors $\mathbf{e}_1$, $\mathbf{e}_2$, and $\mathbf{e}_3$ drawn from the origin of the fixed system XYZ and rotating together as a rigid body with an absolute angular velocity $\boldsymbol{\omega}$ (Fig. 2-5).

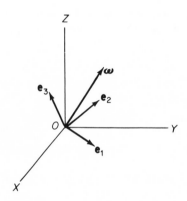

Fig. 2-5. The rotation of a unit vector triad.

From Eq. (2-5), we see that the velocities of the tips of the unit vectors, that is, their time derivatives, are

$$\dot{\mathbf{e}}_1 = \boldsymbol{\omega} \times \mathbf{e}_1$$

$$\dot{\mathbf{e}}_2 = \boldsymbol{\omega} \times \mathbf{e}_2 \qquad (2\text{-}8)$$

$$\dot{\mathbf{e}}_3 = \boldsymbol{\omega} \times \mathbf{e}_3$$

For the case where $\mathbf{e}_1$, $\mathbf{e}_2$, and $\mathbf{e}_3$ form an orthogonal set of unit vectors, that is, $\mathbf{e}_1 \times \mathbf{e}_2 = \mathbf{e}_3$, and so on, we can perform the vector multiplications using the determinant expression of Eq. (1-27). Writing the angular velocity $\boldsymbol{\omega}$ in the form

$$\boldsymbol{\omega} = \omega_1 \mathbf{e}_1 + \omega_2 \mathbf{e}_2 + \omega_3 \mathbf{e}_3 \qquad (2\text{-}9)$$

we obtain

$$\dot{\mathbf{e}}_1 = \begin{vmatrix} \mathbf{e}_1 & \mathbf{e}_2 & \mathbf{e}_3 \\ \omega_1 & \omega_2 & \omega_3 \\ 1 & 0 & 0 \end{vmatrix} = \omega_3 \mathbf{e}_2 - \omega_2 \mathbf{e}_3 \qquad (2\text{-}10)$$

Similarly,

$$\dot{\mathbf{e}}_2 = \begin{vmatrix} \mathbf{e}_1 & \mathbf{e}_2 & \mathbf{e}_3 \\ \omega_1 & \omega_2 & \omega_3 \\ 0 & 1 & 0 \end{vmatrix} = \omega_1 \mathbf{e}_3 - \omega_3 \mathbf{e}_1 \qquad (2\text{-}11)$$

and

$$\dot{\mathbf{e}}_3 = \begin{vmatrix} \mathbf{e}_1 & \mathbf{e}_2 & \mathbf{e}_3 \\ \omega_1 & \omega_2 & \omega_3 \\ 0 & 0 & 1 \end{vmatrix} = \omega_2 \mathbf{e}_1 - \omega_1 \mathbf{e}_2 \qquad (2\text{-}12)$$

In the particular case where we consider a rotating triad of cartesian unit vectors $\mathbf{i}$, $\mathbf{j}$, and $\mathbf{k}$, we can make the substitutions $\mathbf{e}_1 = \mathbf{i}$, $\mathbf{e}_2 = \mathbf{j}$, and $\mathbf{e}_3 = \mathbf{k}$. Noting that

$$\boldsymbol{\omega} = \omega_x \mathbf{i} + \omega_y \mathbf{j} + \omega_z \mathbf{k} \tag{2–13}$$

where ω_x, ω_y, and ω_z are the components of $\boldsymbol{\omega}$ in the directions of $\mathbf{i}$, $\mathbf{j}$, and $\mathbf{k}$, respectively, (not along the X, Y, and Z axes), we obtain from Eqs. (2–8), (2–9), (2–10), and (2–11) that

$$\dot{\mathbf{i}} = \boldsymbol{\omega} \times \mathbf{i} = \omega_z \mathbf{j} - \omega_y \mathbf{k}$$
$$\dot{\mathbf{j}} = \boldsymbol{\omega} \times \mathbf{j} = \omega_x \mathbf{k} - \omega_z \mathbf{i} \tag{2–14}$$
$$\dot{\mathbf{k}} = \boldsymbol{\omega} \times \mathbf{k} = \omega_y \mathbf{i} - \omega_x \mathbf{j}$$

It can be seen that, in each case, the time derivative of a unit vector lies in a plane perpendicular to the vector, in accordance with the definition of a cross product.

Observe that in each case we have calculated the rate of change of a unit vector with respect to a fixed coordinate system but have expressed the result in terms of the unit vectors of the moving system. Since this approach will be used extensively in our later work, the terminology should be made clear. The terms *relative to* or *with respect to* a given system mean as viewed by an observer fixed in that system and moving with it. On the other hand, the term *referred to* a certain system means that the vector is expressed in terms of the unit vectors of that system. For example, the absolute acceleration of a certain particle can be given in terms of the unit vectors of a fixed or of a moving coordinate system. In either event, we are considering the same vector but are merely expressing the result in two different forms. On the other hand, the acceleration of the particle relative to a fixed coordinate system and the acceleration relative to a moving coordinate system would, in general, be quite different vectors.

2–5. VELOCITY AND ACCELERATION OF A PARTICLE IN SEVERAL COORDINATE SYSTEMS

Cartesian Coordinates. Suppose that the position of a particle P relative to the origin O of the cartesian system xyz is given by the vector (Fig. 2–6)

$$\mathbf{r} = x\mathbf{i} + y\mathbf{j} + z\mathbf{k} \tag{2–15}$$

where $\mathbf{i}$, $\mathbf{j}$, and $\mathbf{k}$ are unit vectors fixed in the xyz system.

Differentiating Eq. (2–15) with respect to time, we find that the velocity of point P relative to the xyz system is given by

$$\mathbf{v} = \dot{\mathbf{r}} = \dot{x}\mathbf{i} + \dot{y}\mathbf{j} + \dot{z}\mathbf{k} \tag{2–16}$$

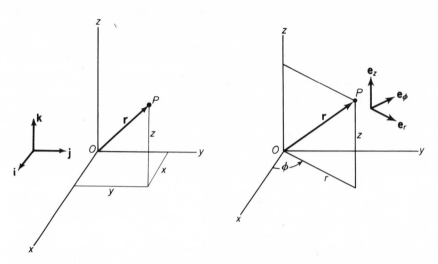

Fig. 2-6. Cartesian coordinates and unit vectors.

Fig. 2-7. Cylindrical coordinates and unit vectors.

where we note that the time derivatives of the unit vectors **i**, **j**, and **k** are all zero since they are fixed in the given cartesian system.

Similarly, differentiating Eq. (2–16) with respect to time, we find that the acceleration of P relative to the xyz system is

$$\mathbf{a} = \ddot{x}\mathbf{i} + \ddot{y}\mathbf{j} + \ddot{z}\mathbf{k} \qquad (2\text{--}17)$$

For the common case where the cartesian system is assumed to be fixed in an inertial frame, the preceding expressions refer to the *absolute* velocity and acceleration.

Cylindrical Coordinates. Now consider the case where the position of point P relative to the xyz system shown in Fig. 2–7 is expressed in terms of the cylindrical coordinates r, ϕ, and z. In this case, the position vector of P is

$$\mathbf{r} = r\mathbf{e}_r + z\mathbf{e}_z \qquad (2\text{--}18)$$

Note that the unit vectors $\mathbf{e}_r$, $\mathbf{e}_\phi$, and $\mathbf{e}_z$ form a mutually orthogonal triad whose directions are given by the directions in which P moves for small increases in r, ϕ, and z, respectively. As before, we obtain the velocity by finding the derivative of **r** with respect to time:

$$\mathbf{v} = \dot{\mathbf{r}} = \dot{r}\mathbf{e}_r + \dot{z}\mathbf{e}_z + r\dot{\mathbf{e}}_r + z\dot{\mathbf{e}}_z \qquad (2\text{--}19)$$

It can be seen that $\mathbf{e}_r$ and $\mathbf{e}_\phi$ change their directions in space as P moves through a general displacement, but $\mathbf{e}_z$ is always parallel to the z axis. Further, the changes in the directions of $\mathbf{e}_r$ and $\mathbf{e}_\phi$ are due solely to changes in ϕ, corresponding to rotations about the z axis. Thus the absolute rotation rate of the unit vector triad is

$$\boldsymbol{\omega} = \dot{\phi}\mathbf{e}_z \qquad (2\text{--}20)$$

and, in accordance with Eq. (2–8), we find that

$$\dot{\mathbf{e}}_r = \boldsymbol{\omega} \times \mathbf{e}_r = \dot{\phi}\mathbf{e}_\phi$$
$$\dot{\mathbf{e}}_\phi = \boldsymbol{\omega} \times \mathbf{e}_\phi = -\dot{\phi}\mathbf{e}_r \qquad (2\text{--}21)$$
$$\dot{\mathbf{e}}_z = \boldsymbol{\omega} \times \mathbf{e}_z = 0$$

So from Eqs. (2–19) and (2–21), we obtain

$$\mathbf{v} = \dot{r}\mathbf{e}_r + r\dot{\phi}\mathbf{e}_\phi + \dot{z}\mathbf{e}_z \qquad (2\text{--}22)$$

Another differentiation with respect to time results in the acceleration

$$\mathbf{a} = \ddot{\mathbf{r}} = \ddot{r}\mathbf{e}_r + (\dot{r}\dot{\phi} + r\ddot{\phi})\mathbf{e}_\phi + \ddot{z}\mathbf{e}_z + \dot{r}\dot{\mathbf{e}}_r + r\dot{\phi}\dot{\mathbf{e}}_\phi + \dot{z}\dot{\mathbf{e}}_z$$

which can be simplified with the aid of Eq. (2–21) to yield

$$\mathbf{a} = (\ddot{r} - r\dot{\phi}^2)\mathbf{e}_r + (r\ddot{\phi} + 2\dot{r}\dot{\phi})\mathbf{e}_\phi + \ddot{z}\mathbf{e}_z \qquad (2\text{--}23)$$

Spherical Coordinates. The position vector of the point P in terms of spherical coordinates and the corresponding unit vectors is simply

$$\mathbf{r} = r\mathbf{e}_r \qquad (2\text{--}24)$$

since, in this case, the unit vector $\mathbf{e}_r$ is defined to be in the direction of the position vector $\mathbf{r}$ (Fig. 2–8). Differentiating Eq. (2–24) with respect to time, we obtain the velocity

$$\mathbf{v} = \dot{\mathbf{r}} = \dot{r}\mathbf{e}_r + r\dot{\mathbf{e}}_r \qquad (2\text{--}25)$$

In order to evaluate $\dot{\mathbf{e}}_r$, $\dot{\mathbf{e}}_\theta$, and $\dot{\mathbf{e}}_\phi$, we note first that changes in the directions of the unit vectors can arise because of changes in θ or ϕ. An increase in θ rotates the unit vector triad about an axis having the direction $\mathbf{e}_\phi$, whereas an increase in ϕ corresponds to a rotation about the z axis. Thus the total rotation rate of the unit vector triad is

$$\boldsymbol{\omega} = \dot{\theta}\mathbf{e}_\phi + \dot{\phi}\mathbf{e}_z \qquad (2\text{--}26)$$

where $\mathbf{e}_z$ is a unit vector in the direction of the positive z axis.

Since $\mathbf{e}_\phi$ and $\mathbf{e}_z$ are perpendicular, it is relatively easy to express $\mathbf{e}_z$ in terms of the remaining unit vectors:

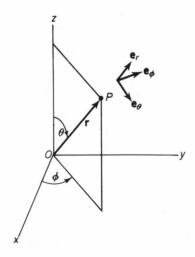

Fig. 2-8. Spherical coordinates and unit vectors.

$$\mathbf{e}_z = \cos\theta\,\mathbf{e}_r - \sin\theta\,\mathbf{e}_\theta \tag{2-27}$$

Therefore, from Eqs. (2–26) and (2–27), we obtain

$$\boldsymbol{\omega} = \dot{\phi}\cos\theta\,\mathbf{e}_r - \dot{\phi}\sin\theta\,\mathbf{e}_\theta + \dot{\theta}\,\mathbf{e}_\phi \tag{2-28}$$

The unit vectors $\mathbf{e}_r$, $\mathbf{e}_\theta$, and $\mathbf{e}_\phi$ are mutually orthogonal, so we can use the determinant method of evaluating the cross products involved in obtaining $\dot{\mathbf{e}}_r$, $\dot{\mathbf{e}}_\theta$, and $\dot{\mathbf{e}}_\phi$. From Eqs. (2–8) and (2–28), we have

$$\dot{\mathbf{e}}_r = \boldsymbol{\omega} \times \mathbf{e}_r = \begin{vmatrix} \mathbf{e}_r & \mathbf{e}_\theta & \mathbf{e}_\phi \\ \dot{\phi}\cos\theta & -\dot{\phi}\sin\theta & \dot{\theta} \\ 1 & 0 & 0 \end{vmatrix} \tag{2-29}$$

and, using a similar procedure to evaluate $\dot{\mathbf{e}}_\theta$ and $\dot{\mathbf{e}}_\phi$, we can summarize as follows:

$$\begin{aligned} \dot{\mathbf{e}}_r &= \dot{\theta}\mathbf{e}_\theta + \dot{\phi}\sin\theta\,\mathbf{e}_\phi \\ \dot{\mathbf{e}}_\theta &= -\dot{\theta}\mathbf{e}_r + \dot{\phi}\cos\theta\,\mathbf{e}_\phi \\ \dot{\mathbf{e}}_\phi &= -\dot{\phi}\sin\theta\,\mathbf{e}_r - \dot{\phi}\cos\theta\,\mathbf{e}_\theta \end{aligned} \tag{2-30}$$

Finally, from Eqs. (2–25) and (2–30), the velocity of point P is

$$\mathbf{v} = \dot{r}\mathbf{e}_r + r\dot{\theta}\mathbf{e}_\theta + r\dot{\phi}\sin\theta\,\mathbf{e}_\phi \tag{2-31}$$

In order to find the acceleration of P, we differentiate again, obtaining

$$\begin{aligned} \mathbf{a} = \dot{\mathbf{v}} &= \ddot{r}\mathbf{e}_r + (\dot{r}\dot{\theta} + r\ddot{\theta})\mathbf{e}_\theta + (\dot{r}\dot{\phi}\sin\theta + r\ddot{\phi}\sin\theta + r\dot{\theta}\dot{\phi}\cos\theta)\mathbf{e}_\phi \\ &\quad + \dot{r}\dot{\mathbf{e}}_r + r\dot{\theta}\dot{\mathbf{e}}_\theta + r\dot{\phi}\sin\theta\,\dot{\mathbf{e}}_\phi \end{aligned} \tag{2-32}$$

Substituting Eq. (2–30) into (2–32) and collecting terms, the result is

$$\begin{aligned} \mathbf{a} &= (\ddot{r} - r\dot{\theta}^2 - r\dot{\phi}^2\sin^2\theta)\mathbf{e}_r + (r\ddot{\theta} + 2\dot{r}\dot{\theta} - r\dot{\phi}^2\sin\theta\cos\theta)\mathbf{e}_\theta \\ &\quad + (r\ddot{\phi}\sin\theta + 2\dot{r}\dot{\phi}\sin\theta + 2r\dot{\theta}\dot{\phi}\cos\theta)\mathbf{e}_\phi \end{aligned} \tag{2-33}$$

Tangential and Normal Components. The velocity and acceleration of a point P as it moves on a curved path in space may be expressed in terms of tangential and normal components. Let us assume that the position of P is specified by its distance s along the curve from a given reference point. From Fig. 2–9, it can be seen that, as P moves an infinitesimal distance ds along the curve, the corresponding change in the position vector $\mathbf{r}$ is

$$d\mathbf{r} = ds\,\mathbf{e}_t$$

or

$$\mathbf{e}_t = \frac{d\mathbf{r}}{ds} \tag{2-34}$$

where $\mathbf{e}_t$ is a unit vector that is tangent to the path at P and points in the direction of increasing s. The velocity of P is

$$\mathbf{v} = \frac{d\mathbf{r}}{dt} = \dot{s}\frac{d\mathbf{r}}{ds} = \dot{s}\mathbf{e}_t \tag{2-35}$$

To find the acceleration of P, we differentiate Eq. (2–35) with respect to time, obtaining

$$\mathbf{a} = \dot{\mathbf{v}} = \ddot{s}\mathbf{e}_t + \dot{s}\dot{\mathbf{e}}_t \qquad (2\text{–}36)$$

Before evaluating $\dot{\mathbf{e}}_t$ explicitly, we note that, in general, $\mathbf{e}_t$ changes direction continuously with increasing s. The direction of the derivative $d\mathbf{e}_t/ds$ is normal to $\mathbf{e}_t$, since the magnitude of $\mathbf{e}_t$ is constant. This is known as the *normal* direction at P and is designated by the unit vector $\mathbf{e}_n$. The plane of $\mathbf{e}_t$ and $\mathbf{e}_n$ at any point P is called the *osculating plane*. Therefore, any motion of $\mathbf{e}_t$ must take place in the osculating plane at

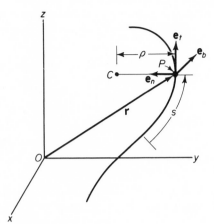

Fig. 2-9. Tangential, normal, and binormal unit vectors for a curve in space.

that point, and hence the vector $\dot{\mathbf{e}}_t$ lies in this plane and points in the $\mathbf{e}_n$ direction. Its magnitude is equal to its rotation rate ω_b about an axis perpendicular to the osculating plane. Thus

$$\dot{\mathbf{e}}_t = \dot{s}\frac{d\mathbf{e}_t}{ds} = \omega_b \mathbf{e}_n \qquad (2\text{–}37)$$

where ω_b is the rotation rate of $\mathbf{e}_t$ about an axis in the *binormal* direction $\mathbf{e}_b$. The binormal unit vector $\mathbf{e}_b$ is perpendicular to the osculating plane and is given by

$$\mathbf{e}_b = \mathbf{e}_t \times \mathbf{e}_n \qquad (2\text{–}38)$$

thus completing the orthogonal triad of unit vectors at P.

The angular rate ω_b can be expressed in terms of the *radius of curvature* ρ and the speed $\dot{s}$ of the point P as it moves along the curve.

$$\omega_b = \frac{\dot{s}}{\rho} \qquad (2\text{–}39)$$

Note that the vector $\rho\mathbf{e}_n$ gives the position of the *center of curvature* C relative to P. Of course, the position of the center of curvature changes, in general, as P moves along the curve.

From Eqs. (2–37) and (2–39), we obtain

$$\dot{\mathbf{e}}_t = \frac{\dot{s}}{\rho} \mathbf{e}_n \qquad (2\text{–}40)$$

and, substituting this expression into Eq. (2–36), we have

$$\mathbf{a} = \ddot{s}\mathbf{e}_t + \frac{\dot{s}^2}{\rho} \mathbf{e}_n \qquad (2\text{–}41)$$

The acceleration component $\ddot{s}$ is the *tangential* acceleration and the component $\dot{s}^2/\rho$ is the *normal or centripetal* acceleration.

The time derivatives of the unit vectors can also be obtained in terms of the angular rotation rate $\boldsymbol{\omega}$ of the unit vector triad. We have

$$\boldsymbol{\omega} = \omega_t \mathbf{e}_t + \omega_b \mathbf{e}_b \tag{2–42}$$

Note that the normal component of $\boldsymbol{\omega}$ is zero. This can be explained by the fact that $\mathbf{e}_n$ was defined to lie in the direction of $\dot{\mathbf{e}}_t$. Any normal component of $\boldsymbol{\omega}$ would result in a binormal component of $\dot{\mathbf{e}}_t$, in conflict with the original assumption. Of course, ω_t does not influence $\dot{\mathbf{e}}_t$.

From Eqs. (2–8) and (2–42), we find that

$$\dot{\mathbf{e}}_t = \boldsymbol{\omega} \times \mathbf{e}_t = \omega_b \mathbf{e}_n$$
$$\dot{\mathbf{e}}_n = \boldsymbol{\omega} \times \mathbf{e}_n = -\omega_b \mathbf{e}_t + \omega_t \mathbf{e}_b \tag{2–43}$$
$$\dot{\mathbf{e}}_b = \boldsymbol{\omega} \times \mathbf{e}_b = -\omega_t \mathbf{e}_n$$

We have seen that ω_b is equal in magnitude to $\dot{\mathbf{e}}_t$ and is a function of the speed $\dot{s}$ and the radius of curvature ρ. Similarly, ω_t is equal in magnitude to $\dot{\mathbf{e}}_b$ and is a function of $\dot{s}$ and $d\mathbf{e}_b/ds$. Thus,

$$\dot{\mathbf{e}}_b = \dot{s}\frac{d\mathbf{e}_b}{ds} \tag{2–44}$$

where the magnitude of $d\mathbf{e}_b/ds$ is called the *torsion* of the curve. It can be seen that ω_t is also the rotation rate of the osculating plane.

2–6. SIMPLE MOTIONS OF A POINT

Circular Motion. Consider the motion of a point P in a circular path of radius a (Fig. 2–10) in the fixed xy plane. The motion will be described in terms of the polar coordinates (r, θ) and the corresponding unit vectors. In this case, the coordinate r is the scalar constant a. Thus, we have

$$\mathbf{r} = a\mathbf{e}_r \tag{2–45}$$

and

$$\mathbf{v} = \dot{\mathbf{r}} = a\dot{\mathbf{e}}_r \tag{2–46}$$

Using the notation that

$$\omega = \dot{\theta} \tag{2–47}$$

we find that the speed along the path is

$$v = a\dot{\theta} = a\omega \tag{2–48}$$

or

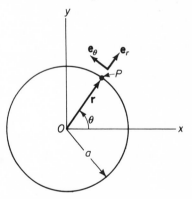

Fig. 2-10. Circular motion of a point.

$$\mathbf{v} = a\omega\mathbf{e}_\theta \tag{2-49}$$

in agreement with Eq. (2-46), since

$$\dot{\mathbf{e}}_r = \omega\mathbf{e}_\theta \tag{2-50}$$

The acceleration is obtained by differentiating Eq. (2-49) with respect to time:

$$\mathbf{a} = \dot{\mathbf{v}} = a\dot{\omega}\mathbf{e}_\theta + a\omega\dot{\mathbf{e}}_\theta \tag{2-51}$$

or, noting that

$$\dot{\mathbf{e}}_\theta = -\omega\mathbf{e}_r \tag{2-52}$$

we obtain

$$\mathbf{a} = -a\omega^2\mathbf{e}_r + a\dot{\omega}\mathbf{e}_\theta \tag{2-53}$$

An alternate form is obtained by substituting for ω from Eq. (2-48), obtaining

$$\mathbf{a} = -\frac{v^2}{a}\,\mathbf{e}_r + \dot{v}\mathbf{e}_\theta \tag{2-54}$$

The first term on the right represents the centripetal acceleration; the second term is the tangential acceleration. Note that the same result is obtained by using the tangential and normal unit vectors of Eq. (2-41) where $\mathbf{e}_t = \mathbf{e}_\theta$, $\mathbf{e}_n = -\mathbf{e}_r$, $\rho = a$, and $\dot{s} = v$.

Helical Motion. Next consider the velocity and acceleration of a particle moving along a helix which is given in terms of cylindrical coordinates by

$$r = a, \qquad z = ka\phi \tag{2-55}$$

where k is the tangent of the helix angle and a is constant (Fig. 2-11). Now let

$$\dot{\phi} = \omega \tag{2-56}$$

Differentiating the expression for z in Eq. (2-55), we obtain

$$\dot{z} = ka\dot{\phi} = ka\omega \tag{2-57}$$

and, from Eqs. (2-22), (2-56), and (2-57), we find that the velocity is

$$\mathbf{v} = \dot{\mathbf{r}} = a\omega\mathbf{e}_\phi + ka\omega\mathbf{e}_z \tag{2-58}$$

The acceleration is found by a similar evaluation of Eq. (2-23), yielding

$$\mathbf{a} = -a\omega^2\mathbf{e}_r + a\dot{\omega}\mathbf{e}_\phi + ka\dot{\omega}\mathbf{e}_z \tag{2-59}$$

For the case where the point P

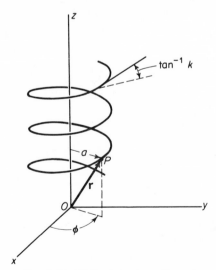

Fig. 2-11. Helical motion of a point.

moves at uniform speed along the helix, the angular acceleration $\dot{\omega}$ is zero and only the radial acceleration remains.

The radius of curvature at a point on the helix can be calculated by equating the magnitude of the radial acceleration found in Eq. (2–59) to the normal or centripetal acceleration previously found in Eq. (2–41). Noting that the speed is

$$\dot{s} = a\omega\sqrt{1 + k^2} \qquad (2\text{–}60)$$

we find that

$$\frac{\dot{s}^2}{\rho} = \frac{a^2\omega^2(1 + k^2)}{\rho} = a\omega^2$$

from which we obtain

$$\rho = a(1 + k^2) \qquad (2\text{–}61)$$

For example, if the helix angle is $45°$, the radius of curvature is $\rho = 2a$ and the center of curvature is diametrically opposite P on the surface of the cylinder $r = a$. In this case, the locus of the centers of curvature is a helix with the same radius and helix angle.

Harmonic Motion. A common type of particle motion is that in which the particle is attracted toward a fixed point by a force that is directly proportional to the distance of the particle from the point. Consider first the one-dimensional case in which the motion takes place along the x axis of a fixed cartesian system because of an attracting center at $x = 0$. From Newton's law of motion as given by Eq. (1–41) or Eq. (1–43), we recall that the acceleration of the particle P is proportional to the applied force. But we have assumed that the applied force is proportional to x and is directed toward the origin. So we can write the differential equation

$$\ddot{x} = -\omega^2 x \qquad (2\text{–}62)$$

where ω^2 is a constant.

The solution can be written in the form

$$x = A\cos(\omega t + \alpha) \qquad (2\text{–}63)$$

where, in general, the constants A and α are evaluated from the values of x and $\dot{x}$ at $t = 0$. If, however, we choose to measure time from the instant when the particle is at the positive extreme of its motion, then $\alpha = 0$, and

$$x = A\cos\omega t \qquad (2\text{–}64)$$

Motion wherein the position oscillates sinusoidally with time is known as *simple harmonic motion* (Fig. 2–12). This motion is periodic with a period

$$T = \frac{2\pi}{\omega} \qquad (2\text{–}65)$$

where ω is usually measured in radians per second and is known as the *circular frequency.*[2]

Differentiating Eq. (2–64), we obtain expressions for the velocity and acceleration.

$$\dot{x} = -A\omega \sin \omega t$$
$$= A\omega \cos \left(\omega t + \frac{\pi}{2} \right) \qquad (2\text{–}66)$$

and

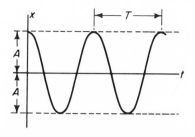

Fig. 2-12. Simple harmonic motion.

$$\ddot{x} = -A\omega^2 \cos \omega t = A\omega^2 \cos (\omega t + \pi) \qquad (2\text{–}67)$$

It can be seen that each differentiation with respect to time results in a sinusoidal oscillation of the same frequency, but the amplitude is multiplied by ω and the phase angle is advanced by $\pi/2$ radians. Thus the extreme values of the displacement and acceleration occur at the moment when the velocity is zero, and vice versa.

Now consider the case of two-dimensional harmonic motion. Suppose the particle P moves in the x direction in accordance with Eq. (2–62), as discussed previously. But it also moves independently in the y direction according to the following equation:

$$\ddot{y} = -\omega^2 y \qquad (2\text{–}68)$$

where the value of ω is assumed to be the same as for the x motion. The solution of this differential equation can be written in the form

$$y = B \cos (\omega t + \beta) \qquad (2\text{–}69)$$

where the angle β is not zero, in general, since the time reference was chosen such that $\alpha = 0$ in Eq. (2–63). Using vector notation, we find that the position of P is

$$\mathbf{r} = x\mathbf{i} + y\mathbf{j} = A \cos \omega t\, \mathbf{i} + B \cos (\omega t + \beta)\, \mathbf{j} \qquad (2\text{–}70)$$

where $\mathbf{i}$ and $\mathbf{j}$ are the usual cartesian unit vectors. Differentiating Eq. (2–70) with respect to time, we see that the velocity is

$$\mathbf{v} = \dot{\mathbf{r}} = -A\omega \sin \omega t\, \mathbf{i} - B\omega \sin (\omega t + \beta)\, \mathbf{j} \qquad (2\text{–}71)$$

and the acceleration is

$$\mathbf{a} = \dot{\mathbf{v}} = -A\omega^2 \cos \omega t\, \mathbf{i} - B\omega^2 \cos (\omega t + \beta)\, \mathbf{j} \qquad (2\text{–}72)$$

The path of the particle P in the xy plane may also be found by first noting from Eq. (2–64) that

---

[2]The term *frequency* refers to the number of cycles per unit time, or T^{-1}. Thus $\omega = 2\pi f$. In this book, we shall, for convenience, generally use the circular frequency ω rather than f and often shall refer to ω as the frequency.

$$\cos \omega t = \frac{x}{A} \tag{2-73}$$

Then, using Eqs. (2–69) and (2–73) and the trigonometric identity for cos $(\omega t + \beta)$, we see that

$$\sin \omega t = (\sin \beta)^{-1} \left(\frac{x}{A} \cos \beta - \frac{y}{B} \right) \tag{2-74}$$

Squaring Eqs. (2–73) and (2–74) and adding, we obtain

$$(\sin \beta)^{-2} \left(\frac{x^2}{A^2} + \frac{y^2}{B^2} - 2 \frac{x}{A} \frac{y}{B} \cos \beta \right) = 1 \tag{2-75}$$

which is the equation of an ellipse (Fig. 2–13).

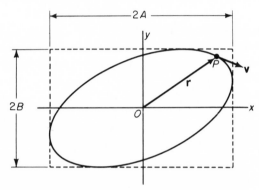

Fig. 2-13. Harmonic motion in two dimensions.

We saw that the x and y motions have the same frequency but are independent of each other. The relative phase angle β between these motions is determined from the initial conditions and the value of β will influence the shape and sense of the motion along the elliptical path. In any event, however, the path is inscribed within the dashed rectangle shown in Fig. 2-13. For example, if $\beta = \pm \pi/2$, Eq. (2–75) reduces to

$$\frac{x^2}{A^2} + \frac{y^2}{B^2} = 1 \tag{2-76}$$

indicating that the x and y axes are principal axes in this case.

On the other hand, if $\beta = 0$ or π, Eq. (2–75) reduces to

$$x = \pm \frac{A}{B} y \tag{2-77}$$

and the path is a straight line along a diagonal of the rectangle of Fig. 2–13. In this case, we again have simple rectilinear harmonic motion.

For the case where $\beta = \pm \pi/2$ and $A = B$, Eq. (2–76) reduces to

$$x^2 + y^2 = A^2 \tag{2-78}$$

which is the equation of a circle of radius A. Also,

$$v^2 = \dot{x}^2 + \dot{y}^2 = A^2\omega^2(\sin^2 \omega t + \cos^2 \omega t) = A^2\omega^2 \qquad (2\text{-}79)$$

Thus the particle P moves at uniform speed about a circular path. The acceleration is also constant in magnitude and, from Eq. (2–53), is

$$\mathbf{a} = -A\omega^2 \mathbf{e}_r \qquad (2\text{-}80)$$

It is interesting to note that elliptical harmonic motion can be considered to be the projection of uniform circular motion onto a plane that is not parallel to the circle. For the case of simple harmonic motion, the two planes are orthogonal and the path reduces to a line. The velocity and acceleration vectors in the case of elliptical motion are found by projecting the corresponding vectors for the case of circular motion. Hence it can be seen that the maximum velocity occurs at the ends of the minor axis and the maximum acceleration occurs at the ends of the major axis. In each of these two cases, the vectors at the given points are parallel to the plane onto which they are projected.

In case the motions in the x and y directions are of *different* frequencies, then a different class of curves known as *Lissajous figures* is generated. Again, the motion remains within the dashed rectangle of Fig. 2–13 but, unless the frequency ratio is a rational number, the curve does not close. For the case of rational frequency ratios, the motion is periodic and retraces itself with a period equal to the least common multiple of the periods of the two component vibrations. We shall not, however, pursue this topic further.

2-7. VELOCITY AND ACCELERATION OF A POINT IN A RIGID BODY

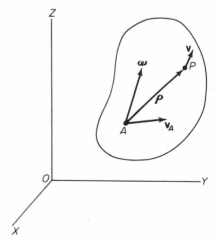

In Sec. 2–3, we calculated the absolute velocity of a point in a rigid body that is rotating about a fixed base point. Now consider the case shown in Fig. 2–14 where the base point A has a velocity $\mathbf{v}_A$ relative to the inertial system XYZ. Also, the body is rotating with an angular velocity $\boldsymbol{\omega}$ relative to this system.

Suppose we define the velocity

$$\mathbf{v}_{PA} = \mathbf{v} - \mathbf{v}_A \qquad (2\text{-}81)$$

where $\mathbf{v}$ and $\mathbf{v}_A$ are the absolute velocities of the points P and A, respectively. Then $\mathbf{v}_{PA}$ is known as the *relative velocity* of P with respect

Fig. 2-14. Motion of a point in a rigid body.

to A as viewed by an observer fixed in the XYZ system. Now suppose the observer is translating in an arbitrary fashion but is not rotating relative to XYZ. The velocities of points P and A as viewed by the translating observer would not be the same as in the previous case. The velocity difference would, however, be the same as before because the apparent velocities of the points P and A would each change by the same amount, namely, by the negative of the observer's translational velocity. Thus the velocity of point P relative to point A is identical for any nonrotating observer.

On the other hand, it is important to realize that the velocity of P relative to A will be different, in general, when the motion is viewed from various reference frames in relative rotational motion, as we shall demonstrate. Hence a statement of the relative velocity of two points should also specify the reference frame. An inertial or a nonrotating frame is assumed if none is stated explicitly.

Returning now to a consideration of the relative velocity $\mathbf{v}_{PA}$ as viewed by a nonrotating observer, we note that it can also be considered to be the velocity of P as viewed by an observer on a nonrotating system that is translating with A. In this case, the base point A would have no velocity relative to the observer, and therefore we could use the results of Sec. 2–3 which apply to the rotation of a rigid body about a fixed point. Proceeding in this fashion, we obtain from Eq. (2–5) that

$$\mathbf{v}_{PA} = \boldsymbol{\omega} \times \boldsymbol{\rho} \qquad (2\text{–}82)$$

where $\boldsymbol{\rho}$ is the position vector of P relative to A and $\boldsymbol{\omega}$ is the absolute angular velocity of the body. (Here we note that if $\boldsymbol{\omega}$ were measured relative to a rotating system, its value would change and so would the value of $\mathbf{v}_{PA}$, in general.) So, from Eqs. (2–81) and (2–82), we find that the absolute velocity of point P is

$$\mathbf{v} = \mathbf{v}_A + \boldsymbol{\omega} \times \boldsymbol{\rho} \qquad (2\text{–}83)$$

The acceleration of P is obtained by differentiating Eq. (2–83) with respect to time:

$$\dot{\mathbf{v}} = \dot{\mathbf{v}}_A + \dot{\boldsymbol{\omega}} \times \boldsymbol{\rho} + \boldsymbol{\omega} \times \dot{\boldsymbol{\rho}} \qquad (2\text{–}84)$$

To evaluate $\dot{\boldsymbol{\rho}}$ we recall that the time rate of change of a vector is the velocity of the tip of the vector when the origin remains fixed. Since $\boldsymbol{\rho}$ is of constant magnitude, we can again use Eq. (2–5), obtaining

$$\dot{\boldsymbol{\rho}} = \boldsymbol{\omega} \times \boldsymbol{\rho} \qquad (2\text{–}85)$$

which, of course, is identical with $\mathbf{v}_{PA}$ obtained previously. Finally, denoting the acceleration $\dot{\mathbf{v}}_A$ by $\mathbf{a}_A$, we obtain from Eqs. (2–84) and (2–85) the absolute acceleration of P.

$$\mathbf{a} = \mathbf{a}_A + \dot{\boldsymbol{\omega}} \times \boldsymbol{\rho} + \boldsymbol{\omega} \times (\boldsymbol{\omega} \times \boldsymbol{\rho}) \qquad (2\text{–}86)$$

2–8. VECTOR DERIVATIVES IN ROTATING SYSTEMS

Suppose that a vector **A** is viewed by an observer on a fixed system XYZ (Fig. 2–15) and also by another observer on a rotating system. The rotating system is designated by the unit vector triad $\mathbf{e}_1$, $\mathbf{e}_2$, and $\mathbf{e}_3$ which is rotating with an angular velocity $\boldsymbol{\omega}$ relative to XYZ. Since **A** is consider-ed to be a free vector during the differentiation process, no generality is lost by taking the point O as the common origin of the unit vector triad and also the vector **A**.

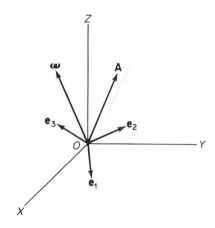

Fig. 2-15. The vector **A** relative to fixed and rotating reference frames.

Now, as each observer views the vector **A**, he might choose to express it in terms of the unit vectors of his own system. Thus each observer would give a different set of compo-nents. Nevertheless, they would be viewing the same vector and a simple coordinate conversion based upon the relative orientation of the co-ordinate systems would provide a check of one observation with the other.

On the other hand, if each observer were to calculate the time rate of change of **A**, the results would, in general, not agree, even after performing the coordinate conversion used previously. To clarify this point, recall from Eq. (1–38) that the absolute rate of change of **A**, written in terms of the unit vectors of the rotating system, is

$$\dot{\mathbf{A}} = \dot{A}_1\mathbf{e}_1 + \dot{A}_2\mathbf{e}_2 + \dot{A}_3\mathbf{e}_3 + A_1\dot{\mathbf{e}}_1 + A_2\dot{\mathbf{e}}_2 + A_3\dot{\mathbf{e}}_3 \qquad (2\text{–}87)$$

But the rate of change of **A**, as viewed by an observer in the rotating system, is

$$(\dot{\mathbf{A}})_r = \dot{A}_1\mathbf{e}_1 + \dot{A}_2\mathbf{e}_2 + \dot{A}_3\mathbf{e}_3 \qquad (2\text{–}88)$$

since the unit vectors are fixed in this system.

Now let us consider the last three terms on the right side of Eq. (2–87). Using Eq. (2–8), we see that

$$A_1\dot{\mathbf{e}}_1 + A_2\dot{\mathbf{e}}_2 + A_3\dot{\mathbf{e}}_3 = A_1\boldsymbol{\omega} \times \mathbf{e}_1 + A_2\boldsymbol{\omega} \times \mathbf{e}_2 + A_3\boldsymbol{\omega} \times \mathbf{e}_3$$

or, since vector multiplication is distributive,

$$A_1\dot{\mathbf{e}}_1 + A_2\dot{\mathbf{e}}_2 + A_3\dot{\mathbf{e}}_3 = \boldsymbol{\omega} \times (A_1\mathbf{e}_1 + A_2\mathbf{e}_2 + A_3\mathbf{e}_3) = \boldsymbol{\omega} \times \mathbf{A} \qquad (2\text{–}89)$$

Therefore, from Eqs. (2–87), (2–88), and (2–89), we find that the absolute rate of change of **A** can be expressed in terms of its value relative to a rotating system as follows:

$$\dot{\mathbf{A}} = (\dot{\mathbf{A}})_r + \boldsymbol{\omega} \times \mathbf{A} \tag{2-90}$$

where $(\dot{\mathbf{A}})_r$ is the rate of change of **A** as viewed from the rotating system and $\boldsymbol{\omega}$ is the absolute angular velocity of the rotating system.

It is important to note that **A** can be any vector whatever. A relatively simple application of Eq. (2–90) would occur for the case where **A** is the position vector of a given point P relative to the common origin O of the fixed and rotating systems. On the other hand, a more complicated situation would arise if, for example, **A** were the velocity of P relative to another point P', as viewed from a third system that is rotating separately. Thus Eq. (2–90) has a wide application and should be studied carefully.

Although we have referred to the coordinate systems in the preceding discussion as *fixed* or *rotating*, the derivation of Eq. (2–90) was based upon mathematics rather than upon physical law. Therefore we need not consider either system as being more fundamental than the other. If we call them system A and system B, respectively, we can write

$$(\dot{\mathbf{A}})_A = (\dot{\mathbf{A}})_B + \boldsymbol{\omega}_{BA} \times \mathbf{A} \tag{2-91}$$

where $\boldsymbol{\omega}_{BA}$ is the rotation rate of system B as viewed from system A. Since the result must be symmetrical with respect to the two systems, we could also write

$$(\dot{\mathbf{A}})_B = (\dot{\mathbf{A}})_A + \boldsymbol{\omega}_{AB} \times \mathbf{A} \tag{2-92}$$

where we note that $\boldsymbol{\omega}_{AB} = -\boldsymbol{\omega}_{BA}$ and that **A** is the same when viewed from either system.

Example 2–1. A turntable rotates with a constant angular velocity $\boldsymbol{\omega}$ about a perpendicular axis through O (Fig. 2–16). The position of a point P which is moving relative to the turntable is given by

$$\mathbf{r} = x\mathbf{i} + y\mathbf{j}$$

where the unit vectors **i** and **j** are fixed in the turntable. Solve for the absolute velocity and acceleration of P in terms of its motion relative to the turntable.

Using Eq. (2–90), we see that the absolute velocity **v** is given by

$$\mathbf{v} = \dot{\mathbf{r}} = (\dot{\mathbf{r}})_r + \boldsymbol{\omega} \times \mathbf{r} \tag{2-93}$$

But the velocity seen by an observer rotating with the turntable is just

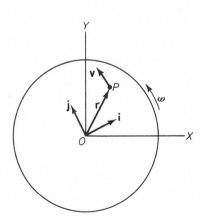

Fig. 2-16. Motion of a point that is moving on a turntable.

$$(\dot{\mathbf{r}})_r = \dot{x}\mathbf{i} + \dot{y}\mathbf{j}$$

Also,

$$\boldsymbol{\omega} \times \mathbf{r} = x\boldsymbol{\omega} \times \mathbf{i} + y\boldsymbol{\omega} \times \mathbf{j} = -y\omega\mathbf{i} + x\omega\mathbf{j}$$

Therefore,

$$\mathbf{v} = (\dot{x} - y\omega)\mathbf{i} + (\dot{y} + x\omega)\mathbf{j} \qquad (2\text{–}94)$$

where we note that the absolute velocity $\mathbf{v}$ is expressed in terms of the unit vectors of the rotating system.

To find the acceleration, we again apply Eq. (2–90), this time to the absolute velocity vector.

$$\mathbf{a} = \dot{\mathbf{v}} = (\dot{\mathbf{v}})_r + \boldsymbol{\omega} \times \mathbf{v} \qquad (2\text{–}95)$$

The term $(\dot{\mathbf{v}})_r$ is found by differentiating Eq. (2–94) with respect to time, assuming that the unit vectors are constant since they are viewed from the rotating system.

$$(\dot{\mathbf{v}})_r = (\ddot{x} - \dot{y}\omega)\mathbf{i} + (\ddot{y} + \dot{x}\omega)\mathbf{j} \qquad (2\text{–}96)$$

Note that $(\dot{\mathbf{v}})_r$ is not the acceleration of P relative to the rotating system. Rather, it is the time rate of change of the absolute velocity vector as viewed from the rotating system. In general, the subscript after a differentiated (dotted) vector refers to the coordinate system from which the rate of change is viewed. If the subscript is omitted, a nonrotating reference frame is assumed.

Next, using the result given in Eq. (2–94), the term $\boldsymbol{\omega} \times \mathbf{v}$ is evaluated in a straightforward fashion, giving

$$\boldsymbol{\omega} \times \mathbf{v} = -\omega(\dot{y} + x\omega)\mathbf{i} + \omega(\dot{x} - y\omega)\mathbf{j} \qquad (2\text{–}97)$$

Finally, from Eqs. (2–95), (2–96), and (2–97) we obtain that the absolute acceleration is

$$\mathbf{a} = (\ddot{x} - x\omega^2 - 2\omega\dot{y})\mathbf{i} + (\ddot{y} - y\omega^2 + 2\omega\dot{x})\mathbf{j} \qquad (2\text{–}98)$$

2–9. MOTION OF A PARTICLE IN A MOVING COORDINATE SYSTEM

Now we shall use the general result of Eq. (2–90) to obtain the equations for the absolute velocity and acceleration of a particle P that is in motion relative to a moving coordinate system. In Fig. 2–17, the XYZ system is fixed in an inertial frame and the xyz system translates and rotates relative to it.

Suppose that $\mathbf{r}$ is the position vector of P and $\mathbf{R}$ is the position vector of O', both relative to point O in the fixed XYZ system. Then

$$\mathbf{r} = \mathbf{R} + \boldsymbol{\rho} \qquad (2\text{–}99)$$

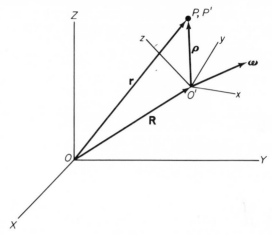

Fig. 2-17. The position vectors of a point P relative to a fixed system and a moving system.

where ρ is the position vector of P relative to O'. Differentiating with respect to time, we obtain the absolute velocity

$$\mathbf{v} = \dot{\mathbf{r}} = \dot{\mathbf{R}} + \dot{\rho} \qquad (2\text{–}100)$$

where both derivatives on the right are calculated from the viewpoint of a fixed observer. Next we express $\dot{\rho}$ in terms of its value relative to the rotating xyz system, using Eq. (2–90), that is,

$$\dot{\rho} = (\dot{\rho})_r + \boldsymbol{\omega} \times \rho \qquad (2\text{–}101)$$

where $\boldsymbol{\omega}$ is the absolute rotation rate of the xyz system. Then from Eqs. (2–100) and (2–101) we obtain

$$\mathbf{v} = \dot{\mathbf{R}} + (\dot{\rho})_r + \boldsymbol{\omega} \times \rho \qquad (2\text{–}102)$$

In order to explain further the meaning of the terms on the right side of Eq. (2–102), let us define the point P' to be coincident with P at the time of observation but fixed in the xyz system. We see that $\dot{\mathbf{R}}$ is the absolute velocity of O' and that $\boldsymbol{\omega} \times \rho$ is the velocity of P' relative to O' as viewed by a nonrotating observer. Thus $\dot{\mathbf{R}} + \boldsymbol{\omega} \times \rho$ represents the absolute velocity of P'. The remaining term $(\dot{\rho})_r$ is the velocity of P relative to O', as viewed by an observer rotating with the xyz system. It is interesting to note that $(\dot{\rho})_r$ can also be interpreted as the velocity of P relative to P', as viewed by a nonrotating observer.[3] Therefore the three terms on the right side of

[3] This result can be seen more clearly if we let ρ' be a vector drawn from P' to P. At the moment in question, $\rho' = 0$, and therefore, from Eq. (2–90), $\dot{\rho}' = (\dot{\rho}')_r$. In other words, at this particular moment, the velocity of P relative to P' is the same, whether viewed from a rotating or nonrotating system. Also, of course, the velocity of P relative to any point fixed in the xyz system, whether it be P' or O', is the same when viewed by an observer moving with the system.

Eq. (2–102) are, respectively, the velocity of O' relative to O, the velocity of P relative to P', and the velocity of P' relative to O', all as viewed by an absolute or nonrotating observer.

To obtain the absolute acceleration of P, we find the rate of change of each of the terms in Eq. (2–102), as viewed by a fixed observer. Thus,

$$\frac{d}{dt}(\dot{\mathbf{R}}) = \ddot{\mathbf{R}} \tag{2–103}$$

Using Eq. (2–90), we obtain

$$\frac{d}{dt}[(\dot{\boldsymbol{\rho}})_r] = (\ddot{\boldsymbol{\rho}})_r + \boldsymbol{\omega} \times (\dot{\boldsymbol{\rho}})_r \tag{2–104}$$

Also, using Eq. (2–101),

$$\frac{d}{dt}(\boldsymbol{\omega} \times \boldsymbol{\rho}) = \dot{\boldsymbol{\omega}} \times \boldsymbol{\rho} + \boldsymbol{\omega} \times \dot{\boldsymbol{\rho}}$$

$$= \dot{\boldsymbol{\omega}} \times \boldsymbol{\rho} + \boldsymbol{\omega} \times (\dot{\boldsymbol{\rho}})_r + \boldsymbol{\omega} \times (\boldsymbol{\omega} \times \boldsymbol{\rho}) \tag{2–105}$$

Finally, adding Eqs. (2–103), (2–104), and (2–105), we obtain an expression for the absolute acceleration of P.

$$\mathbf{a} = \ddot{\mathbf{R}} + \dot{\boldsymbol{\omega}} \times \boldsymbol{\rho} + \boldsymbol{\omega} \times (\boldsymbol{\omega} \times \boldsymbol{\rho}) + (\ddot{\boldsymbol{\rho}})_r + 2\boldsymbol{\omega} \times (\dot{\boldsymbol{\rho}})_r \tag{2–106}$$

Now let us explain the nature of each of the terms. $\ddot{\mathbf{R}}$ is the absolute acceleration of O'. The terms $\dot{\boldsymbol{\omega}} \times \boldsymbol{\rho}$ and $\boldsymbol{\omega} \times (\boldsymbol{\omega} \times \boldsymbol{\rho})$ together represent the acceleration of P' relative to O', as viewed by a nonrotating observer. The term $\dot{\boldsymbol{\omega}} \times \boldsymbol{\rho}$ is similar in nature to the tangential acceleration of Eq. (2–7), whereas the term $\boldsymbol{\omega} \times (\boldsymbol{\omega} \times \boldsymbol{\rho})$ represents a centripetal acceleration since it is directed from P toward, and perpendicular to, the axis of rotation through O'. Thus the first three terms of Eq. (2–106) represent the absolute acceleration of P'. The fourth term $(\ddot{\boldsymbol{\rho}})_r$ is the acceleration of the point P relative to the xyz system, that is, as viewed by an observer moving with the xyz system. The fifth term $2\boldsymbol{\omega} \times (\dot{\boldsymbol{\rho}})_r$ is known as the *Coriolis acceleration.* Note that the Coriolis acceleration arises from two sources, namely, Eqs. (2–104) and (2–105). The term in Eq. (2–104) is due to the changing direction in space of the velocity of P relative to the moving system. The term in Eq. (2–105) represents the rate of change of the velocity $\boldsymbol{\omega} \times \boldsymbol{\rho}$ due to a changing magnitude or direction of the position vector $\boldsymbol{\rho}$ relative to the moving system.

Note also that the last two terms of Eq. (2–106) represent the acceleration of P relative to P', as viewed by a nonrotating observer. Thus we can summarize by saying that the first three terms give the acceleration of P', whereas the last two terms give the acceleration of P relative to P'; the sum, of course, resulting in the absolute acceleration of P.

2–10. PLANE MOTION

If a particle moves so that it remains in a single fixed plane, it is said to move with plane motion. Similarly, plane motion of a rigid body requires that all points of the body move parallel to the same fixed plane. In the latter case, the kinematical aspects of the motion are adequately described in terms of a lamina chosen such that its motion is confined to its own plane. Let us consider, then, the motion of a lamina in its own plane (Fig. 2–18).

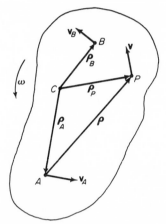

Fig. 2-18. The motion of a lamina in its own plane.

Instantaneous Center. In Eq. (2–83), we found that the velocity of a point P in a rigid body is given by

$$\mathbf{v} = \mathbf{v}_A + \boldsymbol{\omega} \times \boldsymbol{\rho}$$

where $\mathbf{v}_A$ is the velocity of the base point A, $\boldsymbol{\omega}$ is the absolute angular velocity of the body, and $\boldsymbol{\rho}$ is the position vector of P relative to A. For plane motion, the angular velocity vector $\boldsymbol{\omega}$, if it exists, must be perpendicular to the plane of the motion. Furthermore, we recall that a different base point will, in general, have a different velocity. So if we can find a base point location C whose velocity is zero, we see from Eq. (2–83) that

$$\mathbf{v} = \boldsymbol{\omega} \times \boldsymbol{\rho}_P \tag{2–107}$$

where $\boldsymbol{\rho}_P$ is the position vector of P relative to C. Thus the line CP must be perpendicular to $\mathbf{v}$ and of length ρ_P such that

$$v = \omega \rho_P \tag{2–108}$$

Hence, if $\mathbf{v}$ and $\boldsymbol{\omega}$ are known, the point C can be determined.

Similarly, it can be seen that

$$\mathbf{v}_A = \boldsymbol{\omega} \times \boldsymbol{\rho}_A \tag{2–109}$$

or, for another point B in the lamina,

$$\mathbf{v}_B = \boldsymbol{\omega} \times \boldsymbol{\rho}_B \tag{2–110}$$

Thus one concludes that the instantaneous center C is located at the intersection of two or more lines, each of which is drawn from some point in the lamina in a direction perpendicular to its velocity vector.

The point C, whose velocity is instantaneously zero, and about which the lamina appears to be rotating at that moment, is known as the *instan-*

taneous center of the rotation. It may be located in the lamina or in an imaginary extension of it. The instantaneous center is at a unique point for any given instant in time, provided that the motion is not that of pure translation ($\omega = 0$). In the latter case, the instantaneous center is located at an infinite distance from the point in a direction perpendicular to the velocity.

In considering the velocity of various points in the lamina, it should be noted that only certain velocity distributions are possible. In particular, the rigid-body assumption requires that the velocities at all points on any given straight line in the lamina must have equal components along the line. In fact, from Eq. (2–110) we see that the locus of all points having a given speed v_B is a circle centered at C and with a radius $\rho_B = v_B/\omega$, the velocity at each point being tangent to the locus. Thus the velocity distribution has a circular symmetry about the instantaneous center. Furthermore, a knowledge of the angular velocity and the location of the instantaneous center is sufficient to determine the velocity of any point in the lamina, thereby providing a convenient means of calculating velocities for the case of plane motion. It should be emphasized, however, that the *acceleration* of the instantaneous center is not necessarily zero, and therefore the instantaneous center is not particularly useful in performing acceleration calculations.

We have seen that the angular velocity $\boldsymbol{\omega}$ is perpendicular to the lamina and is not a function of position. Hence the angular velocity of any line in the lamina is equal to the angular velocity of the lamina. So if we choose any two points of the lamina, A and P for example, the angular velocity is found by taking the difference of the velocity components at A and P perpendicular to AP and dividing by the distance AP. Using vector notation,

$$\boldsymbol{\omega} = \frac{\boldsymbol{\rho} \times (\mathbf{v} - \mathbf{v}_A)}{\rho^2} \tag{2-111}$$

Space and Body Centrodes. In general, the instantaneous center does not retain the same position, either in space or in the lamina. The locus of the instantaneous centers forms a curve in space known as the *space centrode*. Similarly, the locus of the instantaneous centers relative to the moving lamina is known as the *body centrode*. Both curves are in the same plane, of course, but neither curve need be closed.

Now suppose that, at a given moment, the space centrode S and the body centrode B corresponding to a given motion are as shown in Fig. 2–19. The instantaneous center is a point common to both curves at this moment. Also, we note that curve S is fixed in space and

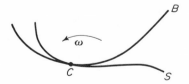

Fig. 2-19. The space centrode S and the body centrode B.

curve B is rotating with angular velocity ω. In general, the curves B and S are tangent at C but have different curvatures at this point. This allows for the instantaneous center C to proceed smoothly along curves B and C because of a finite rotation rate ω of the body centrode. But this is just the condition for rolling motion. Hence *the body centrode rolls on the space centrode*. A further condition for rolling motion is that equal path lengths are traced by point C in curves B and S in any given time interval. If this were not true, slipping would result and pure rolling motion would not occur.

Perhaps a source of confusion can be avoided if we emphasize again that, at a given moment, the point known as the instantaneous center has zero velocity. We may speak of the instantaneous center as moving along the curves B and S but, more accurately, a succession of points on these curves are called the *instantaneous center at succeeding instants of time*. The points on the body centrode are in motion, in general, but when each becomes the instantaneous center of rotation, its velocity is zero at that instant. Each point on the space centrode is at rest at all times.

Now let us take a more general view of the instantaneous center. We saw in the case of the lamina of Fig. 2–18 that the instantaneous center C is instantaneously fixed both in the lamina and in space. So if we consider a lamina A moving with plane motion and lamina B fixed in space, then the instantaneous center is fixed in both laminas at that moment. But, as we emphasized earlier, there are no absolute systems from the viewpoint of kinematics. Therefore, in a more general sense, the instantaneous center relative to any two laminas A and B in plane motion is that point C which is instantaneously at rest, as viewed by observers on both A and B. If more than two laminas are in plane motion, then there is, in general, a separate instantaneous center for each pair. Thus, four laminas would have six instantaneous centers. To avoid confusion, however, we shall always assume that the instantaneous center for a given lamina is defined relative to a fixed frame unless a contrary statement is made.

Example 2–2. As an example of plane motion, consider a wheel of radius a which is rolling along a straight line (Fig. 2–20). Assume that the wheel has a uniform angular velocity ω.

In this case, the instantaneous center is at the point of contact C. The space centrode is the straight line on which the wheel is rolling, whereas the body centrode is the circumference of the wheel.

Considering the instantaneous center C at the given moment as a base point whose velocity is instantaneously zero, we can use Eq. (2–107) to calculate the velocity of any point on the wheel. The velocity of the center O' is constant since its distance from C is constant and the rotation rate is assumed to be constant. Its magnitude is

$$v_{O'} = a\omega \qquad (2\text{-}112)$$

Similarly, the speed of a general point P is

$$v = \rho_P \omega \qquad (2\text{-}113)$$

where ρ_P is the distance from C to P.

The path in space of a point P on the circumference can be shown to be a cycloid. Its speed along its path reaches a maximum when ρ_P is a maximum, that is, when P is diametrically opposite C and at the top point of the cycloid. At this point,

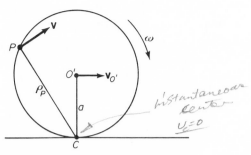

Instantaneous center $v=0$

Fig. 2-20. A wheel rolling on a straight line.

$v = 2v_{O'} = 2a\omega$. On the other hand, the velocity of P goes to zero when P coincides with C and $\rho_P = 0$. In terms of the cycloidal motion, this occurs when P is at the cusp of the cycloid and instantaneously at rest while reversing the direction of its motion. Of course, the acceleration of P is not zero when it coincides with C. In fact, the acceleration of P is of uniform magnitude $a\omega^2$, being at all times directed toward O' for the case of uniform ω.

$\vec{v} = \rho\dot\omega + \dot\rho\omega = \rho\dot\omega + \rho\omega^2$

$\omega = c$ $\dot\omega = 0$

$\Rightarrow \dot v = \rho\omega^2$

Example 2-3. A bar of length l is constrained to move in the xy plane such that end A remains on the x axis and the other end B remains on the y axis. Assuming that the angular velocity ω of the bar is constant, find

instantaneous center

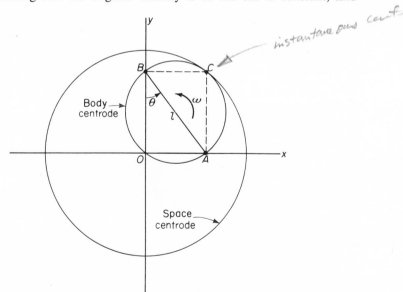

Fig. 2-21. The motion of a bar such that its ends remain on the cartesian axes.

the space and body centrodes and solve for the motion of end A as a function of time.

Let the position of the bar be specified by the angle θ between the bar and y axis at B, as shown in Fig. 2–21. If we measure time from the moment when point A is at the origin and moving in the positive sense along the x axis, we obtain

$$\theta = \omega t \tag{2–114}$$

The instantaneous center of rotation of the bar is at the intersection of lines drawn perpendicular to the velocity vectors at two points on the bar. Choosing the end points, for example, point A always moves along the x axis and point B along the y axis. Therefore, the instantaneous center C is located at the intersection of the dashed lines of Fig. 2–21, that is, at

Roller bearing

$$\begin{aligned} x &= l \sin \theta \\ y &= l \cos \theta \end{aligned} \tag{2–115}$$

indicating that the space centrode is a circle of radius l since the distance OC is equal to $\sqrt{x^2 + y^2} = l$. Furthermore, the line OC makes an angle θ with the positive y axis and, from Eq. (2–114), the point C proceeds around the space centrode at a uniform rate.

We have seen that AB and OC are of equal length l and therefore they bisect each other at the center of the rectangle $OACB$. Thus the instantaneous center C remains at a constant distance $l/2$ from the midpoint of the rod but changes its position relative to the rod. Therefore we conclude that the body centrode is a circle of radius $l/2$ and centered at the midpoint of the rod.

To find the motion of point A, we note that it moves along the x axis and has the same x coordinate as the instantaneous center. So, from Eqs. (2–114) and (2–115), we obtain

$$x = l \sin \omega t \tag{2–116}$$

indicating that point A moves with simple harmonic motion along the x axis. Similarly, point B moves with simple harmonic motion along the y axis.

The instantaneous center C moves clockwise at the same angular rate as the bar rotates counterclockwise. Therefore point C moves completely around the space centrode during one complete rotation of the bar. Since, however, the instantaneous center must traverse the same distance along the space and body centrodes, it must make two complete circuits of the body centrode for every complete rotation of the bar.

2-11. EXAMPLES

Example 2-4. A particle P moves along a straight radial groove in a circular disk of radius a which is pivoted about a perpendicular axis through its center O (Fig. 2-22). The particle moves relative to the disk such that

$$r = \frac{a}{2}(1 + \sin \omega t)$$

and the disk rotates according to

$$\phi = \phi_0 \sin \omega t$$

Find the general expression for the absolute acceleration of P.

First we shall solve the problem using the general acceleration equation in cylindrical coordinates, Eq. (2-23). Setting $\ddot{z} = 0$ since the motion is two-dimensional, we obtain the components in the $\mathbf{e}_r$ and $\mathbf{e}_\phi$ directions as follows:

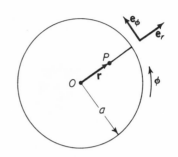

Fig. 2-22. A particle with radial harmonic motion relative to a harmonically oscillating disk.

$$a_r = \ddot{r} - r\dot{\phi}^2 = -\frac{a\omega^2}{2}\sin \omega t - \frac{a\omega^2 \phi_0^2}{2}(1 + \sin \omega t)\cos^2 \omega t$$

$$a_\phi = r\ddot{\phi} + 2\dot{r}\dot{\phi} = -\frac{a\omega^2 \phi_0}{2}(1 + \sin \omega t)\sin \omega t + a\omega^2 \phi_0 \cos^2 \omega t$$

$$(2\text{-}117)$$

Next let us solve the same problem using the general vector expression of Eq. (2-106). Let the rotating coordinate system be fixed in the disk such that the origins O and O' coincide. Then, noting that

$$\boldsymbol{\rho} = r\mathbf{e}_r = \frac{a}{2}(1 + \sin \omega t)\mathbf{e}_r$$

$$\boldsymbol{\omega} = \dot{\phi}\mathbf{e}_z = \omega\phi_0 \cos \omega t \, \mathbf{e}_z$$

we evaluate each of the terms.

$$\ddot{\mathbf{R}} = 0$$

$$\dot{\boldsymbol{\omega}} \times \boldsymbol{\rho} = -\frac{a\omega^2 \phi_0}{2}(1 + \sin \omega t)\sin \omega t \, \mathbf{e}_\phi$$

$$\boldsymbol{\omega} \times (\boldsymbol{\omega} \times \boldsymbol{\rho}) = -\frac{a\omega^2 \phi_0^2}{2}(1 + \sin \omega t)\cos^2 \omega t \, \mathbf{e}_r$$

$$(\ddot{\boldsymbol{\rho}})_r = -\frac{a\omega^2}{2}\sin \omega t \, \mathbf{e}_r$$

$$2\boldsymbol{\omega} \times (\dot{\boldsymbol{\rho}})_r = a\omega^2 \phi_0 \cos^2 \omega t \, \mathbf{e}_\phi$$

In obtaining the foregoing results, the procedure is straightforward except for finding the derivatives of $\boldsymbol{\rho}$ as viewed from the rotating system. Here we differentiate the expression for $\boldsymbol{\rho}$, assuming that $\mathbf{e}_r$ is constant since it is fixed in the rotating system. Adding the components in the $\mathbf{e}_r$ and $\mathbf{e}_\phi$ directions, we obtain agreement with the results found previously in Eq. (2–117).

Example 2–5. Suppose that a particle P moves along a line of longitude on a sphere of radius a rotating at a constant angular velocity $\boldsymbol{\omega}$ about the polar axis. If its speed relative to the sphere is $v = kt$ and if the center of the sphere is fixed, find the absolute acceleration of P in terms of the spherical unit vectors $\mathbf{e}_r$, $\mathbf{e}_\theta$, and $\mathbf{e}_\phi$, (Fig. 2–23). Let $\theta(0) = \theta_0$.

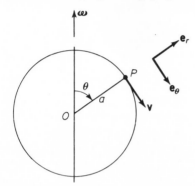

One approach to the problem is to evaluate the acceleration directly in terms of spherical coordinates, using Eq. (2–33). First we note that $r = a$ and $\dot{\phi} = \omega$, where both a and ω are constants. Also,

$$\dot{\theta} = \frac{v}{a} = \frac{kt}{a} \qquad (2\text{--}118)$$

Evaluating the spherical components of the acceleration, we obtain

Fig. 2-23. The motion of a particle P having a velocity $\mathbf{v}$ relative to a sphere.

$$a_r = \ddot{r} - r\dot{\theta}^2 - r\dot{\phi}^2 \sin^2 \theta$$
$$= -\frac{k^2 t^2}{a} - a\omega^2 \sin^2 \theta$$

$$a_\theta = r\ddot{\theta} + 2\dot{r}\dot{\theta} - r\dot{\phi}^2 \sin\theta\cos\theta = k - a\omega^2 \sin\theta\cos\theta \qquad (2\text{--}119)$$

$$a_\phi = r\ddot{\phi} \sin\theta + 2\dot{r}\dot{\phi}\sin\theta + 2r\dot{\theta}\dot{\phi}\cos\theta = 2\omega kt\cos\theta$$

where we note from Eq. (2–118) that

$$\theta = \theta_0 + \frac{kt^2}{2a} \qquad (2\text{--}120)$$

Now let us solve this problem using Eq. (2–106). The moving coordinate system is assumed to be fixed in the sphere and rotating with it at a constant rate $\boldsymbol{\omega}$. The origin O' is at the center O. Evaluating the first two terms, we find that

$$\ddot{\mathbf{R}} = 0 \qquad (2\text{--}121)$$

and

$$\dot{\boldsymbol{\omega}} \times \boldsymbol{\rho} = 0 \qquad (2\text{--}122)$$

Before evaluating the next term, we note that

$$\boldsymbol{\omega} = \omega\cos\theta\,\mathbf{e}_r - \omega\sin\theta\,\mathbf{e}_\theta$$

$$\boldsymbol{\rho} = a\mathbf{e}_r$$

Therefore,

$$\boldsymbol{\omega} \times \boldsymbol{\rho} = a\omega \sin \theta \, \mathbf{e}_\phi$$

and

$$\boldsymbol{\omega} \times (\boldsymbol{\omega} \times \boldsymbol{\rho}) = -a\omega^2(\sin^2 \theta \, \mathbf{e}_r + \sin \theta \cos \theta \, \mathbf{e}_\theta) \qquad (2\text{-}123)$$

It can be seen that the path of P relative to the sphere, that is, the moving system, is just a circular path which is traversed at a constantly increasing speed. From Eq. (2–54), we find that

$$(\ddot{\boldsymbol{\rho}})_r = -\frac{k^2 t^2}{a} \mathbf{e}_r + k\mathbf{e}_\theta \qquad (2\text{-}124)$$

Also, we see that

$$(\dot{\boldsymbol{\rho}})_r = kt\mathbf{e}_\theta$$

so the Coriolis acceleration is

$$2\boldsymbol{\omega} \times (\dot{\boldsymbol{\rho}})_r = 2\omega kt \cos \theta \, \mathbf{e}_\phi \qquad (2\text{-}125)$$

Adding the terms given by Eqs. (2–121) to (2–125), we obtain the same results as were obtained previously in Eq. (2–119).

Example 2-6. Find the acceleration of point P on the circumference of a wheel of radius r_2 which is rolling on the inside of a fixed circular track of radius r_1 (Fig. 2–24). An arm connecting the fixed point O and the wheel hub at O' moves at a constant angular velocity ω. The position of P relative to the arm is given by the angle ϕ.

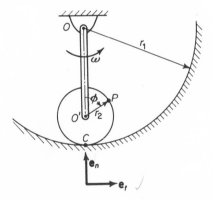

Our procedure will be to solve this example using the general vector equation given in Eq. (2–106). The origin of the fixed system is taken at O. The moving system is fixed in the moving arm and its origin is at the hub O'. Normal and tangential unit vectors will be used, $\mathbf{e}_n$ having the direction of the line from O' to O and $\mathbf{e}_t$ being perpendicular, as shown.

It can be seen that the wheel undergoes plane motion, and since there

Fig. 2-24. A wheel rolling on the inside of a fixed circular track.

is no slipping of the wheel on the track, the instantaneous center of rotation for the wheel is at the contact point C. This enables us to obtain an equation relating $\dot{\phi}$ and ω. First consider the velocity with which the hub moves along its circular path. The wheel is rotating clockwise at an angular rate $\dot{\phi}$ relative to the arm but the arm rotates with an angular velocity ω in a counterclockwise sense. Therefore the wheel rotates clockwise at an absolute angular velocity $(\dot{\phi} - \omega)$. Since the wheel is instantaneously rotating about C, the absolute velocity of the hub is

$$\mathbf{v}_{O'} = r_2(\dot{\phi} - \omega)\mathbf{e}_t \qquad \checkmark \qquad (2\text{-}126)$$

But the velocity of the hub can also be calculated from the motion of the arm. The distance OO' is $(r_1 - r_2)$ and the arm rotates with an absolute angular velocity ω, so

$$\mathbf{v}_{O'} = (r_1 - r_2)\omega\mathbf{e}_t \qquad \checkmark \qquad (2\text{-}127)$$

From Eqs. (2–126) and (2–127), we obtain

$$\dot{\phi} = \frac{r_1}{r_2}\omega \qquad (2\text{-}128)$$

Now let us proceed with the evaluation of the absolute acceleration of P. We find that

$$\ddot{\mathbf{R}} = (r_1 - r_2)\omega^2\mathbf{e}_n \qquad (2\text{-}129)$$

which is just the centripetal acceleration of O' due to its uniform circular motion about O, in accordance with Eq. (2–53). The next term, $\dot{\boldsymbol{\omega}} \times \boldsymbol{\rho}$, is zero because the angular velocity $\boldsymbol{\omega}$ is constant.

In order to evaluate the term $\boldsymbol{\omega} \times (\boldsymbol{\omega} \times \boldsymbol{\rho})$, we note first that

$$\boldsymbol{\omega} = \omega\mathbf{e}_b$$

where $\mathbf{e}_b$ is a unit vector pointing out of the plane of the figure in accordance with Eq. (2–38). Also,

$$\boldsymbol{\rho} = r_2(\sin\phi\,\mathbf{e}_t + \cos\phi\,\mathbf{e}_n) \qquad \checkmark$$

Consequently,

$$\boldsymbol{\omega} \times (\boldsymbol{\omega} \times \boldsymbol{\rho}) = -r_2\omega^2(\sin\phi\,\mathbf{e}_t + \cos\phi\,\mathbf{e}_n) \qquad (2\text{-}130)$$

which is a centripetal acceleration directed from P toward O'.

The point P, as viewed by an observer on the moving coordinate system, moves at a constant angular rate $\dot{\phi}$ in a circle of radius r_2. Thus we find that the acceleration due to this motion is again centripetal, being directed from P toward O'.

$$(\ddot{\boldsymbol{\rho}})_r = -r_2\dot{\phi}^2(\sin\phi\,\mathbf{e}_t + \cos\phi\,\mathbf{e}_n)$$

or, substituting from Eq. (2–128), we obtain

$$(\ddot{\boldsymbol{\rho}})_r = -\frac{r_1^2\omega^2}{r_2}(\sin\phi\,\mathbf{e}_t + \cos\phi\,\mathbf{e}_n) \qquad (2\text{-}131)$$

In order to evaluate the Coriolis acceleration term, we first obtain the relative velocity

$$(\dot{\boldsymbol{\rho}})_r = r_2\dot{\phi}(\cos\phi\,\mathbf{e}_t - \sin\phi\,\mathbf{e}_n)$$
$$= r_1\omega(\cos\phi\,\mathbf{e}_t - \sin\phi\,\mathbf{e}_n)$$

where we note again that an observer on the moving system sees point P moving in a circular path with a uniform speed $r_2\dot{\phi}$. Thus we obtain

$$2\boldsymbol{\omega} \times (\dot{\boldsymbol{\rho}})_r = 2r_1\omega^2(\sin\phi\,\mathbf{e}_t + \cos\phi\,\mathbf{e}_n) \tag{2-132}$$

which is directed radially outward in the direction $O'P$.

Finally, adding the individual acceleration terms given in Eqs. (2-129) to (2-132), we obtain

$$\mathbf{a} = \left(2r_1 - r_2 - \frac{r_1^2}{r_2}\right)\omega^2\sin\phi\,\mathbf{e}_t + \left[\left(2r_1 - r_2 - \frac{r_1^2}{r_2}\right)\omega^2\cos\phi \right.$$
$$\left. + (r_1 - r_2)\omega^2\right]\mathbf{e}_n \tag{2-133}$$

The preceding examples have illustrated the application of Eq. (2-106) which is the general vector equation for the acceleration of a point in terms of its motion relative to a moving coordinate system. Although this equation is valid for an arbitrary motion of the moving coordinate system, this system should be chosen such that the calculations are made as simple as possible. An unfortunate choice at this point can result in a large increase in the required effort. Roughly speaking, the motion of P relative to the moving system should be of about the same complexity as the absolute motion of O', provided that the angular velocity $\boldsymbol{\omega}$ is constant or varies in a simple fashion. Also, the choice of unit vectors in expressing the result should be made for convenience. In general, they should form an orthogonal set.

REFERENCES

1. Banach, S., *Mechanics*, E. J. Scott, trans. Mathematical Monographs, vol. 24, Warsaw, 1951; distributed by Hafner Publishing Co., New York.

2. Halfman, R. L., *Dynamics—Particles, Rigid Bodies, and Systems*. Reading, Mass.: Addison-Wesley Publishing Company, 1962.

3. Kane, T. R., *Analytical Elements of Mechanics—Dynamics*. New York: Academic Press, Inc., 1961.

4. Shames, I. H., *Engineering Mechanics*. Englewood Cliffs, N.J.: Prentice-Hall, Inc., 1960.

5. Synge, J. L., and B. A. Griffith, *Principles of Mechanics*, 3rd ed. New York: McGraw-Hill, Inc., 1959.

6. Yeh, H., and J. I. Abrams, *Principles of Mechanics of Solids and Fluids—Particle and Rigid-Body Mechanics*. New York: McGraw-Hill, Inc., 1960.

PROBLEMS

2-1. A vertical wheel of radius a rolls without slipping along a straight horizontal line. If its angular velocity is given by $\omega = \alpha t$, solve for the acceleration of

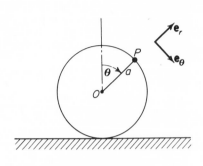

Fig. P2-1

a point P on its rim, assuming that P is initially at the highest point of its path. Express the result in terms of the unit vectors $\mathbf{e}_r$ and $\mathbf{e}_\theta$.

2–2. Solve for the hodograph of the two-dimensional harmonic motion given by Eqs. (2–64) and (2–69). Sketch the result.

2–3. A lamina undergoes general motion in its own xy plane. At a given instant, its angular motion is given by $\boldsymbol{\omega} = \omega \mathbf{k}$, $\dot{\boldsymbol{\omega}} = \dot{\omega} \mathbf{k}$; whereas a base point A in the lamina has an acceleration $\mathbf{a}_A = a_A \mathbf{i}$. Find the position relative to A of a point P whose absolute acceleration is zero.

2–4. A water particle P moves outward along the impeller of a centrifugal pump with a constant tangential velocity of 60 ft/sec relative to the impeller, which is rotating at a uniform rate of 1200 rpm in the direction shown. Find the acceleration of the particle at the point where it leaves the impeller.

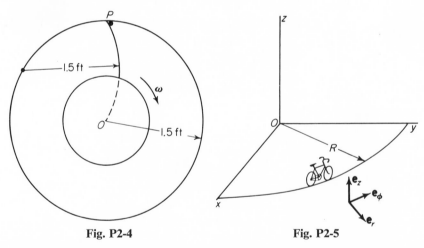

Fig. P2-4 **Fig. P2-5**

2–5. A cyclist rides around a circular track ($R = 100$ ft) such that the point of contact of the wheel on the track moves at a constant speed of 30 ft/sec. He banks his bicycle at 15° with the vertical. Find the acceleration of a tack in the tire (1.25 ft radius) as it passes through the highest point of its path. Use cylindrical unit vectors in expressing the answer.

2–6. An airplane flies with a constant speed v in a level turn to the left at a constant radius R. The propeller is of radius a and rotates about its axis in a clockwise sense (as viewed from the rear) with a constant angular velocity Ω. Find the total acceleration of a point P at the tip of the propeller, assuming that its axis is always aligned with the flight path. Use cylindrical unit vectors and assume that the velocity of P relative to the airplane is vertically upward at $t = 0$.

2-7. A circular disk of radius r_2 rolls in its plane on the inside of a fixed circular cylinder of radius r_1. Find the acceleration of a point P on the wheel at a distance b from its hub O'. Assume that $\dot{\phi}$ is not constant, where the angle ϕ is measured between $O'P$ and the line of centers $O'O$.

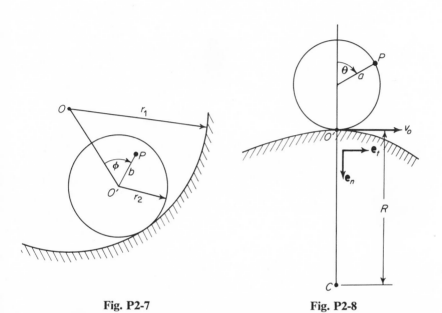

Fig. P2-7   **Fig. P2-8**

2-8. A wheel of radius a rolls along a general convex curve of varying radius of curvature R such that all motion is confined to a single plane. The contact point O' traverses the curve at a constant speed v_0. Find the absolute acceleration of a point P on the rim. Express the result in terms of the unit vectors $\mathbf{e}_n$ and $\mathbf{e}_t$.

2-9. The origin O' of a moving coordinate system has a constant absolute velocity $\mathbf{v}_{O'}$; whereas the point P has a constant absolute velocity $\mathbf{v}_P$. If the moving system rotates with a constant angular velocity $\boldsymbol{\omega}$, and $\boldsymbol{\rho}$ is the position vector of P relative to O', find:

$$\text{(a) } \dot{\boldsymbol{\rho}}; \text{ (b) } (\dot{\boldsymbol{\rho}})_r; \text{ (c) } \ddot{\boldsymbol{\rho}}; \text{ (d) } (\ddot{\boldsymbol{\rho}})_r; \text{ (e) } \frac{d}{dt}[(\dot{\boldsymbol{\rho}})_r].$$

2-10. The points P and P' are each located on the edge of separate circular disks which are in essentially the same plane and rotate at constant but unequal angular velocities ω and ω' about a perpendicular axis through the common center O. Assuming that O is fixed, solve for the velocity and acceleration of P' relative to P at the moment when P and P' have the least separation. Consider the following cases: (a) the observer is nonrotating; (b) the observer is rotating with the unprimed system. Express the results in terms of the unit vectors $\mathbf{e}_n$ and $\mathbf{e}_t$ which rotate with the unprimed system.

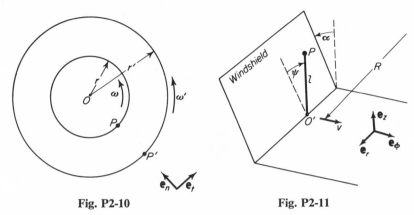

Fig. P2-10   Fig. P2-11

2–11. The plane of the windshield of a certain auto is inclined at an angle α with the vertical. The windshield wiper blade is of length l and oscillates according to the equation $\psi = \psi_0 \sin \beta t$. Assuming that the auto travels with a constant speed v around a circular path of radius R in a counterclockwise sense, (more exactly, the point O' traces out a circle of this radius), solve for the acceleration of the point P at the tip of the wiper. For what value of $\dot{\psi}$ would you expect the largest force of the blade against the windshield?

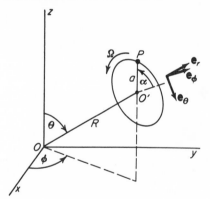

2–12. A gyroscope in the form of a wheel of radius a spins with a constant angular velocity $\dot{\alpha} = \Omega$ about its axis of symmetry. This axis maintains a constant angle θ with the vertical and precesses at a constant rate $\dot{\phi}$. Find the acceleration of a point P on the rim using spherical unit vectors $\mathbf{e}_r, \mathbf{e}_\theta, \mathbf{e}_\phi$. The angle α is measured in the plane of the rim and specifies the location of P relative to a horizontal diameter through O'.

Fig. P2-12

3

DYNAMICS OF A PARTICLE

In Chapter 2, we studied some of the methods of kinematics which can be used to obtain the absolute acceleration of a particle. Then, knowing the acceleration and assuming a knowledge of its mass, one can use Newton's law of motion to determine the total external force acting on the particle. This chapter emphasizes the inverse problem, namely, the problem of calculating the motion of a particle from a knowledge of the external forces acting upon it.

First consider the general case where the force acting on a particle is a function of its position and velocity and also the time. From Eq. (1–41), its differential equation of motion is

$$m\ddot{\mathbf{r}} = \mathbf{F}(\mathbf{r}, \dot{\mathbf{r}}, t) \tag{3-1}$$

Assuming that the function $\mathbf{F}(\mathbf{r}, \dot{\mathbf{r}}, t)$ and the mass m are known, we would like to solve for the position $\mathbf{r}$ as a function of time. Unfortunately, an analytic solution of this equation is impossible except in special cases.

To see the difficulties more clearly, let us write the vector equation in terms of its cartesian components.

$$m\ddot{x} = F_x(x, y, z, \dot{x}, \dot{y}, \dot{z}, t)$$
$$m\ddot{y} = F_y(x, y, z, \dot{x}, \dot{y}, \dot{z}, t) \tag{3-2}$$
$$m\ddot{z} = F_z(x, y, z, \dot{x}, \dot{y}, \dot{z}, t)$$

The force components F_x, F_y, and F_z are, in general, nonlinear functions of the coordinates, velocities, and time; thus the equations are hopelessly complex from the standpoint of obtaining an analytical solution. Nevertheless, it is the thesis of Newtonian mechanics that a complete knowledge of the external forces acting on a particle determines its motion, provided that the initial values of displacement and velocity are known. For example, using cartesian coordinates, initial values of x, y, z and also $\dot{x}$, $\dot{y}$, and $\dot{z}$ would be specified. So a solution to the problem does, in fact, exist. With the aid of modern electronic computers, and using approximate methods, it is possible to obtain solutions to the complete equations that are of sufficient accuracy for engineering purposes.

Any general analytical solution of Eq. (3–2) will contain six arbitrary coefficients which are evaluated from the six initial conditions. One method of obtaining the general solution is to look for integrals or constants of the motion, that is, to attempt to find six functions of the form

$$f_k(x, y, z, \dot{x}, \dot{y}, \dot{z}, t) = \alpha_k \quad (k = 1, 2, \ldots, 6) \tag{3-3}$$

where the α_k are constants. If the functions are all *distinct*, that is, if none is derivable from the others, then, in principle, they may be solved for the displacement and the velocity of the particle as a function of time and the constants α_k.

It is usually not possible to obtain all the α_k by any direct process. One of the principal topics of advanced classical mechanics, however, is the study of coordinate transformations such that the solution for the constants in terms of the new coordinates is a straightforward process.

Sometimes the constants can be given a simple physical interpretation, thereby giving us more insight into the nature of the motion. For example, a constant of the motion might be the total energy or the angular momentum about a given point. Even in cases where we do not solve completely for the motion, a knowledge of some of the constants that are applicable to the given problem may aid in obtaining certain results, such as the limiting values of certain coordinates.

In this chapter, we shall discuss some of the simpler methods and principles to be used in solving for the motion of a particle. As will be seen in the following chapters, these principles can be expanded to apply to systems of particles and to rigid-body motion, thereby forming an important part of our treatment of the subject of dynamics.

3-1. DIRECT INTEGRATION OF THE EQUATIONS OF MOTION

Returning now to the general equation of motion as expressed in Eq. (3–1) or Eq. (3–2), let us consider several cases in which direct integration can be used to find the motion of the particle.

Case 1: Constant Acceleration. The simplest case is that in which the external force on the particle is constant in magnitude and direction. Considering the cartesian components of the motion, we find from Eq. (3–2) that

$$m\ddot{x} = F_x$$
$$m\ddot{y} = F_y \tag{3-4}$$
$$m\ddot{z} = F_z$$

where F_x, F_y, and F_z are each constant.

From Eq. (3–4), we see that the motions in the x, y, and z directions are independent. Therefore, let us consider the one-dimensional case of motion parallel to the x axis. Denoting the velocity and acceleration by v and a, respectively, we find that

$$a = \ddot{x} = \frac{F_x}{m} \tag{3-5}$$

and, by direct integration with respect to time, we obtain that

$$v = v_o + at \tag{3-6}$$

and

$$x = x_o + v_o t + \frac{1}{2} at^2 \tag{3-7}$$

where x_o and v_o are the displacement and velocity at $t = 0$. Of course, similar equations would apply to motion in the direction of the y or z axes.

The time required for the particle to attain a given speed v or displacement x is found by solving Eqs. (3–6) and (3–7) for t.

$$t = \frac{1}{a}(v - v_o) \tag{3-8}$$

Also,

$$t = \frac{1}{a}\left(\sqrt{2a(x - x_o) + v_o^2} - v_o\right) \tag{3-9}$$

An expression relating the speed and displacement can be obtained for this case of constant acceleration by eliminating t between Eqs. (3–8) and (3–9), with the result:

$$v^2 = v_o^2 + 2a(x - x_o) \tag{3-10}$$

Motion of a Particle in a Uniform Gravitational Field. The motion of a particle in a uniform gravitational field is confined to the plane determined by the initial velocity vector of the particle and the gravitational force. This follows from the fact that the force remains in this plane, and therefore there is no component of acceleration normal to the plane; hence there is no tendency for the particle to leave it. Of course, the velocity vector will change direction in the general case, but it will always lie in the same plane.

Let us choose the xy plane as the plane of motion with the y axis directed vertically upward, that is, opposite to the direction of the gravitational force. Assume for convenience that the particle is located at

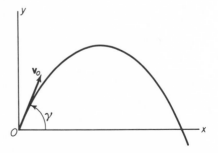

Fig. 3-1. The trajectory of a particle in a uniform gravitational field.

the origin O at time $t = 0$ and is moving with velocity v_o at an angle γ above the horizontal, as shown in Fig. 3–1. The initial values of the velocity components are

$$\dot{x}(0) = v_o \cos \gamma$$
$$\dot{y}(0) = v_o \sin \gamma \tag{3–11}$$

Noting that the acceleration components are

$$a_x = 0$$
$$a_y = -g \tag{3–12}$$

and recalling again that the motion in the x and y directions proceeds independently, we obtain the velocity components from Eq. (3–6).

$$v_x = v_o \cos \gamma \tag{3–13}$$
$$v_y = v_o \sin \gamma - gt \tag{3–14}$$

Similarly, the displacement components are obtained from Eq. (3–7).

$$x = v_o t \cos \gamma \tag{3–15}$$

$$y = v_o t \sin \gamma - \frac{1}{2} g t^2 \tag{3–16}$$

Solving for t from Eq. (3–15) and substituting into Eq. (3–16), we obtain the trajectory

$$y = x \tan \gamma - \frac{g x^2}{2 v_o^2 \cos^2 \gamma} \tag{3–17}$$

which is the equation of a vertical parabola.

The vertex of the parabola occurs at the point of zero slope in this case and is found by differentiating Eq. (3–17) with respect to x and setting the result equal to zero. We obtain

$$x_v = \frac{v_o^2}{g} \sin \gamma \cos \gamma = \frac{v_o^2}{2g} \sin 2\gamma \tag{3–18}$$

and, from Eqs. (3–17) and (3–18),

$$y_v = \frac{v_o^2}{2g} \sin^2 \gamma \tag{3–19}$$

where (x_v, y_v) is the vertex location.

The time required to attain a given value of x is found from Eq. (3–15).

$$t = \frac{x}{v_o \cos \gamma} \tag{3–20}$$

where we note that the velocity component in the x direction is constant. Therefore, from Eqs. (3–18) and (3–20), we find that the time to reach the vertex is

$$t_v = \frac{v_o}{g} \sin \gamma \qquad (3\text{-}21)$$

Of course, this value is also obtained from Eq. (3–14) by setting $v_y = 0$.

For the case of a trajectory over a flat surface, the range is

$$R = 2x_v = \frac{v_o^2}{g} \sin 2\gamma \qquad (3\text{-}22)$$

and the time of flight is

$$t_f = 2t_v = \frac{2v_o}{g} \sin \gamma \qquad (3\text{-}23)$$

It can be seen that the maximum range is achieved for $\gamma = 45°$ and the maximum time of flight for $\gamma = 90°$.

Now let us calculate the initial flight path angle γ such that the particle passes through a given point P at the coodinates (x, y), as shown in Fig. 3–2. Noting that $\sec^2\gamma = 1 + \tan^2\gamma$, we can write Eq. (3–17) in the form

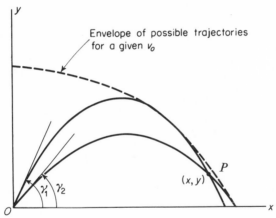

Fig. 3-2. The required initial flight path angle γ in order that the trajectory pass through a given point P.

$$\frac{gx^2}{2v_o^2}(1 + \tan^2 \gamma) - x \tan \gamma + y = 0$$

or

$$\tan^2 \gamma - \frac{2v_o^2}{gx} \tan \gamma + \frac{2v_o^2 y}{gx^2} + 1 = 0 \qquad (3\text{-}24)$$

If the point P is within range for a given initial velocity v_o, then there are two real roots of this quadratic equation in $\tan \gamma$, corresponding to the angles γ_1 and γ_2. Complex roots result for the case where the point is out of range.

The maximum range in a given direction is a point on the dashed envelope of Fig. 3–2 and corresponds to a double root. The value of this double root can be found from Eq. (3–24) by the method of completing the square, that is,

$$\left(\tan \gamma - \frac{v_o^2}{gx}\right)^2 = \tan^2 \gamma - \frac{2v_o^2}{gx} \tan \gamma + \frac{2v_o^2}{gx^2} y + 1 = 0 \qquad (3\text{--}25)$$

from which we obtain that

$$\tan \gamma = \frac{v_o^2}{gx_e} \qquad (3\text{--}26)$$

where γ is the initial flight path angle such that the trajectory just reaches a point (x_e, y_e) on the envelope of possible trajectories for a given v_o. From Eq. (3–25) we obtain the equation of the envelope:

$$\left(\frac{v_o^2}{gx_e}\right)^2 = \frac{2v_o^2}{gx_e^2} y_e + 1$$

or

$$x_e^2 = -\frac{2v_o^2}{g}\left(y_e - \frac{v_o^2}{2g}\right) \qquad (3\text{--}27)$$

which is a parabola with the vertex located at $x = 0$, $y = v_o^2/2g$.

Assuming that the azimuth angle (that is, the direction of the horizontal velocity component) of the trajectory is arbitrary, any point within the paraboloid formed by rotating the envelope about the vertical line $x = 0$ will have at least one trajectory passing through it and will thereby be within the range of a projectile of initial velocity v_o.

Example 3–1. Assuming a given initial speed v_o, find the initial flight path angle γ such that maximum range is achieved in a direction 45° above the horizontal.

Because the maximum range is obtained at 45° above the horizon, the trajectory must pass through a point $x_e = y_e$ on the envelope. Dividing Eq. (3–27) by x_e^2, we obtain

$$1 = -\frac{2v_o^2}{gx_e}\left(\frac{y_e}{x_e} - \frac{v_o^2}{2gx_e}\right)$$

which, by using Eq. (3–26), may be written in the form:

$$-2 \tan \gamma \left(1 - \frac{1}{2} \tan \gamma\right) = 1$$

or

$$\tan^2 \gamma - 2 \tan \gamma - 1 = 0$$

The only root fitting the requirement that the end point (x_e, y_e) lie above the horizontal is

$$\tan \gamma = 1 + \sqrt{2}$$

or

$$\gamma = 67\tfrac{1}{2}°$$

The slant range in this case is

$$R = \sqrt{2}\, x_e = \frac{v_0^2}{g}(2 - \sqrt{2})$$

where x_e is evaluated using Eq. (3–26).

Case 2: F = F(t). Now assume that the external force is a function of time only. Again the general equation of motion can be written in terms of three independent equations giving orthogonal components of the motion. For the case of motion parallel to the x axis, the equation of motion is

$$m\ddot{x} = F_x(t) \tag{3–28}$$

which can be integrated directly to give the velocity

$$v = v_o + \frac{1}{m}\int_o^t F_x(\tau)\, d\tau \tag{3–29}$$

where v_o is the initial velocity. Another integration results in

$$x = x_o + v_o t + \frac{1}{m}\int_o^t \left[\int_o^{\tau_2} F_x(\tau_1)\, d\tau_1\right] d\tau_2 \tag{3–30}$$

where x_o is the initial displacement.

Example 3–2. Solve for the displacement of a particle which is subject to a force of constant magnitude P which is rotating at a uniform angular rate ω in the xy plane such that

$$F_x = P \sin \omega t$$

$$F_y = P \cos \omega t$$

Initially the particle is at the origin and has a velocity v_o in the direction of the positive x axis.

First, we can solve for the velocity component parallel to the x axis from Eq. (3–29).

$$\dot{x} = v_o + \frac{P}{m\omega}(1 - \cos \omega t) \tag{3–31}$$

Another integration results in the displacement

$$x = \left(v_o + \frac{P}{m\omega}\right)t - \frac{P}{m\omega^2}\sin \omega t \tag{3–32}$$

in agreement with Eq. (3–30).

In a similar fashion, one finds that the y components of the velocity and displacement are

$$\dot{y} = \frac{P}{m\omega}\sin \omega t \tag{3–33}$$

$$y = \frac{P}{m\omega^2}(1 - \cos \omega t) \tag{3–34}$$

It is interesting to note from Eqs. (3–32) and (3–34) that the path of the particle consists of a uniform circular motion plus a uniform translation. To see this more clearly, consider a primed coordinate system translating uniformly relative to the unprimed system such that the particle position in the primed system is

$$x' = x - \left(v_o + \frac{P}{m\omega}\right)t = -\frac{P}{m\omega^2}\sin\omega t \qquad (3\text{--}35)$$

$$y' = y = \frac{P}{m\omega^2}(1 - \cos\omega t) \qquad (3\text{--}36)$$

These equations represent a circular path of radius $P/m\omega^2$ centered at $x' = 0$, $y' = P/m\omega^2$ as shown in Fig. 3–3.

Of course, the primed system is an inertial system since it is translating uniformly relative to the fixed xy system. It can be seen that, even in the case

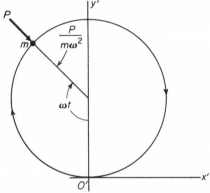

Fig. 3-3. The path of the particle relative to a uniformly translating system.

of general initial conditions, a uniformly translating coordinate system can always be found such that the motion of the particle relative to this system consists of uniform circular motion in a plane.

Case 3: $\mathbf{F} = F_x(x)\mathbf{i} + F_y(y)\mathbf{j} + F_z(z)\mathbf{k}$. For this case, we shall again consider only the x component of the motion, noting that similar results apply to the motion in the y and z directions. The differential equation for the motion is

$$m\ddot{x} = F_x(x) \qquad (3\text{--}37)$$

Because the force F_x is a function of the displacement rather than time, it is desirable to integrate with respect to x. This can be accomplished by making the substitution

$$\dot{x} = v \qquad (3\text{--}38)$$

implying that

$$\ddot{x} = \frac{dv}{dt} = \frac{dv}{dx}\frac{dx}{dt} = v\frac{dv}{dx} \tag{3-39}$$

From Eqs. (3-37) and (3-39), we obtain

$$mv\frac{dv}{dx} = F_x(x) \tag{3-40}$$

which can be integrated directly to give

$$\frac{1}{2}m(v^2 - v_o^2) = \int_{x_o}^{x} F_x(x)\,dx \tag{3-41}$$

This result is an application of the principle of work and kinetic energy which will be discussed in Sec. 3-2.

Equation (3-41) can be integrated again to give a solution of the form

$$t = f(x_o, v_o, x) \tag{3-42}$$

from which one obtains a solution for x of the form

$$x = g(x_o, v_o, t) \tag{3-43}$$

Example 3-3. A mass m and a linear spring of stiffness k are connected as shown in Fig. 3-4. Assuming one-dimensional motion with no friction, solve for the displacement x as a function of time and the initial conditions.

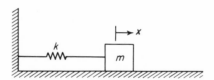

Fig. 3-4. A mass-spring system.

In this example,

$$F_x = -kx \tag{3-44}$$

or, from Eq. (3-37),

$$m\ddot{x} + kx = 0 \tag{3-45}$$

Using Eq. (3-41), we obtain

$$\frac{1}{2}m(v^2 - v_o^2) = -\int_{x_o}^{x} kx\,dx = -\frac{1}{2}k(x^2 - x_o^2) \tag{3-46}$$

and, solving for the velocity, we find that

$$v = \frac{dx}{dt} = \left[v_o^2 - \frac{k}{m}(x^2 - x_o^2)\right]^{1/2}$$

or

$$t = \int_{x_o}^{x}\left[v_o^2 - \frac{k}{m}(x^2 - x_o^2)\right]^{-1/2} dx \tag{3-47}$$

Evaluating the integral, we obtain

$$t = \sqrt{\frac{m}{k}} \left[\sin^{-1} \frac{x}{\sqrt{(m/k)v_o^2 + x_o^2}} - \sin^{-1} \frac{x_o}{\sqrt{(m/k)v_o^2 + x_o^2}} \right]$$

or

$$x = \sqrt{(m/k)v_o^2 + x_o^2} \sin \left(\sqrt{\frac{k}{m}} \, t + \alpha \right) \tag{3-48}$$

where

$$\alpha = \sin^{-1} \frac{x_o}{\sqrt{(m/k)v_o^2 + x_o^2}} \tag{3-49}$$

This is the solution for the free motion of a mass-spring system with arbitrary initial conditions and represents an example of one-dimensional harmonic motion.

Case 4: $\mathbf{F} = F_x(\dot{x})\mathbf{i} + F_y(\dot{y})\mathbf{j} + F_z(\dot{z})\mathbf{k}$. In the analysis of this case, we shall again consider just the x component of the motion because the form of the equation is such that the three component motions are each independent. The equation of motion can be written as

$$m\frac{dv}{dt} = F_x(v) \tag{3-50}$$

from which we obtain

$$t = m \int_{v_o}^{v} \frac{dv}{F_x(v)} \tag{3-51}$$

where, in this case, the velocity v is in the x direction. Evaluating the integral, one can solve for velocity as a function of the initial velocity and time:

$$v = \frac{dx}{dt} = g(v_o, t) \tag{3-52}$$

A second integration results in the displacement in the form

$$x = f(x_o, v_o, t) \tag{3-53}$$

Another approach is to write Eq. (3–50) in the form

$$mv\frac{dv}{dx} = F_x(v) \tag{3-54}$$

Integrating this equation results in

$$m \int_{v_o}^{v} \frac{v\,dv}{F_x(v)} = x - x_o \tag{3-55}$$

Now eliminate v from Eqs. (3–51) and (3–55) and one again obtains

$$x = f(x_o, v_o, t)$$

as in Eq. (3–53).

Example 3–4. Solve for the motion of a particle of mass m that is moving in a uniform gravitational field and is subject to a linear damping force.

The force of the linear damper always acts in a direction opposite to the velocity vector and its magnitude is directly proportional to the velocity. So if we consider the motion to occur in the xy plane with gravity acting in the direction of the negative y axis, then the total force acting on the particle is

$$\mathbf{F} = -c\mathbf{v} - mg\mathbf{j} \tag{3-56}$$

Now the x component of the force is

$$F_x = \frac{\dot{x}}{v}(-cv) = -c\dot{x} \tag{3-57}$$

since $\dot{x}/v$ is the cosine of the angle between the velocity vector and the x axis (Fig. 3-5). Similarly,

$$F_y = -c\dot{y} - mg \tag{3-58}$$

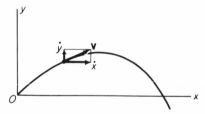

Fig. 3-5. The trajectory of a linearly-damped particle which moves in a uniform gravitational field.

So the differential equations of motion are

$$m\ddot{x} + c\dot{x} = 0 \tag{3-59}$$

and

$$m\ddot{y} + c\dot{y} = -mg \tag{3-60}$$

Consider first the motion in the x direction. Eq. (3–51) can be used directly in this case, resulting in

$$t = -\frac{m}{c}\int_{\dot{x}_o}^{\dot{x}}\frac{d\dot{x}}{\dot{x}} = -\frac{m}{c}\ln\left(\frac{\dot{x}}{\dot{x}_o}\right)$$

or

$$\dot{x} = \dot{x}_o e^{-ct/m} \tag{3-61}$$

where $\dot{x}_o$ is the x component of the initial velocity. Another integration results in

$$x = x_o + \frac{m\dot{x}_o}{c}(1 - e^{-ct/m}) \tag{3-62}$$

where the constant of integration has been evaluated from initial conditions.

Now consider the motion in the y direction. Equation (3–60) can be integrated to obtain

$$t = -\frac{m}{c} \int_{\dot{y}_o}^{\dot{y}} \frac{d\dot{y}}{\dot{y} + (mg/c)} = -\frac{m}{c} \ln \left[\frac{\dot{y} + (mg/c)}{\dot{y}_o + (mg/c)} \right]$$

or

$$\dot{y} = -\frac{mg}{c} + \left(\dot{y}_o + \frac{mg}{c} \right) e^{-ct/m} \tag{3-63}$$

Integrating again, assuming an initial displacement y_o, we obtain

$$y = y_o - \frac{mgt}{c} + \frac{m}{c} \left(\dot{y}_o + \frac{mg}{c} \right)(1 - e^{-ct/m}) \tag{3-64}$$

From Eqs. (3–62) and (3–64), we note that the displacement in the x direction has a limiting value as t approaches infinity, namely,

$$x = x_o + \frac{m\dot{x}_o}{c}$$

On the other hand, the y displacement changes continuously in the negative direction for large t, its limiting velocity being

$$\dot{y} = -\frac{mg}{c}$$

The physical interpretation of this last result is that the gravitational force mg and the friction force $c\dot{y}$ are of equal magnitude and opposite direction in the steady state, resulting in no net external force being applied to the particle.

Another point to notice is that the solutions for the motions in both the x and y directions involve a term of the form $\exp(-ct/m)$. The exponent contains the ratio

$$\tau = \frac{m}{c}$$

which has the dimensions of time and is known as the *time constant* for the system. Now, time constants are often associated with first-order systems, but Eqs. (3–59) and (3–60) are seen to be second-order differential equations. Since they contain no term in x or y, however, they are also first-order in the velocities v_x and v_y as may be observed by writing them in the form:

$$\tau \dot{v}_x + v_x = 0$$

$$\tau \dot{v}_y + v_y = -\tau g$$

The solutions for the velocities given by Eqs. (3–61) and (3–63) are typical of the form of the response of first-order systems to an initial condition or to a constant forcing term.

3–2. WORK AND KINETIC ENERGY

We now present some general principles of particle mechanics. For the cases where these principles apply, they are directly derivable from Newton's laws of motion, and thus they contain no new information. Nevertheless, they promote further insight into particle dynamics and, by providing some integrals of the motion, aid in the solution of many specific problems.

The first of these principles to be presented is the *principle of work and kinetic energy: The increase in the kinetic energy of a particle in going from one point to another is equal to the work done by the external forces acting on the particle as it moves over the given interval.*

To illustrate the meaning of the principle, as well as to define some of the terms, consider a particle of mass m that moves from A to B under the action of an arbitrary external force $\mathbf{F}$, as shown in Fig. 3–6. Starting with

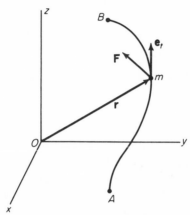

Fig. 3-6. The path of a particle moving under the action of an external force.

Newton's law of motion in the form

$$\mathbf{F} = m\ddot{\mathbf{r}}$$

let us evaluate the line integral of each side of the equation over the given path from A to B.

$$\int_A^B \mathbf{F} \cdot d\mathbf{r} = \int_A^B m\ddot{\mathbf{r}} \cdot d\mathbf{r} \tag{3-65}$$

where $d\mathbf{r}$ is taken in a direction tangent to the curve at each point.

We note that

$$\ddot{\mathbf{r}} \cdot d\mathbf{r} = \frac{1}{2}\frac{d}{dt}(\dot{\mathbf{r}} \cdot \dot{\mathbf{r}})dt = \frac{1}{2}d(v^2)$$

where it is assumed that the infinitesimal changes in position and velocity occur during the same time interval. Thus, we see that

$$\int_A^B m\ddot{\mathbf{r}} \cdot d\mathbf{r} = \frac{m}{2}\int_{v_A}^{v_B} d(v^2) = \frac{m}{2}(v_B^2 - v_A^2) \tag{3-66}$$

where v_A and v_B are the velocities of the particle at points A and B, respectively, of the path. From Eqs. (3–65) and (3–66), we obtain the result:

$$\int_A^B \mathbf{F} \cdot d\mathbf{r} = \frac{1}{2} mv_B^2 - \frac{1}{2} mv_A^2 \qquad (3\text{–}67)$$

The integral on the left side of this equation is the *work* W which is done by the external force $\mathbf{F}$ as it moves along the path from A to B.

$$W = \int_A^B \mathbf{F} \cdot d\mathbf{r} \qquad (3\text{–}68)$$

The *kinetic energy* T of a particle relative to an inertial system is

$$T = \frac{1}{2} mv^2 \qquad (3\text{–}69)$$

where v is the speed of the particle relative to that system. Therefore, the right side of Eq. (3–67) represents the increase of kinetic energy in going from A to B.

Using Eqs. (3–68) and (3–69), we can express the general result of Eq. (3–67) in the form

$$W = T_B - T_A \qquad (3\text{–}70)$$

It should be emphasized that calculations of work and kinetic energy are dependent upon which inertial system is used as a reference frame. Nevertheless, even though the individual terms may vary with the reference frame, the general principle of work and kinetic energy is valid in any inertial frame.

Example 3–5. A whirling particle of mass m is pulled slowly by a string toward a fixed center at O (Fig. 3–7) in such a manner that the radial component of velocity is small compared to the tangential component. Also,

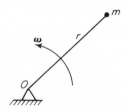

Fig. 3-7. A particle whirling about a fixed point.

$\ddot{r}$ can be neglected relative to the centripetal acceleration. Using energy methods, find the angular velocity ω as a function of r. Initial conditions are $\omega(0) = \omega_o$ and $r(0) = r_o$.

In order to use the principle of work and kinetic energy, we must first determine the total external force acting on the particle. The string provides the only means for exerting an external force and therefore this force must be radial in nature. Thus, in accordance with Newton's law of motion, the acceleration of the particle is entirely radial. Since $\ddot{r}$ is assumed to be negligible, the path at any point is approximately circular and the acceleration is just the centripetal acceleration. From Eqs. (1–41) and (2–53), the total external force is the radial force

$$F_r = -mr\omega^2 \tag{3-71}$$

The work done on the particle in an infinitesimal radial displacement dr is

$$dW = F_r \, dr = -m\omega^2 r \, dr \tag{3-72}$$

But, by the principle of work and kinetic energy, each increment of work done on the particle is accompanied by an equal increase in the kinetic energy. The kinetic energy is

$$T = \frac{1}{2} mv^2 = \frac{1}{2} mr^2 \omega^2 \tag{3-73}$$

since we neglect the radial velocity. Thus,

$$dT = m\omega^2 r \, dr + mr^2 \omega \, d\omega \tag{3-74}$$

Therefore we obtain

$$dW = dT$$

and, from Eqs. (3–72) and (3–74),

$$-2m\omega^2 r \, dr = mr^2 \omega \, d\omega$$

Rearranging and integrating in the interval r_0 to r, corresponding to the angular rates ω_0 and ω, we obtain

$$-2 \int_{r_0}^{r} \frac{dr}{r} = \int_{\omega_0}^{\omega} \frac{d\omega}{\omega}$$

or

$$-2 \ln \left(\frac{r}{r_0} \right) = \ln \left(\frac{\omega}{\omega_0} \right) \tag{3-75}$$

Solving for ω as a function of r, we find that

$$\omega = \left(\frac{r_0}{r} \right)^2 \omega_0 \tag{3-76}$$

This final result can also be written in the form

$$mr^2 \omega = mr_0^2 \omega_0 \tag{3-77}$$

which is in agreement with the principle of conservation of angular momentum to be developed in Sec. 3–6.

3-3. CONSERVATIVE SYSTEMS

Referring again to Fig. 3–6 let us suppose that the force $\mathbf{F}$ acting on the given particle has the following characteristics: (1) it is a function of position only; (2) the line integral

$$\int_A^B \mathbf{F} \cdot d\mathbf{r}$$

is a function of the end points only and is independent of the path taken between A and B.

Now, from characteristic (1),

$$\int_A^B \mathbf{F} \cdot d\mathbf{r} = -\int_B^A \mathbf{F} \cdot d\mathbf{r} \tag{3-78}$$

for the case where the line integral on the right is taken in the reverse direction along the same path. But, by characteristic (2), it is also true for an integration along any two paths connecting A and B. Therefore,

$$\oint \mathbf{F} \cdot d\mathbf{r} = 0 \tag{3-79}$$

where the integral is taken around any closed path.

A force with the characteristics just cited is said to be a *conservative force*, that is, it forms a conservative force field. Practically speaking, this means that the force is not dissipative in nature and that any mechanical process taking place under its influence is reversible. The property of reversibility can be clarified as follows: if, at a certain moment, the velocities of all moving parts are reversed, then, following the same physical laws, a reversible mechanical process will retrace its former sequence of positions and accelerations in reverse order, as though time were running backwards.

If the work W, as given by Eq. (3–68), is found to depend only upon the location of the end points, then the integrand must be an exact differential.

$$\mathbf{F} \cdot d\mathbf{r} = -dV \tag{3-80}$$

where the minus sign has been chosen for convenience in the statement of later results. Therefore we find that

$$W = \int_A^B \mathbf{F} \cdot d\mathbf{r} = -\int_A^B dV = V_A - V_B \tag{3-81}$$

Equation (3–81) states that the decrease in the *potential energy* V in moving the particle from A to B is equal to the work done on the particle *by* the conservative force field. Conversely, the increase in potential energy in moving the particle between two points is equal to the work done *against* the conservative field forces by the particle.

The potential energy V is a scalar function of position only, for a given particle. The sum of the potential and kinetic energies is known as the *total energy* E, and from Eqs. (3–70) and (3–81), we find that this sum is constant for a conservative system.

$$V_A + T_A = V_B + T_B = E \tag{3-82}$$

Equation (3–82) is a mathematical statement of the *principle of conservation of mechanical energy*. It applies to systems in which the only forces that do work on the particle are those arising from a conservative force field. Note that workless forces, such as those due to frictionless, fixed constraints, do not change the applicability of the principle.

3-4. POTENTIAL ENERGY

Let us consider again the potential energy V. We saw that V is a function of position only, for the case of a conservative force field. Therefore, if we express the position of a particle in terms of its cartesian coordinates, we find that

$$dV = \frac{\partial V}{\partial x} dx + \frac{\partial V}{\partial y} dy + \frac{\partial V}{\partial z} dz \tag{3-83}$$

Also,

$$d\mathbf{r} = dx\mathbf{i} + dy\mathbf{j} + dz\mathbf{k}$$

and therefore

$$\mathbf{F} \cdot d\mathbf{r} = F_x \, dx + F_y \, dy + F_z \, dz \tag{3-84}$$

From Eqs. (3-80), (3-83), and (3-84), we obtain

$$F_x = -\frac{\partial V}{\partial x}$$

$$F_y = -\frac{\partial V}{\partial y} \tag{3-85}$$

$$F_z = -\frac{\partial V}{\partial z}$$

since we note that Eq. (3-80) is applicable for an arbitrary infinitesimal displacement; hence the coefficients of dx, dy, and dz in Eq. (3-83) must be equal to the negative of the corresponding coefficients in Eq. (3-84). Thus we can write

$$\mathbf{F} = F_x\mathbf{i} + F_y\mathbf{j} + F_z\mathbf{k} = -\frac{\partial V}{\partial x}\mathbf{i} - \frac{\partial V}{\partial y}\mathbf{j} - \frac{\partial V}{\partial z}\mathbf{k} \tag{3-86}$$

Now the *gradient* of the scalar function V is

$$\nabla V = \frac{\partial V}{\partial x}\mathbf{i} + \frac{\partial V}{\partial y}\mathbf{j} + \frac{\partial V}{\partial z}\mathbf{k} \tag{3-87}$$

and therefore we find that the force exerted by a conservative force field acting on the particle is[1]

$$\mathbf{F} = -\nabla V \tag{3-88}$$

This means that the force is in the direction of the largest spatial rate of decrease of V and is equal in magnitude to that rate of decrease.

[1] The converse is not necessarily true, that is, a force derivable from a potential function V by using Eq. (3-88) may not be conservative, as in the case of a time-varying field.

Equation (3–88) is a general vector equation and therefore the gradient need not be expressed in terms of cartesian coordinates. For a general coordinate system, the component of the force **F** in the direction of the unit vector $\mathbf{e}_1$ is given by

$$F_1 = -\frac{\partial V}{\partial x_1} \tag{3-89}$$

where x_1 is the linear displacement in the direction $\mathbf{e}_1$ at the point under consideration.

Inverse-square Attraction. As an example of a potential energy calculation, consider the case of an inverse-square attraction of a particle of mass m toward a fixed point. The force exerted by the attracting field is entirely radial and is equal to

$$F_r = -\frac{\partial V}{\partial r} = -\frac{K}{r^2} \tag{3-90}$$

where K is a constant and r is the distance of the particle from the attracting center. Since the force is a function of r only, we can integrate Eq. (3–90) directly to give

$$V = -\frac{K}{r} + C \tag{3-91}$$

where C is an arbitrary constant of integration. In the usual case of inverse-square attraction, we choose $C = 0$, implying that the potential energy is always negative and approaches zero as r approaches infinity.

In the general case, the fact that the potential energy contains an arbitrary constant would seem to require that all measurable quantities of the motion such as velocity, acceleration, and so on, should be independent of the choice of C, since the motion in a given situation is not arbitrary. That this is actually true is confirmed further by noting that the potential energy enters all computations as a potential energy *difference*, in which case the constant C cancels out. As a result, the choice of C, that is, the choice of the datum or reference point of zero potential energy, is made for convenience in solving the problem at hand.

Gravitational Potential Energy. The most commonly encountered inverse-square force in the study of mechanics is the force of gravitational attraction, and as we have seen, the corresponding gravitational potential must be of the form given by Eq. (3–91).

Now let us consider the particular case of gravitational attraction by the earth. Assuming that the mass distribution of the earth is spherically symmetrical about the center, it can be shown (Sec. 5–1) that the attractive force on an external particle is the same as if the entire mass of the earth were concentrated at its center. Hence we see from Eq. (3–90) that the gravitational force on a particle of mass m outside the earth's surface is the radial force

$$F_r = -\frac{K}{r^2} \quad (r \geq R) \tag{3-92}$$

where R is the radius of the earth (Fig. 3-8). The constant K can be evaluated

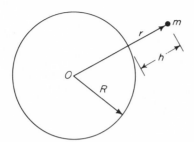

Fig. 3-8. The earth and a particle of mass m.

from the knowledge (Sec. 1–3) that, assuming a nonrotating earth, the weight w of a particle is the force of the gravitational attraction of the earth on the particle, measured at the earth's surface. Thus

$$w = -F_r = \frac{K}{R^2} \tag{3-93}$$

But, in this instance, the weight is also given by

$$w = mg_0 \tag{3-94}$$

where g_0 is the acceleration of gravity at the surface of a spherical nonrotating earth.

From Eqs. (3–93) and (3–94), we obtain that

$$K = mg_0 R^2 \tag{3-95}$$

Therefore, from Eq. (3–92),

$$F_r = -\frac{mg_0 R^2}{r^2} \quad (r \geq R) \tag{3-96}$$

and, from Eq. (3–91),

$$V = -\frac{mg_0 R^2}{r} \quad (r \geq R) \tag{3-97}$$

But we can also write the potential energy in terms of the height h above the earth's surface. Letting

$$r = R + h \tag{3-98}$$

we find from Eq. (3–97) that

$$V = -\frac{mg_0 R}{1 + (h/R)}$$

For motion near the surface of the earth, that is, for $h \ll R$, this equation is approximated by

$$\left(1 - \frac{h}{R}\right)^{-1} = 1 - \frac{h}{R} + 3\left(\frac{h}{R}\right)^2 \cdots$$

$$V \cong -mg_oR\left(1 - \frac{h}{R}\right)$$

Now we eliminate the constant term $-mg_oR$ by choosing the zero reference for potential energy at the earth's surface. Then we obtain

$$V \cong mg_oh \quad (h \ll R) \tag{3-99}$$

In the calculation of local trajectories near the earth's surface, a reference frame is often chosen that is fixed in the earth and the value of the acceleration of gravity includes the effects at that point of the earth's rotation. In this case, the gravitational field is essentially uniform and we use the symbol g for the local acceleration of gravity. Hence we find that

$$V = mgh \tag{3-100}$$

Potential Energy of a Linear Spring. Another commonly encountered form of potential energy is that due to elastic deformation. As an example of elastic potential energy, consider a particle P which is attached by a linear spring of stiffness k to a fixed point O, as shown in Fig. 3–9. If the elonga-

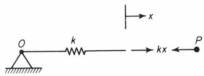

Fig. 3-9. The interaction forces of a particle P and a linear spring.

tion x of the spring is measured from its unstressed position, the particle will experience a force

$$F_x = -\frac{\partial V}{\partial x} = -kx \tag{3-101}$$

Direct integration of Eq. (3–101) with respect to x, choosing the zero reference for potential energy at $x = 0$, results in

$$V = \frac{1}{2}kx^2 \tag{3-102}$$

Note again that, in the preceding development, the force F_x is the force exerted by the spring on the particle and not the force of the opposite sign that is applied to the spring by the particle. In other words, the emphasis here is on the potential ability of the spring to do work on its surroundings, and not vice versa.

Example 3–6. A particle of mass m is suspended vertically by a spring of stiffness k in the presence of a uniform gravitational field, the direction of the gravitational force being as shown in Fig. 3–10. If the vertical displacement y of the mass is measured from its position when the spring is unstressed, solve for y as a function of time. The initial conditions are

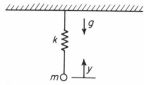

Fig. 3-10. A mass suspended by a linear spring.

$y(0) = y_o$ and $\dot{y}(0) = 0$. Also find the maximum values of kinetic energy and potential energy in the ensuing motion.

The equation of motion can be written with the aid of the free-body diagram shown in Fig. 3–11, where we note that the spring force and the

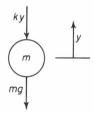

Fig. 3-11. A free-body diagram showing the external forces which act on mass m.

gravity force are the only forces acting on the particle. Using Newton's law of motion we can write

$$m\ddot{y} = -ky - mg$$

or

$$m\ddot{y} + ky = -mg \qquad (3\text{–}103)$$

Rather than solve this equation by direct integration in a fashion similar to Example 3–3, we shall obtain the solution as the sum of two parts; that is,

$$y = y_t + y_s \qquad (3\text{–}104)$$

where the *transient solution* y_t (also called the *complementary function*) is the solution to the homogeneous equation

$$m\ddot{y} + ky = 0 \qquad (3\text{–}105)$$

and the *steady-state solution* y_s (also called the *particular integral*) is a solution which satisfies the complete differential equation.

In general, for this case of an *ordinary differential equation with constant coefficients*, one assumes that the transient solution contains terms of the form $Ce^{\lambda t}$, where C and λ are constants.[2] The exponential function is chosen, since it retains the same variable part upon differentiation. For the problem at hand, the substitution of $y = Ce^{\lambda t}$ into Eq. (3–105) results in

[2]This assumption is valid except for the case of repeated roots, a situation which will not be considered here.

$$(m\lambda^2 + k)Ce^{\lambda t} = 0$$

Since this result must hold for all values of time and we rule out the trivial case, $C = 0$, we obtain the so-called *characteristic equation*

$$m\lambda^2 + k = 0 \qquad (3\text{–}106)$$

The roots are imaginary in this case:

$$\lambda_{1,2} = \pm i\sqrt{\frac{k}{m}} \qquad (3\text{–}107)$$

where $i = \sqrt{-1}$.

Thus we assume a transient solution of the form

$$y_t = C_1 e^{i\sqrt{k/m}\,t} + C_2 e^{-i\sqrt{k/m}\,t} \qquad (3\text{–}108)$$

Substituting for the exponential functions in terms of sine and cosine functions, we use the general formula

$$e^{i\theta} = \cos\theta + i\sin\theta \qquad (3\text{–}109)$$

to obtain

$$y_t = (C_1 + C_2)\cos\sqrt{\frac{k}{m}}\,t + i(C_1 - C_2)\sin\sqrt{\frac{k}{m}}\,t \qquad (3\text{–}110)$$

Now if we let

$$A = C_1 + C_2 \quad \text{and} \quad B = i(C_1 - C_2)$$

we obtain the transient solution in the form

$$y_t = A\cos\sqrt{\frac{k}{m}}\,t + B\sin\sqrt{\frac{k}{m}}\,t \qquad (3\text{–}111)$$

where A and B are arbitrary constants. In general, there must be as many arbitrary constants as the order of the differential equation, and as we shall see, the constants are ultimately evaluated on the basis of initial conditions.

The steady-state solution can be found by several methods. For the common case where the forcing function is represented by a constant or by terms of the form t^n (n a positive integer), e^{kt}, $\sin\omega t$, $\cos\omega t$, or by sums and products of these terms, the method of *undetermined coefficients* is applicable. (See a standard textbook on differential equations for details of that method.) Briefly, however, the usual form of the steady-state solution consists of terms of the form of the forcing function plus terms containing any different variable parts obtainable by successive differentiations of the forcing function. In this particular case, the forcing function is the constant term $-mg$ and the steady-state solution is simply

$$y_s = -\frac{mg}{k} \qquad (3\text{–}112)$$

as may be verified by substitution into the differential equation. Therefore the total solution is of the form

$$y = y_t + y_s = -\frac{mg}{k} + A \cos \sqrt{\frac{k}{m}} t + B \sin \sqrt{\frac{k}{m}} t \qquad (3\text{-}113)$$

Now we find the constants A and B from the two given initial conditions. Upon evaluating Eq. (3–113) at $t = 0$, we obtain

$$y(0) = y_o = A - \frac{mg}{k}$$

Differentiating Eq. (3–113) with respect to time and setting $t = 0$, we have

$$\dot{y}(0) = 0 = \sqrt{\frac{k}{m}} B$$

Hence

$$A = y_o + \frac{mg}{k}$$
$$\tag{3-114}$$
$$B = 0$$

From Eqs. (3–113) and (3–114), the complete solution is

$$y = -\frac{mg}{k} + \left(y_o + \frac{mg}{k}\right) \cos \sqrt{\frac{k}{m}} t \qquad (3\text{-}115)$$

It can be seen that the solution consists of a sinusoidal oscillation of the mass about the position $y = -mg/k$, which is the position of static equilibrium for the particle.

The terms *transient* and *steady-state* should not be taken literally as describing the nature of certain terms in the solution, although in many instances the transient solution amplitude decreases exponentially with time and the steady-state solution persists with undiminished magnitude. In this example, the transient solution oscillates with constant amplitude. For the case of a system with one or more roots with positive real parts, it would actually show an amplitude which increases with time. On the other hand, the steady-state solution could have a magnitude which increases or decreases with time, depending upon the nature of the forcing function.

Now let us calculate the kinetic and potential energies for this system. From Eq. (3–69), the kinetic energy is

$$T = \frac{1}{2} m \dot{y}^2 \qquad (3\text{-}116)$$

The potential energy arises from both the gravitational force and the spring force, since they are conservative in nature. Adding the individual potential energies to obtain the total, we have from Eqs. (3–100) and (3–102),

$$V = mgy + \frac{1}{2} ky^2 \qquad (3\text{-}117)$$

the reference point of zero potential energy being taken at $y = 0$. We are considering a conservative system and therefore the total energy E is a constant. Evaluating E from its initial value, we obtain

$$E = T + V = mgy_o + \frac{1}{2} ky_o^2 \tag{3-118}$$

It can be seen that the kinetic energy is maximum when the potential energy is minimum and vice versa. From Eq. (3-116), we see that

$$T_{min} = 0 \tag{3-119}$$

and therefore

$$V_{max} = E \tag{3-120}$$

On the other hand, we see by setting $dV/dy = 0$ that V_{min} occurs at

$$y = -\frac{mg}{k}$$

and is equal to

$$V_{min} = -\frac{1}{2} \frac{m^2 g^2}{k} \tag{3-121}$$

Therefore,

$$T_{max} = E + \frac{1}{2} \frac{m^2 g^2}{k} \tag{3-122}$$

Thus we see that

$$V_{max} - V_{min} = T_{max} - T_{min} \tag{3-123}$$

even though the extreme values of V are not equal to the corresponding extreme values of T.

Now let us consider the case where the displacement is measured from its equilibrium position. Calling this vertical displacement z, we can write

$$z = y + \frac{mg}{k} \tag{3-124}$$

From Eqs. (3-115) and (3-124), we find that the complete solution can be written in the form

$$z = z_o \cos \sqrt{\frac{k}{m}} t \tag{3-125}$$

where

$$z_o = y_o + \frac{mg}{k} \tag{3-126}$$

Now let us choose the zero reference for potential energy to be at $z = 0$, that is, we choose a constant such that $V = 0$ when $z = 0$. Then,

$$V = mgz + \frac{1}{2} ky^2 - \frac{1}{2} \frac{m^2 g^2}{k}$$

$$= mgz + \frac{1}{2} k \left(z - \frac{mg}{k} \right)^2 - \frac{1}{2} \frac{m^2 g^2}{k}$$

which reduces to

$$V = \frac{1}{2} kz^2 \qquad (3\text{--}127)$$

Therefore, if we use the static equilibrium position as the zero reference for potential energy, we find that the total energy is

$$E = \frac{1}{2} m\dot{z}^2 + \frac{1}{2} kz^2 = \frac{1}{2} kz_o^2 \qquad (3\text{--}128)$$

Also,

$$T_{\max} = V_{\max} = E \qquad (3\text{--}129)$$

and

$$T_{\min} = V_{\min} = 0 \qquad (3\text{--}130)$$

In summary, it can be seen that the analysis is simplified by measuring displacements from the position of static equilibrium and setting the potential energy equal to zero at this point. In this case, the total force acting on the particle is $-kz$ and, from Eq. (3–88), we obtain

$$-\frac{\partial V}{\partial z} = -kz$$

or

$$V = \frac{1}{2} kz^2$$

in agreement with Eq. (3–127), provided that the reference for zero potential energy is again taken at the static equilibrium position. Note that this expression includes the gravitational as well as the elastic potential energy.

3-5. LINEAR IMPULSE AND MOMENTUM

In Sec. 1–2, we saw that the *linear momentum* $\mathbf{p}$ of a particle is just the product of its mass and velocity.

$$\mathbf{p} = m\mathbf{v} \qquad (3\text{--}131)$$

Thus Newton's law of motion can be written in the form

$$\mathbf{F} = \dot{\mathbf{p}} \qquad (3\text{--}132)$$

where it is assumed that the particle mass is constant and its velocity is measured relative to an inertial frame.

Now let us integrate both sides of Eq. (3–132) over the time interval t_1 to t_2.

$$\int_{t_1}^{t_2} \mathbf{F}\, dt = \int_{t_1}^{t_2} \dot{\mathbf{p}}\, dt = \mathbf{p}_2 - \mathbf{p}_1 \qquad (3\text{--}133)$$

where $\mathbf{p}_1$ and $\mathbf{p}_2$ are the values of the linear momentum vector at times t_1

and t_2, respectively. The time integral of the force $\mathbf{F}$ is known as the *impulse* $\mathscr{F}$ of the force, that is,

$$\mathscr{F} = \int_{t_1}^{t_2} \mathbf{F}\, dt \tag{3-134}$$

So, from Eqs. (3–133) and (3–134) we obtain a statement of the *principle of linear impulse and momentum:*

> *The change in the linear momentum of a particle during a given interval is equal to the total impulse of the external forces acting on the particle over the same interval.*

$$\mathscr{F} = \mathbf{p}_2 - \mathbf{p}_1 = m\mathbf{v}_2 - m\mathbf{v}_1 \tag{3-135}$$

The length of time over which the integration proceeds does not influence the result. Of course, we assume that $t_2 > t_1$, but the interval may approach zero. If the duration of the external force approaches zero, but its amplitude becomes very large in such a manner that the time integral of the force remains finite, then the force is known as an *impulsive force*. The effect of an impulsive force in changing the motion of a particle is expressed entirely by its total impulse $\mathscr{F}$, in accordance with Eq. (3–135). We see that the velocity changes instantaneously, but the position cannot change instantaneously for a finite impulse because the velocity remains finite.

It is convenient to express an impulsive force of total impulse $\mathscr{F}$ occurring at time $t = \tau$ in the form

$$\mathbf{F} = \mathscr{F}\, \delta(t - \tau) \tag{3-136}$$

where $\delta(t)$ is the *Dirac delta function* defined as follows:

$$\delta(t) = \begin{cases} 0 & t \neq 0 \\ \infty & t = 0 \end{cases} \tag{3-137}$$

such that

$$\int_{-\infty}^{\infty} \delta(t)\, dt = 1 \tag{3-138}$$

It can be seen that the total impulse due to $\mathbf{F}$ is

$$\int_{-\infty}^{\infty} \mathbf{F}\, dt = \mathscr{F} \int_{-\infty}^{\infty} \delta(t - \tau)\, dt = \mathscr{F}$$

in agreement with the original assumption.

Equation (3–135) can also be written in terms of its scalar components. Choosing an inertial cartesian system in which to express the motion of the particle, we find that

$$\mathscr{F}_x = \int_{t_1}^{t_2} F_x\, dt = m(\dot{x})_2 - m(\dot{x})_1$$

$$\mathscr{F}_y = \int_{t_1}^{t_2} F_y\, dt = m(\dot{y})_2 - m(\dot{y})_1 \tag{3-139}$$

$$\mathscr{F}_z = \int_{t_1}^{t_2} F_z\, dt = m(\dot{z})_2 - m(\dot{z})_1$$

where the subscripts 1 and 2 indicate that the evaluations of the given velocity components are to be made at t_1 and t_2, respectively. These equations are obtained directly from Eq. (3–135) or by integrating Eq. (1–43) with respect to time.

From Eq. (3–132) we see that if the external force **F** equals zero, then the linear momentum **p** is constant. This is the *principle of conservation of linear momentum*. Similarly, we note from Eq. (3–139) that if any component of **F** is zero for a certain interval of time, then the corresponding component of the momentum is conserved during that interval.

A comparison of Eq. (3–135), which equates the total impulse to the change of momentum, and Eq. (3–70), which equates the work done on the particle to the change in kinetic energy, reveals some interesting qualitative differences. First, Eq. (3–135) is a vector equation, whereas Eq. (3–70) is a scalar equation. The vector equation is advantageous at times because it gives the direction as well as the magnitude of the velocity. On the other hand, scalar equations are often easier to use. If the force **F** is given as a function of *time*, impulse and momentum methods are usually called for; but if **F** is given as a function of *position*, then work and energy methods can be used to advantage.

Another point of interest is that the total impulse $\mathscr{F}$, and consequently the change in linear momentum, is independent of which inertial frame is chosen for viewing the motion of the particle. This is in contrast to the calculations of work and the change of kinetic energy which, as we have seen, are dependent upon a specific frame of reference.

Example 3–7. A particle m slides along a frictionless wire that is fixed in inertial space, as shown in Fig. 3–12. Assuming that a known external

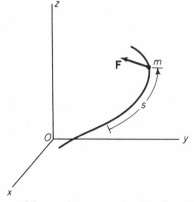

Fig. 3-12. A particle moving over a fixed path under the action of an external force.

force **F** acts on the system, find a general method for calculating speed changes of the particle.

If at each instant we take components of force and acceleration along the path, we find from Eqs. (1–41) and (2–41) that

$$F_s = m\ddot{s} \tag{3-140}$$

where F_s is the component, along the tangential direction, of the external force $\mathbf{F}$, and $\ddot{s}$ is the tangential component of the acceleration. Note that the frictionless constraint forces are normal to the path and do not contribute to F_s.

Integrating Eq. (3–140) with respect to time, we obtain

$$\int_{t_1}^{t_2} F_s \, dt = m(\dot{s})_2 - m(\dot{s})_1 \tag{3-141}$$

Thus we obtain an essentially one-dimensional or scalar version of the equation of impulse and momentum that applies to motion along a fixed curve in space.

Of course, the equation of work and kinetic energy, Eq. (3–67), can be applied directly, resulting in

$$\int_{A}^{B} F_s \, ds = \frac{1}{2} m(\dot{s})_B^2 - \frac{1}{2} m(\dot{s})_A^2 \tag{3-142}$$

3–6. ANGULAR MOMENTUM AND ANGULAR IMPULSE

Angular Momentum. We have seen that the linear momentum of a particle with respect to a fixed reference frame is the vector $m\mathbf{v}$, where m is the mass of the particle and $\mathbf{v}$ is its absolute velocity. Now let us consider the momentum vector as a sliding vector whose line of action passes through the particle, as shown in Fig. 3–13. If $\mathbf{r}$ is the position vector of the particle

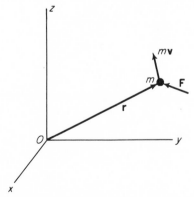

Fig. 3-13. A particle moving relative to a fixed point O.

with respect to a fixed reference point O, then the *moment of momentum* or *angular momentum* about O is given by

$$\mathbf{H} = \mathbf{r} \times m\mathbf{v} \tag{3-143}$$

Let us differentiate this equation with respect to time and note that $\mathbf{v} = \dot{\mathbf{r}}$. We obtain

$$\dot{\mathbf{H}} = \mathbf{r} \times m\ddot{\mathbf{r}} + \dot{\mathbf{r}} \times m\dot{\mathbf{r}}$$

and, since the cross product of a vector with itself is zero, this reduces to

$$\dot{\mathbf{H}} = \mathbf{r} \times m\ddot{\mathbf{r}} \tag{3-144}$$

Now let us consider again Newton's law of motion written in the form

$$\mathbf{F} = m\ddot{\mathbf{r}}$$

and take the cross product of each side with the position vector $\mathbf{r}$. We find that

$$\mathbf{r} \times \mathbf{F} = \mathbf{r} \times m\ddot{\mathbf{r}} \tag{3-145}$$

and immediately identify the left side of the equation as the *moment* $\mathbf{M}$ of the total external force $\mathbf{F}$ about the fixed point O.

$$\mathbf{M} = \mathbf{r} \times \mathbf{F} \tag{3-146}$$

Then, from Eqs. (3-144), (3-145), and (3-146), we obtain

$$\mathbf{M} = \dot{\mathbf{H}} \tag{3-147}$$

which is a statement of the important principle that the moment about a fixed point of the total external force applied to a particle is equal to the time rate of change of the angular momentum of the particle about the same fixed point.

We see from Eq. (3-147) that for the case where the external moment $\mathbf{M}$ is zero, the angular momentum $\mathbf{H}$ must be constant in magnitude and direction. This is known as the *principle of conservation of angular momentum*.

The general vector relationship given by Eq. (3-147) can also be written in terms of its components. Thus, choosing a fixed cartesian coordinate system, one obtains the scalar equations:

$$M_x = \dot{H}_x$$
$$M_y = \dot{H}_y \tag{3-148}$$
$$M_z = \dot{H}_z$$

When written in this manner, each equation can be interpreted as relating the moment and the rate of change of angular momentum about the corresponding fixed *axis* passing through the point O. So we can see that even though the *total* angular momentum is not conserved in a given case, one of the *components* of $\mathbf{M}$ might vanish. This would require the angular momentum about the corresponding axis to be conserved.

A similar situation occurs when the motion of a particle is confined to a plane (Fig. 3-14). In this case, the angular momentum is essentially scalar

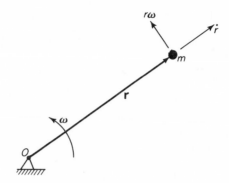

Fig. 3-14. The motion of a particle in a plane.

in nature since its direction is fixed. Let us assume that the velocity of the particle m has radial and tangential components given by

$$v_r = \dot{r}$$
$$v_\phi = r\omega \tag{3-149}$$

where ω is the angular velocity of the radius vector as it moves in the plane of the particle motion. The angular momentum is of magnitude

$$H = mrv_\phi = mr^2\omega \tag{3-150}$$

and, in accordance with the right-hand rule, is directed out of the page. We note that H is independent of v_r since the line of action of the corresponding component of linear momentum passes through the reference point O.

Now let us assume that the particle is acted upon by radial forces only. Then, regardless of the manner of their variation, the applied moment M will be zero at all times, and therefore the angular momentum will be conserved. From Eq. (3-150), we obtain that

$$r_1^2\omega_1 = r_2^2\omega_2 \tag{3-151}$$

where the subscripts 1 and 2 refer to the values of the variable at the arbitrary times t_1 and t_2, respectively.

Returning briefly to a more general discussion of the angular momentum of a particle, let us consider what latitude is permissible in the application of Eq. (3-147). Specifically, in the analysis of the general motion of a single particle, what constitutes a proper choice of (1) a *reference frame* for calculating the linear momentum vector; (2) a *reference point* O? Eliminating at the outset the trivial case where the reference point O is chosen to coincide with the particle, we can state that the reference frame must be inertial and the reference point must be fixed in that frame. These requirements arise because the derivation of Eq. (3-147) is based upon Newton's law of motion and also because $\dot{\mathbf{r}}$ and $\mathbf{v}$ are assumed to be equal.

The question reduces, then, to the choice of a reference point fixed in an inertial frame. Nevertheless, the proper choice of the reference point can greatly clarify the analysis of a problem and simplify its solution. Quite frequently, one attempts to choose a reference point such that one or more components of the total angular momentum are conserved.

Angular Impulse. The equation of linear impulse and momentum was obtained by integrating Newton's equation of motion with respect to time. In a similar fashion, starting with Eq. (3–147), namely,

$$\mathbf{M} = \dot{\mathbf{H}}$$

we can integrate each side with respect to time over an arbitrary interval t_1 to t_2. The time integral of the moment $\mathbf{M}$ is called the *angular impulse* and is designated by $\mathcal{M}$.

$$\int_{t_1}^{t_2} \mathbf{M}\, dt = \mathcal{M} \tag{3–152}$$

Also,

$$\int_{t_1}^{t_2} \dot{\mathbf{H}}\, dt = \mathbf{H}_2 - \mathbf{H}_1 \tag{3–153}$$

So, equating the right-hand sides of Eqs. (3–152) and (3–153), we obtain the *principle of angular impulse and momentum:*

$$\mathcal{M} = \mathbf{H}_2 - \mathbf{H}_1 \tag{3–154}$$

This equation states that the change in the angular momentum of a particle over an arbitrary time interval is equal to the total angular impulse of the external forces acting on the particle during that interval, the reference point being the same fixed point in each computation.

Example 3–8. A particle of mass m slides along a frictionless horizontal track in the form of a logarithmic spiral

$$r = r_o e^{-a\theta}$$

If its initial speed is v_o when $\theta = 0$, find the speed of the particle as a function of θ and also the magnitude of the track force acting on the particle (Fig. 3–15).

In this example, the only force acting on the particle is the track force

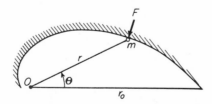

Fig. 3-15. A particle moving along a logarithmic spiral.

F which is normal to the direction of motion. Therefore, from Eq. (3–140), the speed along the track is constant:

$$v = v_o$$

Also, we find from the equation of the spiral that

$$\frac{\dot{r}}{r\dot{\theta}} = -a \tag{3–155}$$

implying that the ratio of radial to transverse velocities is constant. Therefore the angle between the radial line from *O* and the tangent to the curve is constant. Let this angle be α, where

$$\tan \alpha = \frac{1}{a} \tag{3–156}$$

as can be seen from Fig. 3–16 and Eq. (3–155).

The track force is calculated from the rate of change of angular momentum about *O*.

$$H = mrv_o \sin \alpha$$

and therefore

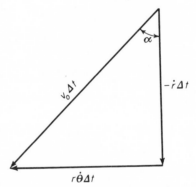

Fig. 3-16. Radial and transverse displacements in an infinitesimal interval Δt.

$$\dot{H} = m\dot{r}v_o \sin \alpha = -mv_o^2 \sin \alpha \cos \alpha \tag{3–157}$$

We see from Eq. (3–147) that the moment of the track force about *O* must be constant, since $\dot{H}$ about that point is constant. The moment is seen to be

$$M = -Fr \cos \alpha \tag{3–158}$$

Equating the right-hand sides of Eqs. (3–157) and (3–158), we obtain

$$F = \frac{mv_o^2}{r} \sin \alpha \tag{3–159}$$

or, writing the result in terms of θ and *a*, using Eq. (3–156) and the equation of the curve, we obtain

$$F = \frac{mv_o^2}{r_0\sqrt{1 + a^2}} e^{a\theta} \tag{3-160}$$

3-7. THE MASS-SPRING-DAMPER SYSTEM

Let us consider the one-dimensional motion of the system shown in Fig. 3-17(a). We can treat the mass m as though it were a particle in this case,

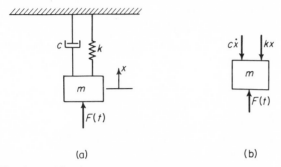

(a)   (b)

Fig. 3-17. (a) The mass-spring-damper system. (b) A free-body diagram showing the forces acting on the mass m.

since the motion consists of pure translation in which all points in the mass have identical values of velocity and acceleration, and undergo identical displacements relative to their initial positions. The forces acting on the mass are shown in Fig. 3-17(b). The force $c\dot{x}$, which is directed opposite to the velocity vector at all times, is dissipative in nature, and is due to a linear viscous damper with a damping coefficient c. The spring force kx is directed opposite to the displacement x, where x is measured from the unstressed position of the spring.[3]

Writing the differential equation for motion in the x direction using Eq. (1-43), we obtain

$$m\ddot{x} = -c\dot{x} - kx + F(t)$$

or

$$m\ddot{x} + c\dot{x} + kx = F(t) \tag{3-161}$$

This second-order linear differential equation with constant coefficients is quite important in the analysis of mechanical vibrations. Many simple vibrating systems can be adequately represented using only a single degree of freedom. Even for the case of certain more complex systems, the analysis

---

[3] In case gravity acts on the system in a direction parallel to the spring, then we can measure the displacement x from the equilibrium position. In this case, the equations are unchanged but the term kx corresponds to the sum of the gravitational and spring forces (see Example 3-6).

can be accomplished in terms of the motion of independent vibrating systems, each with a single degree of freedom, as will be shown in Chapter 9. Finally, the mathematical methods used here are readily extended to the more complicated linear systems to be considered later.

Transient Solution. Proceeding with the solution of Eq. (3–161), we first find the transient solution x_t which is the solution to the homogeneous equation

$$m\ddot{x} + c\dot{x} + kx = 0 \qquad (3\text{–}162)$$

It is convenient to introduce at this point the *undamped natural frequency*[4]

$$\omega_n = \sqrt{\frac{k}{m}} \qquad (3\text{–}163)$$

and the *damping ratio*

$$\zeta = \frac{c}{2\sqrt{km}} \qquad (3\text{–}164)$$

Then, dividing Eq. (3–162) by m and using Eqs. (3–163) and (3–164), we obtain the homogeneous equation in the following form:

$$\ddot{x} + 2\zeta\omega_n\dot{x} + \omega_n^2 x = 0 \qquad (3\text{–}165)$$

The transient solution is obtained by using the same procedure that was used in Example 3–6. We assume a solution of the form $x = Ce^{\lambda t}$ and substitute it into Eq. (3–165), thereby obtaining the characteristic equation

$$\lambda^2 + 2\zeta\omega_n\lambda + \omega_n^2 = 0 \qquad (3\text{–}166)$$

where we have omitted the common factor $Ce^{\lambda t}$ since it cannot be zero for cases of interest. The roots of the characteristic equation are

$$\lambda_{1,2} = -\zeta\omega_n \pm \omega_n\sqrt{\zeta^2 - 1} \qquad (3\text{–}167)$$

and the transient solution is

$$x_t = C_1 e^{\lambda_1 t} + C_2 e^{\lambda_2 t} \qquad (3\text{–}168)$$

It is convenient to express the transient solution in different forms, depending upon the value of the damping ratio. (We shall assume throughout this discussion that ζ is zero or positive.) For the *undamped case*, $\zeta = 0$, the roots are

$$\lambda_{1,2} = \pm i\omega_n$$

The transient solution is of the form

$$x_t = C_1 \cos \omega_n t + C_2 \sin \omega_n t \qquad (3\text{–}169)$$

in agreement with the result found in Example 3–6.

More generally, we find for the *underdamped case*, $0 \leq \zeta < 1$, that the roots are

[4] Note that this is actually circular frequency and the units are radians per unit time.

$$\lambda_{1,2} = -\zeta \omega_n \pm i\omega_n\sqrt{1 - \zeta^2}$$

and the transient solution is

$$x_t = e^{-\zeta\omega_n t}(C_1 \cos \omega_n\sqrt{1 - \zeta^2}\, t + C_2 \sin \omega_n\sqrt{1 - \zeta^2}\, t) \quad (0 \leq \zeta < 1)$$
$$(3\text{-}170)$$

The constant $\omega_n\sqrt{1 - \zeta^2}$ is the *damped natural frequency* ω_d. An alternative form of Eq. (3–170) is

$$x_t = C e^{-\zeta\omega_n t} \cos (\omega_n\sqrt{1 - \zeta^2}\, t + \theta) \quad (0 \leq \zeta < 1) \qquad (3\text{-}171)$$

where C and θ are the two arbitrary constants required in the solution of the second-order equation.

For the *critically damped case*, $\zeta = 1$, the roots are identical.

$$\lambda_{1,2} = -\omega_n$$

According to the theory of ordinary differential equations, the transient solution for this case is of the form

$$x_t = (C_1 + C_2 t)e^{-\omega_n t} \quad (\zeta = 1) \qquad (3\text{-}172)$$

Finally, for the *overdamped case*, $\zeta > 1$, a pair of negative real roots occurs, as shown in Eq. (3–167). The transient solution is of the form

$$x_t = C_1 e^{-(\zeta+\sqrt{\zeta^2-1})\omega_n t} + C_2 e^{-(\zeta-\sqrt{\zeta^2-1})\omega_n t} \quad (\zeta > 1) \qquad (3\text{-}173)$$

For the case of *free motion* wherein $F(t) = 0$, the arbitrary constants can be evaluated immediately from the initial conditions. Let us assume general initial conditions $x(0) = x_o$ and $\dot{x}(0) = v_o$. Noting that the transient solution constitutes the entire solution for this case, we can solve for C_1 and C_2 in Eq. (3–170), obtaining

$$C_1 = x_o \quad \text{and} \quad C_2 = \frac{v_o + \zeta\omega_n x_o}{\omega_n\sqrt{1 - \zeta^2}}$$

Then we can write

$$x = e^{-\zeta\omega_n t}\left(x_o \cos \omega_n\sqrt{1 - \zeta^2}\, t + \frac{v_o + \zeta\omega_n x_o}{\omega_n\sqrt{1 - \zeta^2}} \sin \omega_n\sqrt{1 - \zeta^2}\, t\right)$$
$$(0 \leq \zeta < 1) \qquad (3\text{-}174)$$

If the alternative form of Eq. (3–171) is used, we find in a similar manner that

$$C = \frac{x_o}{\sqrt{1 - \zeta^2}}\left(1 + 2\zeta\frac{v_o}{x_o\omega_n} + \frac{v_o^2}{x_o^2\omega_n^2}\right)^{1/2} \quad (0 \leq \zeta < 1) \quad (3\text{-}175)$$

and

$$\theta = \tan^{-1}\left[-\frac{\zeta + (v_o/x_o\omega_n)}{\sqrt{1 - \zeta^2}}\right] \quad (0 \leq \zeta < 1) \qquad (3\text{-}176)$$

From Eq. (3–172), we obtain that the free motion of a *critically damped system* with arbitrary initial conditions is

$$x = e^{-\omega_n t}[x_0(1 + \omega_n t) + v_0 t] \quad (\zeta = 1) \tag{3-177}$$

Similarly, for the free motion of an *overdamped system*, we obtain from Eq. (3–173) that

$$x = \frac{e^{-\zeta\omega_n t}}{2\sqrt{\zeta^2 - 1}}\left\{\left[(\sqrt{\zeta^2 - 1} - \zeta)x_0 - \frac{v_0}{\omega_n}\right]e^{-\sqrt{\zeta^2 - 1}\omega_n t}\right.$$

$$\left. + \left[(\sqrt{\zeta^2 - 1} + \zeta)x_0 + \frac{v_0}{\omega_n}\right]e^{\sqrt{\zeta^2 - 1}\omega_n t}\right\} \quad (\zeta > 1) \tag{3-178}$$

Typical plots showing the transient response of a mass-spring-damper system are shown in Fig. 3–18 for various values of the damping ratio ζ. The system is passive and therefore the total energy, that is, the kinetic energy of the mass plus the potential energy of the spring, must decrease or

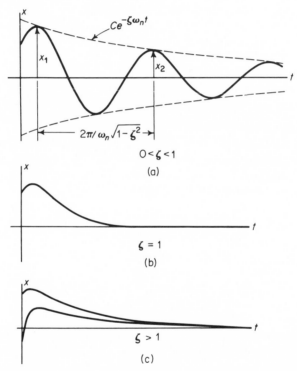

Fig. 3-18. Typical transient solutions for a mass-spring-damper system which is (a) under-damped, (b) critically damped, (c) over-damped.

remain constant. For $0 < \zeta < 1$, we have seen that the transient response has the form of an oscillation whose envelope decreases exponentially. Thus, a plot of the displacement versus time crosses the time axis at equal intervals for an indefinite period of time. On the other hand, for $\zeta \geq 1$, a similar plot can cross the time axis only once at the most. After the mass stops and reverses its direction of motion, it can never again pass through $x = 0$.

For the underdamped case $(0 \leq \zeta < 1)$, we can see from Eq. (3–174) that the ratio of the displacements at any two instants separated by one period of the damped vibration, that is, by $\Delta t = 2\pi/\omega_n\sqrt{1 - \zeta^2}$, is constant. The natural logarithm of this amplitude ratio is called the *logarithmic decrement* δ.

$$\delta = \ln\left[\frac{e^{-\zeta\omega_n t}}{e^{-\zeta\omega_n(t + 2\pi/\omega_n\sqrt{1-\zeta^2})}}\right]$$

or

$$\delta = \frac{2\pi\zeta}{\sqrt{1 - \zeta^2}} \tag{3–179}$$

It is convenient to think of the logarithmic decrement in terms of the ratio of successive positive *peak amplitudes*. But we also recall that the total energy is proportional to x_{max}^2 since the kinetic energy is zero at these moments of zero velocity. Therefore the ratio of total energies, measured at successive displacement maximums, is just the square of the amplitude ratio. It can be seen, then, that 2δ is the natural logarithm of this energy ratio.

Steady-state Solution for a Sinusoidal Input. Now let us consider the steady-state response of a mass-spring-damper system which is being driven by the sinusoidal force

$$F(t) = F_o \cos \omega t \tag{3–180}$$

In accordance with the method of undetermined coefficients (Example 3–6), let us assume a solution of the form

$$x_s = C_1 \cos \omega t + C_2 \sin \omega t$$

or, equivalently,

$$x_s = A \cos(\omega t + \phi) \tag{3–181}$$

where A and ϕ are the coefficients to be evaluated. Substituting into the differential equation of motion, Eq. (3–161), we obtain

$$A(k - m\omega^2)\cos(\omega t + \phi) - Ac\omega \sin(\omega t + \phi) = F_o \cos \omega t$$

Expanding $\cos(\omega t + \phi)$ and $\sin(\omega t + \phi)$ and collecting terms, we find that

$$[(k-m\omega^2)\cos\phi - c\omega \sin\phi]\cos \omega t$$
$$- [(k - m\omega^2)\sin\phi + c\omega \cos\phi]\sin \omega t = \frac{F_o}{A}\cos \omega t$$

Now we equate the coefficients of the same variable parts on the left and right sides of the equality. From the $\sin \omega t$ terms we obtain

$$(k - m\omega^2)\sin\phi + c\omega \cos\phi = 0$$

or

$$\phi = \tan^{-1}\left(\frac{-c\omega}{k - m\omega^2}\right) = \tan^{-1}\left[\frac{-2\zeta(\omega/\omega_n)}{1 - (\omega^2/\omega_n^2)}\right] \qquad (3\text{-}182)$$

In a similar fashion, we obtain from the cos ωt terms that

$$(k - m\omega^2)\cos\phi - c\omega\sin\phi = \frac{F_o}{A}$$

or

$$A = \frac{F_o}{\sqrt{(k - m\omega^2)^2 + c^2\omega^2}} = \frac{F_o/k}{\sqrt{[1 - (\omega^2/\omega_n^2)]^2 + [2\zeta(\omega/\omega_n)]^2}} \qquad (3\text{-}183)$$

From Eqs. (3–181) and (3–183), we can write the *steady-state solution:*

$$x_s = \frac{(F_o/k)\cos(\omega t + \phi)}{\sqrt{[1 - (\omega^2/\omega_n^2)]^2 + [2\zeta(\omega/\omega_n)]^2}} \qquad (3\text{-}184)$$

where ϕ is given by Eq. (3–182). Plots of the *amplification factor* $A/(F_o/k)$ and the *phase angle* ϕ are shown in Figs. 3–19 and 3–20. The amplification factor is the ratio of the amplitude A of the steady-state solution to the static deflection F_o/k due to a constant force of magnitude F_o. It can be seen from Fig. 3–19 that the amplification factor is unity for $\omega = 0$, regardless of the damping ratio ζ. Also, when the forcing frequency ω is equal to the

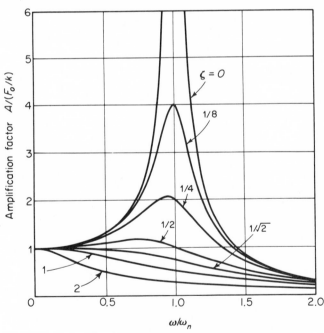

Fig. 3-19. Amplification factor versus frequency.

undamped natural frequency ω_n, the amplification factor is $1/(2\zeta)$. The *resonant frequency* ω_r is the frequency of maximum amplification for a given value of ζ. By differentiating Eq. (3–183) with respect to ω/ω_n and setting this derivative equal to zero, we can solve for the resonant frequency. The result is

$$\omega_r = \omega_n\sqrt{1 - 2\zeta^2} \tag{3–185}$$

It can be seen that a resonance or peaking effect occurs in all curves for $\zeta < 1/\sqrt{2}$. The peak amplitude is found from Eqs. (3–183) and (3–185).

$$A_{\max} = \left(\frac{F_o}{k}\right)\frac{1}{2\zeta\sqrt{1 - \zeta^2}} \tag{3–186}$$

From Eq. (3–182) or Fig. 3–20 we note that the phase angle ϕ lies in the range $0 \leq \phi \leq -\pi$. In other words, the steady-state displacement lags the

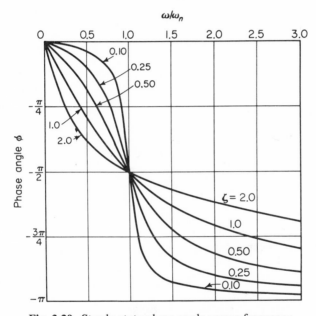

Fig. 3-20. Steady-state phase angle versus frequency.

input force by from $0°$ to $180°$. Furthermore, for $\omega = 0$, the phase angle is zero and for $\omega \gg \omega_n$, the phase angle is approximately $-180°$, for any finite value of ζ. Of particular interest is the point that $\phi = -90°$ at $\omega = \omega_n$, regardless of the damping ratio ζ. This can be used in vibration testing as a sensitive criterion to determine the undamped natural frequency ω_n. It indicates that the velocity and the force are in phase and therefore that work is being done on the system throughout the cycle.

The complete solution for the case of certain given initial conditions is the sum of the transient solution as given by Eq. (3–170), (3–172), or (3–173) and the steady-state solution of Eq. (3–184). The two arbitrary constants in the transient solution are evaluated from the initial conditions upon velocity and displacement which are applied to the *complete* solution.

Unit Step and Unit Impulse Responses. Let us consider next the response of a mass-spring-damper system to a force of unit magnitude applied at $t = 0$. We use the notation

$$F(t) = u(t)$$

where the *unit step function* $u(t)$ is plotted in Fig. 3–21(a). It can be seen that the steady-state solution in this case is simply

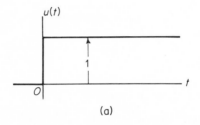

(a)

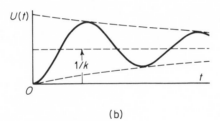

(b)

Fig. 3-21. A unit step function and the corresponding response of an underdamped system.

$$x_s = \frac{1}{k} \tag{3–187}$$

Assuming that the initial velocity and displacement are zero for the complete solution, we can immediately solve for the initial conditions on the transient portion of the solution. They are just the negative of the steady-state values at $t = 0$. In this particular example of a unit step forcing function, we obtain

$$x_t(0) = -x_s(0) = -\frac{1}{k}$$

$$\dot{x}_t(0) = -\dot{x}_s(0) = 0 \tag{3–188}$$

So we can write down the transient solution directly from results obtained previously. For example, for the *underdamped case*, we find from Eq. (3–

174) that the transient solution is

$$x_t = e^{-\zeta\omega_n t}\left(-\frac{1}{k}\cos\omega_n\sqrt{1-\zeta^2}\,t - \frac{\zeta}{k\sqrt{1-\zeta^2}}\sin\omega_n\sqrt{1-\zeta^2}\,t\right)$$

$$(0 \le \zeta < 1) \qquad (3\text{-}189)$$

Adding Eqs. (3-187) and (3-189), we obtain the complete response. Designating this response to a unit step function by $U(t)$, we can write

$$U(t) = \frac{1}{k}\left[1 - e^{-\zeta\omega_n t}\left(\cos\omega_n\sqrt{1-\zeta^2}\,t + \frac{\zeta}{\sqrt{1-\zeta^2}}\sin\omega_n\sqrt{1-\zeta^2}\,t\right)\right]$$

$$(0 \le \zeta < 1) \qquad (3\text{-}190)$$

For the critically damped or overdamped cases, the complete solution $U(t)$ is obtained in a similar manner.

The unit step response for the underdamped case is plotted in Fig. 3-21(b). Note that it is identical in shape to the transient solution for the case of an initial displacement, but the oscillations occur about the new static equilibrium position, namely, $x = 1/k$.

Now let us find the response to a unit impulse at $t = 0$, that is, the case where the forcing function is

$$F(t) = \delta(t)$$

Recall from Eqs. (3-137) and (3-138) that the delta function is a pulse of infinite amplitude and infinitesimal width occurring at $t = 0$. Then, since the total impulse acting on the mass due to this force is unity, the change in linear momentum must also be unity, in accordance with Eq. (3-135). The velocity changes instantaneously because of the unit impulse, but the displacement is unchanged since a finite time is required for the mass to move a finite distance. Hence the conditions immediately after the impulse are

$$x(0+) = 0$$

$$\dot{x}(0+) = \frac{1}{m} \qquad (3\text{-}191)$$

These serve as initial conditions for the transient solution. The transient and complete solutions are identical in this case because the steady-state solution is zero. As an example, we again consider the *underdamped case*. From Eqs. (3-174) and (3-191), and noting that $\omega_n = \sqrt{k/m}$, we find that the response to a unit impulse at $t = 0$ is

$$h(t) = \frac{\omega_n e^{-\zeta\omega_n t}}{k\sqrt{1-\zeta^2}}\sin\omega_n\sqrt{1-\zeta^2}\,t \quad (0 \le \zeta < 1) \qquad (3\text{-}192)$$

The response $h(t)$ to a unit impulse input is known as the *weighting function* for the system.

Convolution Integral. Perhaps the principal reason for introducing the unit step and unit impulse responses in the discussion of the mass-spring-damper is that this system is an example of a *linear system*, that is, it is a system described by linear differential equations, and the *principle of superposition* applies. The principle of superposition can be stated as follows:

> If $x_1(t)$ *is the response of a linear system to an input* $F_1(t)$ *for initial conditions* $x_1(0)$, $\dot{x}_1(0)$, *and so on, and if* $x_2(t)$ *is the response of the same system to an input* $F_2(t)$ *for initial conditions* $x_2(0)$, $\dot{x}_2(0)$, *and so on, then* $x_1(t) + x_2(t)$ *is the response of that system to the input* $F_1(t) + F_2(t)$, *assuming the initial conditions are* $x_1(0) + x_2(0)$, $\dot{x}_1(0) + \dot{x}_2(0)$, *and so on.*

Of course, the superposition principle can be extended to more than two inputs or forcing functions. Also, the input need not be a force, but could be a given velocity, displacement, and so on, or some combination of these. It is interesting to note that we have already applied the principle of superposition in obtaining the complete solution to a differential equation as the sum of the transient and steady-state solutions.

Now let us apply the principle of superposition to the problem of finding the response of a linear time-invariant system to an arbitrary forcing function $F(t)$. We consider $F(t)$ to be composed of a sequence of infinitesimal impulses. The response of the system at time t due to a unit impulse at time τ is just $h(t - \tau)$ since $(t - \tau)$ is the interval between the time of the impulse and the time of observation, and the various system parameters do not change with time. In other words, the response due to a given impulse is determined solely by the time interval since that impulse, not by the absolute

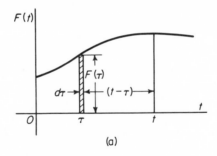

(a)

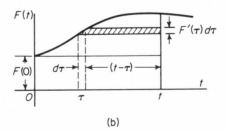

(b)

Fig. 3-22. Quantities used in the convolution integral.

time. Therefore the response of the system at time t due to an impulse of magnitude $F(\tau)\,d\tau$ at time τ (Fig. 3–22a) is given by

$$h(t - \tau)F(\tau)\,d\tau$$

Now, by the principle of superposition, the total response at time t due to the effect of all previous portions of the forcing function, considered as a series of impulses, is just the sum of the individual responses. In the limiting case of infinitesimal impulses, the following integral results:

$$x(t) = \int_{-\infty}^{t} h(t - \tau)F(\tau)\,d\tau \tag{3-193}$$

For the common situation where the response is desired for $t \geq 0$ with *all initial conditions equal to zero*, we obtain

$$x(t) = \int_{0}^{t} h(t - \tau)F(\tau)\,d\tau \tag{3-194}$$

If the initial conditions are not all zero, we add to the convolution integral of Eq. (3–194) the transient solution for $F(t) = 0$, assuming the actual initial conditions.

An alternative form of Eq. (3–194) can be written by interchanging the roles of τ and $(t - \tau)$. The result is

$$x(t) = \int_{0}^{t} h(\tau)F(t - \tau)\,d\tau \tag{3-195}$$

where τ is now interpreted as the interval between the impulse and the time of observation.

Additional forms of the convolution integral can be obtained in terms of the unit step function response. Considering the forcing function $F(t)$ to be composed of a series of step functions applied in sequence—Fig. 3–22(b), we can see that the response at time t due to a step function of amplitude $F'(\tau)\,d\tau$ applied at time τ is

$$U(t - \tau)\,F'(\tau)\,d\tau$$

where the prime indicates differentiation of the function with respect to its argument. Assuming that the initial conditions are all zero, we superimpose the effects of all the infinitesimal step functions as previously given and, in addition, include the term $F(0)U(t)$ which is the response at time t due to an initial step function of magnitude $F(0)$. The result is

$$x(t) = F(0)\,U(t) + \lim_{\epsilon \to 0} \int_{\epsilon}^{t} U(t - \tau)F'(\tau)\,d\tau$$

where we use as the lower limit on the integral a small positive quantity ϵ to emphasize that the value of $F'(0)$ is normally finite and is measured to the right of any discontinuity of $F(\tau)$ at the origin. Of course, $F(0)$ is determined in a similar manner. Now we can write

$$x(t) = F(0)U(t) + \int_0^t U(t - \tau)F'(\tau)\,d\tau \qquad (3\text{–}196)$$

An alternate form is

$$x(t) = F(0)U(t) + \int_0^t U(\tau)F'(t - \tau)\,d\tau \qquad (3\text{–}197)$$

where τ has the same meaning as in Eq. (3–195).

As an example of the use of the convolution integral, let us calculate the response of an underdamped mass-spring-damper system to a unit step forcing function, that is,

$$F(t) = u(t)$$

Using the impulse response of Eq. (3–192) and the convolution integral in the form given by Eq. (3–195), we obtain

$$U(t) = \int_0^t \frac{\omega_n e^{-\zeta\omega_n\tau}}{k\sqrt{1 - \zeta^2}} \sin \omega_n\sqrt{1 - \zeta^2}\,\tau\, u(t - \tau)\,d\tau$$

$$= \frac{\omega_n}{k\sqrt{1 - \zeta^2}} \int_0^t e^{-\zeta\omega_n\tau} \sin \omega_n\sqrt{1 - \zeta^2}\,\tau\,d\tau$$

Evaluating this integral, the result is

$$U(t) = \frac{1}{k}\left[1 - e^{-\zeta\omega_n t}\left(\cos \omega_n\sqrt{1 - \zeta^2}\,t + \frac{\zeta}{\sqrt{1 - \zeta^2}} \sin \omega_n\sqrt{1 - \zeta^2}\,t\right)\right]$$

in agreement with the result obtained in Eq. (3–190) by other methods.

Continuing this discussion of the application of superposition methods to linear systems, using the mass-spring-damper as an example, let us consider further the unit step and unit impulse responses. From the definition of the delta function given by Eqs. (3–137) and (3–138) we see that the step function $u(t)$ can be expressed as

$$u(t) = \int_{-\infty}^t \delta(\tau)\,d\tau \qquad (3\text{–}198)$$

Differentiating both sides of this equation with respect to t, noting that t occurs in the upper limit of the integral, we obtain

$$\frac{du(t)}{dt} = \delta(t) \qquad (3\text{–}199)$$

Now let us calculate the unit impulse response of a system in terms of its unit step response. First, we can use Eq. (3–195) to obtain

$$U(t) = \int_0^\infty h(\tau)u(t - \tau)\,d\tau \qquad (3\text{–}200)$$

where we note that $u(t - \tau) = 0$ for $\tau > t$. But $U(t)$ can also be written in the form

$$U(t) = U(0)u(t) + \int_0^t \frac{dU}{dt}(\tau)\,d\tau \qquad (3\text{–}201)$$

where we take

$$\frac{dU}{dt}(0) = \lim_{\epsilon \to 0} \frac{dU}{dt}(\epsilon) \quad (\epsilon > 0)$$

in order to avoid difficulties with discontinuities in the derivative at the origin.

Equating the right-hand sides of Eqs. (3–200) and (3–201) and differentiating with respect to time, we obtain

$$\int_0^\infty h(\tau)\,\delta(t - \tau)\,d\tau = U(0)\,\delta(t) + \frac{dU}{dt}$$

where we recall from Eq. (3–199) that the derivative of a unit step function is a unit impulse. Now, it can be seen from Eq. (3–138) that

$$\int_0^\infty h(\tau)\,\delta(t - \tau)\,d\tau = h(t) \tag{3–202}$$

since $h(\tau)$ is essentially constant in the infinitesimal interval around $\tau = t$ and the integrand is zero for $\tau \neq t$. Hence we obtain the important result:

$$h(t) = U(0)\,\delta(t) + \frac{dU}{dt} \tag{3–203}$$

This means that the unit impulse response is the time derivative of the unit step response. In case there is a discontinuity in $U(t)$ at $t = 0$, we must include an impulse of magnitude $U(0)$ to account for the sharp leading edge. This last situation is not common in mechanical systems because it implies that the response magnitude does not drop off to zero in the limit for an increasingly high input frequency.

We observe, then, that the unit impulse response is the time derivative of the unit step response and, furthermore, that the unit impulse function is the time derivative of the unit step function. Thus a differentiation of the input function has resulted in the differentiation of the response. Since a general input or output function can be considered to be composed of a sequence of small superimposed step functions, or as a sequence of impulses, and since the superposition principle applies to these linear systems, we conclude that the differentiation of a general input to a linear system results in the differentiation of the output. In other words, if $x(t)$ is the response of a given system to an input $F(t)$, then $\dot{x}(t)$ is the response of the same system to an input $\dot{F}(t)$, where, of course, allowance is made for non-zero initial values of $x(t)$ or $F(t)$ in the manner previously described.

Example 3–9. Find the steady-state response of an undamped mass-spring system to an input force which is a square wave of amplitude A and period T (see Fig. 3–23).

Our approach will be to represent the square wave by its Fourier series and to sum the steady-state responses to the individual terms in order to obtain the complete steady-state solution.

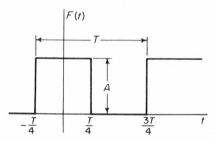

Fig. 3-23. A square wave of amplitude A, period T.

In general, a periodic function $f(t)$ of period T can be represented by the Fourier series

$$f(t) = \frac{a_o}{2} + \sum_{n=1}^{\infty} a_n \cos \frac{2n\pi t}{T} + \sum_{n=1}^{\infty} b_n \sin \frac{2n\pi t}{T} \qquad (3\text{-}204)$$

where

$$a_n = \frac{2}{T} \int_{-T/2}^{T/2} f(t) \cos \frac{2n\pi t}{T} \, dt \quad (n = 0, 1, 2, \ldots)$$

$$b_n = \frac{2}{T} \int_{-T/2}^{T/2} f(t) \sin \frac{2n\pi t}{T} \, dt \quad (n = 1, 2, 3, \ldots) \qquad (3\text{-}205)$$

In this instance, we have taken the origin for the square wave such that it is an even function of time. Consequently, it can be seen from Eq. (3–205) that all the b_n are zero. Solving for the a_n, we obtain

$$a_n = \frac{2A}{T} \int_{-T/4}^{T/4} \cos \frac{2n\pi t}{T} \, dt = \begin{cases} A & (n = 0) \\ \dfrac{2A}{n\pi} \sin \dfrac{n\pi}{2} & (n = 1, 2, 3 \ldots) \end{cases}$$

Thus the Fourier series representation of the square wave is

$$F(t) = \frac{A}{2} \left[1 + \frac{4}{\pi} \left(\cos \frac{2\pi t}{T} - \frac{1}{3} \cos \frac{6\pi t}{T} + \frac{1}{5} \cos \frac{10\pi t}{T} - \cdots \right) \right] \qquad (3\text{-}206)$$

where we note that the constant term is the average value of the function. For convenience we shall use the notation

$$\omega = \frac{2\pi}{T}$$

Then, obtaining the steady-state response for each term from Eq. (3–184) and setting $\zeta = 0$, we obtain the total response

$$x_s = \frac{A}{2k} \left[1 + \frac{4}{\pi} \sum_{n \text{ odd}} \frac{\cos n\omega t}{1 - (\omega/\omega_n)^2} \right] \qquad (3\text{-}207)$$

The procedure used here to calculate the steady-state response of a mass-spring system to a square wave input can be extended to calculate the steady-state response of a more general linear time-invariant system which is excited by a periodic function. One merely represents the input function by its Fourier series and sums the solutions resulting from the sinusoidal or constant input terms considered separately.

3-8. COULOMB FRICTION

We have seen that a lumped viscous damper is a linear element, that is, the force which it exerts on the external system varies linearly with the relative velocity of its ends. It is also dissipative because the force always opposes this relative motion, and thereby continuously absorbs energy so long as the relative velocity exists.

Now let us consider the force associated with *Coulomb* or *sliding friction*. This sort of force is dissipative, like other frictional forces, but is not linear. As an illustration of Coulomb friction, suppose block A slides with a velocity $\mathbf{v}_r$ relative to block B, as shown in Fig. 3–24. The force of block B acting

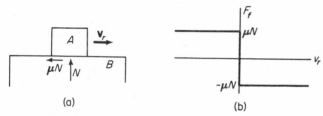

Fig. 3-24. The force of Coulomb friction.

on block A has a component N which is normal to the flat contact surface and a component μN parallel to that surface and opposing the relative motion. The classical law of sliding friction states that this frictional force is directly proportional to the normal force N, but is independent of the contact area and the magnitude of the relative velocity, so long as sliding exists. Thus we can state that

$$F_f = -\mu N \operatorname{sgn}(v_r) \qquad (3\text{–}208)$$

where the *coefficient of sliding friction* μ depends only upon the roughness of the sliding surfaces and the materials used. Also, the function $\operatorname{sgn}(v_r)$ has the value ± 1, depending upon the *sign* of its argument; in this case, the sign of v_r. The Coulomb friction force F_f is plotted against the relative velocity v_r in Fig. 3–24(b) for the case of a constant normal force N, assuming the positive directions for F_f and v_r are the same.

The case of $v_r = 0$, that is, no sliding, deserves some comment. The force of friction in this case can have any magnitude less than that required to initiate sliding, the actual magnitude normally being obtained from the equations of statics. Actually, the force required to initiate sliding is somewhat larger than that required to sustain it. We shall assume, however, that the maximum frictional force magnitude is μN, as shown in Fig. 3–24(b). With these assumptions, it can be seen that the *direction* of the force of block B acting on block A must lie within an angle ϕ from the normal to the surface, where

$$\tan \phi = \mu \tag{3–209}$$

Furthermore, when slipping actually occurs, the angle between this force and the normal is exactly ϕ, since μ is the ratio of the tangential and normal components. Thus the force vector lies on, or within, a cone of semivertex angle ϕ whose symmetry axis is normal to the contact surface.

For a more general case, it is helpful to consider the force of sliding friction as arising from a frictional shear stress at the contact area that is equal to μ times the normal pressure. This gives the same results for the simpler cases, such as sliding blocks, but aids in the analysis of more complicated situations, such as those involving curved contact surfaces or nonuniform pressure or velocity distributions.

As an example of this approach, consider the frictional moment arising from the flat end of a circular rotating shaft of radius a being pressed against a plane surface with a total force N. Assuming a uniform normal pressure or compressive stress at the contact area A, we obtain a uniform frictional stress of magnitude

$$\tau_f = \mu \frac{N}{A} = \frac{\mu N}{\pi a^2}$$

which is everywhere in a direction normal to a radial line drawn from the center of the circular contact area. It can be seen (Fig. 3–25) that the moment

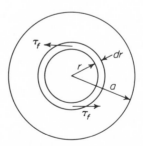

Fig. 3-25. Frictional stresses on a rotating circular surface.

due to an annular element of width dr and area $2\pi r \, dr$ is

$$dM = \frac{\mu N}{\pi a^2} 2\pi r^2 \, dr$$

The total frictional moment is

$$M = \frac{2\mu N}{a^2} \int_0^a r^2 \, dr = \frac{2}{3} \mu N a$$

indicating that the average moment arm is $2a/3$.

If $\mu = 0$, the contact surface is said to be *perfectly smooth*. On the other hand, a *perfectly rough* surface corresponds to $\mu = \infty$, and implies that no slipping can occur. Both bodies must have the same velocity at any points of contact in the latter case, and the relative motion consists of rolling motion at the contact points.

The energy lost due to friction can be calculated by multiplying the frictional force on a given elemental area by the *relative* velocity at that point and integrating over the contact area and over the required time interval. It can be seen that no energy is dissipated in friction for either a perfectly smooth or a perfectly rough surface. In the former case, the friction force is zero; in the latter case, the relative velocity at the point of contact is zero.

Example 3–10. Block A can slide relative to block B which, in turn, can slide on a perfectly smooth horizontal plane (Fig. 3–26). If the initial velocities are $v_A(0) = v_0$ and $v_B(0) = 0$, find the final velocities of the two blocks and the distance that A slides relative to B.

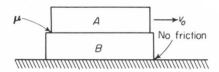

Fig. 3-26. Coulomb friction between sliding blocks.

The final velocity v_f is the same for both blocks and can be obtained by applying the principle of conservation of linear momentum, since there is no net external force acting on the system consisting of the two blocks. Equating the total linear momenta before and after the sliding of A on B, we obtain

$$m_A v_0 = (m_A + m_B) v_f$$

or

$$v_f = \frac{m_A v_0}{m_A + m_B} \tag{3–210}$$

The distance d that A slides relative to B can be obtained by equating the change of kinetic energy during the sliding process to the work done against friction; that is,

$$\frac{1}{2} m_A v_0^2 - \frac{1}{2} (m_A + m_B) v_f^2 = \mu m_A g d \tag{3–211}$$

where we note that the normal force at the contact surface is $m_A g$. Solving for the distance d from Eqs. (3–210) and (3–211), we obtain

$$d = \left(m_A - \frac{m_A^2}{m_A + m_B} \right) \frac{v_0^2}{2\mu m_A g} = \frac{m_B v_0^2}{2\mu g(m_A + m_B)} \qquad (3\text{–}212)$$

Another approach to this problem is to use the principle of work and kinetic energy on each mass separately. For block A, we equate the work done *against* the friction force to the loss in kinetic energy.

$$\mu m_A g x_A = \frac{1}{2} m_A v_0^2 - \frac{1}{2} m_A v_f^2 \qquad (3\text{–}213)$$

where x_A is the *absolute* displacement of A during the sliding process. Similarly, the work done *by* the friction force acting on block B is equated to the increase in its kinetic energy.

$$\mu m_A g x_B = \frac{1}{2} m_B v_f^2 \qquad (3\text{–}214)$$

The displacement of A relative to B is, from Eqs. (3–213) and (3–214),

$$d = x_A - x_B = \frac{1}{2\mu g} \left[v_0^2 - \left(1 + \frac{m_B}{m_A} \right) v_f^2 \right]$$

Substituting for v_f from Eq. (3–210), we obtain

$$d = \frac{v_0^2}{2\mu g} \left[1 - \left(\frac{1}{1 + m_B/m_A} \right) \right]$$

$$= \frac{m_B v_0^2}{2\mu g(m_A + m_B)}$$

in agreement with Eq. (3–212).

Example 3–11. A mass-spring system is connected as shown in Fig. 3–27(a). There is Coulomb friction between the mass and the horizontal surface on which it slides. If the mass is released from a positive initial displacement x_0, measured from the unstressed position of the spring, solve for x as a function of time.

Assuming that the initial spring force $-kx_0$ is greater in magnitude than the friction force μmg, the initial acceleration is negative. So long as the velocity $\dot{x}$ is negative, the differential equation for the motion is

$$m\ddot{x} + kx = \mu mg \qquad (\dot{x} < 0) \qquad (3\text{–}215)$$

Conversely, if $\dot{x}$ is positive, the corresponding equation is

$$m\ddot{x} + kx = -\mu mg \qquad (\dot{x} > 0) \qquad (3\text{–}216)$$

Although the Coulomb friction force is nonlinear in nature, it can be seen that Eqs. (3–215) and (3–216) are both *linear*. Thus, the nonlinearity consists of switching between the two linear equations, the choice being determined by the sign of $\dot{x}$. Superposition does not apply to this system because the switching does not occur as an explicit function of time, but is deter-

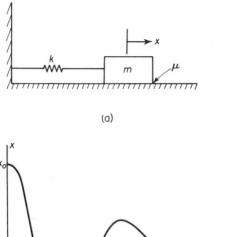

(a)

(b)

Fig. 3-27. A mass-spring system with Coulomb damping.

mined by the response. Thus it could occur at various times, depending upon the forcing function and initial conditions.

The solution can be obtained in the same manner as for an undamped mass-spring system that is excited by a constant force (see Example 3-6), except that the sign of the forcing function must be reversed after each half-cycle of the response. We have seen that the response to a step function input consists of an oscillation about the static equilibrium position. From Eqs. (3-215) and (3-216), we can see that the static equilibrium positions (in this case, the steady-state solutions) are

$$x_s = \pm \frac{\mu m g}{k} \tag{3-217}$$

for $\dot{x} < 0$ or $\dot{x} > 0$, respectively. Hence the solution for the first half-cycle is

$$x = \frac{\mu m g}{k} + \left(x_0 - \frac{\mu m g}{k}\right) \cos \omega_n t \quad \left(0 < t < \frac{\pi}{\omega_n}\right)$$

where $\omega_n = \sqrt{k/m}$. After a half-cycle the displacement is

$$x = -x_0 + 2\frac{\mu m g}{k}$$

The second half-cycle is an oscillation about $-\mu mg/k$, as given by

$$x = -\frac{\mu mg}{k} + \left(x_0 - 3\frac{\mu mg}{k}\right)\cos \omega_n t \quad \left(\frac{\pi}{\omega_n} < t < \frac{2\pi}{\omega_n}\right)$$

In a similar fashion, the solution for the succeeding half-cycles consists of oscillations about one or the other of the two dashed lines of Fig. 3–27(b) and, for the nth half-cycle, is found to be

$$x = (-1)^{n-1}\frac{\mu mg}{k} + \left[x_0 - (2n - 1)\frac{\mu mg}{k}\right]\cos \omega_n t$$

$$\left[\frac{(n - 1)\pi}{\omega_n} < t < \frac{n\pi}{\omega_n}\right]$$

(3–218)

provided, of course, that $x_0 > (2n - 1)\mu mg/k$. If the latter inequality does not hold, then the oscillation has stopped permanently at

$$x = (-1)^n\left(x_0 - 2n\frac{\mu mg}{k}\right) \quad \left(\frac{n\pi}{\omega_n} < t\right)$$

(3–219)

where n in this case is the total number of half-cycles which occur.

We have assumed that x_0 is positive. It is clear, however, that, if x_0 were negative, the solution would proceed in a manner similar to the foregoing solution after the first half-cycle.

It can be seen from Eq. (3–217) that the oscillation amplitude decreases by $2\mu mg/k$ per half-cycle. Therefore a straight line can be drawn through successive positive (or negative) peaks of the oscillation. This is in contrast to the exponential envelope that occurs with linear damping. Another point of interest is that the period is $2\pi/\omega_n$ whenever oscillations occur, regardless of μ; whereas the period for the case of a linear damper increases with the damping ratio.

3-9. THE SIMPLE PENDULUM

As another important example of particle motion, consider a simple pendulum of length l whose mass m moves on a vertical circular path centered

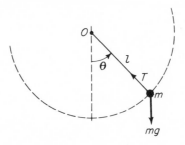

Fig. 3-28. A simple pendulum showing the external forces acting on the particle of mass m.

at the fixed support point O (Fig. 3–28). This system has one degree of freedom, the corresponding coordinate θ being the angular position of the mass m measured from the downward vertical through O.

The differential equation of motion can be obtained by applying Newton's law of motion in terms of the tangential components of force and acceleration.

$$F_\theta = -mg \sin \theta = ml\ddot{\theta}$$

or

$$\ddot{\theta} + \frac{g}{l} \sin \theta = 0 \tag{3-220}$$

Consider first the case of small motion in which $|\theta| \ll 1$, implying that

$$\sin \theta \cong \theta$$

Equation (3–220) is linearized by this assumption and becomes

$$\ddot{\theta} + \frac{g}{l} \theta = 0 \tag{3-221}$$

This equation is of the familiar form associated with harmonic oscillators such as the undamped mass-spring system discussed earlier (Example 3–6). The solution can be seen to be of the form

$$\theta = \theta_0 \sin \omega_n(t - t_0) \tag{3-222}$$

where

$$\omega_n = \sqrt{\frac{g}{l}} \tag{3-223}$$

and where θ_0 is the amplitude of the oscillation and t_0 is the time of a passage through $\theta = 0$ in the positive direction.

Next, let us consider the general case of motions of the simple pendulum where θ is not necessarily small. Rather than integrating Eq. (3–220) directly, we can obtain the first integral of the motion by using the principle of conservation of energy. In this case, taking point O as the reference level for potential energy, the total energy is found to be

$$E = \frac{1}{2} ml^2 \dot{\theta}^2 - mgl \cos \theta \tag{3-224}$$

Again assuming that θ_0 is the maximum value of θ, we can evaluate E at this point, obtaining

$$E = -mgl \cos \theta_0 \tag{3-225}$$

Therefore, from Eqs. (3–224) and (3–225), we can write

$$\dot{\theta}^2 = \frac{2g}{l} (\cos \theta - \cos \theta_0) \tag{3-226}$$

or, separating the variables and integrating,

$$t - t_0 = \sqrt{\frac{l}{2g}} \int_0^\theta \frac{d\theta}{\sqrt{\cos \theta - \cos \theta_0}} \tag{3-227}$$

where $t - t_0$ is the time interval from the passage through $\theta = 0$ in the positive direction until reaching the angle θ. This is an elliptic integral. It can be put in a standard form by making the substitutions

$$k = \sin \frac{\theta_0}{2} \quad \text{and} \quad \sin \phi = \frac{\sin (\theta/2)}{\sin (\theta_0/2)}$$

Noting that

$$\cos \theta_0 = 1 - 2 \sin^2 \frac{\theta_0}{2} = 1 - 2k^2$$

and

$$\cos \theta = 1 - 2 \sin^2 \frac{\theta}{2} = 1 - 2k^2 \sin^2 \phi$$

we obtain by differentiation that

$$d\theta = \frac{2k \cos \phi \, d\phi}{\sqrt{1 - k^2 \sin^2 \phi}}$$

Hence we can write Eq. (3–227) in the following form:

$$t - t_0 = \sqrt{\frac{l}{g}} \int_0^\phi \frac{d\phi}{\sqrt{1 - k^2 \sin^2 \phi}}$$

or

$$t - t_0 = \sqrt{\frac{l}{g}} \, F(\phi, k) \tag{3–228}$$

where

$$F(\phi, k) = \int_0^\phi \frac{d\phi}{\sqrt{1 - k^2 \sin^2 \phi}} \tag{3–229}$$

This integral is known as *Legendre's elliptic integral of the first kind*. It can be seen that the time required to reach a certain angle θ, or its corresponding value of ϕ, depends upon maximum amplitude θ_0 which is expressed in terms of the modulus k.

The period T of the oscillation is four times the interval required for the movement from $\theta = 0$ to $\theta = \theta_0$, or from $\phi = 0$ to $\phi = \pi/2$.

$$T = 4\sqrt{\frac{l}{g}} \int_0^{\pi/2} \frac{d\phi}{\sqrt{1 - k^2 \sin^2 \phi}} = 4\sqrt{\frac{l}{g}} \, K(k) \tag{3–230}$$

where $K(k) = F(\pi/2, k)$ is called the *complete elliptic integral of the first kind*. For example, if $\theta_0 = \pi/2$, then $k = 1/\sqrt{2}$ and we find from the tables that $K = 1.8541$, resulting in a period $T = 7.4164\sqrt{l/g}$. This is approximately 18 per cent longer than the period $2\pi\sqrt{l/g}$ which applies for the case of small motion. In general, the period increases for increasing θ_0 and becomes infinite for $\theta_0 = \pi$.

Equation (3–228) can be expressed in the form

$$\sin \phi = \text{sn} \sqrt{\frac{g}{l}} (t - t_0) \tag{3–231}$$

where the *sn* function is known as the *Jacobian elliptic function of the first kind*. It is periodic in its argument with a period $4K$, the shape of the function depending upon the modulus k.

For the case $k \ll 1$, that is, for small motion, we can use the approximation

$$\text{sn } x \cong \sin x \qquad (3\text{-}232)$$

and, from Eq. (3–231), we obtain

$$\sin \phi = \frac{\theta}{\theta_0} = \sin \sqrt{\frac{g}{l}} (t - t_0)$$

in agreement with the expression given in Eq. (3–222) for the linearized equations. This result can also be obtained directly from Eq. (3–227) by expanding the cosine functions and keeping the first two terms.

If $k = 1$, corresponding to $\theta_0 = \pi$, the elliptic function becomes

$$\text{sn } x = \tanh x \qquad (3\text{-}233)$$

This equation can be checked by making the substitution

$$y = \sin \phi$$

in Eq. (3–229) and evaluating the integral. We obtain

$$F = \int_0^y \frac{dy}{1 - y^2} = \tanh^{-1} y$$

which, when substituted into Eq. (3–228), results in

$$y = \sin \phi = \tanh \sqrt{\frac{g}{l}} (t - t_0) \qquad (3\text{-}234)$$

Comparing Eqs. (3–231) and (3–234), we see that Eq. (3–233) applies.

The previous solutions for the motion of a simple pendulum have assumed that $\dot{\theta} = 0$ whenever $\theta = \pm\theta_0$. Now we consider the case where the particle proceeds continuously around the circle in the same direction, corresponding to a total energy $E > mgl$. Suppose that $\dot{\theta} = \dot{\theta}_0$ when $\theta = 0$ and $t = t_0$. Then, using Eq. (3–224), we find that the total energy is

$$E = \frac{1}{2} ml^2 \dot{\theta}_0^2 - mgl$$

and, from the principle of conservation of energy, we obtain

$$\frac{1}{2} ml^2 \dot{\theta}^2 - mgl \cos \theta = \frac{1}{2} ml^2 \dot{\theta}_0^2 - mgl$$

or

$$\dot{\theta}^2 = \dot{\theta}_0^2 - \frac{2g}{l} (1 - \cos \theta)$$

$$= \dot{\theta}_0^2 \left(1 - \frac{4g}{l\dot{\theta}_0^2} \sin^2 \frac{\theta}{2}\right) \qquad (3\text{-}235)$$

We see from energy considerations that $\dot{\theta}_0^2 > 4g/l$. So let us define

$$k^2 = \frac{4g}{l\dot\theta_0^2} < 1 \quad \text{and} \quad \phi = \tfrac{1}{2}\theta$$

Then Eq. (3–235) can be written in the form

$$\dot\phi^2 = \tfrac{1}{4}\dot\theta_0^2(1 - k^2 \sin^2 \phi)$$

from which we obtain

$$t - t_0 = \frac{2}{\dot\theta_0} \int_0^\phi \frac{d\phi}{\sqrt{1 - k^2 \sin^2 \phi}} = \frac{2}{\dot\theta_0} F(\phi, k) \qquad (3\text{–}236)$$

Thus the solution is again expressed in terms of elliptic integrals of the first kind. Again we use the *sn* function to express the motion of the angle θ as a function of time.

$$\sin \frac{\theta}{2} = \operatorname{sn} \frac{\dot\theta_0(t - t_0)}{2} \qquad (3\text{–}237)$$

The period of a complete revolution is twice the time required for θ to go from 0 to π. Hence, from Eq. (3–236),

$$T = \frac{4}{\dot\theta_0} F\!\left(\frac{\pi}{2}, k\right) = \frac{4}{\dot\theta_0} K(k) \qquad (3\text{–}238)$$

It can be seen that the period decreases continuously with increasing $\dot\theta_0$ if $\dot\theta_0^2 > 4g/l$. This effect arises from two sources: (1) the $\dot\theta_0$ term in the denominator of the right-hand side of Eq. (2–238), (2) the inverse variation of k with $\dot\theta_0$, causing the value of the complete elliptic integral K to decrease with increasing $\dot\theta_0$. As $\dot\theta_0$ becomes very large, the modulus k approaches zero and, from Eqs. (3–232) and (3–237),

$$\theta \cong \dot\theta_0(t - t_0)$$

indicating nearly uniform motion around the circular path.

3–10. EXAMPLES

Now let us consider several examples whose solutions illustrate the application of various principles of particle dynamics.

Example 3–12. A projectile is fired from the top of a cliff onto a level plain which is 1000 ft below. Its initial velocity is 500 ft/sec and it lands at a point such that the trajectory makes an angle of 70° with respect to the horizontal. What is its slant range, that is, the distance between the firing point and the impact point?

The principal difficulty encountered in the analysis of this problem is that the speed is given at one point of the trajectory and the direction at another. This is in contrast to the usual situation where complete initial conditions are given. Nevertheless, we can proceed by first calculating the

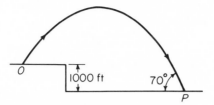

Fig. 3-29. The trajectory of a projectile fired onto a lower target.

impact velocity at P (Fig. 3–29). Using the principle of conservation of energy and assuming that the reference level for potential energy is at P, we see that the kinetic energy at P is equal to the kinetic energy plus the potential energy at O. Thus,

$$\frac{1}{2} m v_P^2 = 1000 \, mg + \frac{1}{2} m(500)^2$$

or

$$v_P^2 = 2000g + 250{,}000 = 3.144 \times 10^5 \text{ ft}^2/\text{sec}^2$$

$$v_P = 560.7 \text{ ft/sec}$$

Now we can use the *reversibility* property of conservative systems and calculate the trajectory due to firing a projectile backwards from point P to O with an initial velocity $v_0 = 560.7$ ft/sec and an initial angle $\gamma = 70°$. The horizontal range x can be obtained from Eq. (3–17) by multiplying through by $2v_0^2 \cos^2 \gamma/g$ and setting $y = 1000$ ft. We find that

$$x^2 - \frac{v_0^2 \sin 2\gamma}{g} x + \frac{2v_0^2 \cos^2 \gamma}{g} y = 0$$

or, assuming $g = 32.2$ ft/sec^2,

$$x^2 - 6.28 \times 10^3 x + 2.28 \times 10^6 = 0$$

The roots are

$$x_{1,2} = 390 \text{ ft}, \ 5890 \text{ ft}$$

From Fig. 3–29, it can be seen that the larger root is the one giving the horizontal distance of point O from point P, the smaller root corresponding to the other crossing of the line $y = 1000$ ft. Finally, the total range is

$$R = \sqrt{x^2 + y^2} = \sqrt{5890^2 + 1000^2} = 5980 \text{ ft}$$

A second method of solution again involves the reverse trajectory from P to O, but after calculating v_P, we compute the vertex altitude and the corresponding time, using Eqs. (3–19) and (3–21). The result is

$$y_v = \frac{v_0^2}{2g} \sin^2 \gamma = 4320 \text{ ft}$$

$$t_v = \frac{v_0}{g} \sin \gamma = 16.40 \text{ sec}$$

Next we calculate the time to fall from the vertex a distance of $4320 - 1000 = 3320$ ft. From Eq. (3–9), we obtain

$$t_2 = \sqrt{\frac{2(y_v - 1000)}{g}} = 14.35 \text{ sec}$$

Hence, the total time of flight is

$$t = t_v + t_2 = 30.75 \text{ sec}$$

and, from Eq. (3–15), the horizontal range is

$$x = v_0 t \cos \gamma = 5890 \text{ ft}$$

in agreement with the previous result.

Example 3–13. A particle of mass m starts from rest and slides on a frictionless track around a vertical circular loop of radius a, as shown in Fig. 3–30(a). Find the minimum starting height h above the bottom of the loop in order that the particle will not leave the track at any point.

First consider the free-body diagram shown in Fig. 3–30(b). Since the track is fixed and frictionless, the force R which it exerts on the particle is

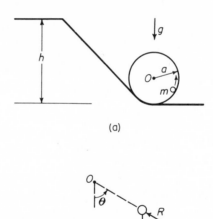

(a)

(b)

Fig. 3-30. A particle sliding around a smooth vertical loop.

normal to the direction of motion at that point and does no work. The only external force that works on the particle is the force of gravity and so we can use conservation of energy principles in solving for its speed at any point. Choosing the reference level for potential energy at the bottom of the loop, we find that

$$mgh = \frac{1}{2}mv^2 + mga(1 - \cos\theta)$$

or

$$v^2 = 2g(h - a + a\cos\theta) \tag{3-239}$$

for any position on the loop.

Next let us determine the conditions such that the particle will just fail to leave the track. It is clear that the particle cannot leave on the straight portions and therefore we shall consider only the circular loop. We assume that sliding occurs on the interior surface, implying that R must be positive at all times when the particle is in contact with the track. Hence the required conditions for the particle to leave the track are that $\dot{R} < 0$ and $R = 0$. The borderline case occurs when the minimum value of R is zero, that is, when the conditions $R = 0$ and $\dot{R} = 0$ occur simultaneously.

Applying Newton's law of motion to the radial component of the force and the acceleration, we see that

$$mg\cos\theta - R = ma_r$$

or, substituting for a_r from Eq. (2-54),

$$R = \frac{mv^2}{a} + mg\cos\theta \tag{3-240}$$

From Eqs. (3-239) and (3-240), we obtain

$$R = 3mg\cos\theta + 2mg\left(\frac{h}{a} - 1\right) \tag{3-241}$$

and it follows that

$$\dot{R} = -3mg\sin\theta \tag{3-242}$$

Setting R and $\dot{R}$ equal to zero simultaneously we find from Eqs. (3-241) and (3-242) that

$$\theta = \pi \text{ rad} \qquad h = \frac{5}{2}a$$

Hence the particle does not leave the track if $h \geq 5a/2$. It may be seen that the centripetal acceleration at $\theta = \pi$ is just equal to the acceleration of gravity for the borderline case where $R = 0$ at this point. Also note from Eq. (3-241) that the minimum value of R always occurs at $\theta = \pi$ if the particle remains on the track.

Example 3-14. A particle having a mass m and a velocity v_m in the y direction is projected onto a horizontal belt that is moving with a uniform velocity v_b in the x direction, as shown in Fig. 3-31(a). There is a coefficient of sliding friction μ between the particle and the belt. Assuming that the particle first touches the belt at the origin of the fixed xy coordinate system and remains on the belt, find the coordinates (x, y) of the point where sliding stops.

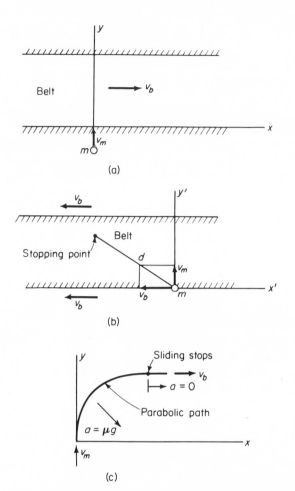

Fig. 3-31. The path of a particle sliding on a moving belt.

In the analysis of this problem we first note that the fixed xy coordinate system and also an $x'y'$ coordinate system which translates uniformly with the belt are *both* inertial systems. Because the frictional force depends upon the motion of the particle relative to the belt, it is more convenient to consider first the motion relative to the $x'y'$ system. As viewed by an observer riding with the belt, the particle moves initially with a velocity component v_m in the positive y' direction and a component v_b in the negative x' direction, as shown in Fig. 3–31(b). The path of the particle relative to the belt is a straight line since the frictional force directly opposes the motion and there are no horizontal forces perpendicular to the path.

Let us use the principle of work and kinetic energy to find the stopping point in the $x'y'$ frame. The work done against friction is equal to the loss of kinetic energy, or

$$\mu mgd = \frac{1}{2} m(v_m^2 + v_b^2)$$

where d is the stopping distance and the normal force is mg. Hence the stopping distance is

$$d = \frac{1}{2\mu g}(v_m^2 + v_b^2)$$

Assuming that the particle moved onto the belt at the origin of the $x'y'$ system, we can take components of d in the x' and y' directions to find the stopping point in the primed system. Thus we obtain that

$$x' = -d\frac{v_b}{\sqrt{v_m^2 + v_b^2}} = \frac{-v_b}{2\mu g}\sqrt{v_m^2 + v_b^2}$$

$$y' = d\frac{v_m}{\sqrt{v_m^2 + v_b^2}} = \frac{v_m}{2\mu g}\sqrt{v_m^2 + v_b^2}$$

are the coordinates of the stopping point.

We have seen that the force of friction during sliding is μmg. Therefore, the particle has a constant deceleration of magnitude μg as it moves along its straight path. It follows that the time required to stop sliding is just the original speed in the primed system divided by μg.

$$t = \frac{1}{\mu g}\sqrt{v_m^2 + v_b^2}$$

To convert the stopping position back to the fixed xy coordinate system, let us use the following equations:

$$x = x' + v_b t \qquad y = y'$$

Evaluating these equations at the time when sliding stops, we obtain

$$x = \frac{v_b}{2\mu g}\sqrt{v_m^2 + v_b^2}, \qquad y = \frac{v_m}{2\mu g}\sqrt{v_m^2 + v_b^2}$$

A *second method* of solution is to work entirely in the fixed xy frame. From this viewpoint, it is no longer obvious that the acceleration is constant in direction, although its magnitude must be μg as long as sliding continues. The direction of the acceleration, however, is always opposite to the *relative* velocity, so we can write

$$a_x = \frac{dv_x}{dt} = -\mu g\frac{v_x - v_b}{\sqrt{(v_x - v_b)^2 + v_y^2}}$$

$$a_y = \frac{dv_y}{dt} = -\mu g\frac{v_y}{\sqrt{(v_x - v_b)^2 + v_y^2}}$$

and therefore

$$\frac{dv_x}{dv_y} = \frac{v_x - v_b}{v_y}$$

or

$$\int \frac{dv_x}{v_x - v_b} = \int \frac{dv_y}{v_y}$$

Integrating, we obtain

$$\ln (v_x - v_b) = \ln v_y + \ln C$$

or

$$v_x = C v_y + v_b$$

Evaluating the constant C from initial conditions we find that

$$v_x = -\frac{v_b}{v_m} v_y + v_b$$

This allows the acceleration components to be evaluated, with the result that

$$a_x = \mu g \frac{v_b}{\sqrt{v_m^2 + v_b^2}}$$

$$a_y = -\mu g \frac{v_m}{\sqrt{v_m^2 + v_b^2}}$$

thereby confirming that the acceleration is constant in direction.

The time required for the velocity v_y to reach zero is

$$t = -\frac{v_m}{a_y} = \frac{1}{\mu g} \sqrt{v_m^2 + v_b^2}$$

at which time the velocity v_x is

$$v_x = a_x t = v_b$$

which is just the belt velocity. So the sliding in the x and y directions stops simultaneously. From Eq. (3–7), the position of the particle is

$$x = \frac{1}{2} a_x t^2 = \frac{v_b}{2\mu g} \sqrt{v_m^2 + v_b^2}$$

$$y = v_m t + \frac{1}{2} a_y t^2 = \frac{v_m}{2\mu g} \sqrt{v_m^2 + v_b^2}$$

in agreement with our previous results. The path of the particle relative to the fixed xy system is parabolic while sliding occurs, as shown in Fig. 3–31(c). This results from the fact that the force of sliding friction acts like a uniform force field in this case.

Comparing the two methods of solution, it can be seen that the proper choice of a reference frame can result in a considerable simplification in the analysis of a problem. In this example, both the xy and $x'y'$ systems are inertial, but the motion relative to the belt, and hence the direction of the frictional force, are much more easily visualized when viewed from the primed system.

REFERENCES

1. Ames, J. S., and F. D. Murnaghan, *Theoretical Mechanics*. New York: Dover Publications, Inc., 1957.
2. Banach, S., *Mechanics*, E. J. Scott, trans. Mathematical Monographs, vol. 24, Warsaw, 1951; distributed by Hafner Publishing Co., New York.
3. Becker, R. A., *Introduction to Theoretical Mechanics*. New York: McGraw-Hill, Inc., 1954.
4. Synge, J. L., and B. A. Griffith, *Principles of Mechanics*, 3rd ed. New York: McGraw-Hill, Inc., 1959.
5. Webster, A. G., *The Dynamics of Particles and of Rigid, Elastic and Fluid Bodies*, 2nd ed. New York: Dover Publications, Inc., 1959.

PROBLEMS

3–1. A smooth sphere of mass m and radius r is squeezed between two massless levers, each of length l, which are inclined at an angle ϕ with the vertical. If a force P is applied between the ends of the levers, as shown, what is the vertical acceleration of the sphere when $\phi = 30°$? Consider the mass of the sphere to be concentrated at its center.

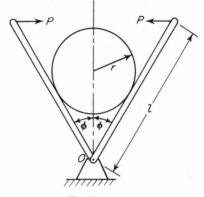

Fig. P3-1

3–2. A tube rotates in the horizontal xy plane with a constant angular velocity ω about the z axis. A particle of mass m is released from a radial distance of 1 ft when the tube is in the

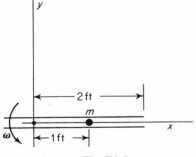

Fig. P3-2

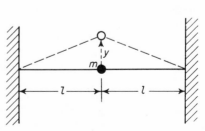

Fig. P3-3

position shown. If the tube is frictionless, find the direction and magnitude of the velocity of the particle as it leaves the tube.

3-3. A particle of mass m is supported by two elastic cords, each having a nominal length l and tension T. Assuming that the transverse displacement y is much smaller than either l or T/k, where k is the stiffness coefficient of each cord, solve for the natural frequency of the transverse motion.

3-4. A particle of mass m is projected vertically upward with an initial velocity v_0. If the drag force of the air is bv^2, where b is constant, solve for the upward velocity v as a function of time. Also solve for v as a function of the vertical displacement z. What is z_{max}? Assume v is positive.

3-5. Consider the free motion of an underdamped mass-spring-damper system such as that in Sec. 3-7. For each cycle of oscillation, what is the interval between the time of maximum velocity and the time at which the displacement is zero?

3-6. Consider an underdamped mass-spring-damper system. Using the weighting function and the method of convolution, find the steady-state response to the forcing function $F(t) = F_0 \cos \omega t$.

3-7. A mass m and a spring of stiffness k are connected in series and lie on a horizontal floor. At $t = 0$, the free end of the initially unstressed spring is moved at a constant speed v_0 in a straight line directly away from the mass. Assuming a friction coefficient μ between the mass and the floor, solve for the displacement of the mass as a function of time.

3-8. Suppose that the cartesian components of the force $\mathbf{F}$ acting upon a certain particle are each given as a function of its position (x,y,z). Show that if $\partial F_x/\partial y = \partial F_y/\partial x$, $\partial F_x/\partial z = \partial F_z/\partial x$, and $\partial F_y/\partial z = \partial F_z/\partial y$ in a given region, where these partial derivatives are continuous functions, then the force field is conservative.

3-9. Given that the force acting on a particle has the following components: $F_x = -x + y$, $F_y = x - y + y^2$, $F_z = 0$. Solve for the potential energy V.

3-10. Calculate the energy necessary to put a satellite of mass m into a circular orbit at an altitude h above the earth's surface. Assume a spherical earth of radius R and neglect the effects of atmospheric drag and the spin of the earth.

3-11. A particle of mass m is fastened to an inextensible wire of length l to form a spherical pendulum. Using spherical coordinates to describe the position of the particle, where θ is measured from the upward vertical, the initial angular velocities are $\dot{\theta}_0$ and $\dot{\phi}_0$. Find the initial values of: (a) the angular accelerations $\ddot{\phi}$ and $\ddot{\theta}$; (b) the force in the wire.

3-12. A particle of mass m moves under the action of gravity on the inner surface of a smooth inverted right circular cone of vertex angle 2θ. Using a spherical coordinate system with its origin at the vertex of the cone, find the maximum value of the distance r of the particle from the vertex, assuming that $\theta = 30°$ and the initial conditions of the motion are $r(0) = a$, $\dot{r}(0) = 0$, and $\dot{\phi}(0) = 4\sqrt{g/a}$.

3-13. A thin flexible rope of negligible mass is wrapped around a cylinder of radius a that is rotating with a constant angular velocity of Ω rad/sec. A particle of mass m is attached to the end of the rope. Assuming that the rope does not slip relative to the cylinder, but can unwind such that a straight portion of length l is produced, write a differential equation of motion for the particle in terms of the single dependent variable l. If the initial conditions are $l(0) = 0$, $\dot{l}(0) = a\Omega$, solve for l and the tensile force in the rope as functions of time.

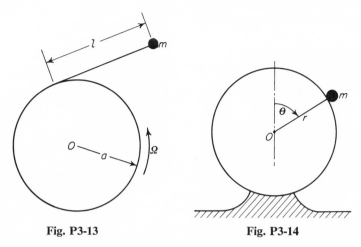

Fig. P3-13   **Fig. P3-14**

3–14. A particle of mass m is displaced slightly from its equilibrium position at the top of a smooth fixed sphere of radius r, and it slides because of gravity. Through what angular displacement θ does the particle move before it leaves the sphere?

3–15. A small fixed tube is shaped in the form of a vertical helix of radius a and helix angle γ, that is, the tube always makes an angle γ with the horizontal. A particle of mass m slides down the tube under the action of gravity. If there is a coefficient of friction μ between the tube and the particle, what is the steady-state speed of the particle? Let $\gamma = 30°$ and assume that $\mu < 1/\sqrt{3}$.

3–16. A particle of mass m is embedded at a distance $\frac{1}{4}R$ from the center of a massless circular disk of radius R which can roll without slipping on the inside surface of a fixed circular cylinder of radius $3R$. The disk is released with zero velocity from the position shown and rolls because of gravity, all motion taking place in the same vertical plane. Find: (a) the maximum velocity of the particle during the resulting motion; (b) the reaction force acting on the disk at the point of contact when it is at its lowest position.

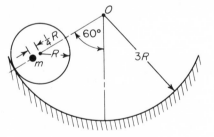

Fig. P3-16

4

DYNAMICS OF A SYSTEM OF PARTICLES

In the previous chapter, we discussed some of the more important principles and techniques to be used in the analysis of the motion of a single particle. When one considers the dynamics of a group of interacting particles, one may still look at individual particles, but in addition, certain over-all aspects of the motion of the system may be calculated without specifically solving for the individual motions. It is with these extensions and generalizations of previously discussed principles that this chapter will be primarily concerned. In addition, the chapter will discuss particular applications, such as the collision and rocket propulsion problems.

4–1. THE EQUATIONS OF MOTION

Consider first a system of n particles, of which three are shown in Fig. 4–1. The forces applied to a given particle may be classified as external or internal, according to their source. The *total* force on the ith particle arising from sources external to the system of n particles is designated by $\mathbf{F}_i$ and is known as an *external force*. All interaction forces among the particles are known as *internal forces* and are designated by individual force vectors of the form $\mathbf{f}_{ij}$, where the first subscript indicates the particle on which the force acts and the second subscript indicates the acting particle.

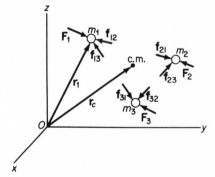

Fig. 4-1. The forces acting on a system of particles.

From Newton's law of action and reaction, we know that the interaction forces between any two particles are equal and opposite:

$$\mathbf{f}_{ij} = -\mathbf{f}_{ji} \tag{4-1}$$

Also, these forces are assumed to be collinear, that is, they act along the straight line connecting the particles. It can be seen that

$$\mathbf{f}_{ii} = 0$$

in agreement with Eq. (4–1), indicating the fact that a particle cannot exert a force on itself which affects its motion.

Now let us sum all the forces acting on the ith particle. Including both external and internal forces, we can write the equation of motion in the form:

$$m_i \ddot{\mathbf{r}}_i = \mathbf{F}_i + \sum_{j=1}^{n} \mathbf{f}_{ij} \quad (i = 1, 2, \ldots, n) \tag{4-2}$$

where m_i is the mass of the ith particle and $\mathbf{r}_i$ is its position vector relative to the fixed point O. Next we sum Eq. (4–2) over all n particles, obtaining

$$\sum_{i=1}^{n} m_i \ddot{\mathbf{r}}_i = \sum_{i=1}^{n} \mathbf{F}_i + \sum_{i=1}^{n} \sum_{j=1}^{n} \mathbf{f}_{ij} \tag{4-3}$$

But, from Eq. (4–1), we see that

$$\sum_{i=1}^{n} \sum_{j=1}^{n} \mathbf{f}_{ij} = 0 \tag{4-4}$$

since the internal forces always occur in equal and opposite pairs. Also, we note that the total mass m is

$$m = \sum_{i=1}^{n} m_i \tag{4-5}$$

and the *center of mass* location is given by

$$\mathbf{r}_c = \frac{1}{m} \sum_{i=1}^{n} m_i \mathbf{r}_i \tag{4-6}$$

The total external force is

$$\mathbf{F} = \sum_{i=1}^{n} \mathbf{F}_i \tag{4-7}$$

Therefore Eq. (4–3) can be written in the form

$$\mathbf{F} = m \ddot{\mathbf{r}}_c \tag{4-8}$$

This result has the familiar form of a force being equal to the product of a mass and an acceleration. It indicates that the motion of the center of mass of a system of particles is the same as if the entire mass of the system were concentrated at the center of mass and were driven by the sum of all the forces external to the system.

To illustrate this point, the trajectory of a bomb in a vacuum is sometimes used. If the bomb explodes, then the center of mass of all the fragments continues on the same path as the bomb would have taken had it not exploded. This result is accurate if the total external force acting on particles of the bomb is not changed by the explosion. It is clear, however, that for the case of motion through the air, the external drag force would be vastly increased after the explosion, thereby altering the path of the center of mass. Also, for the case of certain conservative forces, a nonuniform gravitational field, for example, the total external force is altered by replacing a system of particles by a single particle at the center of mass and calculating the force on this particle.

It should be emphasized, then, that although Eq. (4–8) has general validity, the force **F** is calculated for the actual system of particles and not, in general, on the basis of a single particle of total mass m located at the center of mass.

4–2. WORK AND KINETIC ENERGY

We have seen that Eq. (4–6) is identical in mathematical form to the equation of motion for a single particle. It is apparent, then, that a set of principles similar to those for a single particle also applies to the motion of the center of mass of a system of particles.

Let us assume that the center of mass of the system moves from A_c to B_c under the action of the external forces $\mathbf{F}_i$ whose sum is **F**. Then, taking the line integral of each side of Eq. (4–8) over the path of the center of mass, we obtain a result similar to that obtained previously in Eq. (3–67). It is

$$\int_{A_c}^{B_c} \mathbf{F} \cdot d\mathbf{r}_c = \frac{1}{2}\, mv_c^2 \Big|_{A_c}^{B_c} \tag{4–9}$$

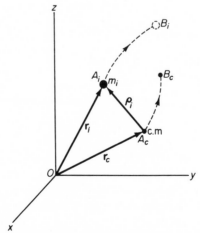

Fig. 4-2. Position vectors of the center of mass and of a typical particle during an arbitrary displacement.

where v_c is the absolute velocity of the center of mass. This equation states that if one considers the sum of the external forces to be acting at the center of mass of the system, then the work done by the total external force in moving over the path of the center of mass is equal to the change in the translational kinetic energy associated with the motion of the center of mass.

Note that the work expressed by the integral on the left side of Eq. (4-9) does not include that work done by the internal forces, nor is it even the total work of the external forces. To clarify this point, let us calculate the total work done by all the forces, external as well as internal. Considering first the ith particle, the work done by all the forces acting on m_i as it moves from A_i to B_i is

$$W_i = \int_{A_i}^{B_i} \left(\mathbf{F}_i + \sum_{j=1}^{n} \mathbf{f}_{ij} \right) \cdot d\mathbf{r}_i \qquad (4\text{-}10)$$

But, as shown in Fig. 4-2, we can express the position of the ith particle as the sum

$$\mathbf{r}_i = \mathbf{r}_c + \boldsymbol{\rho}_i \qquad (4\text{-}11)$$

where $\boldsymbol{\rho}_i$ is the position vector of the ith particle relative to the center of mass. So, substituting Eq. (4-11) into (4-10) and summing over all n particles, we obtain the total work

$$W = \sum_{i=1}^{n} W_i = \sum_{i=1}^{n} \int_{A_i}^{B_i} \left(\mathbf{F}_i + \sum_{j=1}^{n} \mathbf{f}_{ij} \right) \cdot (d\mathbf{r}_c + d\boldsymbol{\rho}_i)$$

or

$$W = \sum_{i=1}^{n} \int_{A_c}^{B_c} \left(\mathbf{F}_i + \sum_{j=1}^{n} \mathbf{f}_{ij} \right) \cdot d\mathbf{r}_c + \sum_{i=1}^{n} \int_{A_i}^{B_i} \left(\mathbf{F}_i + \sum_{j=1}^{n} \mathbf{f}_{ij} \right) \cdot d\boldsymbol{\rho}_i \qquad (4\text{-}12)$$

where the limits on the second integral refer, in this case, to the position of the ith particle relative to the position of the center of mass of the system.

The first integral on the right side of Eq. (4-12) can be simplified by using Eqs. (4-4) and (4-7). We note that $\mathbf{r}_c$ and the limits of integration are not dependent upon the summation index i, and therefore the summations can be carried out before the integration. The second integral cannot be simplified, since $\boldsymbol{\rho}_i$ is a function of i. With these changes, we obtain

$$W = \int_{A_c}^{B_c} \mathbf{F} \cdot d\mathbf{r}_c + \sum_{i=1}^{n} \int_{A_i}^{B_i} \left(\mathbf{F}_i + \sum_{j=1}^{n} \mathbf{f}_{ij} \right) \cdot d\boldsymbol{\rho}_i \qquad (4\text{-}13)$$

Thus we see that the total work can be considered as the sum of two parts: (1) the work done by the total external force acting through the displacement of the center of mass; (2) the summation of the work done on all particles by both the external and internal forces on each particle acting through the displacement of that particle relative to the center of mass.

Now, for each particle, the principle of work and kinetic energy applies, so we can write

$$W_i = \frac{1}{2} m_i \dot{\mathbf{r}}_i \cdot \dot{\mathbf{r}}_i \Big|_{A_i}^{B_i} = \frac{1}{2} m_i (\dot{\mathbf{r}}_c \cdot \dot{\mathbf{r}}_c + 2\dot{\mathbf{r}}_c \cdot \dot{\boldsymbol{\rho}}_i + \dot{\boldsymbol{\rho}}_i \cdot \dot{\boldsymbol{\rho}}_i) \Big|_{A_i}^{B_i}$$

where we have substituted for $\mathbf{r}_i$ from Eq. (4–11). Summing over all particles, we find that

$$W = \sum_{i=1}^{n} W_i = \frac{1}{2} m v_c^2 \Big|_{A_c}^{B_c} + \dot{\mathbf{r}}_c \cdot \sum_{i=1}^{n} m_i \dot{\boldsymbol{\rho}}_i \Big|_{A_i}^{B_i} + \frac{1}{2} \sum_{i=1}^{n} m_i \dot{\boldsymbol{\rho}}_i^2 \Big|_{A_i}^{B_i} \qquad (4\text{–}14)$$

where $\dot{\boldsymbol{\rho}}_i$ is the relative velocity of m_i, as viewed by a nonrotating observer moving with the center of mass. Here we have used the notation:

$$\dot{\boldsymbol{\rho}}_i^2 = \dot{\boldsymbol{\rho}}_i \cdot \dot{\boldsymbol{\rho}}_i$$

But we see from Eq. (4–6) that

$$m\mathbf{r}_c = \sum_{i=1}^{n} m_i \mathbf{r}_i = \sum_{i=1}^{n} m_i (\mathbf{r}_c + \boldsymbol{\rho}_i)$$

$$= m\mathbf{r}_c + \sum_{i=1}^{n} m_i \boldsymbol{\rho}_i$$

and, consequently, that

$$\sum_{i=1}^{n} m_i \boldsymbol{\rho}_i = 0 \qquad (4\text{–}15)$$

in agreement with the original assumption that $\boldsymbol{\rho}_i$ is measured from the center of mass. Also, of course,

$$\sum_{i=1}^{n} m_i \dot{\boldsymbol{\rho}}_i = 0 \qquad (4\text{–}16)$$

and therefore Eq. (4–14) reduces to

$$W = \frac{1}{2} m v_c^2 \Big|_{A_c}^{B_c} + \sum_{i=1}^{n} \frac{1}{2} m_i \dot{\boldsymbol{\rho}}_i^2 \Big|_{A_i}^{B_i} \qquad (4\text{–}17)$$

The right-hand side of Eq. (4–17) represents the sum of the increases in kinetic energy of the individual particles, or, in other words, the increase in the total kinetic energy of the system. Thus we can write

$$W = T_B - T_A \qquad (4\text{–}18)$$

in a manner similar to Eq. (3–70) which was derived for a single particle. In this case, T_A and T_B represent the total kinetic energy of the system at the beginning and at the end, respectively, of the line integrations. It is apparent from Eq. (4–17) that the total kinetic energy is

$$T = \frac{1}{2} m v_c^2 + \sum_{i=1}^{n} \frac{1}{2} m_i \dot{\boldsymbol{\rho}}_i^2 \qquad (4\text{–}19)$$

Now Eq. (4–18) is valid for an arbitrary interval. So if we consider the case where the work and the change in kinetic energy are evaluated during an infinitesimal time interval Δt, we can write

$$\Delta W = \Delta T$$

or, in the limit as Δt approaches zero, we obtain

$$\dot{W} = \dot{T} \tag{4-20}$$

Thus the rate of increase of the total kinetic energy is equal to the rate at which work is done on the system, that is, it is the instantaneous *power* associated with the external and internal forces.

Returning now to a further consideration of work and kinetic energy relationships, we find from Eqs. (4–9), (4–13), and (4–17) that

$$\sum_{i=1}^{n} \int_{A_i}^{B_i} \left(\mathbf{F}_i + \sum_{j=1}^{n} \mathbf{f}_{ij} \right) \cdot d\boldsymbol{\rho}_i = \sum_{i=1}^{n} \frac{1}{2} m_i \dot{\boldsymbol{\rho}}_i^2 \bigg|_{A_i}^{B_i} \tag{4-21}$$

Hence the work done by the external and internal forces in moving through displacements relative to the center of mass is equal to the increase in the kinetic energy of relative motion. It is important to note that the relative velocity $\dot{\boldsymbol{\rho}}_i$ can arise from rigid body rotations in which the particle separations do not change with time, as well as in the more obvious case of changing particle separations.

Referring again to Eqs. (4–9), (4–19), and (4–21), we can summarize the results of this section as follows:

1. The total kinetic energy is equal to that due to the total mass moving with velocity of the center of mass plus that due to the motions of the individual particles relative to the center of mass.

2. The work done by the external forces in moving through the displacement of the center of mass is equal to the increase in the kinetic energy due to the total mass moving with the velocity of the center of mass.

3. The work done by the external plus the internal forces in moving through displacements relative to the center of mass is equal to the increase in the kinetic energy associated with the relative motions.

4–3. CONSERVATION OF MECHANICAL ENERGY

The principle of conservation of mechanical energy was developed in Sec. 3–3 for the case of a single particle. Now let us extend this principle to apply to a system of particles. First, recall from Eq. (4–8) that the motion of the center of mass is the same as though it were a particle with a mass equal to the total mass m and acted upon by the total external force $\mathbf{F}$. Therefore, if the total external force is derivable from a potential function involving the center of mass position only, then, by analogy to the results obtained previously for a single particle, the center of mass moves such that the energy E_c is constant.

$$E_c = T_c + V_c \tag{4-22}$$

T_c is the kinetic energy due to the translational motion of the center of mass, and V_c is the potential energy associated with the position of the center of mass. It can be seen that it is possible for conservation of the energy E_c to occur even in the case of dissipative internal forces.

For the case where the internal as well as the external forces are conservative, that is, they are derivable from a potential function involving the coordinates only, we find that the total energy E is conserved, where

$$E = T + V \qquad (4\text{-}23)$$

In this case, T is the sum of the kinetic energies of the individual particles, or it can be considered to be the sum of the portion due to the motion of the center of mass plus the portion due to motion relative to the center of mass, as in Eq. (4–19). The potential energy V is often just the sum of the potential energy due to gravity and that due to the deformations of elastic elements such as springs. In any event, the potential energy of n particles in a three-dimensional space can be written in the form:

$$V = V(x_1, x_2, \ldots, x_{3n}) \qquad (4\text{-}24)$$

since we assume that the system may have as many as $3n$ degrees of freedom. Thus, as in Eq. (3–89), we find that if a small increase in x_k results in a small displacement of a certain particle in a given direction, then the force

$$F_k = -\frac{\partial V}{\partial x_k} \qquad (4\text{-}25)$$

acts on the given particle in the direction of increasing x_k. This is the total force and includes, in general, both internal and external forces.

For the case of a system in which both conservative and nonconservative forces are acting, one can use work and energy concepts in place of a strict

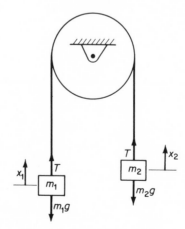

Fig. 4-3. Atwood's machine.

conservation of energy. For example, if W_n represents the work done *on* the system by *nonconservative* internal and external forces in going from configuration A to configuration B, then

$$W_n = E_B - E_A \tag{4-26}$$

where the total energies E_B and E_A include potential energy terms corresponding to all the forces doing work on the system except for the nonconservative forces.

Example 4-1. Atwood's Machine. Two masses m_1 and m_2 are connected by a massless, inextensible rope which passes over a pulley, as shown in Fig. 4-3. Neglecting the mass and the bearing friction of the pulley, find the acceleration of m_1 and the tension in the rope as the system moves under the action of gravity.

Let us use the coordinates x_1 and x_2 to designate the vertical displacements of m_1 and m_2, respectively, measured from some nominal position. Since the rope is inextensible, we see that

$$x_1 = -x_2 \tag{4-27}$$

Also, the pulley exerts no inertial or frictional forces on the rope and therefore the tension T is uniform throughout its length.

First we shall solve the problem using Newton's law of motion for each mass separately. Each mass has two forces acting on it, namely, the rope force T and the gravitational force. Thus we can write

$$m_1 \ddot{x}_1 = T - m_1 g$$
$$m_2 \ddot{x}_2 = T - m_2 g$$

Subtracting the second equation from the first and noting from Eq. (4-27) that $\ddot{x}_1 = -\ddot{x}_2$, we obtain

$$(m_1 + m_2)\ddot{x}_1 = (m_2 - m_1)g$$

or

$$\ddot{x}_1 = \left(\frac{m_2 - m_1}{m_1 + m_2}\right) g \tag{4-28}$$

The tension T is found by substituting this result into the first equation of motion, yielding

$$T = \left(\frac{2m_1 m_2}{m_1 + m_2}\right) g \tag{4-29}$$

Next, let us solve for $\ddot{x}_1$, using the principle of work and kinetic energy. Assuming that m_1 and m_2 comprise the system under consideration, we see that the work done on the system in an infinitesimal displacement is

$$dW = (T - m_1 g)\, dx_1 + (T - m_2 g)\, dx_2$$
$$= (m_2 - m_1)g\, dx_1$$

since $dx_1 = -dx_2$. Now, the total work done in a certain displacement x_1 is equal to the total kinetic energy of the system at the end of that interval, assuming that the system started from rest. Noting that the force $(m_2 - m_1)g$ is constant, we can write

$$(m_2 - m_1)gx_1 = \frac{1}{2} m_1 \dot{x}_1^2 + \frac{1}{2} m_2 \dot{x}_2^2 = \frac{1}{2}(m_1 + m_2)\dot{x}_1^2 \qquad (4\text{-}30)$$

where we recall that $\dot{x}_1 = -\dot{x}_2$. Differentiating this equation with respect to time and dividing by $\dot{x}_1$ (which is not zero, in general), we obtain

$$(m_2 - m_1)g = (m_1 + m_2)\ddot{x}_1$$

in agreement with the previous result of Eq. (4–28).

It is interesting to note that the rope tension T does not enter into the expression for the work on the left side of Eq. (4–30). Although the rope does positive or negative work on m_1 and m_2 individually, the total work done by the rope is zero for any displacement consistent with the constraint that $x_1 = -x_2$. Thus the rope and pulley system constitute what is known as a *workless constraint*.

One can always neglect the forces due to workless constraints when applying energy methods. For example, the potential energy function includes just those terms corresponding to gravitational forces when applying conservation of energy principles to Atwood's machine. Assuming that the system starts from rest at $x_1 = x_2 = 0$ and that the potential energy reference is chosen such that $V = 0$ initially, we see that the total energy is zero, or

$$m_1 gx_1 + m_2 gx_2 + \frac{1}{2} m_1 \dot{x}_1^2 + \frac{1}{2} m_2 \dot{x}_2^2 = 0$$

Substituting $x_2 = -x_1$ and $\dot{x}_2 = -\dot{x}_1$, we obtain

$$(m_1 - m_2)gx_1 + \frac{1}{2}(m_1 + m_2)\dot{x}_1^2 = 0$$

in agreement with Eq. (4–30).

Example 4–2. Masses m_1 and m_2 are connected by a spring of stiffness k. If a constant force F is applied to m_1 at $t = 0$ as shown in Fig. 4–4, and assuming that the masses can slide without friction on the horizontal surface, solve for the displacement x_1 as a function of time. Initially, the system

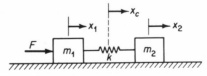

Fig. 4-4. A mass-spring system in rectilinear motion.

starts from rest with $x_1 = x_2 = 0$ and with the spring unstressed. ($m_1 = m_2 = m$.)

Let the displacement of the center of mass of the system, measured from its initial position, be x_c.

$$x_c = \frac{1}{2}(x_1 + x_2) \tag{4-31}$$

From Eq. (4-8), we see that the motion of the center of mass is the same as if the total force F were applied to a particle of mass $2m$. Thus

$$F = 2m\ddot{x}_c \tag{4-32}$$

Integrating twice and evaluating the constants of integration from the initial conditions $\dot{x}_c(0) = 0$ and $x_c(0) = 0$, we obtain

$$x_c = \frac{F}{4m} t^2 \tag{4-33}$$

in agreement with the general result previously given in Eq. (3-7).

Now let us write the equation of motion for m_1. The spring force is $k(x_1 - x_2)$ acting to the left on m_1, and so we obtain

$$F - k(x_1 - x_2) = m\ddot{x}_1$$

or, substituting for x_2 from Eq. (4-31),

$$F - 2k(x_1 - x_c) = m\ddot{x}_1 \tag{4-34}$$

Assuming for convenience that the center of the spring coincides with the center of mass, we can interpret the term $2k(x_1 - x_c)$ in the last equation as being the force on m_1 due to the compression $(x_1 - x_c)$ of the half-spring of stiffness $2k$.[1]

Next, we divide Eq. (4-32) by 2 and subtract it from Eq. (4-34). We obtain the equation

$$m(\ddot{x}_1 - \ddot{x}_c) + 2k(x_1 - x_c) = \frac{1}{2} F \tag{4-35}$$

which is in the standard form of a mass-spring system being excited by a step function of magnitude $\frac{1}{2}F$. We can write down the solution immediately by analogy to the results previously obtained in Eq. (3-115) or in Eq. (3-190).

$$x_1 - x_c = \frac{F}{4k}\left(1 - \cos\sqrt{\frac{2k}{m}}\, t\right) \tag{4-36}$$

Finally, from Eqs. (4-33) and (4-36), the solution for x_1 is seen to be

---

[1] In general, if we think of dividing springs or of connecting several similar springs in series, the stiffness varies inversely with the unstressed length. On the other hand, if n similar springs of stiffness k are connected in parallel, the over-all stiffness is nk. These rules can be checked easily by noting that springs connected in series all have the same force applied but the extensions are additive, whereas similar springs connected in parallel all have the same extension but the forces are additive.

$$x_1 = \frac{F}{4m} t^2 + \frac{F}{4k} \left(1 - \cos \sqrt{\frac{2k}{m}} t\right)$$ (4-37)

It is instructive to use the known solution for this example to illustrate some of the ideas which have been presented concerning work and energy. Let us evaluate the kinetic energy using Eq. (4-19). From Eq. (4-33), we see that

$$v_c = \dot{x}_c = \frac{F}{2m} t$$

and from Eq. (4-36), we obtain

$$\dot{\boldsymbol{\rho}}_1^2 = \dot{\boldsymbol{\rho}}_2^2 = (\dot{x}_1 - \dot{x}_c)^2 = \frac{F^2}{8mk} \sin^2 \sqrt{\frac{2k}{m}} t$$

Thus the total kinetic energy is

$$T = \frac{F^2}{4m} t^2 + \frac{F^2}{8k} \sin^2 \sqrt{\frac{2k}{m}} t$$

The potential energy of the spring is found by using Eq. (3-102).

$$V = \frac{1}{2} k(x_1 - x_2)^2 = 2k(x_1 - x_c)^2$$

$$= \frac{F^2}{8k} \left(1 - \cos \sqrt{\frac{2k}{m}} t\right)^2$$

Therefore the total energy is

$$E = T + V = \frac{F^2}{4m} t^2 + \frac{F^2}{4k} \left(1 - \cos \sqrt{\frac{2k}{m}} t\right)$$

Now the work done by the external force F is just

$$W = Fx_1 = \frac{F^2}{4m} t^2 + \frac{F^2}{4k} \left(1 - \cos \sqrt{\frac{2k}{m}} t\right)$$

Therefore we see that $W = E$, which agrees with Eq. (4-26) for the present case where the initial total energy is zero.

4-4. LINEAR IMPULSE AND MOMENTUM

Consider again Eq. (4-8) which relates the total external force to the motion of the center of mass.

$$\mathbf{F} = m\ddot{\mathbf{r}}_c$$

Both sides of this equation can be integrated with respect to time over the interval t_1 to t_2, yielding

$$\int_{t_1}^{t_2} \mathbf{F} \, dt = m(\mathbf{v}_{c2} - \mathbf{v}_{c1})$$ (4-38)

where $\mathbf{v}_{c1}$ and $\mathbf{v}_{c2}$ are the velocities of the center of mass at times t_1 and t_2, respectively. The integral on the left is the total impulse $\mathscr{F}$ of the external forces during the given interval. The right side of the equation represents the change in the total linear momentum $\mathbf{p}$ during the same interval, as we shall demonstrate.

The total linear momentum of a system of particles is just the vector sum of the individual momenta.

$$\mathbf{p} = \sum_{i=1}^{n} m_i \mathbf{v}_i \tag{4-39}$$

where $\mathbf{v}_i$ is the absolute velocity of m_i. But from Eq. (4-6), we see that

$$m\dot{\mathbf{r}}_c = \sum_{i=1}^{n} m_i \dot{\mathbf{r}}_i = \sum_{i=1}^{n} m_i \mathbf{v}_i$$

and therefore

$$\mathbf{p} = m\mathbf{v}_c \tag{4-40}$$

From Eqs. (4-38) and (4-40), we obtain the result:

$$\mathscr{F} = \mathbf{p}_2 - \mathbf{p}_1 \tag{4-41}$$

which is the *principle of linear impulse and momentum* for a system of particles. Note that because the internal forces occur in equal and opposite pairs, they do not contribute to the impulse $\mathscr{F}$, and hence have no influence on the total linear momentum of the system.

Equation (4-38) or Eq. (4-41) could have been written in terms of cartesian components as follows:

$$\mathscr{F}_x = \int_{t_1}^{t_2} F_x \, dt = m(\dot{x}_{c2} - \dot{x}_{c1})$$

$$\mathscr{F}_y = \int_{t_1}^{t_2} F_y \, dt = m(\dot{y}_{c2} - \dot{y}_{c1}) \tag{4-42}$$

$$\mathscr{F}_z = \int_{t_1}^{t_2} F_z \, dt = m(\dot{z}_{c2} - \dot{z}_{c1})$$

where (x_c, y_c, z_c) is the center of mass location.

It can be seen that if any component of the total impulse is zero, then the momentum is conserved in this direction. Furthermore, if there are no external forces acting on the system, whatever the nature of the internal forces, then the total momentum is constant. This is the *principle of conservation of linear momentum* as it applies to a system of particles. It is particularly useful in the analysis of problems in which the internal forces are not accurately known, as in collision and explosion problems. Examples of the use of linear impulse and momentum methods will be found in Secs. 4-7 and 4-8 where problems involving collisions and also rocket propulsion will be discussed.

4–5. ANGULAR MOMENTUM

In Sec. 3–6, we developed equations for the angular momentum of a single particle and also its time rate of change, using a point fixed in inertial space as a reference. In this section, we extend this development to apply to a system of particles, and in doing so several possible reference points will be considered.

Fixed Reference Point. Consider the total angular momentum of a system of n particles, taking as a reference the fixed point O, as shown in Fig. 4–5. Using Eq. (3–143), we find that the angular momentum of the particle m_i is

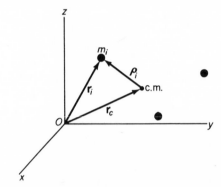

Fig. 4-5. Position vectors for a system of particles.

$$\mathbf{H}_i = \mathbf{r}_i \times m\dot{\mathbf{r}}_i \qquad (4\text{–}43)$$

The total angular momentum of the system is found by summing the angular momenta of the individual particles. Thus

$$\mathbf{H} = \sum_{i=1}^{n} \mathbf{H}_i = \sum_{i=1}^{n} \mathbf{r}_i \times m_i\dot{\mathbf{r}}_i \qquad (4\text{–}44)$$

Now let us differentiate Eq. (4–44) with respect to time, obtaining

$$\dot{\mathbf{H}} = \sum_{i=1}^{n} \mathbf{r}_i \times m_i\ddot{\mathbf{r}}_i + \sum_{i=1}^{n} \dot{\mathbf{r}}_i \times m_i\dot{\mathbf{r}}_i$$

$$= \sum_{i=1}^{n} \mathbf{r}_i \times m_i\ddot{\mathbf{r}}_i \qquad (4\text{–}45)$$

since $\dot{\mathbf{r}}_i \times \dot{\mathbf{r}}_i = 0$. The equation of motion can be written for each particle, as in Eq. (4–2). Then, substituting for each $m_i\ddot{\mathbf{r}}_i$ in terms of the applied forces, we obtain

$$\dot{\mathbf{H}} = \sum_{i=1}^{n} \mathbf{r}_i \times \mathbf{F}_i + \sum_{i=1}^{n} \sum_{j=1}^{n} \mathbf{r}_i \times \mathbf{f}_{ij} \qquad (4\text{–}46)$$

We have assumed that the internal forces occur in equal, opposite, and collinear pairs. Therefore,

$$\sum_{i=1}^{n} \sum_{j=1}^{n} \mathbf{r}_i \times \mathbf{f}_{ij} = 0 \qquad (4\text{-}47)$$

since, for each $\mathbf{f}_{ij}$, there is an $\mathbf{f}_{ji} = -\mathbf{f}_{ij}$ having the same line of action and therefore an equal and opposite moment about O. The expression for the total moment about O due to the external forces acting on the system is thus reduced to

$$\mathbf{M} = \sum_{i=1}^{n} \mathbf{r}_i \times \mathbf{F}_i \qquad (4\text{-}48)$$

From Eqs. (4–46), (4–47), and (4–48), we obtain

$$\mathbf{M} = \dot{\mathbf{H}} \qquad (4\text{-}49)$$

Hence we see that, for a system of particles, *the time rate of change of the angular momentum about a fixed point O is equal to the total moment about O of the external forces acting on the system.*

Reference Point at the Center of Mass. If we substitute $\mathbf{r}_i = \mathbf{r}_c + \boldsymbol{\rho}_i$ into the expression given in Eq. (4–44) for the angular momentum about the fixed point O, we obtain

$$\mathbf{H} = \sum_{i=1}^{n} (\mathbf{r}_c + \boldsymbol{\rho}_i) \times m_i(\dot{\mathbf{r}}_c + \dot{\boldsymbol{\rho}}_i)$$

$$= \mathbf{r}_c \times m\dot{\mathbf{r}}_c + \mathbf{r}_c \times \sum_{i=1}^{n} m_i\dot{\boldsymbol{\rho}}_i + \sum_{i=1}^{n} m_i\boldsymbol{\rho}_i \times \dot{\mathbf{r}}_c + \sum_{i=1}^{n} \boldsymbol{\rho}_i \times m_i\dot{\boldsymbol{\rho}}_i$$

From Eqs. (4–15) and (4–16), which assume that $\boldsymbol{\rho}_i$ is measured from the center of mass, we see that the two middle terms on the right are zero. Hence the preceding expression for $\mathbf{H}$ can be written in the form:

$$\mathbf{H} = \mathbf{r}_c \times m\dot{\mathbf{r}}_c + \sum_{i=1}^{n} \boldsymbol{\rho}_i \times m_i\dot{\boldsymbol{\rho}}_i \qquad (4\text{-}50)$$

or

$$\mathbf{H} = \mathbf{r}_c \times m\dot{\mathbf{r}}_c + \mathbf{H}_c \qquad (4\text{-}51)$$

where

$$\mathbf{H}_c = \sum_{i=1}^{n} \boldsymbol{\rho}_i \times m_i\dot{\boldsymbol{\rho}}_i \qquad (4\text{-}52)$$

Here we have the important result that *the total angular momentum of a system of particles about a fixed point O is equal to the angular momentum of a particle of mass* m *moving with the velocity of the center of mass plus the angular momentum* $\mathbf{H}_c$ *about the center of mass.* More explicitly, we see from Eq. (4–52) that $\mathbf{H}_c$ is the angular momentum of the system with respect to the center of mass, as viewed by a nonrotating observer moving with the center of mass.

Now let us differentiate Eq. (4–50) with respect to time, obtaining

$$\dot{\mathbf{H}} = \mathbf{r}_c \times m\ddot{\mathbf{r}}_c + \sum_{i=1}^{n} \boldsymbol{\rho}_i \times m_i\ddot{\boldsymbol{\rho}}_i \qquad (4\text{-}53)$$

where again we have used the fact that all terms involving the cross product of a vector with itself are zero. Of course, $\mathbf{F} = m\ddot{\mathbf{r}}_c$ and, from Eq. (4–52),

$$\dot{\mathbf{H}}_c = \sum_{i=1}^{n} \boldsymbol{\rho}_i \times m_i \ddot{\boldsymbol{\rho}}_i \tag{4–54}$$

so Eq. (4–53) can be written in the form

$$\dot{\mathbf{H}} = \mathbf{r}_c \times \mathbf{F} + \dot{\mathbf{H}}_c \tag{4–55}$$

The moment about O due to external forces is

$$\mathbf{M} = \sum_{i=1}^{n} \mathbf{r}_i \times \mathbf{F}_i = \sum_{i=1}^{n} (\mathbf{r}_c + \boldsymbol{\rho}_i) \times \mathbf{F}_i$$

$$= \mathbf{r}_c \times \mathbf{F} + \sum_{i=1}^{n} \boldsymbol{\rho}_i \times \mathbf{F}_i \tag{4–56}$$

But $\mathbf{M} = \dot{\mathbf{H}}$, as stated in Eq. (4–49), and therefore we obtain from Eqs. (4–55) and (4–56) that

$$\mathbf{M}_c = \dot{\mathbf{H}}_c \tag{4–57}$$

where

$$\mathbf{M}_c = \sum_{i=1}^{n} \boldsymbol{\rho}_i \times \mathbf{F}_i \tag{4–58}$$

It is seen that $\mathbf{M}_c$ is the moment of the external forces about the center of mass.

Comparing Eqs. (4–49) and (4–57) we see that they are of identical form. So we can state that *if the chosen reference point is either (1) fixed in inertial space or (2) at the center of mass of the system, then the time rate of change of the angular momentum of a system of particles about the given reference point is equal to the moment about that point of the external forces acting on the system.* It will be shown that this equation can apply to certain other reference points as well, but these are associated with rather special cases.

The equation $\mathbf{M} = \dot{\mathbf{H}}$ implies equality of each of its components. So we can apply Eq. (3–148) to a system of particles.

$$M_x = \dot{H}_x$$
$$M_y = \dot{H}_y \tag{4–59}$$
$$M_z = \dot{H}_z$$

where the cartesian reference frame either is fixed or is translating with the center of mass. More generally, we can write

$$M_k = \dot{H}_k \tag{4–60}$$

where the components are taken in the direction of an arbitrary unit vector $\mathbf{e}_k$ that is fixed in space.

From Eq. (4–49) or Eq. (4–57) we see that, if the moment of the external forces about either a fixed point or about the center of mass is zero, then the angular momentum about the same point is conserved. This is the *principle of conservation of angular momentum* as it applies to a system of particles. Similarly, if a certain component M_k remains zero, then the corresponding angular momentum about a fixed axis through the reference point and parallel to $\mathbf{e}_k$ is conserved.

It can be seen that the use of the center of mass as a reference point has some interesting and important consequences. We note that the equation $\mathbf{M}_c = \dot{\mathbf{H}}_c$ does not involve in its detailed representation given by Eqs. (4–54) and (4–58) any terms involving $\mathbf{r}_c$ or its derivatives. Similarly, the equation $\mathbf{F} = m\ddot{\mathbf{r}}_c$ does not contain explicit terms in $\boldsymbol{\rho}_i$ or its derivatives. Therefore, if the total force $\mathbf{F}$ does not depend upon the relative positions of the particles, and if the moment $\mathbf{M}_c$ does not depend upon the motion of the center of mass, then the translational and rotational ($\mathbf{H}_c$) portions of the total motion are independent and can be calculated separately. Furthermore, if in addition *all* forces are independent of the motion of the center of mass, then the work and kinetic energy relationship of Eq. (4–21) for *any* motion relative to the center of mass can be evaluated separately from that of Eq. (4–9) for the motion of the center of mass.

Another interesting characteristic of the choice of a nonrotating reference frame which translates with the center of mass is that, in general, it is a *noninertial frame*. Hence, it can be seen that the rotational equation of motion, $\mathbf{M} = \dot{\mathbf{H}}$, is frequently written with respect to a noninertial reference. Being able to use the center of mass as a reference point for the rotational equations simplifies considerably the analysis of rigid-body motion, as we shall see in Chapters 7 and 8.

Arbitrary Reference Point. Consider the angular momentum of a system of particles relative to an arbitrary point P, that is, as viewed by a nonrotating observer who is moving with P. Let the xyz frame of Fig. 4–6 be fixed

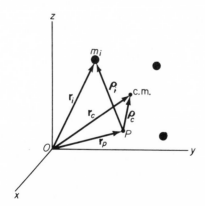

Fig. 4-6. Position vectors for a system of particles, using an arbitrary reference point.

in inertial space and let the point P move in an arbitrary manner relative to this frame. Noting that the vectors $\boldsymbol{\rho}_i$ and $\boldsymbol{\rho}_c$ originate at the reference point P, we see that

$$\mathbf{r}_i = \mathbf{r}_p + \boldsymbol{\rho}_i \tag{4-61}$$

$$\mathbf{r}_c = \mathbf{r}_p + \boldsymbol{\rho}_c \tag{4-62}$$

Also, the center of mass location relative to P is

$$\boldsymbol{\rho}_c = \frac{1}{m} \sum_{i=1}^{n} m_i \boldsymbol{\rho}_i \tag{4-63}$$

and the angular momentum about P is

$$\mathbf{H}_p = \sum_{i=1}^{n} \boldsymbol{\rho}_i \times m_i \dot{\boldsymbol{\rho}}_i \tag{4-64}$$

Equation (4–51) expresses the angular momentum about O in terms of its value relative to the center of mass. In a similar fashion, the angular momentum about the reference point P can be written in the form:

$$\mathbf{H}_p = \boldsymbol{\rho}_c \times m\dot{\boldsymbol{\rho}}_c + \mathbf{H}_c \tag{4-65}$$

From Eqs. (4–51) and (4–65), we obtain the following equation relating $\mathbf{H}$ and $\mathbf{H}_p$:

$$\mathbf{H} = \mathbf{H}_p - \boldsymbol{\rho}_c \times m\dot{\boldsymbol{\rho}}_c + \mathbf{r}_c \times m\dot{\mathbf{r}}_c \tag{4-66}$$

where we recall that $\mathbf{H}$ is the angular momentum with respect to the fixed point O. Differentiating with respect to time, we obtain

$$\dot{\mathbf{H}} = \dot{\mathbf{H}}_p - \boldsymbol{\rho}_c \times m\ddot{\boldsymbol{\rho}}_c + \mathbf{r}_c \times m\ddot{\mathbf{r}}_c$$

$$= \dot{\mathbf{H}}_p - \boldsymbol{\rho}_c \times m(\ddot{\mathbf{r}}_c - \ddot{\mathbf{r}}_p) + (\mathbf{r}_p + \boldsymbol{\rho}_c) \times m\ddot{\mathbf{r}}_c$$

Noting that $\mathbf{F} = m\ddot{\mathbf{r}}_c$ and collecting terms, we find that

$$\dot{\mathbf{H}} = \dot{\mathbf{H}}_p + \mathbf{r}_p \times \mathbf{F} + \boldsymbol{\rho}_c \times m\ddot{\mathbf{r}}_p \tag{4-67}$$

But the moment about O is

$$\mathbf{M} = \sum_{i=1}^{n} \mathbf{r}_i \times \mathbf{F}_i = \mathbf{r}_p \times \mathbf{F} + \sum_{i=1}^{n} \boldsymbol{\rho}_i \times \mathbf{F}_i$$

or, noting that

$$\mathbf{M}_p = \sum_{i=1}^{n} \boldsymbol{\rho}_i \times \mathbf{F}_i \tag{4-68}$$

we obtain

$$\mathbf{M} = \mathbf{r}_p \times \mathbf{F} + \mathbf{M}_p \tag{4-69}$$

Now, taking the origin O as the reference point, $\mathbf{M} = \dot{\mathbf{H}}$, and therefore from Eqs. (4–67) and (4–69), we see that

$$\mathbf{M}_p - \boldsymbol{\rho}_c \times m\ddot{\mathbf{r}}_p = \dot{\mathbf{H}}_p \tag{4-70}$$

A comparison of this result with Eqs. (4–49) and (4–57) reveals that the choice of an arbitrary reference point has resulted in the additional term $-\boldsymbol{\rho}_c \times m\ddot{\mathbf{r}}_p$. If P is fixed or moves with constant velocity, then $\ddot{\mathbf{r}}_p = 0$ and the term is zero. Also, if P is at the center of mass, then $\boldsymbol{\rho}_c = 0$ and the term is zero, in agreement with our earlier results. Note that if $\boldsymbol{\rho}_c$ and $\ddot{\mathbf{r}}_p$ are parallel, the term again disappears. This last case is not common but can be applied in certain situations, such as in the rolling motion of a wheel, the reference point being taken at the instantaneous center.

The physical interpretation of this term is that it is the moment about P of the inertial force $-m\ddot{\mathbf{r}}_p$ which arises because the nonrotating reference frame moving with P is not an inertial frame. This inertial force can be considered as the resultant of individual inertial forces $-m_i\ddot{\mathbf{r}}_p$ acting at each of the particles. Thus, we can use the standard form of the equation, $\mathbf{M} = \dot{\mathbf{H}}$, even for the case of an accelerating reference point, if we include the inertial forces due to $\ddot{\mathbf{r}}_p$, as well as the actual external forces, in calculating $\mathbf{M}$. This applies to all systems of particles and to a general motion of the reference point P, provided that the internal forces occur in equal, opposite, and collinear pairs.

In a similar fashion, the equation of motion $\mathbf{F} = m\ddot{\mathbf{r}}_c$ can be applied with respect to a nonrotating but accelerating frame if the inertial force $-m\ddot{\mathbf{r}}_p$ is included in $\mathbf{F}$ as an additional external force. Thus we are led to the following general rule: *All the results and principles derivable from Newton's laws of motion relative to an inertial frame can be extended to apply to an accelerating but nonrotating frame if the inertial forces associated with the acceleration of the frame are considered as additional external forces acting on the system.*

This rule applies to all calculations, including those of work, kinetic energy, linear or angular momentum, and so on. We see, for example, that the original introduction of inertial forces in connection with d'Alembert's principle (Sec. 1–5) can be viewed as the special case where the reference frame translates with the particle. Hence there is no motion relative to this frame and the particle is in static equilibrium, with the true external forces being balanced by the inertial force.

When should one choose an inertial frame and when should one choose a noninertial frame in solving for the motions of a system of particles? For the most part, it is advisable to consider dynamics problems from the viewpoint of an observer fixed in an inertial frame and to include only the true external forces acting on a particle or set of particles when applying Newton's equation of motion. On the other hand, certain problems are made simpler by adopting the viewpoint of an accelerating observer. An important example of the latter is the choice of the center of mass as a reference point for the analysis of the rotational aspects of the motion, as in Eq. (4–57).

The arbitrary reference point P is also very convenient for certain problems. One example is the case where the motion of a given point in the system is a known function of time. Choosing this point as the reference point

P, one can calculate immediately the inertial forces $-m_i \ddot{\mathbf{r}}_p$. Furthermore, the external force acting at P has no moment about P and therefore it need not be determined in solving for the rotational motion. Another example is in the formulation of the dynamical equations for a space vehicle with moving parts. It is convenient to choose a reference point that is fixed in one member of the system and to specify the location of the various parts relative to this reference, as they would appear to an observer riding with the vehicle at the reference point. As a simple illustration, suppose we wish to calculate the total kinetic energy of a system of particles, using the arbitrary point P as a reference. Referring to Fig. 4–6, it can be seen that the kinetic energy is

$$T = \frac{1}{2} \sum_{i=1}^{n} m_i \dot{\mathbf{r}}_i \cdot \dot{\mathbf{r}}_i = \frac{1}{2} \sum_{i=1}^{n} m_i (\dot{\mathbf{r}}_p + \dot{\boldsymbol{\rho}}_i) \cdot (\dot{\mathbf{r}}_p + \dot{\boldsymbol{\rho}}_i)$$

$$= \frac{1}{2} \sum_{i=1}^{n} m_i \dot{\mathbf{r}}_p^2 + \frac{1}{2} \sum_{i=1}^{n} m_i \dot{\boldsymbol{\rho}}_i^2 + \sum_{i=1}^{n} m_i \dot{\boldsymbol{\rho}}_i \cdot \dot{\mathbf{r}}_p$$

which can be simplified further using Eqs. (4–5) and (4–63) to yield

$$T = \frac{1}{2} m \dot{\mathbf{r}}_p^2 + \frac{1}{2} \sum_{i=1}^{n} m_i \dot{\boldsymbol{\rho}}_i^2 + m \dot{\mathbf{r}}_p \cdot \dot{\boldsymbol{\rho}}_c \tag{4–71}$$

Thus we have the *theorem of König:* The total kinetic energy of a system is equal to the sum of (1) the kinetic energy due to a particle of total mass m moving with the reference point P, (2) the kinetic energy due to the motion of the system relative to P, (3) the scalar product of the linear momentum vector of a particle of mass m moving with P, and the velocity vector of the center of mass motion relative to P. For the case where P coincides with the center of mass, $\boldsymbol{\rho}_c = 0$ and the expression for the total kinetic energy reduces to that of Eq. (4–19).

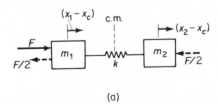

(a)

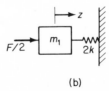

(b)

Fig. 4-7. A mass-spring system as viewed from a reference frame translating with the center of mass.

Example 4-3. Let us solve for the motion of mass m_1 relative to the center of mass in the system of Example 4-2, using a noninertial reference frame which is translating with the center of mass.

First recall that the acceleration of the center of mass is equal to the total external force divided by the total mass.

$$\ddot{x}_c = \frac{F}{2m}$$

Therefore the inertial force acting on each mass due to the acceleration of the reference frame is $-m\ddot{x}_c = -\frac{1}{2}F$, as shown in Fig. 4-7(a). Adding the external and the inertial forces on each mass, we see that a symmetrical situation results with respect to the center of mass position, that is, each mass has a resultant force of magnitude $\frac{1}{2}F$ applied to it and both forces are directed toward the center of mass. Therefore, since the physical system and also the forces exhibit a symmetry about the center of mass, we need to analyze only one-half of the system, as shown in Fig. 4-7(b). Letting $z = x_1 - x_c$, we can immediately write the equation of motion

$$m\ddot{z} + 2kz = \frac{1}{2}F$$

corresponding to a simple mass-spring system with spring stiffness $2k$ and with a constant force of magnitude $\frac{1}{2}F$ applied at $t = 0$. Using the results previously derived in Eq. (3-190) for a unit step response, we can solve for z, obtaining

$$z = \frac{F}{4k}\left(1 - \cos\sqrt{\frac{2k}{m}}\,t\right)$$

which is identical to the solution for $(x_1 - x_c)$ given in Example 4-2.

Because we have chosen the center of mass as the reference, and also in accordance with the symmetry about this point, we find that

$$x_1 - x_c = x_c - x_2$$

in agreement with Eq. (4-31). Hence, the solution for x_2 can be obtained directly from the solutions for x_1 and x_c.

Example 4-4. A particle of mass m_1 moves on a frictionless, horizontal plane. It is connected to a second particle of mass m_2 by an inextensible string which passes through a small hole at point O, as shown in Fig. 4-8. If the second mass moves only along a vertical line through O, and if the initial conditions are $r(0) = r_0$, $\dot{r}(0) = 0$, $\dot{\theta}(0) = \omega_0$, find the minimum value of r and the maximum force in the string during the ensuing motion. Assume $m_1 = m_2 = m$ and $r_0\omega_0^2 = g/3$.

This system is conservative, so let us first find an expression for the total energy. The velocity of mass m_1 is given by $v_1^2 = \dot{r}^2 + r^2\dot{\theta}^2$ and the velocity of m_2 is $v_2 = \dot{r}$. If we measure the potential energy of m_2 from its value when

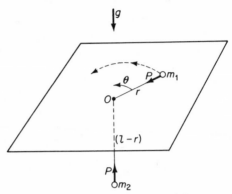

Fig. 4-8. A system consisting of two particles connected by an inextensible string, one particle moving on a horizontal plane and the other along a vertical axis.

$r = 0$, we can write the total energy in the form:

$$E = T + V = m\dot{r}^2 + \frac{1}{2}mr^2\dot{\theta}^2 + mgr \qquad (4\text{-}72)$$

But, from the initial conditions, we see that

$$E = \frac{1}{2}mr_0^2\omega_0^2 + mgr_0 \qquad (4\text{-}73)$$

Also, the angular momentum of m_1 about O is constant since the total moment **M** about O is zero. This results because the total external force acting on m_1 arises from the string tension, and its line of action passes through O. Hence, using the principle of conservation of angular momentum, we can write

$$H = mr^2\dot{\theta} = mr_0^2\omega_0$$

from which we obtain

$$\dot{\theta} = \left(\frac{r_0}{r}\right)^2 \omega_0 \qquad (4\text{-}74)$$

Equating the right-hand sides of Eqs. (4–72) and (4–73) and substituting for $\dot{\theta}$ from Eq. (4–74), we obtain

$$m\dot{r}^2 + m\frac{r_0^4\omega_0^2}{2r^2} + mgr = \frac{1}{2}mr_0^2\omega_0^2 + mgr_0 \qquad (4\text{-}75)$$

To find the minimum value of r, we set $\dot{r} = 0$, and after collecting and rearranging terms, the roots are found from

$$r^3 - \left(r_0 + \frac{r_0^2\omega_0^2}{2g}\right)r^2 + \frac{r_0^4\omega_0^2}{2g} = 0$$

Dividing out the known factor $(r - r_0)$, since it is given that $\dot{r} = 0$ and $r = r_0$ initially, we obtain

$$r^2 - \frac{\omega_0^2 r_0^2}{2g} r - \frac{\omega_0^2 r_0^3}{2g} = 0$$

which has a single positive root, corresponding to r_{min} in this case.

$$r_{min} = \frac{\omega_0^2 r_0^2}{4g} \left(1 + \sqrt{1 + \frac{8g}{\omega_0^2 r_0}} \right) = \frac{1}{2} r_0$$

If the initial values of r and $\dot{\theta}$ are such that $r_0 \omega_0^2 > g$, the preceding general expression is valid but corresponds to r_{max}, in which case $r_{min} = r_0$.

To find the maximum force in the string, let us differentiate Eq. (4–75) with respect to time and solve for $\ddot{r}$. The result is

$$\ddot{r} = \frac{r_0^4 \omega_0^2}{2r^3} - \frac{1}{2} g$$

Now, the vertical acceleration of m_2 is $\ddot{r}$, and therefore from Newton's law of motion, we obtain

$$P - mg = m\ddot{r} \quad \text{or} \quad P = \frac{1}{2} mg + \frac{mr_0^4 \omega_0^2}{2r^3}$$

It is clear that P is a maximum for a minimum length r. Therefore, substituting $r = \frac{1}{2}r_0$ into the foregoing expression, we obtain

$$P_{max} = \frac{11}{6} mg$$

4–6. ANGULAR IMPULSE

An equation relating the change in the angular momentum to the total angular impulse can be derived for the case of a system of particles in a manner similar to that used in obtaining Eq. (3–154) for a single particle. We have seen from Eqs. (4–49) and (4–57) that

$$\mathbf{M} = \dot{\mathbf{H}}$$

applies to a system of particles when the reference point is fixed or is at the center of mass. Integration of this equation with respect to time over the interval t_1 to t_2 results in the *principle of angular impulse and momentum:*

$$\mathcal{M} = \mathbf{H}_2 - \mathbf{H}_1 \tag{4–76}$$

where the total angular impulse acting on the system due to external forces is

$$\mathcal{M} = \int_{t_1}^{t_2} \mathbf{M} \, dt \tag{4–77}$$

As we have seen, the internal forces occur in equal, opposite, and collinear pairs, and therefore they cancel out in this calculation.

If we perform a similar integration on Eq. (4–70), we obtain

$$\mathbf{H}_{p2} - \mathbf{H}_{p1} = \mathscr{M}_p - \int_{t_1}^{t_2} \boldsymbol{\rho}_c \times m\ddot{\mathbf{r}}_p \, dt \qquad (4\text{–}78)$$

The integral on the right can be interpreted as the angular impulse about P due to the inertial forces arising from the acceleration of the reference frame translating with P.

4–7. COLLISIONS

One of the important types of interactions between two bodies occurs during a collision. In order to be able to treat the colliding bodies as particles, we shall consider the special case of impact between smooth spheres. Nevertheless, the analysis of this case will involve many of the important concepts used in the solution of more complex problems.

Consider, then, the case of two smooth spheres whose centers of mass coincide with their geometrical centers. At the moment of impact, the interaction forces are normal to the common tangent plane at the first point of contact, and therefore the line of action of these forces is coincident with the line of centers. Hence no rotational motion is introduced and the problem reduces to one of particle motion. Furthermore, let us assume that the spheres move in the same plane before impact, that is, the velocity vectors and the line of centers at impact all lie in the same plane. As shown in Fig. 4–9, v_n represents a velocity component along the line of centers just before impact, and v_t represents a velocity component perpendicular to the line of centers, the velocities being relative to an inertial frame.

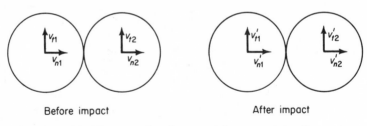

Before impact   After impact

Fig. 4-9. The normal and tangential components of velocity of colliding spheres.

Let us assume that the forces between the colliding spheres at the moment of impact are impulsive in nature. In other words, the forces are very large compared to other forces acting at the same time and are of very short duration. Thus they are best described by specifying their total impulse rather than by making any attempt to give their variation with time. So let $\mathscr{F}$ be the total impulse at the moment of impact in this case. It acts to the left on m_1 and to the right on m_2, in each case being normal to the common tangent plane at the point of contact. Using the equation of linear impulse and momentum on each particle, we see that

$$\mathscr{F} = m_1(v_{n1} - v'_{n1}) = m_2(v'_{n2} - v_{n2})$$

or

$$m_1 v_{n1} + m_2 v_{n2} = m_1 v'_{n1} + m_2 v'_{n2} \tag{4-79}$$

where the primes refer to velocities after impact. There is no impulse on the spheres in the tangential direction, and therefore the velocity components in this direction are unchanged.

$$v_{t1} = v'_{t1}$$
$$v_{t2} = v'_{t2} \tag{4-80}$$

Hence we obtain that

$$m_1 v_{t1} + m_2 v_{t2} = m_1 v'_{t1} + m_2 v'_{t2} \tag{4-81}$$

We see from Eqs. (4–79) and (4–81) that the normal and tangential components of the linear momentum are conserved during impact and therefore the total linear momentum of the system is conserved. This result is to be expected, since there are no external forces acting on the system.

For the case where other forces are acting on the system, Eqs. (4–79), (4–80), and (4–81) still apply, provided that these other forces are not impulsive forces applied at the instant of impact. The other forces can be neglected because their total impulse during the infinitesimal interval of impact is itself infinitesimal. Of course, they cannot be neglected, in general, during the longer time interval before and after impact.

In order to solve for the motion of the system, we must solve for the two tangential and the two normal components of the velocity after impact. The tangential components are given by Eq. (4–80). Another equation, however, is required in addition to Eq. (4–79) in order to be able to solve for both v'_{n1} and v'_{n2}. This additional equation relates the normal components of the relative velocities of approach and separation. It is

$$v'_{n2} - v'_{n1} = e(v_{n1} - v_{n2}) \quad (0 \le e \le 1) \tag{4-82}$$

where e is known as the *coefficient of restitution*. The coefficient of restitution depends upon the composition of the bodies and also, to some extent, upon their shape, size, and impact velocities. For our purposes, we shall consider the coefficient of restitution to be a constant for a given pair of substances.

Solving for v'_{n1} and v'_{n2} from Eqs. (4–79) and (4–82), we obtain

$$v'_{n1} = \frac{m_1 - em_2}{m_1 + m_2} v_{n1} + \frac{(1 + e)m_2}{m_1 + m_2} v_{n2}$$

$$v'_{n2} = \frac{(1 + e)m_1}{m_1 + m_2} v_{n1} + \frac{m_2 - em_1}{m_1 + m_2} v_{n2} \tag{4-83}$$

If $e = 0$, the normal velocities after impact are equal and the collision is said to be *inelastic*. If $e = 1$, then, in accordance with Eq. (4–82), the relative velocities of approach and separation have the same magnitude. This case is known as *perfectly elastic* impact.

We have seen that the total linear momentum is conserved during impact. For all cases except perfectly elastic collisions, however, there is a loss of kinetic energy. This loss in the total mechanical energy appears as heat resulting from plastic deformations or as internal vibrations that are not included in the analysis and are finally dissipated as heat due to internal friction losses.

It is interesting to view the collision process from the standpoint of an inertial observer translating with the center of mass. We assume that no external forces are acting on the system of two colliding spheres and therefore the center of mass moves in a straight line at constant speed throughout the collision. Because we have taken the center of mass as the reference, we find that

$$v_{n1} = -\frac{m_2}{m_1} v_{n2}$$

$$v_{t1} = -\frac{m_2}{m_1} v_{t2}$$
(4–84)

Then, substituting Eq. (4–84) into Eq. (4–83), we obtain

$$v'_{n1} = -e v_{n1}$$

$$v'_{n2} = -e v_{n2}$$
(4–85)

Of course, the tangential components are unchanged in accordance with Eq. (4–80).

From this point of view, it is clear that, for perfectly elastic collisions ($e = 1$), the normal velocity components are reversed in direction but unchanged in magnitude. Therefore, there is no loss of kinetic energy associated with motion relative to the center of mass. Furthermore, there is no change in the velocity of the center of mass; hence the total kinetic energy relative to a fixed system is also unchanged by the collision, as can be seen by referring to Eq. (4–19).

On the other hand, for the case of inelastic impact ($e = 0$), we obtain $v'_{n1} = v'_{n2} = 0$. Therefore, we see that the kinetic energy due to normal velocity components relative to the center of mass is entirely lost as a result of the collision.

In general, we can write the total kinetic energy of two colliding spheres in the form

$$T = T_c + T_t + T_n$$
(4–86)

where T_c is the kinetic energy due to the motion of the center of mass, T_t is the kinetic energy due to the tangential components of the velocity relative to the center of mass, and T_n is the kinetic energy due to the normal components of the velocity relative to the center of mass. Using primes to indicate the corresponding quantities after the collision, we find that

$$T'_c = T_c$$
$$T'_t = T_t \tag{4-87}$$
$$T'_n = e^2 T_n$$

Another approach which gives a better insight into the physical nature of the collision process is obtained by considering the impact to occur in two phases. First is the *compression* phase, during which elastic deformations of the spheres occur and the relative normal velocity is reduced to zero. At this point, the second or *restitution* phase begins and lasts until the spheres separate. Now, a portion of the total collision impulse $\mathscr{F}$ which is applied to a given sphere occurs during the compression phase and the remainder occurs during the restitution. Calling these impulses $\mathscr{F}_c$ and $\mathscr{F}_r$, respectively, we see that

$$\mathscr{F} = \mathscr{F}_c + \mathscr{F}_r \tag{4-88}$$

All impulses on a given sphere occur in the same direction (normal to the common tangent plane), so we need to consider only their magnitudes. Furthermore, since the change in the velocity of a given mass is directly proportional to the applied impulse, we conclude from Eq. (4-85) that

$$\mathscr{F}_r = e\mathscr{F}_c \tag{4-89}$$

Thus we could equally well define the coefficient of restitution as the ratio of the restitution impulse to the compression impulse.

We assumed originally that both spheres move in the same plane before impact. We saw from the results of the analysis that no forces are produced during the collision which would cause the motion to deviate from that plane. Hence the entire motion takes place within the given plane. But upon further study, it becomes apparent that the original requirement was not necessary since an inertial reference can always be found such that the initial velocities and the line of centers at impact all lie in the same plane. An example of such a reference frame is one which translates with the center of mass. Another example is a reference frame which translates with the initial velocity of one of the spheres. Thus the results which we have obtained are valid for the general case of colliding spheres if we choose a proper reference frame from which to view the motion.

Example 4-5. A ball is dropped from a height of 10 ft onto a level floor. If the coefficient of restitution $e = 0.9$, how long will it take the ball to come to rest? What is the total distance traveled by the ball?

Let us use the notation that v_n is the speed with which the ball hits the floor on the nth bounce. The floor is assumed to have zero velocity at all times; therefore, from Eq. (4-82),

$$v_{n+1} = ev_n$$

since the speed with which the ball hits the floor on a given bounce is just the speed with which it rebounded from the previous bounce. The time interval between the nth and $(n + 1)$st bounce is

$$t_n = \frac{2v_{n+1}}{g} = \frac{2ev_n}{g}$$

since it is twice the time required for the ball to stop momentarily at the peak height after rebounding with a speed ev_n. Now, the time required to hit the floor the first time after being dropped from a height h is, from Eq. (3–9),

$$t_0 = \sqrt{\frac{2h}{g}}$$

and the speed at the first impact is

$$v_1 = gt_0 = \sqrt{2gh}$$

Thus the total time required for the ball to come to rest is

$$
\begin{aligned}
t &= t_0 + t_1 + t_2 + \cdots \\
&= t_0(1 + 2e + 2e^2 + \cdots) \\
&= \sqrt{\frac{2h}{g}} \left(\frac{1 + e}{1 - e}\right) = 15.0 \text{ sec}
\end{aligned}
$$

The distance traveled between the nth and $(n + 1)$st bounce is $2h_n$, where h_n is the rebound height after the nth bounce.

$$h_n = \frac{v_{n+1}^2}{2g} = \frac{e^2 v_n^2}{2g}$$

So we see that the total distance traveled along the path is

$$
\begin{aligned}
s &= h + 2h_1 + 2h_2 + \cdots \\
&= h(1 + 2e^2 + 2e^4 + \cdots) \\
&= h\left(\frac{1 + e^2}{1 - e^2}\right) = 95.3 \text{ ft}
\end{aligned}
$$

Note that the number of bounces is, in theory, infinite, even though the total time and distance are finite.

Example 4–6. Mass m_1, moving along the x axis with velocity v, hits m_2 and sticks to it (Fig. 4–10). If all three particles are of equal mass m, and if m_2 and m_3 are connected by a rigid, massless rod, as shown, find the motion of the particles after impact. All particles can move without friction on the horizontal xy plane.

First, we notice that the total linear momentum of the system of three particles is conserved, since no external forces act on the system. Next, we

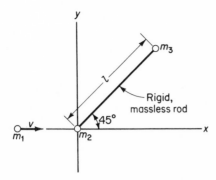

Fig. 4-10. Impact of a partially constrained system.

see that the rod can transmit axial forces only. It is impossible for bending moments to occur at either end of the rod because a moment cannot be applied to a point mass. This follows, because a point mass has no means of resisting with an equal and opposite inertial moment. The lack of bending moments at either end requires that no shear forces or bending moments occur in the rod.

Now consider the inertial properties of the system consisting of masses m_2 and m_3 and the rod. Suppose a unit *axial* impulse is applied at m_2 in a direction along the rod. In this case, masses m_2 and m_3 move in the axial direction with velocities equal to $1/2m$, as though the impulse were applied to a single particle of mass $2m$. Next consider a unit *transverse* impulse applied at m_2. In this case, the rod exerts no forces at the instant of the impulse, and mass m_2 moves with a transverse velocity $1/m$.

We can apply these observations by writing expressions for the axial and transverse components of the linear momentum. We note that the velocity component along the rod after impact is the same for all three masses. Calling this velocity v'_a, we can write the equation for conservation of the axial component of linear momentum, obtaining

$$\frac{mv}{\sqrt{2}} = 3mv'_a$$

or

$$v'_a = \frac{v}{3\sqrt{2}}$$

Similarly, we find that the equation of momentum conservation in the transverse direction is

$$\frac{mv}{\sqrt{2}} = 2mv'_t$$

or

$$v'_t = \frac{v}{2\sqrt{2}}$$

where v'_t is the transverse component of the velocity of m_1 and m_2 immediately after impact. The velocity of m_3 just after impact is entirely axial, since it receives an impulse in this direction through the rod.

The velocities of $(m_1 + m_2)$ and m_3 immediately after impact are as shown in Fig. 4–11. Throughout the whole problem, that is, before, during, and

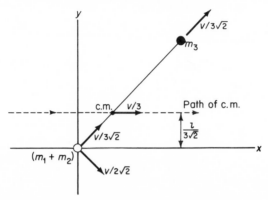

Fig. 4-11. Velocity components immediately after impact.

after impact, the center of mass of the system moves with a constant velocity $v/3$ in the direction of the positive x axis along a line $y = l/3\sqrt{2}$. After impact, the system rotates at a constant angular rate ω which is found by dividing the difference in the transverse components of velocity at the two ends by the rod length l. In this case,

$$\omega = \frac{v}{2\sqrt{2}\,l}$$

Note that the total angular momentum about the center of mass is constant throughout the problem. Its magnitude is

$$H = \frac{mvl}{3\sqrt{2}}$$

Example 4–7. Solve Example 4–6 for the case of perfectly elastic impact between m_1 and m_2.

We assume that the spheres m_1 and m_2 are perfectly smooth. Therefore the impulse occurring during impact must be normal to the tangent plane at the point of contact, that is, it is along the x axis. If we designate the magnitude of this impulse by $\mathscr{F}$, we can solve for the velocities of the three masses immediately after impact in terms of $\mathscr{F}$. Using the equation of linear impulse and momentum, Eq. (4–41), we see that the velocity of m_1 after

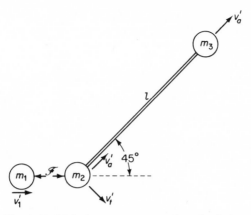

Fig. 4-12. Velocity components after elastic impact.

impact is

$$v'_1 = v - \frac{\mathscr{F}}{m}$$

as shown in Fig. 4–12. Taking axial and transverse components of the total impulse $\mathscr{F}$ which is acting on $(m_2 + m_3)$, we find that the axial component of velocity for m_2 and m_3 is

$$v'_a = \frac{\mathscr{F}}{2\sqrt{2}\ m}$$

since the effective mass resisting this impulse is $2m$. In a similar manner, we find that the transverse velocity of m_2 is

$$v'_t = \frac{\mathscr{F}}{\sqrt{2}\ m}$$

since the effective mass in this case is m.

Now we use Eq. (4–82), the defining equation for the coefficient of restitution in terms of normal velocity components. Setting $e = 1$, corresponding to perfectly elastic impact, we obtain

$$\frac{v'_a}{\sqrt{2}} + \frac{v'_t}{\sqrt{2}} - v'_1 = v$$

or

$$\left(\frac{1}{4} + \frac{1}{2} + 1 \right) \frac{\mathscr{F}}{m} = 2v$$

from which we obtain

$$\mathscr{F} = \frac{8}{7}\ mv$$

Therefore,

$$v'_1 = -\frac{1}{7}\ v, \qquad v'_a = \frac{4}{7\sqrt{2}}\ v, \qquad v'_t = \frac{8}{7\sqrt{2}}\ v$$

Let us calculate the total kinetic energy after impact. Designating the kinetic energy of masses m_1, m_2, and m_3 by T_1', T_2', and T_3', respectively, we find that

$$T_1' = \frac{1}{2} m v_1'^2 = \frac{mv^2}{98}$$

$$T_2' = \frac{1}{2} m(v_a'^2 + v_t'^2) = \frac{20mv^2}{49}$$

$$T_3' = \frac{1}{2} m v_a'^2 = \frac{4mv^2}{49}$$

giving

$$T' = T_1' + T_2' + T_3' = \frac{1}{2} mv^2$$

We see that the total kinetic energy T' is equal to the original kinetic energy, thereby illustrating once again that energy is conserved in a perfectly elastic collision.

An interesting aspect of this example is the calculation of the effective mass of the system composed of m_2 and m_3, insofar as it affects m_1 during the collision. We have seen that the effective mass for an axial collision is $2m$, whereas the effective mass for a transverse collision is m. Hence we might expect the effective mass in this case to lie somewhere between these values. Using Eq. (4-83) and letting $e = 1$, we find that

$$v_1' = \frac{m - m_{\text{eff}}}{m + m_{\text{eff}}} v = -\frac{v}{7}$$

from which we obtain

$$m_{\text{eff}} = \frac{4}{3} m$$

for this case in which the impact occurs at 45° from the axis of the rod.

More generally, we can think of a unit impulse being applied to m_2, and noting that the effective mass is inversely proportional to the velocity change in the direction of the impulse. Now, the velocity changes in the direction of the impulse due to the axial and transverse components of the impulse are additive. Therefore, assuming that the impulse occurs at an angle α measured from the axial direction, we find that

$$\frac{1}{m_{\text{eff}}} = \frac{\cos^2 \alpha}{m_a} + \frac{\sin^2 \alpha}{m_t} \tag{4-90}$$

where m_a and m_t are the effective masses for axial and transverse impulses, respectively. The factors $\cos^2 \alpha$ and $\sin^2 \alpha$ occur because first the axial or transverse components of the impulse are used, and then components of the corresponding velocities must be taken in the direction of the impulse. In this particular case where $\alpha = 45°$, we obtain

$$\frac{1}{m_{\text{eff}}} = \frac{\frac{1}{2}}{2m} + \frac{\frac{1}{2}}{m} = \frac{3}{4m}$$

in agreement with our previous results.

It should be emphasized again that the effective mass of the system composed of m_2 and m_3 refers to its effect on m_1 during a collision. In this respect, it (or the reciprocal, $1/m_{\text{eff}}$) resembles an input impedance of electrical circuit theory. Note that the actual mass $2m$ is used in calculations of the change in the velocity of the center of mass of m_2 and m_3.

4-8. THE ROCKET PROBLEM

Particle Mechanics Approach. An important application of the dynamical principles which we have obtained for a system of particles is in the analysis of rocket propulsion, that is, propulsion by means of reaction forces due to the ejection of mass. Of particular importance in this analysis are two equations: the equation of motion and the equation of linear impulse and momentum. From Eqs. (4–3) and (4–4), we can write the equation of motion in the form

$$\mathbf{F} = \sum_{i=1}^{n} \mathbf{F}_i = \sum_{i=1}^{n} m_i \dot{\mathbf{v}}_i \tag{4-91}$$

Also, from Eqs. (4–38), (4–39), and (4–41), we see that the equation of linear impulse and momentum can be written in the form

$$\mathscr{F} = \int_{t_1}^{t_2} \mathbf{F}\, dt = \sum_{i=1}^{n} m_i \mathbf{v}_i \Big|_{t_1}^{t_2} \tag{4-92}$$

where it is understood that the system must consist of the same set of particles throughout the interval from t_1 to t_2.

As a simplified example of the rocket problem, consider the system shown in Fig. 4–13. Assume that the rocket is operating in a vacuum in the absence of gravitational forces. The area of the nozzle exit is A_e and the average pressure of the exhaust gases at this area is p_e. The average exit velocity of the exhaust gases *relative to the rocket* is v_e. Assume also that the mass of the unburned fuel plus the rocket structure is given by

$$m = m_0 - bt \tag{4-93}$$

where b is the rate at which fuel is burned and ejected from the rocket. The burning rate b is normally assumed to be constant during the burning interval, but, in general,

$$b = \rho_e A_e v_e \tag{4-94}$$

where ρ_e is the average density of the discharged gases at the exit area A_e.

Let us take as the system of particles under consideration the total mass m of the rocket at time t. We think of the distributed mass as a large number

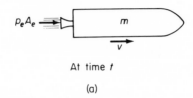

At time t

(a)

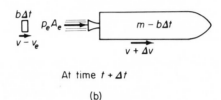

At time $t + \Delta t$

(b)

Fig. 4-13. The rocket system at successive instants of time.

of discrete particles. As the fuel burns, some of these particles are ejected through the nozzle. Although the mass of the unburned fuel is decreasing, we assume that this is due entirely to the ejection of mass, the masses of the individual particles remaining constant throughout the burning process. Thus, when we consider the same system again after an interval Δt, we must include the mass $b \, \Delta t$ which has been ejected during that interval.

Now let us use the equation of impulse and momentum to calculate the change in the rocket velocity. The total impulse acting to the right on the system during the interval Δt is of magnitude

$$\mathscr{F} = p_e A_e \, \Delta t \tag{4-95}$$

since the only external force on the system is due to the pressure p_e acting at the exit area. (Although the rocket is moving in a vacuum, the pressure p_e exists because the exhaust gases are not fully expanded at the exit of the nozzle.) The total momentum of the system at time t is mv. On the other hand, the total momentum at time $t + \Delta t$ is

$$(m - b \, \Delta t)(v + \Delta v) + b \, \Delta t(v - v_e)$$

where the positive direction is to the right. The first term in this last expression is the momentum of the rocket structure and unburned fuel and the second term is the momentum of the mass $b \, \Delta t$ which has been ejected during the interval Δt and which has an absolute velocity $v - v_e$. Therefore, equating the total impulse $\mathscr{F}$ to the change in the total linear momentum in that direction, we obtain

$$p_e A_e \, \Delta t = (m - b \, \Delta t)(v + \Delta v) + b \, \Delta t(v - v_e) - mv$$

which simplifies to

$$p_e A_e \, \Delta t = m \, \Delta v - b v_e \, \Delta t - b \, \Delta v \, \Delta t \qquad (4\text{-}96)$$

Dividing by Δt and taking the limit as Δt approaches zero, we find that

$$m\dot{v} = p_e A_e + b v_e \qquad (4\text{-}97)$$

where the last term of Eq. (4–96) has been neglected because Δv approaches zero as Δt approaches zero.

Equation (4–97) can also be written in the form

$$F_s = m\dot{v} \qquad (4\text{-}98)$$

where F_s is known as the *static thrust* of the rocket.

$$F_s = p_e A_e + b v_e \qquad (4\text{-}99)$$

To see that this is actually the static thrust, suppose that the rocket of Fig. 4–13 is held fixed by a test stand, as shown in Fig. 4–14. The static thrust F_s is the force that is transmitted by the test stand to the earth and is also the force on the rocket that is required to keep it stationary. In this case, the

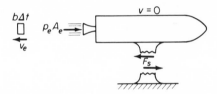

Fig. 4-14. The forces acting on a rocket in a test stand.

total external force acting on the system is $(F_s - p_e A_e)$ in a rearward direction. The rocket itself has no momentum; the only momentum is that of the exhaust gases. So, if we use the principle of linear impulse and momentum, we obtain

$$(F_s - p_e A_e) \, \Delta t = b v_e \, \Delta t$$

or

$$F_s = p_e A_e + b v_e$$

in agreement with Eq. (4–99). Note that we have again been careful to include the same particles in the system at the beginning and at the end of the interval Δt.

In our discussion of the rocket problem, we have assumed that the total momentum of the gases inside the rocket is constant, as viewed by an observer traveling with the rocket. In other words, we have assumed a situation of steady internal flow relative to the rocket. Hence we have represented changes in the linear momentum of the total mass contained within the rocket as changes in the product mv. Although this assumption is not strictly true, it is accurate enough for most purposes. Also, we have previously considered that the rocket is moving in a vacuum. In case it is moving in the atmosphere, then p_e can be viewed as the average gauge pressure at the exit. Any additional external forces due to aerodynamic drag, gravity, and so on, can be included in the analysis by adding the corresponding force components to the right side of Eq. (4–97).

Returning now to a consideration of Eq. (4–98), it can be seen that the static thrust F_s is the total force acting on the mass of the rocket structure and unburned fuel for the original case of free flight in a vacuum. This total force F_s consists of the exit pressure force $p_e A_e$ and the jet reaction force bv_e. Thus we see that the total force is equal to the product of the instantaneous mass and the corresponding absolute acceleration, even for the case of a system whose mass is changing. Hence Newton's equation of motion

$$\mathbf{F} = m\mathbf{a}$$

applies *instantaneously* to a group of particles of total mass m whose center of mass has an absolute acceleration $\mathbf{a}$. This is true even though the system consists of a slightly different set of particles at successive instants of time due, in this particular case, to the expulsion of mass from the rocket. The principal precaution to be taken here is to remember that $\mathbf{a}$ *is the acceleration of the center of mass of the particles composing the system at the given instant*. It is not necessarily equal to $\ddot{\mathbf{r}}_c$, since the position vector $\mathbf{r}_c$ may change

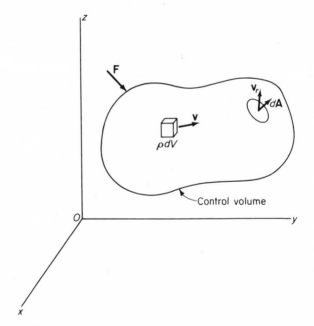

Fig. 4-15. A general control volume showing a mass element ρdV and a surface element $d\mathbf{A}$.

because different particles compose the system at different times. For example, the center of mass of a rocket on a test stand may move without the rocket itself moving. More generally, we can use Eq. (4–91), remembering that it applies instantaneously to any system so long as the summations of external forces and momenta are over the same set of particles.

It should be noted in passing that the equation

$$F = \frac{d}{dt}(m\mathbf{v})$$

applies to systems whose particles have a changing mass due to relativistic effects, but does not apply to systems whose total mass is changing because particles are continuously entering or leaving the system.

Control Volume Approach. Another approach to the derivation of the rocket acceleration equation is provided by an analysis of the material within a *control volume* bounded by a closed mathematical surface (Fig. 4–15). The control volume may move or change its shape and, in general, there is a mass flow through its surface. In this rather general approach, the problem is to express the time rate of change of the linear momentum of the material within the control volume in terms of the external forces acting on it.

As the system under consideration, let us take the material within the control volume at time t. The total linear momentum of this system is

$$\mathbf{p} = \int_V \rho \mathbf{v}\, dV \qquad (4\text{-}100)$$

where ρ and $\mathbf{v}$ are the mass density and the absolute velocity, respectively, of the material within the volume element dV. The integral is taken over the entire control volume V.

Now, if we evaluate the same integral at time $t + \Delta t$, we find that a slightly different set of particles is within the control volume. So, in order to follow the original set of particles, we calculate the momentum of all particles within the control volume at $t + \Delta t$, and then we must add the momentum of all the "native" particles that left the control volume in the infinitesimal interval Δt, and subtract the momentum of all the "foreign" particles that entered during the same interval. Therefore, at time $t + \Delta t$, the momentum of the original set of particles is

$$\left[\int_V \rho \mathbf{v}\, dV\right]_{t+\Delta t} + \Delta t \int_A \rho \mathbf{v}(\mathbf{v}_r \cdot d\mathbf{A})$$

where the second integral accounts for the momentum of the particles entering or leaving the control volume in the interval Δt. This integral is over the entire surface of the control volume, $d\mathbf{A}$ being a surface element whose orientation is specified by the outward-pointing normal vector. The vector $\mathbf{v}_r$ is the velocity, *relative to the surface*, of the particles that are entering or leaving. Thus $\rho(\mathbf{v}_r \cdot d\mathbf{A})\, \Delta t$ represents the mass crossing the surface element

$d\mathbf{A}$ in the interval Δt, with a positive $\mathbf{v}_r \cdot d\mathbf{A}$ referring to leaving native particles and a negative $\mathbf{v}_r \cdot d\mathbf{A}$, referring to entering foreign particles. Multiplying each mass element crossing the boundary surface by its absolute velocity $\mathbf{v}$ and integrating over all elements, we obtain a correction term which can be interpreted as the net momentum outflow from the control volume during the interval Δt. Changes in $\mathbf{v}$ or $\mathbf{v}_r$ during this interval can be neglected.

Thus we see that the change in the linear momentum of the original system of particles in the interval Δt is given by

$$\Delta \mathbf{p} = \left[\int_V \rho \mathbf{v} \, dV \right]_{t+\Delta t} - \left[\int_V \rho \mathbf{v} \, dV \right]_t + \Delta t \int_A \rho \mathbf{v}(\mathbf{v}_r \cdot d\mathbf{A})$$

Dividing this expression by Δt and taking the limit as Δt approaches zero, we obtain the time rate of change of the total linear momentum of the particles included within the control volume at the given time t. This result, by Newton's law of motion, must equal the total external force $\mathbf{F}$ applied to the mass within the control volume. Therefore we see that

$$\mathbf{F} = \frac{d}{dt} \int_V \rho \mathbf{v} \, dV + \int_A \rho \mathbf{v}(\mathbf{v}_r \cdot d\mathbf{A}) \qquad (4\text{-}101)$$

The volume integral may be interpreted from the Eulerian viewpoint of considering $\mathbf{v}$ (or $\rho \mathbf{v}$) as a vector field which is a function of position and time. The integration is taken over the control volume at each instant without following the path of any particular particle as time proceeds. Thus the integral represents the instantaneous value of the total linear momentum of all material within the control volume. This momentum may change because of changes in the flow field or changes in the control volume, or both. Generally, however, the control volume is chosen such that certain aspects of the flow are steady, thereby simplifying the analysis.

Note that the force $\mathbf{F}$ may include field forces as well as contact forces so long as the source is external to the control volume. Internal forces, such as those due to viscosity and internal pressures, do not enter the problem, except possibly as a cause of external contact forces at the boundary.

Now let us apply Eq. (4–101) to the rocket problem which we analyzed previously by the particle mechanics approach. We take a control volume which includes the rocket and moves with it, as shown in Fig. 4–16. Note

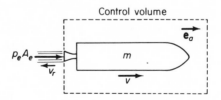

Fig. 4-16. A control volume enclosing the rocket.

that the exit area of the rocket is a portion of the surface of the control volume. The total external force acting on the material within the control volume is due to the exit pressure p_e acting over the flat exit area A_e. Hence,

$$\mathbf{F} = p_e A_e \mathbf{e}_a \qquad (4\text{-}102)$$

where the force acts to the right in the direction of the unit vector $\mathbf{e}_a$. Evaluating the other terms of Eq. (4–101), we obtain

$$\int_V \rho \mathbf{v}\, dV = mv\mathbf{e}_a$$

or

$$\frac{d}{dt} \int_V \rho \mathbf{v}\, dV = (m\dot{v} + \dot{m}v)\mathbf{e}_a = (m\dot{v} - bv)\mathbf{e}_a \qquad (4\text{-}103)$$

where $\dot{m}$ is evaluated from Eq. (4–93). Again we neglect changes in the momentum of the jet relative to the rocket as it flows within the control volume. The jet exhaust furnishes the only particles crossing the surface of the control volume, their relative velocity being v_e to the left, corresponding to an absolute velocity $v - v_e$ to the right. Therefore, using Eq. (4–94), we find that

$$\int_A \rho \mathbf{v}(\mathbf{v}_r \cdot d\mathbf{A}) = \rho_e(v - v_e)v_e A_e \mathbf{e}_a = b(v - v_e)\mathbf{e}_a \qquad (4\text{-}104)$$

Now we can evaluate Eq. (4–101), obtaining

$$p_e A_e \mathbf{e}_a = [m\dot{v} - bv + b(v - v_e)]\mathbf{e}_a$$

or

$$m\dot{v} = p_e A_e + bv_e \qquad (4\text{-}105)$$

in agreement with Eq. (4–97).

Another approach to this problem which again makes use of the control volume concept is obtained directly from the equation of motion for a system of particles, as given by Eq. (4–91). Taking the system to consist of all the material within the control volume at the given instant, we can replace the summation by an integration for this case of distributed mass, obtaining

$$\mathbf{F} = \int_V \rho \dot{\mathbf{v}}\, dV \qquad (4\text{-}106)$$

where $\dot{\mathbf{v}}$ is the absolute acceleration of the mass in the given volume element. In the language of fluid mechanics, it is a *substantial derivative*.

This result appears to be more simple than Eq. (4–101), but is frequently more difficult to apply because the integral involves an acceleration field rather than a velocity field. Thus we need to know the acceleration of all particles within the control volume instead of being able to integrate the velocity or momentum distribution over the control volume first and then differentiate with respect to time.

To illustrate the use of Eq. (4–106), let us calculate the acceleration of the rocket of Fig. 4–16, using the same control volume. Again we obtain that the total external force is

$$\mathbf{F} = p_e A_e \mathbf{e}_a$$

Now, the acceleration $\dot{\mathbf{v}}$ of any particle is equal to the acceleration of the rocket plus the acceleration of the particle relative to the rocket. This relative acceleration is zero everywhere except for those gas particles which are being accelerated rearward until they achieve the relative velocity $-v_e$ at the exit of the nozzle. Regardless of the precise acceleration history of a certain gas particle, but assuming that the internal flow is steady, there is, in effect, a mass b being accelerated from zero relative velocity to $-v_e$ during each unit of time. Therefore, the integral of the relative acceleration term is $-bv_e$ and the entire integral becomes

$$\int_V \rho \dot{\mathbf{v}} \, dV = (m\dot{v} - bv_e)\mathbf{e}_a \tag{4–107}$$

Hence we again obtain

$$m\dot{v} = p_e A_e + bv_e$$

in agreement with our earlier results.

We have seen that it might be convenient for the absolute acceleration $\dot{\mathbf{v}}$ to be written in terms of an acceleration relative to a noninertial reference frame such as the rocket body. The reason for choosing such a reference frame is that the flow, or a portion of it, is steady as viewed by an observer in this frame. For a general noninertial frame which is rotating at an angular rate $\boldsymbol{\omega}$, as well as translating, we can use the notation of Eq. (2–106) and obtain

$$\dot{\mathbf{v}} = \ddot{\mathbf{R}} + \dot{\boldsymbol{\omega}} \times \boldsymbol{\rho} + \boldsymbol{\omega} \times (\boldsymbol{\omega} \times \boldsymbol{\rho}) + 2\boldsymbol{\omega} \times (\dot{\boldsymbol{\rho}})_r + (\ddot{\boldsymbol{\rho}})_r \tag{4–108}$$

The expression for $\dot{\mathbf{v}}$ is then integrated term by term in order to evaluate the volume integral of Eq. (4–106).

Again choosing the rocket problem as an example, suppose we let the moving frame and the control volume be fixed in the rocket. Then $\boldsymbol{\omega} = 0$, and

$$\dot{\mathbf{v}} = \ddot{\mathbf{R}} + (\ddot{\boldsymbol{\rho}})_r$$

We see that

$$\int_V \rho \ddot{\mathbf{R}} \, dV = m\dot{v}\mathbf{e}_a \tag{4–109}$$

Also,

$$\int_V \rho(\ddot{\boldsymbol{\rho}})_r \, dV = \int_V \rho \dot{\mathbf{v}}_r \, dV$$

where, in this case of a control volume of constant dimensions, $\mathbf{v}_r$ is the velocity relative to the control volume of the mass element $\rho \, dV$. But, from Eqs. (4–101) and (4–106), we obtain

$$\int_V \rho \dot{\mathbf{v}} \, dV = \frac{d}{dt} \int_V \rho \mathbf{v} \, dV + \int_A \rho \mathbf{v}(\mathbf{v}_r \cdot d\mathbf{A}) \qquad (4\text{--}110)$$

This equality holds for a general vector field $\mathbf{v}$ and does not require that the reference frame be inertial. Therefore, we can apply it to the case of the relative velocity $\mathbf{v}_r$, obtaining

$$\int_V \rho \dot{\mathbf{v}}_r \, dV = \frac{d}{dt} \int_V \rho \mathbf{v}_r \, dV + \int_A \rho \mathbf{v}_r(\mathbf{v}_r \cdot dA)$$

$$= -\rho_e A_e v_e^2 \mathbf{e}_a = -b v_e \mathbf{e}_a \qquad (4\text{--}111)$$

where the volume integral on the right is constant for this case of steady flow and hence its time derivative is zero. Thus, equating the external force $\mathbf{F}$ to the sum of the expressions obtained in Eqs. (4–109) and (4–111), we once again find that

$$m\dot{v} = p_e A_e + b v_e$$

Now let us extend the control volume approach to rotational motion about an arbitrary point P. We assume that the control volume can move and change shape. The angular momentum $\mathbf{H}_p$ is defined in the manner of Eq. (4–64) and refers to the total angular momentum about P of the material within the control volume, as viewed by a nonrotating observer translating with P. Writing Eq. (4–70) in an integral form analogous to that of Eq. (4–101), we obtain

$$\mathbf{M}_p - \boldsymbol{\rho}_c \times m\ddot{\mathbf{r}}_p = \frac{d}{dt} \int_V \rho \boldsymbol{\rho} \times \mathbf{v}_p \, dV + \int_A \rho \boldsymbol{\rho} \times \mathbf{v}_p(\mathbf{v}_r \cdot d\mathbf{A}) \qquad (4\text{--}112)$$

where $\mathbf{M}_p$ is the total moment about P due to external forces acting on the mass within the control volume. Also, $\boldsymbol{\rho}$ is the position vector of the mass element relative to P, $\boldsymbol{\rho}_c$ is the position vector of the center of mass relative to P, and $\mathbf{v}_p$ is velocity relative to P of a particle at the position $\boldsymbol{\rho}$. As before, $\mathbf{v}_r$ refers to the velocity relative to the surface element $d\mathbf{A}$ of the particles crossing it. Note that $\boldsymbol{\rho}_c \times \ddot{\mathbf{r}}_p$ is zero if we choose a fixed point or the center of mass as the reference point P.

In a similar fashion, we can obtain the rotational counterpart of Eq. (4–106). Again choosing an arbitrary reference point P, we obtain

$$\mathbf{M}_p - \boldsymbol{\rho}_c \times m\ddot{\mathbf{r}}_p = \int_V \rho \boldsymbol{\rho} \times \mathbf{a}_p \, dV \qquad (4\text{--}113)$$

where $\mathbf{a}_p$ is the acceleration relative to P of the particle at the position $\boldsymbol{\rho}$, as viewed by a nonrotating observer. In case the flow is more easily observed by a rotating observer, $\mathbf{a}_p$ may be expressed in terms of a general vector equation such as Eq. (4–108).

Example 4–8. Water of density ρ enters a pipe along a vertical transverse axis through O, about which the pipe rotates at a constant angular rate ω (Fig. 4–17). The water leaves with a relative velocity v_e through a nozzle

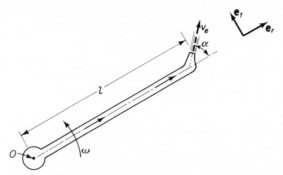

Fig. 4-17. A pipe and nozzle whirling in the horizontal plane about a vertical axis through O.

of exit area A_e at a distance l from the axis at O. If the nozzle axis is horizontal and at an angle α from the radial direction, find the moment **M** applied to the system.

Let us apply Eq. (4–112), choosing as a reference point the fixed point O. The control volume includes the pipe and nozzle and rotates with them. We assume that the water enters the control volume with no angular momentum about O. It leaves the control volume through the nozzle exit. Consider first the term, $\boldsymbol{\rho} \times \mathbf{v}_p$. We see that only transverse components of $\mathbf{v}_p$ contribute to this vector product. Furthermore, it is directed vertically at all times. Since the flow is steady relative to the pipe, we see that

$$\frac{d}{dt} \int_V \rho \boldsymbol{\rho} \times \mathbf{v}_p \, dV = 0$$

At the nozzle, we obtain

$$\boldsymbol{\rho} \times \mathbf{v}_p = l(v_e \sin \alpha + l\omega)\mathbf{e}_z$$

where $\mathbf{e}_z = \mathbf{e}_r \times \mathbf{e}_t$. Therefore, integrating over the nozzle exit only, since the surface integral is zero elsewhere, we find that

$$\int_A \rho \boldsymbol{\rho} \times \mathbf{v}_p(\mathbf{v}_r \cdot d\mathbf{A}) = \rho v_e A_e l(v_e \sin \alpha + l\omega)\mathbf{e}_z$$

Of course, $\ddot{\mathbf{r}}_p = 0$ because the reference point is fixed. Hence an evaluation of Eq. (4–112) yields the following result:

$$\mathbf{M} = \rho v_e A_e l(v_e \sin \alpha + l\omega)\mathbf{e}_z$$

In case the external moment is zero, then we find that the steady-state rotation rate is

$$\omega = \frac{-v_e \sin \alpha}{l}$$

Integration of the Rocket Equation. Let us return now to the differential equation for rocket flight in a vacuum with no gravitational forces. We found in Eqs. (4–98) and (4–99) that

$$m\dot{v} = p_e A_e + b v_e = F_s$$

where the thrust F_s is assumed to be constant. We wish to integrate with respect to mass rather than time, so let us make the substitution

$$m\frac{dv}{dt} = m\frac{dv}{dm}\frac{dm}{dt} = -bm\frac{dv}{dm}$$

which changes the differential equation to the following form:

$$\frac{dv}{dm} = -\frac{F_s}{bm} \tag{4–114}$$

Now we integrate between the initial and final (usually burnout) conditions, using the subscripts 0 and f, respectively, as the corresponding designations. Thus,

$$\int_{v_0}^{v_f} dv = -\frac{F_s}{b}\int_{m_0}^{m_f}\frac{dm}{m} \tag{4–115}$$

or

$$v_f - v_0 = \frac{F_s}{b}\ln\frac{m_0}{m_f} \tag{4–116}$$

It can be seen for this case in which no gravitational or aerodynamic forces are acting that the velocity gain $v_f - v_0$ is independent of the burning time for a given *mass ratio* m_0/m_f, assuming that F_s/b is constant. The coefficient F_s/b has the units of impulse per unit mass. Usually, it is specified in terms of the *specific impulse*

$$I_{\text{sp}} = \frac{F_s}{bg} \tag{4–117}$$

which is the total impulse per pound of propellant, the weight being measured at the earth's surface. It can be seen that if the pressure force is small compared to F_s, then

$$I_{\text{sp}} \cong \frac{v_e}{g}$$

indicating that the specific impulse is a measure of the effective exhaust velocity. Typical values of the specific impulse of chemical propellants are in the range from 200 to 350 sec.

From Eqs. (4–116) and (4–117), we see that the velocity change, written in terms of the specific impulse is

$$v_f - v_0 = I_{\text{sp}}g\ln\frac{m_0}{m_f} \tag{4–118}$$

Similarly, at any time during burning, the velocity v and mass m are related by the equation:

$$v = v_0 + I_{sp}g \ln \frac{m_0}{m} \tag{4-119}$$

For the case in which the rocket is fired vertically upward in a *uniform* gravitational field, the effect of the gravitational force can be superimposed upon the solution of Eq. (4–119) since we are concerned with a linear differential equation. Thus we obtain

$$v = v_0 - gt + I_{sp}g \ln \frac{m_0}{m} \tag{4-120}$$

which applies for any time during burning. After burnout, we let $m = m_f$ and the equation remains valid for this case also.

Example 4–9. A two-stage sounding rocket is to be fired vertically from the earth (Fig. 4–18). Each stage, individually, has a mass ratio equal to 9, the total weight of the first stage being 360 lb and that of the second stage being 90 lb. Assuming that an instrument package weighing 30 lb is mounted atop the second stage and that a 20-sec coasting period occurs between first-stage burnout and the ignition of the second stage, find the maximum altitude reached by the second stage plus instruments. The burning time is assumed to be very short and the change in the acceleration of gravity is neglected during the coasting period before second-stage ignition. The specific impulse of the propellant is $I_{sp} = 250$ sec. Assume a nonrotating earth and neglect atmospheric drag.

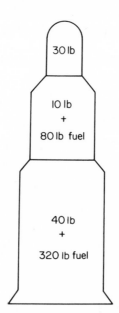

Fig. 4-18. A two-stage rocket and payload.

It can be seen that the effective mass ratio for each stage is the same, namely,

$$\frac{m_0}{m_f} = \frac{480}{160} = \frac{120}{40} = 3$$

where we assume that the first stage is detached before the second stage is ignited. Using Eq. (4–118), we find that the velocity gained per stage is

$$v_f - v_0 = (250)(32.2) \ln 3 = 8850 \text{ ft/sec}$$

From Eqs. (3–6) and (3–7), we see that the velocity and altitude after the 20-sec coasting period are

$$v = 8850 - (32.2)(20) = 8206 \text{ ft/sec}$$

$$h = (8850)(20) - \tfrac{1}{2}(32.2)(20)^2 = 170,600 \text{ ft}$$

After second-stage burnout, this velocity has increased to $8206 + 8850 = 17,056 \text{ ft/sec}$. Now we can find the maximum altitude by using conservation of energy. From Eqs. (3–69) and (3–97), we see that the total energy per unit mass is

$$\frac{1}{2}(1.706)^2 \times 10^8 - (32.2)(3960)(5280)\left[1 + \frac{1.706 \times 10^5}{(3960)(5280)}\right]^{-1}$$
$$= -5.223 \times 10^8 \text{ ft}^2/\text{sec}^2$$

where we have assumed that the earth's radius $R = 3960$ miles. Using Eq. (3–97) again for the case where the kinetic energy is zero, we set the potential energy per unit mass equal to the total energy per unit mass previously calculated, and thereby we are able to solve for the maximum altitude.

$$h_{\text{max}} = (3960)(5280)\left[\frac{(32.2)(3960)(5280)}{5.223 \times 10^8} - 1\right] = 6.030 \times 10^6 \text{ ft} \cong 1140 \text{ mi}$$

It is interesting to note that if a constant acceleration of gravity $g = 32.2$ ft/sec^2 had been used throughout, the maximum height would have been calculated to be only 890 miles. So the variation of the gravitational force with altitude has a significant effect in this example. Note also that the coasting before second-stage ignition resulted in a delay of 20 sec in obtaining the velocity increase due to this second stage, that is, the velocity was 8850 ft/sec lower than it could have been for a period of 20 sec. Therefore the maximum altitude was reduced by 1.77×10^5 ft or 33.5 miles. In the practical case, of course, atmospheric drag forces which are strongly dependent upon velocity would occur. Thus the coasting to a higher altitude with its lower atmospheric density before igniting the second stage actually results in smaller losses than if the stages were fired in quick succession.

4–9. EXAMPLES

Example 4–10. Mass m_1 hits m_2 with inelastic impact ($e = 0$) while sliding horizontally with velocity v along the common line of centers of the three equal masses—Fig. 4–19(a). Initially, masses m_2 and m_3 are stationary and the spring is unstressed. Find (a) the velocities of m_1, m_2, and m_3 imme-

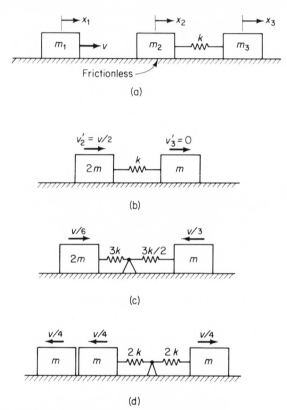

(a)

(b)

(c)

(d)

Fig. 4-19. An inelastic collision and subsequent motion, as viewed from three inertial reference frames.

diately after impact, (b) the maximum kinetic energy of m_3, (c) the minimum kinetic energy of m_2, (d) the maximum compression of the spring, (e) the final motion of m_1.

This example is one-dimensional in nature and no external forces act on the system in this direction. Therefore linear momentum is conserved throughout the problem. Because $e = 0$, we see from Eq. (4–82) that masses m_1 and m_2 will move at the same velocity immediately after impact, resulting in the equivalent system shown in Fig. 4–19(b). Calling this velocity v_2', we can use the principle of conservation of linear momentum to obtain

$$mv = 2mv_2'$$

or

$$v_2' = \frac{1}{2}v$$

where m is the mass of each particle. Mass m_3 does not move at the time of impact because the spring is initially unstressed and therefore $(m_1 + m_2)$ must move through a finite displacement before any force acts on m_3.

At all times after the impact, the total energy is conserved. It is equal to the kinetic energy just after impact, since no potential energy is stored in the spring at this time. So, writing the general expression for the total energy and setting it equal to the kinetic energy just after impact, we obtain

$$mv_2^2 + \frac{1}{2}mv_3^2 + \frac{1}{2}k(x_2 - x_3)^2 = \frac{mv^2}{4}$$

where v_2 and v_3 are the velocities of $(m_1 + m_2)$ and m_3, respectively. Also, using the principle of conservation of linear momentum and assuming that m_1 and m_2 continue to move together, we see that

$$2mv_2 + mv_3 = mv$$

from which we obtain

$$v_2 = \frac{1}{2}(v - v_3)$$

Because the total energy is conserved, the total kinetic energy will be maximum when the potential energy is minimum, that is, zero. This occurs when $x_2 = x_3$, corresponding to an unstressed spring. So, setting $x_2 = x_3$ and substituting for v_2 in the energy equation, we find that

$$\frac{1}{4}(v - v_3)^2 + \frac{1}{2}v_3^2 = \frac{1}{4}v^2$$

or

$$v_3(3v_3 - 2v) = 0$$

from which we obtain the roots

$$v_3 = 0, \frac{2}{3}v$$

Now, to find the extreme values of the kinetic energy associated with the individual particles, we note that these values occur when the individual velocities reach extreme values, that is, when the accelerations are zero. But this occurs for a zero spring force, or $x_2 = x_3$. Therefore, it can be seen that the extreme values of the kinetic energy of the individual particles occur when the total kinetic energy is maximum. It is clear, then, that the maximum kinetic energy of m_3 occurs for $v_3 = 2v/3$, that is, for the largest root.

$$(T_3)_{\text{max}} = \frac{1}{2} m \left(\frac{2v}{3}\right)^2 = \frac{2}{9} mv^2$$

Because the total linear momentum is conserved, we see that $(v_2)_{\text{min}}$ occurs at the same time as $(v_3)_{\text{max}}$. At this time, the total momentum is

$$2m(v_2)_{\text{min}} + \frac{2}{3} mv = mv$$

or

$$(v_2)_{\text{min}} = \frac{v}{6}$$

The corresponding kinetic energy of m_2 is

$$(T_2)_{\text{min}} = \frac{1}{2} m \left(\frac{v}{6}\right)^2 = \frac{mv^2}{72}$$

The maximum compression of the spring occurs when the relative velocity of its two ends is zero, in which case all three particles are moving with the same velocity. Again, we use the conservation of linear momentum to solve for the common velocity at this time.

$$v_2 = v_3 = \frac{v}{3}$$

Now we substitute these values into the general energy equation, obtaining

$$m \left(\frac{v}{3}\right)^2 + \frac{1}{2} m \left(\frac{v}{3}\right)^2 + \frac{1}{2} k(x_2 - x_3)^2 = \frac{mv^2}{4}$$

from which we find that the maximum spring compression is

$$(x_2 - x_3)_{\text{max}} = \sqrt{\frac{m}{6k}} v$$

We have omitted the other sign on the square root because it implies tension in the spring.

In order to solve for the final motion of m_1, we recall that m_1 hit m_2 inelastically, but this does not imply that they move together indefinitely. In fact, m_1 cannot accelerate to the right, since we assume that m_2 cannot exert a pull on it. So, at the first instant in which m_2 accelerates to the right, m_1 and m_2 will separate. This occurs when m_2 accelerates from its minimum velocity $v/6$. From this time onward, m_1 moves at a constant velocity $v/6$. Meanwhile, the center of mass of m_2 and m_3 translates uniformly at a velocity $5v/12$, as can be seen from momentum considerations. So m_1 is left behind permanently. Masses m_2 and m_3 continue to oscillate relative to their center of mass, and since the initial separation occurs at the moment of $(v_2)_{\text{min}}$ and $(v_3)_{\text{max}}$, these extremes are never exceeded.

Before leaving this example, note that it could have been solved by considering the whole process from the viewpoint of an observer translating uniformly with the center of mass of the system. In this case, the system immediately after impact appears as shown in Fig. 4–19(c). The center of mass is fixed in this system and is located at one-third of the distance between $(m_1 + m_2)$ and m_3. Using a procedure similar to that of Example 4–2, we see that the center of mass divides the spring into two portions with the spring constant of each being inversely proportional to its unstressed length. Furthermore, it can be seen that two mass-spring systems are formed, each with the natural frequency $\omega_n = \sqrt{3k/2m}$. Thus the motion of each mass relative to the center of mass is sinusoidal in nature, the initial velocities being as shown in Fig. 4–19(c).

But we have seen that only one half-cycle of this motion occurs before m_1 and m_2 separate. After this separation, we might choose an inertial reference which translates uniformly with the center of mass of m_2 and m_3. From this reference frame, the situation at the time of separation appears as shown in Fig. 4–19(d). We see that m_2 and m_3 oscillate at a frequency $\omega_n = \sqrt{2k/m}$. Meanwhile, m_1 is being left behind with an average separation velocity of $v/4$. In each case, absolute velocities are found by adding the velocity of the reference frame.

Example 4–11. Two equal masses slide on a smooth horizontal plane and are connected by a parallel spring-damper combination, as shown in Fig. 4–20. The separation of the two masses is $(l_0 + x)$, the coordinate x being the extension of the spring from its unstressed length l_0. Initially, $x = 0$, $\dot{x} = 0$, and one mass is motionless, whereas the other mass has a transverse velocity v_0. After reaching the steady-state motion, we find that the spring has stretched to twice its original length. Solve for v_0 as a function of the other system parameters and compute the fraction of the original energy

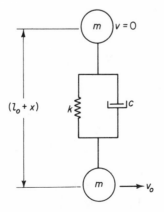

Fig. 4-20. The initial motion of the system of Example 4-11.

which has been dissipated in the damper. Consider the masses as particles.

We can solve for the angular rotation rate as a function of x by writing the equation of conservation of angular momentum about the center of mass. First, we note from Eq. (4–43) that the angular momentum of a single particle in plane motion is

$$H = mr^2 \omega \tag{4–121}$$

where ω is the absolute angular rate with which the particle moves about the reference point and r is its radial distance from that point. So, taking the center of mass as the reference, we see that the total angular momentum of the system is

$$H = \frac{m(l_0 + x)^2}{2} \omega = \frac{m l_0 v_0}{2}$$

where the constant term on the right was evaluated from initial conditions. Hence we find that

$$\omega = \frac{l_0 v_0}{(l_0 + x)^2} \tag{4–122}$$

It can be seen that the damper is dissipating energy except when $\dot{x} = 0$. So, in the steady-state motion, $\dot{x} = 0$ and $x = l_0$. From Eq. (4–122), we see that the steady-state value of ω is

$$\omega_s = \frac{v_0}{4 l_0}$$

But in the steady state the spring force is equal to the mass times the centripetal acceleration. From Eqs. (2–53) and (3–101), we see that

$$m l_0 \omega_s^2 = \frac{m v_0^2}{16 l_0} = k l_0$$

Therefore,

$$v_0 = 4 l_0 \sqrt{\frac{k}{m}}$$

To find the energy loss in the damper, we calculate first the initial energy E_i of the system. Using a fixed reference frame, we see that

$$E_i = \frac{1}{2} m v_0^2 = 8 k l_0^2$$

since all the energy is associated with one particle as kinetic energy. The final potential energy stored in the spring is

$$V_f = \frac{1}{2} k l_0^2$$

The final kinetic energy is the sum of the kinetic energy due to the motion of the center of mass and that due to the motion relative to the center of mass.

$$T_f = m \left(\frac{v_0}{2} \right)^2 + m l_0^2 \omega_s^2 = 5 k l_0^2$$

Therefore the final energy is

$$E_f = T_f + V_f = \frac{11}{2} k l_0^2$$

The fractional loss of energy is

$$\frac{E_i - E_f}{E_i} = \frac{5}{16}$$

It is interesting to note that if an inertial frame had been chosen which translates with the center of mass, then the fractional energy loss would have been greater, namely, $\frac{5}{8}$. For either reference frame, however, the magnitude of the energy loss is the same.

Example 4-12. The point of support O of a simple pendulum of length l moves with a horizontal displacement

$$x = A \sin \omega t$$

in the plane of the motion in θ [see Fig. 4-21(a)]. Find the differential equation of motion.

We shall analyze this problem from the point of view of an observer in a noninertial reference frame which translates with the support point O. Using the principles discussed in Sec. 4-5 for the case of an arbitrary refer-

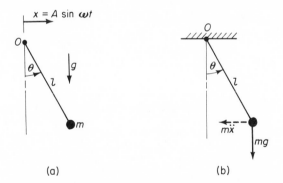

(a)   (b)

Fig. 4-21. A simple pendulum with a moving support.

ence point, we see that we must introduce a horizontal inertial force $-m\ddot{x}$ in addition to the usual external forces. Now we consider the point O to be fixed and solve the dynamical problem shown in Fig. 4-21(b).

The angular momentum about O is

$$H = m l^2 \dot{\theta}$$

and therefore

$$\dot{H} = ml^2 \ddot{\theta}$$

where the positive direction for **H** is taken out of the page. Also, the moment due to the gravitational and inertial forces shown in Fig. 4–21(b) is

$$M = -m\ddot{x}l \cos \theta - mgl \sin \theta$$
$$= ml(A\omega^2 \cos \theta \sin \omega t - g \sin \theta)$$

Now we set $M = \dot{H}$ and obtain the differential equation:

$$\ddot{\theta} + \frac{g}{l} \sin \theta = \frac{A\omega^2}{l} \cos \theta \sin \omega t$$

Note that the foregoing procedure is equivalent to using Eq. (4–70) directly.

REFERENCES

1. Becker R. A., *Introduction to Theoretical Mechanics.* New York: McGraw-Hill, Inc., 1954.

2. Halfman R. L., *Dynamics—Particles, Rigid Bodies, and Systems.* Reading, Mass.: Addison-Wesley Publishing Company, 1962.

3. Housner, G. W., and D. E. Hudson, *Applied Mechanics—Dynamics*, 2nd ed. Princeton, N.J.: D. Van Nostrand Co., Inc., 1959.

4. Shames, I. H., *Engineering Mechanics.* Englewood Cliffs, N.J.: Prentice-Hall, Inc., 1960.

5. Yeh, H., and J. I. Abrams, *Principles of Mechanics of Solids and Fluids—Particle and Rigid-Body Mechanics.* New York: McGraw-Hill, Inc., 1960.

PROBLEMS

4–1. What is the minimum spring compression δ necessary to cause m_2 to leave the floor after m_1 is suddenly released with zero velocity? Measure δ from

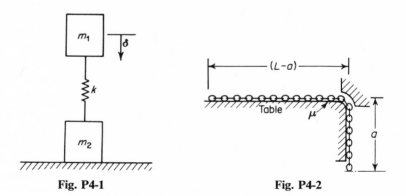

Fig. P4-1   Fig. P4-2

the unstressed length of the spring and assume that all motion is in the vertical direction.

4-2. A chain of length L and mass m rests on a horizontal table. If there is a coefficient of friction μ between the chain and the table top, find the velocity of the chain as it leaves the table, assuming that it is released from the position shown. The initial overhang is a, where $a > \mu L/(1 + \mu)$, and the chain is guided without friction around the corner.

4-3. A system consists of two particles of equal mass m which slide on a rigid massless rod that rotates freely about a vertical axis at O. Solve for the radial interaction force between the particles which will cause their distance from the axis to vary according to the formula, $r = r_0 + A \sin \beta t$, where $A < r_0$. The initial angular velocity is ω_0.

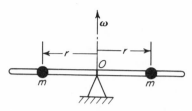

Fig. P4-3

4-4. A tennis ball bouncing on a concrete surface is found to have a coefficient of restitution $e = 0.90$. If such a ball of mass m is lying motionless on a horizontal concrete surface, what downward impulse must be applied to it by a racket to cause it to rise one foot into the air? Give assumptions.

4-5. A ball is thrown with an initial velocity v_0 at an angle γ_0 above the horizontal. It hits a smooth vertical wall at a horizontal distance s and rebounds to the point from which it was thrown. Assuming a coefficient of restitution e at the wall, and neglecting all forces exerted by the air, solve for the required angle γ_0.

4-6. A smooth sphere A of mass m, traveling with velocity $v = v_0\mathbf{i}$, hits a similar sphere B such that the angle between the line of centers and the negative x axis is $45°$. Assuming that the coefficient of restitution is $e = 0.90$, find the velocities of A and B after impact.

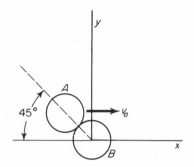

Fig. P4-6

4-7. Bodies B_1 and B_2 each consist of two small spheres of equal mass m connected by a rigid massless rod of length l. Initially, B_2 is motionless and

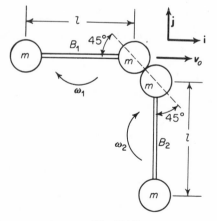

Fig. P4-7

B_1 is moving such that $\omega_1 = 0$ and the velocity of its mass center is $\mathbf{v}_0 = v_0\mathbf{i}$. Two of the smooth spheres hit, as shown in the figure, with their line of centers at $45°$ to each rod during impact. Assuming that all motion takes place in a plane and the coefficient of restitution is $e = 0.5$, solve for the linear and angular velocities of B_1 and B_2 immediately after impact. Consider the spheres as particles.

4–8. Two particles of mass m_1 and m_2, respectively, are connected by a string of length l. Initially, the particles are separated by a distance l and are motionless on a smooth horizontal plane. Then m_2 is given a horizontal impulse of magnitude $\mathscr{F} = m_2v_0$ in a direction perpendicular to the string. Solve for the motion of the system and the tension in the string.

4–9. Masses m_2 and m_3 are initially at rest and the spring of length l_0 is unstressed. Then mass m_1, traveling with velocity v_0 in a direction perpendicular to the spring, hits m_2 inelastically and sticks to it. In the ensuing motion the spring stretches to a maximum length $3l_0$. Solve for v_0, assuming that the masses are equal and can be considered as particles.

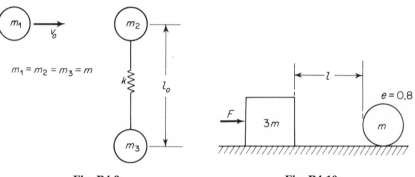

Fig. P4-9   Fig. P4-10

4–10. A block of mass $3m$ and a sphere of mass m are initially motionless and are separated by a distance l on a smooth horizontal plane, as shown. At $t = 0$, a constant horizontal force F is applied to the block, causing it to hit the sphere repeatedly. Assuming a coefficient of restitution $e = 0.8$ for collisions between the block and the sphere, solve for the velocity of the system when the bouncing stops.

4–11. A particle of mass m can slide without friction along a fixed horizontal wire coinciding with the x axis. Another particle of mass m_0 moves with a constant speed v_0 along the line $y = h$ from $x = -\infty$ to $x = \infty$. If the particle m is initially at the origin and if an attractive force of magnitude K/r^2 exists between the two particles, where r is their separation, solve for the maximum speed of m.

4–12. Solve for the maximum speed of particle m in problem 4–11 if m_0 starts at $x = -\infty$ with velocity $v = v_0$ and both particles can slide freely along their respective wires which are separated by a distance h.

4–13. A coiled flexible rope of uniform mass ρ per unit length lies on a horizontal surface. Then, at $t = 0$, one end is raised vertically with a constant velocity v. Find the force F which is required to lift the end of the rope.

4–14. A flexible rope of mass ρ per unit length and total length l is suspended so that its upper end is at a height h above a horizontal floor, where $h > l$.

Suddenly it is released, hitting the floor inelastically. Find the force on the floor as a function of time.

4–15. Material is fed at a rate bg lb/sec and with essentially zero absolute velocity from a hopper onto a belt of total mass m. It sticks to the belt until it reaches a small pulley at B, at which point it falls off. If the belt is inclined at 30° to the horizontal, and if the material is put on the belt at a distance l from where it falls off, find the steady-state speed v of the belt motion. The pulleys are frictionless.

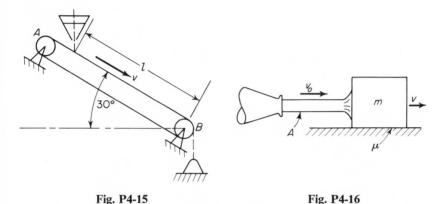

Fig. P4-15   **Fig. P4-16**

4–16. A jet of water of cross-section A, density ρ, and absolute velocity v_0, moves horizontally and hits a block of mass m inelastically, that is, the water leaves with a zero horizontal component of velocity relative to the block. Find the terminal velocity of the block, assuming a coefficient of friction μ between the block and the horizontal plane upon which it slides.

4–17. A rocket of mass ratio 10 (exclusive of payload) is used to carry a 50 lb payload to an altitude $R/4$ above the earth's surface. (a) Assuming a specific impulse $I_{sp} = 250$ sec and the earth radius $R = 3960$ miles, find the minimum total weight necessary to accomplish this result. Neglect the effects of atmospheric drag and the earth's rotation, and assume a very short burning time. (b) Suppose that a two-stage rocket is used to accomplish the same result. Each stage has mass ratio of 10 (exclusive of payload). Assuming that the stages are fired in quick succession and the first stage is five times as heavy as the second, what is the minimum total weight in this case? (c) If we now assume a 10 sec burning time for each rocket, solve for the maximum accelerations of the rocket systems of parts (a) and (b).

4–18. A particle of mass m strikes a horizontal floor while traveling with velocity v_0 at an angle of 45° from the vertical. Assuming that the coefficient of restitution for normal impact is $e = 1$ and the coefficient of friction is $\mu = 0.10$, find: (a) the rebound angle and speed immediately after the first bounce and (b) the position and motion of the particle after a long time.

4–19. Spheres A and B are slightly separated and move to the right with velocity v, as shown. Sphere C moves to the left with velocity v along the common line of centers. If each sphere has the same mass m, and if the coefficient of restitution for collisions between spheres is unity; find the velocities of A, B, and C after all impacts have occurred for the following cases: (a) A and B are not connected; (b) A and B are connected by a short inelastic string of negligible mass; (c) A and B are connected by a rigid massless rod.

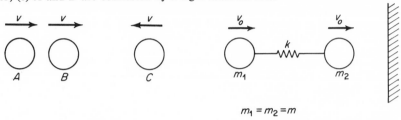

$$m_1 = m_2 = m$$

Fig. P4-19   **Fig. P4-20**

4-20. A system consisting of two equal spherical masses connected by a spring is sliding on a smooth horizontal floor with velocity v_0 along its line of centers. It hits a vertical wall and rebounds along its path of approach. Suppose that the coefficient of restitution e for the system is defined as the ratio of the speed of the center of mass as the system moves away from the wall compared to its approach speed v_0. Solve for e for the cases where the coefficient of restitution e' for a single mass is: (a) zero; (b) unity.

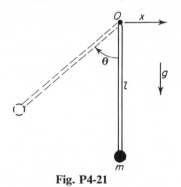

Fig. P4-21

4-21. The support point O of a simple pendulum of mass m and length l is given a uniform horizontal acceleration $\ddot{x} = a$. If the initial conditions are $\theta(0) = 0$, $\dot{\theta}(0) = 0$, $\dot{x}(0) = 0$, find the limits on θ in the ensuing motion. Assuming that $a = \sqrt{3}\,g$, solve for the period of the motion.

4-22. Suppose that the support point O of the simple pendulum in the previous problem is given a constant horizontal acceleration $a = 2g$. What is the angular velocity when the pendulum first becomes horizontal?

4-23. Particles m_1 and m_2, each of mass m, are connected by an inextensible weightless string of length $4l$. Initially, each particle is sliding with velocity v along parallel paths separated by a distance $4l$ on a smooth horizontal floor. Suddenly a point on the string at a distance l from particle m_1 strikes a fixed vertical nail of negligible diameter that projects from the floor. Assuming that the particles proceed to whirl in opposite directions without colliding, and the string can slide freely on the nail, find for the ensuing motion: (a) the maximum distance of m_1 from the nail; (b) the minimum tensile force in the string.

4-24. Consider again the gyroscope of problem 2–12. Assuming that the rotor mass m is uniformly distributed along the rim, solve for the uniform precession rate $\dot{\phi}$ under the influence of gravity by using d'Alembert's principle and equating the magnitudes of the inertial and gravitational moments about O.

5

ORBITAL MOTION

This chapter is primarily concerned with the calculation of some of the important characteristics of the path in space followed by a particle as it moves in a gravitational field. We studied the motion of a particle in a uniform gravitational field in Sec. 3–1. Here we shall emphasize the path or *orbit* of a particle which experiences an inverse-square attraction toward a point.

Although the development of the subject is made in the context of particle motion, the results are applicable to many cases of translational motion of extended bodies of arbitrary shape, the principal restriction being that the external gravitational field be essentially uniform in the region occupied by the body at any given moment. For locally homogeneous fields, it follows directly from Eq. (4–8) that the translational motion is the same as though the entire mass were concentrated at the mass center.

5–1. KEPLER'S LAWS AND NEWTON'S LAW OF GRAVITATION

Kepler's Laws of Planetary Motion. The principal factors affecting the gravitational force on a body were originally discovered by studying planetary motions. Johannes Kepler, after a careful analysis of the observational data of Tycho Brahe, found that he could predict the motions of the planets if he assumed the following laws:

I. *The orbit of each planet is an ellipse with the sun at one focus.*

II. *The radius vector drawn from the sun to a planet sweeps over equal areas in equal times.*

III. *The squares of the periods of the planets are proportional to the cubes of the semimajor axes of their respective orbits.*

These are Kepler's laws. The first two laws were published in 1609 and the third in 1619. Thus, they preceded Newton's laws of motion by nearly seventy years. As a matter of fact, Newton deduced the law of gravitation from Kepler's laws and his own laws of motion.

As the chapter proceeds, we shall obtain Kepler's laws as a natural result of our analysis of the two-body problem and indicate the reasons for the small inaccuracies which occur when they are applied to the planetary orbits of the solar system.

Newton's Law of Gravitation. One of Newton's principal contributions to the development of mechanics is his law of gravitation which may be stated as follows:

Any two particles attract each other with a force which acts along the line joining them and which has a magnitude that is directly proportional to the product of the masses and inversely proportional to the square of the distance between them.

The quantitative aspects of the law of gravitation may be stated in the form of the equation:

$$F_r = -G \frac{m_1 m_2}{r^2} \tag{5-1}$$

where m_1 and m_2 are the masses of the particles, r is their separation, and G is a universal constant which is independent of the nature of the masses or their location in space. In terms of commonly used units,

$$G = 6.67 \times 10^{-8} \mathrm{cm^3/gm\,sec^2} = 3.44 \times 10^{-8} \mathrm{ft^4/lb\,sec^4}$$

The minus sign signifies that F_r is an attracting force.

Gravitational Potential and Potential Energy. We saw from Eq. (3–91) that if the force acting on a body is given by $-K/r^2$, then the potential energy is of the form

$$V = -\frac{K}{r} \tag{5-2}$$

where we have chosen the zero reference for potential energy at $r = \infty$, that is, for an infinite separation of the particles. So, for the foregoing case of the gravitational attraction between two particles, we find that

$$K = Gm_1 m_2 \tag{5-3}$$

Now let us consider the force acting on a *unit mass* due to the presence of a second mass m_2. Setting $m_1 = 1$, we see from Eq. (5–1) that the force can be written in the form

$$\mathbf{f} = -\frac{Gm_2}{r^2} \mathbf{e}_r \tag{5-4}$$

where $\mathbf{e}_r$ is a unit vector directed from m_2 toward the unit mass. By moving the unit mass, the vector $\mathbf{f}$ can be evaluated at all points in space (except at m_2). The force at each point is a measure of the intensity of the gravitational field due to m_2 and is called the *gravitational field strength*. We note that the dimensions of $\mathbf{f}$ are those of force per unit mass, or acceleration. In fact, $\mathbf{f}$ is just the absolute acceleration which the unit mass would have if it were released at the given point. Hence it can be considered as the acceleration of gravity at that point.

We saw in Eq. (3–88) that the force exerted by a conservative force field is

$$\mathbf{F} = -\nabla V$$

It is clear, then, that for this case in which $m_1 = 1$, we can write

$$\mathbf{f} = -\nabla \phi \tag{5-5}$$

where

$$\phi = -\frac{Gm_2}{r} \tag{5-6}$$

This expression represents the potential energy per unit mass, that is, for $m_1 = 1$, and is known as the *gravitational potential*.

If we wish to find the force on a unit mass due to the presence of n other particles, then we sum the individual forces. For example, if $\mathbf{r}_i$ is the position vector of the unit mass relative to m_i, then, from Eq. (5–4), we see that

$$\mathbf{f} = \sum_{i=1}^{n} \mathbf{f}_i = -G \sum_{i=1}^{n} \frac{m_i}{r_i^3} \mathbf{r}_i \tag{5-7}$$

where $\mathbf{f}_i$ is the force on the unit mass due to m_i. Note that the forces act independently and no shielding or other effects occur whereby the interaction of one pair of masses is influenced by the presence of other masses.

In a similar fashion, we find that the total potential at a given point due to the n particles is

$$\phi = \sum_{i=1}^{n} \phi_i = -G \sum_{i=1}^{n} \frac{m_i}{r_i} \tag{5-8}$$

where ϕ_i is the potential due to m_i at a distance r_i. Since the potential ϕ is a scalar function of position, it is often preferable to calculate the field strength $\mathbf{f}$ due to several particles by first calculating ϕ as the sum of the ϕ_i and then finding $\mathbf{f}$ as the negative gradient of ϕ, rather than performing the vector additions directly according to Eq. (5–7).

Earlier, for the case of two particles, we defined the potential ϕ at a given point as the potential energy per unit mass. The potential energy, however, can also be considered as the work required to move the system from a standard configuration to the given one (Sec. 3–3). So in this case, the potential ϕ is just the work done against the field in bringing a unit mass from infinity to the given point. Note that it does not include the work required to assemble the n masses in their given positions and hence is not directly related to the total potential energy of the system.

To clarify this point, let us calculate the total potential energy of a system of n particles. For any two particles, m_i and m_j, the potential energy is given by

$$V_{ij} = -\frac{Gm_i m_j}{r_{ij}} \qquad (5\text{--}9)$$

where r_{ij} is the distance between m_i and m_j. Summing over all pairs of particles, we obtain the total potential energy of the system:

$$V = \frac{1}{2} \sum_{i=1}^{n} \sum_{j=1}^{n} V_{ij} = -\frac{G}{2} \sum_{i=1}^{n} \sum_{j=1}^{n} \frac{m_i m_j}{r_{ij}} \qquad (i \neq j) \qquad (5\text{--}10)$$

where the factor $\frac{1}{2}$ occurs because the double summation includes each term twice. It can be seen that the expressions given in Eqs. (5–8) and (5–10) are quite different.

The Gravitational Field of a Sphere. Now let us find the gravitational field due to a sphere whose mass is distributed with spherical symmetry, that is, the density depends only upon the distance from the center and is essentially constant over a thin spherical shell whose center coincides with that of the sphere. If we calculate first the potential due to a thin spherical shell, we can obtain the potential for the case of a sphere by superimposing the results for shells whose radii vary from zero to the radius of the sphere.

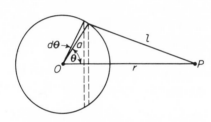

Consider, then, a thin spherical shell (Fig. 5–1) having a radius a and a constant mass σ per unit area. Taking an area element in the form of a ring of width $a\,d\theta$ and mass

$$dm = 2\pi a^2 \sigma \sin\theta\,d\theta$$

we see that all points on the ring are a constant distance l from a point P lying on its axis of symmetry. Therefore, from Eq. (5–8), we find that the potential at P due to the ring is

Fig. 5-1. A spherical shell, showing an area element used in calculating the potential at P.

$$d\phi = -\frac{G\,dm}{l}$$

Noting that

$$l^2 = a^2 + r^2 - 2ar\cos\theta$$

and integrating over the entire shell, we obtain an expression for the potential at P in the following form:

$$\phi = -\int_0^\pi \frac{2\pi a^2 G\sigma \sin\theta\,d\theta}{\sqrt{a^2 + r^2 - 2ar\cos\theta}}$$

$$= -\frac{2\pi aG\sigma}{r}\left[a + r \mp (r - a) \right]$$

where we choose the upper sign for $r > a$ and the lower sign for $r < a$. The total mass m of the shell is the surface density σ times the area, or

$$m = 4\pi a^2 \sigma$$

Therefore the potential due to a thin shell is

$$\phi = -\frac{Gm}{r} \qquad (r > a) \tag{5-11}$$

$$\phi = -\frac{Gm}{a} \qquad (r < a) \tag{5-12}$$

Using Eq. (5–5), we see that the gravitational force per unit mass is

$$\mathbf{f} = -\frac{Gm}{r^2}\mathbf{e}_r \qquad (r > a) \tag{5-13}$$

for points outside the shell and $\mathbf{f} = 0$ for all points inside the shell. Hence we find that the external gravitational field of a thin spherical shell is the same as if its entire mass were concentrated at the center. On the other hand, the shell exerts no gravitational force on a particle which is located at any point within the shell.

If we consider a solid sphere to be composed of concentric spherical shells, then it can be seen that the external field is just that due to the sum of the masses of the shells concentrated at their common center. In other words, the potential ϕ external to the sphere can be written in the form

$$\phi = -\frac{Gm}{r} \qquad (r > a) \tag{5-14}$$

where m is the total mass of the sphere and r is the distance from the center to the given point. The corresponding gravitational field strength is

$$\mathbf{f} = -\frac{Gm}{r^2}\mathbf{e}_r \qquad (r > a) \tag{5-15}$$

where $\mathbf{e}_r$ is a radial unit vector. For a point P inside an isolated *homogeneous* sphere, we see that the gravitational field is due to that portion of the mass which lies inside a spherical surface through P. Assuming a uniform density ρ, we find that this mass is

$$m = \frac{4}{3}\pi\rho r^3$$

and therefore

$$\mathbf{f} = -\frac{4}{3}\pi G\rho r\mathbf{e}_r \qquad (r < a) \tag{5-16}$$

Hence, as one proceeds radially outward from the center of a homogeneous sphere, the gravitational field strength increases linearly with distance until the surface of the sphere is reached. Then it decreases as the inverse square of the distance.

We have seen that the force of a sphere on an external particle is identical to that exerted by the mass of the sphere concentrated at its center. By the

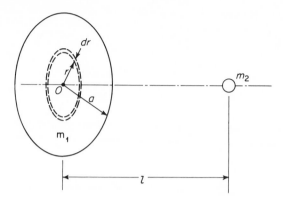

Fig. 5-2. A circular disk of mass m_1 with a particle m_2 on its axis of symmetry.

law of action and reaction, the force of the particle on the sphere can be calculated in the same fashion. Thus it follows that the interaction forces between two or more spheres are calculated by considering the mass of each to be concentrated at its center. This is true, of course, even for the case of close spacing in which the distance between centers is not much greater than the sum of the radii.

Example 5–1. Given a thin homogeneous disk of radius a and mass m_1. A particle of mass m_2 is placed at a distance l from the disk on its axis of symmetry (Fig. 5–2). Initially both are motionless in free space, but they ultimately collide, because of the gravitational attraction. Find the relative velocity at the time of collision. Assume $a \ll l$.

First let us calculate the potential ϕ at m_2. The potential due to a circular ring of radius r and width dr is

$$d\phi = -\frac{G\,dm}{\sqrt{r^2 + l^2}} = -\frac{2Gm_1 r\,dr}{a^2 \sqrt{r^2 + l^2}}$$

since all points on the ring are at the same distance from m_2 and we assume a uniform surface density $\sigma = m_1/\pi a^2$. Integrating, we obtain the potential on the axis of symmetry at a distance l from the disk.

$$\phi = -\int_0^a \frac{2Gm_1 r\,dr}{a^2 \sqrt{r^2 + l^2}} = -\frac{2Gm_1}{a^2}\left(\sqrt{a^2 + l^2} - l\right) \tag{5-17}$$

So we find that the initial potential energy is ϕm_2, or[1]

$$V_0 = -\frac{2Gm_1 m_2}{a^2}\left(\sqrt{a^2 + l^2} - l\right)$$

[1]We do not consider the potential energy of the disk itself since it undergoes no changes in this example.

The final potential energy V_f, occurring just before impact, is found by setting $l = 0$ in the foregoing expression, since, from symmetry considerations, the impact must occur at the center of the disk.

$$V_f = -\frac{2Gm_1m_2}{a}$$

We assumed that $a \ll l$ and therefore we can expand the square root in powers of a^2/l^2 to obtain the approximation

$$V_0 \cong -\frac{Gm_1m_2}{l}$$

Hence the change in potential energy is

$$V_0 - V_f = Gm_1m_2\left(\frac{2}{a} - \frac{1}{l}\right) \tag{5-18}$$

From the principle of conservation of energy, we see that the total kinetic energy just before impact is equal to $V_0 - V_f$, that is,

$$\frac{1}{2}m_1v_1^2 + \frac{1}{2}m_2v_2^2 = Gm_1m_2\left(\frac{2}{a} - \frac{1}{l}\right)$$

Also, since there are no external forces on the system, the total linear momentum is conserved. Therefore,

$$m_1v_1 + m_2v_2 = 0 \quad \text{or} \quad v_2 = -\frac{m_1}{m_2}v_1$$

Substituting this expression for v_2 into the energy equation, we obtain

$$\frac{1}{2}m_1v_1^2\left(1 + \frac{m_1}{m_2}\right) = Gm_1m_2\left(\frac{2}{a} - \frac{1}{l}\right)$$

or

$$v_1 = m_2\left[\frac{2G}{m_1 + m_2}\left(\frac{2}{a} - \frac{1}{l}\right)\right]^{1/2}$$

So the relative velocity at impact is

$$v_{rel} = v_1 - v_2 = \left(1 + \frac{m_1}{m_2}\right)v_1 = \left[2G(m_1 + m_2)\left(\frac{2}{a} - \frac{1}{l}\right)\right]^{1/2}$$

Note that this relative velocity does not depend upon the magnitude of the individual masses, but only on their sum.

5-2. THE TWO-BODY PROBLEM

Absolute and Relative Motions. Let us consider the mutual gravitational attraction between two spherical masses m_1 and m_2. We shall assume that the mass distributions are spherically symmetric and therefore that each can be considered as a particle in calculating the motions of their centers.

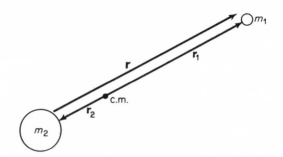

Fig. 5-3. A two-body system showing position vectors.

Furthermore, let us assume an isolated system, implying that we can choose an inertial frame of reference which translates with the center of mass. Let $\mathbf{r}_1$ and $\mathbf{r}_2$ be the position vectors of m_1 and m_2, respectively, relative to the center of mass (Fig. 5–3).

Using Newton's law of gravitation, we see that the mutual force between m_1 and m_2 is

$$f_{12} = f_{21} = -\frac{Gm_1 m_2}{(r_1 + r_2)^2} \tag{5–19}$$

where we use the convention of Fig. 4–1 that internal forces of repulsion are considered to be positive. Since r_1 and r_2 are measured from the center of mass, it is apparent that

$$\frac{r_1}{r_2} = \frac{m_2}{m_1} \tag{5–20}$$

and therefore the force on m_1 can be written in the form:

$$f_{12} = -\frac{K_1}{r_1^2} \tag{5–21}$$

where

$$K_1 = \frac{Gm_1 m_2}{[1 + (m_1/m_2)]^2} \tag{5–22}$$

Similarly,

$$f_{21} = -\frac{K_2}{r_2^2} \tag{5–23}$$

where

$$K_2 = \frac{Gm_1 m_2}{[1 + (m_2/m_1)]^2} \tag{5–24}$$

It can be seen from Eqs. (5–21) and (5–23) that the force on each mass varies inversely as the square of its distance from the center of mass. But the center of mass is fixed in an inertial frame. Therefore the two-body problem reduces to the problem of finding the motion of a particle which is attracted by an inverse-square force toward a fixed point.

Another approach which is often convenient is to solve for the motion of one mass relative to the other. Usually we solve for the motion of the smaller mass relative to the larger. So let the position vector of m_1 relative to m_2 be

$$\mathbf{r} = \mathbf{r}_1 - \mathbf{r}_2 \qquad (5\text{-}25)$$

Now, using Eq. (5-19) and Newton's equation of motion, we see that

$$\mathbf{f}_{12} = m_1 \ddot{\mathbf{r}}_1 = -m_2 \ddot{\mathbf{r}}_2 = -\frac{Gm_1 m_2}{r^3} \mathbf{r} \qquad (5\text{-}26)$$

From Eqs. (5-25) and (5-26) we have

$$\ddot{\mathbf{r}} = \ddot{\mathbf{r}}_1 - \ddot{\mathbf{r}}_2 = -\frac{G(m_1 + m_2)}{r^3} \mathbf{r} \qquad (5\text{-}27)$$

and therefore we obtain

$$\frac{m_1 m_2}{m_1 + m_2} \ddot{\mathbf{r}} = \mathbf{f}_{12} \qquad (5\text{-}28)$$

We can interpret this result as follows: The motion of m_1 relative to m_2 is the same as if m_2 were fixed and m_1 were replaced by a *reduced mass*

$$m_1' = \frac{m_1 m_2}{m_1 + m_2} \qquad (5\text{-}29)$$

the force $\mathbf{f}_{12}$ being unchanged.

Now let us calculate the kinetic energy of the system relative to an inertial frame moving with the center of mass. Assuming that m_1 and m_2 each move about the center of mass at an angular rate ω, we find that

$$T = \frac{1}{2} m_1 (\dot{r}_1^2 + r_1^2 \omega^2) + \frac{1}{2} m_2 (\dot{r}_2^2 + r_2^2 \omega^2) \qquad (5\text{-}30)$$

Using Eq. (5-20) we obtain

$$r = r_1 + r_2 = r_1 \left(\frac{m_1 + m_2}{m_2} \right) = r_2 \left(\frac{m_1 + m_2}{m_1} \right) \qquad (5\text{-}31)$$

Therefore the kinetic energy can be expressed as follows:

$$T = \frac{1}{2} m_1 \left(\frac{m_2}{m_1 + m_2} \right)^2 (\dot{r}^2 + r^2 \omega^2) + \frac{1}{2} m_2 \left(\frac{m_1}{m_1 + m_2} \right)^2 (\dot{r}^2 + r^2 \omega^2)$$

which reduces to

$$T = \frac{1}{2} m_1' (\dot{r}^2 + r^2 \omega^2) \qquad (5\text{-}32)$$

where m_1' is the reduced mass given by Eq. (5-29).

This result shows that the correct kinetic energy is obtained by assuming m_2 is fixed and using the reduced mass m_1' in place of m_1 in the standard equation for the kinetic energy. Also, we see that the potential energy is

$$V = -\frac{Gm_1 m_2}{r} = -\frac{Gm_1' m_2'}{r} \tag{5-33}$$

where

$$m_2' = (m_1 + m_2) \tag{5-34}$$

Hence, if we assume that m_2 is fixed and replace the masses m_1 and m_2 by m_1' and m_2', respectively, then the calculation of the forces, relative motions, and energies of the system will be correct. Furthermore, the proper angular momentum is obtained by using these assumptions. We assume that standard methods are used, such as Newton's laws of motion and gravitation, and the conservation of energy and angular momentum.

What has been accomplished in the foregoing development is to perform the analysis with respect to a *noninertial* frame translating with m_2. It can be seen that the same relative motion could have been obtained by changing the system parameters in several other ways if that were the only objective. By making the additional requirement that the gravitational force be unchanged, however, we insure that the energy relationships are preserved as well, since the work done by gravity involves a relative displacement.

The Orbit Equation. We have shown that the two-body problem can be reduced to the problem of a mass being attracted by an inverse-square force to a fixed point. So let us consider the case of a particle of mass m being attracted to a fixed attracting focus F, as shown in Fig. 5–4.

First we notice that the motion is confined to the plane described by the radius vector and the velocity vector of the particle, the reason being that the only force on the particle is in the negative radial direction; hence the acceleration also is in this plane at all times. Using polar coordinates, where r is measured from a fixed focus and θ is measured from a fixed reference line, we can write the equation of motion in terms of the r and θ components of force and acceleration.

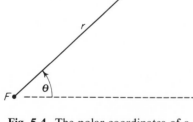

Fig. 5-4. The polar coordinates of a particle m which undergoes an inverse-square attraction toward a fixed attracting focus F.

$$F_r = ma_r = m(\ddot{r} - r\dot{\theta}^2) \tag{5-35}$$

and

$$F_\theta = ma_\theta = m(r\ddot{\theta} + 2\dot{r}\dot{\theta}) \tag{5-36}$$

It is convenient to consider the force per unit mass. Thus we can write

$$\frac{F_r}{m} = -\frac{\mu}{r^2} \tag{5-37}$$

where μ is a positive constant. It can be seen from Eq. (5-34) that, for the case of the two-body problem in which m_2 is considered to be fixed,

$$\mu = G(m_1 + m_2) \tag{5-38}$$

The value of μ could be obtained in a similar fashion for the cases of the motion of m_1 or of m_2 about their common center of mass.

From Eqs. (5-35) and (5-37), we obtain

$$\ddot{r} - r\dot{\theta}^2 = -\frac{\mu}{r^2} \tag{5-39}$$

Since $F_\theta = 0$, we see from Eq. (5-36) that

$$r\ddot{\theta} + 2\dot{r}\dot{\theta} = 0$$

or, equivalently,

$$\frac{d}{dt}(r^2\dot{\theta}) = 0 \tag{5-40}$$

Equations (5-39) and (5-40) are the differential equations of motion for the particle m.

By integrating Eq. (5-40) directly, we obtain

$$h = r^2\dot{\theta} \tag{5-41}$$

where h is a constant. It can be seen from Eq. (3-150) that h is the angular momentum per unit mass; hence Eq. (5-41) states that the angular momentum of the particle with respect to the attracting focus F is conserved.

Note that Eq. (5-39) is a nonlinear differential equation as it stands. Also, the independent variable is time, whereas we would prefer to solve for r directly as a function of θ. In order to avoid these difficulties, let us make the substitution

$$r = \frac{1}{u} \tag{5-42}$$

and note from Eq. (5-41) that

$$\dot{\theta} = hu^2 \tag{5-43}$$

In addition, we find that

$$\dot{r} = -\frac{1}{u^2}\frac{du}{d\theta}\frac{d\theta}{dt} = -h\frac{du}{d\theta} \tag{5-44}$$

and

$$\ddot{r} = -h\frac{d^2u}{d\theta^2}\frac{d\theta}{dt} = -h^2u^2\frac{d^2u}{d\theta^2} \tag{5-45}$$

Making these substitutions into Eq. (5-39) we obtain

$$-h^2 u^2 \frac{d^2 u}{d\theta^2} - h^2 u^3 = -\mu u^2$$

or

$$\frac{d^2 u}{d\theta^2} + u = \frac{\mu}{h^2} \tag{5-46}$$

This differential equation has the same mathematical form as the equation for an undamped mass-spring system with a constant forcing term, such as we discussed in Example 3-6. Hence the general solution can be written immediately. It is

$$u = \frac{\mu}{h^2} + C \cos(\theta - \theta_0) \tag{5-47}$$

where C and θ_0 are constants.

The constants can be evaluated by considering the system when $\theta = \theta_0$. From Eqs. (5-44) and (5-47), we see that $\dot{r} = 0$ at this time; therefore the total energy e per unit mass is

$$e = \frac{1}{2} r^2 \dot{\theta}^2 - \frac{\mu}{r} = \frac{1}{2} h^2 u^2 - \mu u \tag{5-48}$$

Substituting for u from Eq. (5-47), we obtain

$$e = \frac{1}{2} h^2 \left(\frac{\mu}{h^2} + C \right)^2 - \mu \left(\frac{\mu}{h^2} + C \right)$$

$$= \frac{1}{2} h^2 C^2 - \frac{\mu^2}{2h^2}$$

Solving for C, we find that

$$C = \sqrt{\frac{\mu^2}{h^4} + \frac{2e}{h^2}} \tag{5-49}$$

where we arbitrarily choose the positive square root, since any further sign changes can be handled by the choice of θ_0. It is apparent, however, that the choice of θ_0 specifies the direction of the reference line from which θ is measured. So let us set $\theta_0 = 0$ for convenience. Then we obtain the orbit equation

$$u = \frac{\mu}{h^2} \left(1 + \sqrt{1 + \frac{2eh^2}{\mu^2}} \cos \theta \right) \tag{5-50}$$

or

$$r = \frac{h^2/\mu}{1 + \sqrt{1 + (2eh^2/\mu^2)} \cos \theta} \tag{5-51}$$

Note that $r = r_{\min}$ when $\theta = 0$; therefore θ is measured from a line drawn from F to the closest point of the orbit.

Now e and h are independent of the particle position in a given orbit and are known as the *dynamical constants* of the orbit. It can be seen that, for a given gravitational coefficient μ, the size and shape of an orbit are entirely determined by these two constants.

It is interesting to calculate the rate at which the line from the attracting center to the particle sweeps out area. This so-called *areal velocity* is

$$\dot{A} = \frac{1}{2}r^2\dot{\theta} = \frac{1}{2}h \tag{5-52}$$

and is constant for any given orbit. Hence Kepler's second law, stating that the areal velocity of each planet is constant, is equivalent to the statement that the angular momentum due to the orbital motion of each planet about the sun is constant.

5-3. THE GEOMETRY OF CONIC SECTIONS

The equation for the orbit given in Eq. (5–50) or (5–51) is the polar coordinate representation of a class of curves known as *conic sections*. First, we shall present the geometrical characteristics of conic sections that are most important for our purposes. Then, with this background, we shall be able to correlate the geometrical characteristics of the various possible orbits with the dynamical constants e and h.

The equation of a general conic section, written in terms of polar coordinates, is

$$r = \frac{l}{1 + \epsilon \cos\theta} \tag{5-53}$$

where l is the *semilatus rectum* and ϵ is the *eccentricity*. It can be seen that

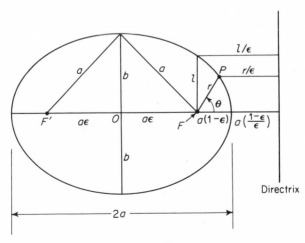

Fig. 5-5. An ellipse.

l is the parameter governing the *size* of the conic section and represents the value of r for $\theta = \pm \pi/2$. On the other hand, the eccentricity ϵ determines its *shape*, as we will show.

Ellipse. The ellipse is a conic section for which $0 \leqslant \epsilon < 1$. (We consider the circle as the special case for which $\epsilon = 0$.) An ellipse can be defined as the locus of points whose distance from a given point F is a constant factor ϵ multiplied by its distance from a straight line known as the *directrix* (Fig. 5–5). It can also be defined as the locus of points such that the sum of the distances to two foci, F and F', is a constant length $2a$ which is also the length of the major axis. The distance between the foci is just ϵ times the length of the major axis. It can be seen from Fig. 5–5 that the *semiminor axis b* is related to the *semimajor axis a* by the equation

$$b = a\sqrt{1 - \epsilon^2} \tag{5-54}$$

From Eq. (5–53) and Fig. 5–5 we obtain that the minimum value of r, that is, the *perigee distance*,[2] is

$$r_p = \frac{l}{1 + \epsilon} = a(1 - \epsilon) \tag{5-55}$$

In a similar fashion, the maximum value of r, the *apogee distance*, is found to be

$$r_a = \frac{l}{1 - \epsilon} = a(1 + \epsilon) \tag{5-56}$$

From Eqs. (5–55) and (5–56) we see that

$$a = \frac{1}{2}(r_p + r_a) \tag{5-57}$$

indicating the reason why a is sometimes known as the *mean distance*.[3] Also, we obtain from these equations an expression for the eccentricity ϵ, that is,

$$\epsilon = \frac{1}{2a}(r_a - r_p) \tag{5-58}$$

From either Eq. (5–55) or (5–56) we find that

$$l = a(1 - \epsilon^2) \tag{5-59}$$

and therefore the general equation of an ellipse, as given in Eq. (5–53), can also be written in the following form:

$$r = \frac{a(1 - \epsilon^2)}{1 + \epsilon \cos \theta} \tag{5-60}$$

This is the equation for an ellipse in terms of its *geometrical constants a* and ϵ.

[2]We shall use the terms *perigee* and *apogee* with reference to general orbits. Strictly, they apply only to orbits about the earth.

[3]Note that a is not the average distance with respect to time.

The area of an ellipse can be calculated by noting that it is just the projected area of a circle of radius a onto a nonparallel plane, resulting in a foreshortening ratio b/a for all lines parallel to the minor axis. Hence the total area of the ellipse is

$$A = \pi a^2 \left(\frac{b}{a}\right) = \pi ab \qquad (5\text{--}61)$$

Parabola. For a given perigee distance r_p, we can see from Eq. (5–55) that, as the eccentricity ϵ of an ellipse approaches unity, the semimajor axis a approaches infinity. Also, the *vacant focus F'* recedes toward infinity. In the limit, $\epsilon = 1$, and the conic section has the form of a parabola, as shown in Fig. 5–6.

Setting $\epsilon = 1$ in Eq. (5–53), we see that the equation of the parabola is

$$r = \frac{l}{1 + \cos \theta} \qquad (5\text{--}62)$$

where, in this case,

$$l = 2r_p \qquad (5\text{--}63)$$

in accordance with Eq. (5–55).

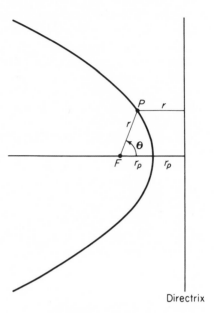

Fig. 5-6. A parabola.

Hyperbola. If we set $\epsilon > 1$ in Eq. (5–53), the resulting curve is a hyperbola. The principal geometrical parameters are shown in Fig. 5–7. The

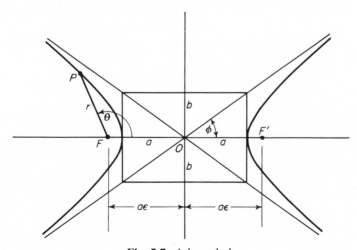

Fig. 5-7. A hyperbola.

hyperbola can be defined as the locus of points such that the difference of their distances from two fixed foci F and F' is a constant length $2a$ where, as in the case of an ellipse, the separation of the foci is $2a\epsilon$. It can be seen that the denominator of the right-hand side of Eq. (5–53) can change sign, with the result that the hyperbola has two parts or *branches*. These branches are separated by a distance $2a$, and a line of this length joining the vertices is known as the major axis. The two branches approach asymptotes making an angle ϕ with the major axis, where

$$\phi = \cos^{-1}\frac{1}{\epsilon} \tag{5–64}$$

as we can verify by noting that the magnitude of r approaches infinity as θ approaches $\cos^{-1}(-1/\epsilon)$.

The semiminor axis b is defined as the distance from the vertex to an asymptote, measured in a direction perpendicular to the major axis. From Fig. 5–7, we see that

$$b = a \tan \phi = a\sqrt{\epsilon^2 - 1} \tag{5–65}$$

Also, setting $\theta = 0$ in Eq. (5–53), we obtain

$$r_p = \frac{l}{1 + \epsilon} = a(\epsilon - 1) \tag{5–66}$$

Therefore

$$l = a(\epsilon^2 - 1) \tag{5–67}$$

and hence the general equation of the hyperbola can be written in the form:

$$r = \frac{a(\epsilon^2 - 1)}{1 + \epsilon \cos \theta} \tag{5–68}$$

We should note that if F is the attracting focus, then the actual path of a particle will be along the nearer branch of the hyperbola. For the case of inverse-square repulsion by the same focus, the particle moves along the other branch.

5–4. ORBITAL RELATIONSHIPS

Dynamical and Geometrical Constants. We saw as a result of Eq. (5–51) that the size and shape of the orbit are determined by the dynamical constants e and h. On the other hand, we found that a conic section can be expressed in terms of the geometrical constants a and ϵ. Now let us determine how the dynamical and geometrical constants are related.

First we compare Eqs. (5–51) and (5–53) and find that

$$\epsilon = \sqrt{1 + \frac{2eh^2}{\mu^2}} \tag{5–69}$$

and

$$l = \frac{h^2}{\mu} \tag{5-70}$$

Also, we see from Eqs. (5-59), (5-67) and (5-70) that the semimajor axis can be written in the form

$$a = \pm \frac{h^2}{\mu(1 - \epsilon^2)} \tag{5-71}$$

or, substituting for ϵ from Eq. (5-69),

$$a = \mp \frac{\mu}{2e} \tag{5-72}$$

where the choice of sign refers to the elliptic and hyperbolic cases, respectively. Incidentally, this sign convention will be continued throughout this chapter, that is, the upper sign refers to an elliptic orbit and the lower sign refers to a hyperbolic orbit. Notice that for a certain mass m and a given gravitational coefficient μ, the *semilatus rectum* l is a function of the angular momentum only, and the semimajor axis a is a function of the total energy only.

We have defined a and μ to be positive quantities. Therefore we see from Eq. (5-72) that the total energy must be negative for an elliptic orbit and positive for a hyperbolic orbit. For the borderline case of a parabolic orbit, the total energy is zero.

In general, the total energy per unit mass is just the sum of the kinetic and potential energies:

$$e = \frac{1}{2} v^2 - \frac{\mu}{r} \tag{5-73}$$

where v is the speed of the particle in orbit. It can be seen that energy is conserved; therefore e can be considered as the residual kinetic energy as r approaches infinity. For the hyperbolic case, e is positive and the particle retains a finite speed at infinity and therefore *escapes* from the influence of the attracting center. On the other hand, e is negative in the case of an elliptic orbit, and the particle cannot escape because the kinetic energy must be positive or zero at all times. For the parabolic case, $e = 0$ and the particle has zero velocity at infinity.

From Eqs. (5-72) and (5-73), we obtain the following equation:

$$v^2 = \mu \left(\frac{2}{r} \mp \frac{1}{a} \right) \tag{5-74}$$

This result, which is known as the *vis viva* integral, is essentially a statement of the conservation of energy. It indicates a remarkable fact, namely, that if a particle at a distance r from the focus F has a certain speed v, then the semimajor axis of its orbit is the same, regardless of the *direction* in which the particle is moving.

For the particular case of a *circular orbit*, we note that $r = a$, and obtain from Eq. (5-74) that

$$v^2 = \frac{\mu}{r} \qquad (\epsilon = 0) \tag{5-75}$$

and from Eq. (5-73),

$$e = -\frac{\mu}{2r} \qquad (\epsilon = 0) \tag{5-76}$$

Finally, it is convenient at times to write the equation for the general conic in terms of r_p. From Eqs. (5-53), (5-55), and (5-66), we find that

$$r = \frac{r_p(1 + \epsilon)}{1 + \epsilon \cos \theta} \tag{5-77}$$

this equation being valid for both the elliptic and hyperbolic cases. Now let us define r_a as the value of r corresponding to $\theta = \pi$. From Eq. (5-77), we obtain

$$r_a = \frac{r_p(1 + \epsilon)}{1 - \epsilon}$$

or

$$\epsilon = \frac{r_a - r_p}{r_a + r_p} \tag{5-78}$$

Note that r_a is negative for the hyperbolic case, its magnitude being the distance from F to the vertex of the second branch.

Orbital Period. The period T of the motion in an elliptical orbit is found by dividing the total area by the areal velocity. From Eqs. (5-52), (5-54), and (5-61), we find that

$$T = \frac{\pi ab}{\dot{A}} = \frac{2\pi a^2 \sqrt{1 - \epsilon^2}}{h}$$

But from Eq. (5-71) we see that

$$h = \sqrt{\mu a(1 - \epsilon^2)}$$

and therefore the period is

$$T = 2\pi \sqrt{\frac{a^3}{\mu}} \tag{5-79}$$

This is a statement of Kepler's third law. We note that it applies exactly for the case of a particle being attracted to a fixed point by an inverse-square force. But for the two-body problem in which the motion of m_1 is calculated relative to m_2, we recall that $\mu = G(m_1 + m_2)$. Hence, when we consider the solar system, the period depends to some extent upon the mass m_1 of the planet. This dependence turns out to be quite small because $m_2 \gg m_1$, that is, the sun's mass is many times larger than that of any of the planets.

Eccentric Anomaly. We have discussed some of the most important factors influencing the form and size of the various orbits associated with an inverse-square attracting force. When one introduces the *time* element into the analysis, either to find the position as function of time or vice versa, then the calculations are aided by introducing another geometrical parameter E known as the *eccentric anomaly* (Fig. 5–8).

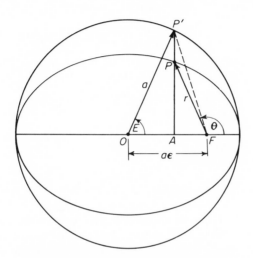

Fig. 5-8. True and eccentric anomalies.

Recall that an ellipse can be considered as the projection of a circle of radius a onto a nonparallel plane, resulting in a foreshortening factor b/a for all lines parallel to the minor axis. Reversing this process, we find that the point P on the ellipse corresponds to P' on the auxiliary circle. The angle E measured at the center O, giving the position of P' relative to the perigee, is known as the *eccentric anomaly*. On the other hand, the familiar angle θ giving the actual position P, as seen from the focus F, is known as the *true anomaly*. An equation relating E and θ is found from Fig. 5–8 by noting that

$$AP = \frac{b}{a} AP'$$

where b/a is the foreshortening ratio. Hence

$$r \sin \theta = \frac{b}{a} (a \sin E) = a\sqrt{1 - \epsilon^2} \sin E \qquad (5\text{--}80)$$

and, using the general equation for an ellipse given in Eq. (5–60), we obtain

$$\sin E = \frac{\sqrt{1 - \epsilon^2} \sin \theta}{1 + \epsilon \cos \theta} \qquad (5\text{--}81)$$

Also, we see from Fig. 5–8 that

$$r \cos \theta = a(\cos E - \epsilon) \tag{5-82}$$

Now, if we square Eqs. (5–80) and (5–82) and add, we obtain the orbit equation in terms of the eccentric anomaly, that is,

$$r = a(1 - \epsilon \cos E) \tag{5-83}$$

From Eqs. (5–82) and (5–83), we obtain

$$\cos E = \frac{\epsilon + \cos \theta}{1 + \epsilon \cos \theta} \tag{5-84}$$

Finally, from Eqs. (5–81) and (5–84), another equation relating E and θ is obtained.

$$\tan \frac{E}{2} = \frac{\sin E}{1 + \cos E} = \sqrt{\frac{1 - \epsilon}{1 + \epsilon}} \tan \frac{\theta}{2} \tag{5-85}$$

5–5. TIME AND POSITION

Elliptic Orbit. Now let us obtain an expression relating the time to the position in an elliptical orbit. We shall use the convention that time t is measured from the instant of perigee passage. First, we recall from Kepler's second law that the radius vector FP of Fig. 5–8 sweeps out equal areas in equal times. Similarly, the line FP' sweeps over equal areas in equal times since, as we have seen, FP is the projection of FP' onto the orbital plane and all areas are reduced by the constant factor b/a as a result of this projection. So the area swept over by FP' is the total area multiplied by the fractional period since perigee passage, or $\pi a^2 t / T$. But this area is just the area of a sector of vertex angle E minus the area of the triangle OFP'. So we can write

$$\frac{\pi a^2 t}{T} = \frac{1}{2} a^2 E - \frac{1}{2} a^2 \epsilon \sin E$$

or, using Eq. (5–79),

$$t = \sqrt{\frac{a^3}{\mu}} (E - \epsilon \sin E) \tag{5-86}$$

This equation can also be written in the form

$$M = \frac{2\pi t}{T} = E - \epsilon \sin E \tag{5-87}$$

where the *mean anomaly M* is the angular displacement from perigee of a line moving at the *average* angular rate $2\pi/T$. This is known as *Kepler's equation.*

In order to calculate the time required between any two points θ_1 and θ_2 in an elliptical orbit, we merely evaluate $t_2 - t_1$, where t_1 and t_2 are the corresponding times since perigee. This is accomplished in each case by first evaluating the eccentric anomaly E from a knowledge of ϵ and θ; then the time is obtained from Kepler's equation.

The reverse problem, namely, that of finding the position if the time since perigee is given, is more difficult. It involves solving Kepler's equation for the eccentric anomaly when the mean anomaly is given, and this requires the solution of a transcendental equation. The computation is often accomplished by a method of successive approximations.[4]

Parabolic Orbit. For a parabolic orbit, a knowledge of the perigee distance r_p and either r or $\dot{r}$ is sufficient to establish the true anomaly θ at that time. This can be seen by noting from Eqs. (5–62) and (5–63) that

$$\cos \theta = \frac{2r_p}{r} - 1 \qquad (5\text{–}88)$$

Differentiating Eq. (5–88) with respect to time and multiplying by r^2, we obtain

$$r^2 \dot{\theta} \sin \theta = 2r_p \dot{r}$$

But we recall from Eqs. (5–41), (5–63), and (5–70) that

$$h = r^2 \dot{\theta} = \sqrt{2r_p \mu} \qquad (5\text{–}89)$$

and therefore

$$\sin \theta = \sqrt{\frac{2r_p}{\mu}} \, \dot{r} \qquad (5\text{–}90)$$

Hence we obtain

$$\tan \frac{\theta}{2} = \frac{\sin \theta}{1 + \cos \theta} = \frac{\dot{r}}{r\dot{\theta}} \qquad (5\text{–}91)$$

indicating that θ can also be found from a knowledge of the distance r and the radial and transverse components of velocity.

To obtain the time since perigee for a parabolic orbit, we note from Eq. (5–89) and the general equation of the parabola that

$$\frac{d\theta}{dt} = \frac{1}{2} \sqrt{\frac{\mu}{2r_p^3}} (1 + \cos \theta)^2$$

or

$$t = \frac{1}{2} \sqrt{\frac{2r_p^3}{\mu}} \int_0^\theta \frac{d\theta}{\cos^4 (\theta/2)}$$

Evaluating this integral, we obtain

$$t = \sqrt{\frac{2r_p^3}{\mu}} \left(\tan \frac{\theta}{2} + \frac{1}{3} \tan^3 \frac{\theta}{2} \right) \qquad (5\text{–}92)$$

[4]For a discussion of the solution of Kepler's equation, see F. R. Moulton, *An Introduction to Celestial Mechanics* (2nd rev. ed.; New York: The Macmillan Company, 1914), pp. 160*ff*.

Of course, an exactly parabolic orbit will not exist in practice. On the other hand, the use of Eq. (5–86) in calculating times for elliptical orbits where $\epsilon \cong 1$ is not satisfactory because the period T becomes very large and the mean anomaly M becomes very small, resulting in an inaccurate product. These difficulties can be largely avoided if we expand Eq. (5–86) in terms of the deviation δ of the eccentricity from unity and the true anomaly θ. Using Eq. (5–85) and neglecting terms of order δ^2 or higher, we obtain

$$t \cong \sqrt{\frac{2r_p^3}{\mu}} \left[\left(1 + \frac{\delta}{4}\right) \tan\frac{\theta}{2} + \frac{1}{3}\left(1 - \frac{3\delta}{4}\right) \tan^3\frac{\theta}{2} - \frac{\delta}{6} \tan^5\frac{\theta}{2} \right] \quad (5\text{–}93)$$

where

$$\delta = 1 - \epsilon \quad (5\text{–}94)$$

This equation gives results with less than 1 per cent error if $-0.2 < \delta < 0.2$ and $-(\pi/2) < \theta < (\pi/2)$. Note that it applies to both the hyperbolic and elliptic cases.

Hyperbolic Orbit. We have seen that the equations applying to hyperbolic orbits are usually quite similar to those for elliptic orbits. This similarity applies to the time equations as well. So rather than presenting an explicit derivation for hyperbolic orbits, we shall transform the results obtained previously for the elliptical case, using Table 5–1 which lists corresponding quantities.

TABLE 5–1. CORRESPONDING QUANTITIES

Elliptic Orbit	Hyperbolic Orbit
a	$-a$
b	ib
θ	θ
r	r
t	t
ϵ	ϵ
r_p	r_p
μ	μ
e	e
h	h
v	v
E	$-iF$

First we define an auxiliary variable F which is analogous to the eccentric anomaly for elliptic orbits. Corresponding to Eq. (5–85), we obtain

$$\tanh\frac{F}{2} = \sqrt{\frac{\epsilon - 1}{\epsilon + 1}} \tan\frac{\theta}{2} \quad (5\text{–}95)$$

and, corresponding to Eq. (5–86), the time since perigee is found to be

$$t = \sqrt{\frac{a^3}{\mu}} \left(\epsilon \sinh F - F \right) \tag{5-96}$$

Also, corresponding to Eq. (5–83), we obtain

$$r = a(\epsilon \cosh F - 1) \tag{5-97}$$

The time equations for hyperbolic orbits can also be written in terms of circular functions. If we let

$$\sinh F = \tan H \tag{5-98}$$

where H is another auxiliary variable, then, using Eqs. (5–95) and (5–96), it can be shown that

$$\tan \frac{H}{2} = \sqrt{\frac{\epsilon - 1}{\epsilon + 1}} \tan \frac{\theta}{2} \tag{5-99}$$

and

$$t = \sqrt{\frac{a^3}{\mu}} \left[\epsilon \tan H - \ln \tan \left(\frac{\pi}{4} + \frac{H}{2} \right) \right] \tag{5-100}$$

Also, corresponding to Eq. (5–97) we obtain

$$r = a \left(\frac{\epsilon}{\cos H} - 1 \right) \tag{5-101}$$

5–6. SATELLITE ORBITS ABOUT THE EARTH

In recent years, the advent of artificial satellites has greatly increased the interest in orbits about the earth. In this brief treatment of the subject, we shall review previously derived material in the context of earth orbits and restate some of the equations in terms of dimensionless ratios which are particularly applicable to this case. Note, however, that the results of this section can also be used for orbits about the other planets or the sun if proper adjustments are made in the numerical values of various coefficients.

Circular Orbit. We recall from Eq. (5–38) that if the motion of a satellite is desired relative to the earth, then we use a gravitational coefficient

$$\mu = G(m_e + m_s) \tag{5-102}$$

where m_e is the mass of the earth and m_s is the mass of the satellite. For the case of artificial satellites, we see that $m_s \ll m_e$ and therefore we can neglect m_s in calculating μ. Also, the reduced mass in this case is just the satellite mass, as can be seen from Eq. (5–29).

Another approach is to express μ in terms of the acceleration of gravity g_0 at the surface of a spherical nonrotating earth. Setting the gravitational force per unit mass equal to the acceleration at the earth's surface, we obtain from Eq. (5–37) that

$$-\frac{\mu}{R^2} = -g_0$$

or

$$\mu = g_0 R^2 \qquad (5\text{-}103)$$

where R is the radius of the earth.

$$R = 20.91 \times 10^6 \text{ ft} = 3960 \text{ miles}$$

$$g_0 = 32.2 \text{ ft/sec}^2 = 7.90 \times 10^4 \text{ miles/hr}^2$$

$$\mu = 1.407 \times 10^{16} \text{ ft}^3/\text{sec}^2 = 1.239 \times 10^{12} \text{ miles}^3/\text{hr}^2$$

Now suppose that a satellite is in a circular orbit at the earth's surface, assuming no atmospheric drag. Noting that the gravitational attraction per unit mass is equal to the centripetal acceleration, we obtain

$$g_0 = \frac{v_c^2}{R}$$

or

$$v_c = \sqrt{g_0 R} \qquad (5\text{-}104)$$

where v_c is the speed of a body in a circular orbit at the earth's surface.

$$v_c = 25,950 \text{ ft/sec} = 17,690 \text{ miles/hr}$$

From Eqs. (5-103) and (5-104) we obtain μ in terms of v_c.

$$\mu = R v_c^2 \qquad (5\text{-}105)$$

Similarly, for a general circular orbit about the earth, the speed v is found by equating the magnitudes of the gravitational and centrifugal forces per unit mass. Thus, in accordance with Eq. (5-75),

$$\frac{v^2}{r} = \frac{\mu}{r^2}$$

or, using Eqs. (5-104) and (5-105), we obtain

$$v = v_c \sqrt{\frac{R}{r}} = R \sqrt{\frac{g_0}{r}} \qquad (5\text{-}106)$$

The total energy is found from Eqs. (5-76) and (5-105). It is

$$e = -\frac{R v_c^2}{2r} = -\frac{g_0 R^2}{2r} \qquad (5\text{-}107)$$

The period of a circular orbit at the earth's surface is

$$T_c = \frac{2\pi R}{v_c} = 1.407 \text{ hr} = 84.4 \text{ min}$$

For a general circular orbit of radius r, we obtain from Kepler's third law that the period is

$$T = \left(\frac{r}{R}\right)^{3/2} T_c \qquad (5\text{-}108)$$

General Orbit. Suppose that an earth satellite has a speed v and a flight path angle γ at a certain point in its orbit. The angle γ is measured positive upward from the local horizontal to the velocity vector **v**, as shown in Fig. 5–9. We will show that a knowledge of r, v, and γ at a given instant is sufficient to calculate the principal parameters of the orbit, aside from the orientation of the orbital plane in space.

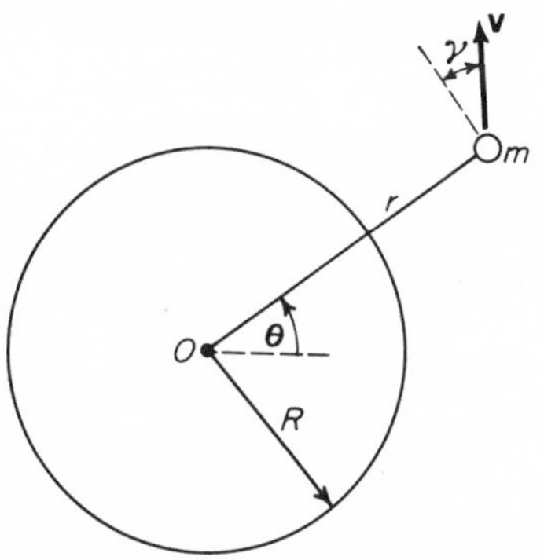

First we note that the angular momentum per unit mass is

$$h = rv \cos \gamma \qquad (5\text{–}109)$$

where we see from Fig. 5–9 that

Fig. 5-9. The flight path angle of an earth satellite.

$$\tan \gamma = \frac{\dot{r}}{r\dot{\theta}} \qquad (5\text{–}110)$$

Also, from Eqs. (5–73) and (5–105) we observe that

$$e = \frac{1}{2} v_c^2 \left[\left(\frac{v}{v_c} \right)^2 - 2\frac{R}{r} \right] \qquad (5\text{–}111)$$

Now from Eqs. (5–72), (5–105), and (5–111), we obtain an expression for the semimajor axis, namely,

$$a = \pm \frac{R}{2(R/r) - (v/v_c)^2} \qquad (5\text{–}112)$$

where the upper and lower signs again refer to elliptic and hyperbolic orbits, respectively. The eccentricity ϵ is found from Eq. (5–69) by substituting the foregoing expressions for e and h. The result is

$$\epsilon = \sqrt{1 - \left(\frac{r}{R} \right)^2 \left(\frac{v}{v_c} \right)^2 \left(2\frac{R}{r} - \frac{v^2}{v_c^2} \right) \cos^2 \gamma} \qquad (5\text{–}113)$$

or

$$\epsilon = \sqrt{\sin^2\gamma + \left(1 - \frac{rv^2}{Rv_c^2} \right)^2 \cos^2 \gamma} \qquad (5\text{–}114)$$

Hence we have computed the dynamical constants h and e and also the geometrical constants a and ϵ from a knowledge of r, v, and γ at a given time, such information being obtained from tracking data in a practical case.

Another parameter of importance is the true anomaly or, equivalently, the location of perigee. From the equation for the general conic, we obtain

$$\cos \theta = \frac{1}{\epsilon} \left[\pm \frac{a}{r} (1 - \epsilon^2) - 1 \right] \qquad (5\text{–}115)$$

which can be differentiated with respect to time to yield

$$\sin \theta = \frac{\pm a}{\epsilon r^2} (1 - \epsilon^2) \frac{\dot{r}}{\dot{\theta}} \tag{5-116}$$

Dividing Eq. (5–116) by (5–115) and noting that $\tan \gamma = \dot{r}/r\dot{\theta}$, we find that

$$\tan \theta = \frac{\pm (a/r)(1 - \epsilon^2)}{\pm (a/r)(1 - \epsilon^2) - 1} \tan \gamma$$

But, from Eqs. (5–112) and (5–113), we see that

$$\pm \frac{a}{r} (1 - \epsilon^2) = \frac{r}{R} \left(\frac{v}{v_c} \right)^2 \cos^2 \gamma$$

and therefore

$$\tan \theta = \frac{\sin \gamma \cos \gamma}{\cos^2 \gamma - \frac{R}{r} \left(\frac{v_c}{v} \right)^2} \tag{5-117}$$

Finally, it is convenient to write the expression for orbital speed in terms of dimensionless ratios. From Eqs. (5–74) and (5–105), we obtain

$$v^2 = R v_c^2 \left(\frac{2}{r} \mp \frac{1}{a} \right) \tag{5-118}$$

It is interesting to calculate the minimum speed which is required at the earth's surface in order to escape permanently from its attraction, in the absence of other bodies. Setting $a = \infty$ and $r = R$ in Eq. (5–118), we find that the *escape velocity* is

$$v_e = \sqrt{2} \, v_c \tag{5-119}$$

Its numerical value is

$$v_e = 36{,}700 \text{ ft/sec} = 25{,}000 \text{ miles/hr}$$

Example 5–2. An earth satellite has a perigee speed v_p and an apogee speed $v_a = v_p/3$. Determine r_p, r_a, ϵ, and a.

First we note that $\gamma = 0$ at perigee and apogee. Therefore, since the angular momentum is conserved, we obtain from Eq. (5–109) that

$$r_p v_p = r_a v_a$$

or

$$\frac{r_p}{r_a} = \frac{1}{3}$$

Also, the total energy is the same at perigee and apogee, so we obtain from Eq. (5–111) that

$$e = \frac{1}{2} v_p^2 - \frac{R v_c^2}{r_p} = \frac{1}{2} v_a^2 - \frac{R v_c^2}{r_a}$$

Solving this equation for r_p, we obtain

$$r_p = \frac{3}{2} R \left(\frac{v_c}{v_p}\right)^2$$

and, of course,

$$r_a = 3r_p = \frac{9}{2} R \left(\frac{v_c}{v_p}\right)^2$$

The eccentricity is found from Eq. (5–78) :

$$\epsilon = \frac{r_a - r_p}{r_a + r_p} = \frac{1}{2}$$

The semimajor axis is just

$$a = \frac{1}{2}(r_a + r_p) = 2r_p = 3R \left(\frac{v_c}{v_p}\right)^2$$

Example 5–3. Immediately after the last-stage burnout of a rocket, the following data are received concerning its motion:

$$r = 4200 \text{ miles} = 22.18 \times 10^6 \text{ ft}$$

$$\dot{r} = 12,000 \text{ ft/sec}$$

$$\dot{\theta} = 1.410 \times 10^{-3} \text{ rad/sec}$$

where the reference frame is an inertial system with its origin at the earth's center (Fig. 5–10). Find the apogee distance r_a and the time, measured from burnout, at which the rocket returns to earth.

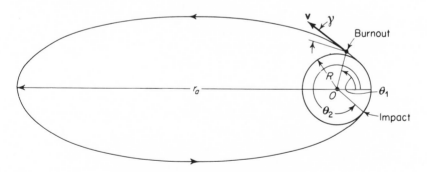

Fig. 5-10. The trajectory of a rocket.

First, we calculate the initial flight path angle γ from Eq. (5–110), obtaining

$$\tan \gamma = \frac{\dot{r}}{r\dot{\theta}} = \frac{12,000}{31,270} = 0.384$$

or

$$\gamma = 21.0°$$

The burnout velocity is

$$v = \sqrt{\dot{r}^2 + r^2 \dot{\theta}^2} = 33{,}500 \text{ ft/sec}$$

We find that

$$\frac{v}{v_c} = \frac{33{,}500}{25{,}950} = 1.291$$

$$\frac{r}{R} = \frac{4200}{3960} = 1.061$$

Thus, from Eq. (5–113) we obtain

$$\epsilon = \sqrt{1 - 0.359} = 0.801$$

We note that $\epsilon < 1$ and therefore the orbit is an ellipse. The semimajor axis is obtained from Eq. (5–112).

$$a = \frac{3960}{0.220} = 18{,}000 \text{ miles} = 9.50 \times 10^7 \text{ ft}$$

Hence the apogee distance is

$$r_a = a(1 + \epsilon) = 32{,}400 \text{ miles} = 1.712 \times 10^8 \text{ ft}$$

In order to calculate the total time, we need first to obtain the angles θ_1 and θ_2. From Eq. (5–117), we find that

$$\theta_1 = \tan^{-1} 1.095 = 47.6°$$

The corresponding eccentric anomaly is obtained by using Eq. (5–85).

$$\tan \frac{E_1}{2} = \sqrt{\frac{1 - \epsilon}{1 + \epsilon}} \tan \frac{\theta_1}{2} = 0.1466$$

or

$$E_1 = 16.7° = 0.291 \text{ rad}$$

Therefore we find from Eq. (5–86) that, at burnout, the time since perigee is

$$t_1 = \sqrt{\frac{a^3}{\mu}} (E_1 - \epsilon \sin E_1) = 0.132 \text{ hr}$$

Next, we can solve for θ_2 from Eq. (5–115) by noting that $r = R$ at impact. We obtain

$$\cos \theta_2 = 0.785$$

and, taking the value of θ_2 in the fourth quadrant, the result is

$$\theta_2 = 321.7°$$

Again we use Eq. (5–85) and find that

$$E_2 = 346.8° = 6.053 \text{ rad}$$

and, from Eq. (5–86), we obtain

$$t_2 = 13.53 \text{ hr}$$

Finally, the total time of flight from burnout until impact at the earth's surface is

$$t = t_2 - t_1 = 13.40 \text{ hr}$$

5-7. ELEMENTARY PERTURBATION THEORY

In the analysis of dynamical systems with respect to the *stability* of the motion and the *sensitivity* of the solution to small disturbances and parameter variations, the ideas and procedures of perturbation theory are widely used. We shall present in this section a few of these basic ideas and illustrate their application to the calculation of orbits. In later chapters, we shall use the same theory in the analysis of other types of motion.

In general, we shall be concerned with small deviations from a reference solution. So let us consider first the expansion of a function $F(\alpha, \beta, \gamma, \ldots)$ about a reference value $F_0(\alpha_0, \beta_0, \gamma_0, \ldots)$. Assuming that the function has derivatives of all orders everywhere in this region, we can write

$$F = F_0 + \delta F = F_0 + \left(\frac{\partial F}{\partial \alpha}\right)_0 \delta\alpha + \left(\frac{\partial F}{\partial \beta}\right)_0 \delta\beta + \left(\frac{\partial F}{\partial \gamma}\right)_0 \delta\gamma + \cdots$$

$$+ \frac{1}{2!}\left[\left(\frac{\partial^2 F}{\partial \alpha^2}\right)_0 (\delta\alpha)^2 + 2\left(\frac{\partial^2 F}{\partial \alpha\, \partial \beta}\right)_0 \delta\alpha\, \delta\beta + \left(\frac{\partial^2 F}{\partial \beta^2}\right)_0 (\delta\beta)^2 + \cdots\right] + \cdots$$

$$(5\text{-}120)$$

where the zero subscript refers to the reference value of the variable, and $\alpha = \alpha_0 + \delta\alpha$, $\beta = \beta_0 + \delta\beta$, and so on. Now, making the assumption that the deviations $\delta\alpha$, $\delta\beta$, and so on, are small, and neglecting all terms involving products of these small quantities, we obtain

$$\delta F = \left(\frac{\partial F}{\partial \alpha}\right)_0 \delta\alpha + \left(\frac{\partial F}{\partial \beta}\right)_0 \delta\beta + \left(\frac{\partial F}{\partial \gamma}\right)_0 \delta\gamma + \cdots \qquad (5\text{-}121)$$

where the partial derivatives are coefficients expressing the sensitivity of function F to small changes in the variables. In some instances, these coefficients are implicit functions of another variable, such as time. Hence they need not be constant.

As an illustrative example, let us calculate the sensitivity of the semimajor axis a and the period T of an elliptical orbit about the earth to small deviations in the burnout velocity. We assume that the independent variables are the values of r, v, and γ at burnout. We recall from Eq. (5-112) that

$$a = \frac{R}{2\dfrac{R}{r} - (v/v_c)^2}$$

Hence

$$\frac{\partial a}{\partial v} = \frac{2Rv}{v_c^2\,[2(R/r) - (v/v_c)^2]^2} = \frac{2a^2 v}{\mu} \qquad (5\text{-}122)$$

where we note from Eq. (5–105) that $\mu = R v_c^2$. So we see that a change in orbital speed is most effective in changing the semimajor axis if it occurs at the point of maximum speed, that is, at perigee.

Next we see from Eq. (5–79) that

$$T = 2\pi \sqrt{\frac{a^3}{\mu}}$$

and therefore

$$\frac{\partial T}{\partial v} = \frac{\partial T}{\partial a} \frac{\partial a}{\partial v} = 6\pi a^{5/2} \mu^{-3/2} v = \frac{3aTv}{\mu} \tag{5–123}$$

Comparing the results of Eqs. (5–122) and (5–123), we see that a given small change in orbital speed will cause a percentage change in the period that is larger than the percentage change in a by a factor $\frac{3}{2}$, as we might have shown directly from Kepler's third law.

Now let us consider the case of small perturbations from a circular reference orbit by looking at the differential equations of motion. We consider the differential equation itself as the function which undergoes small perturbations in its variables. Thus from Eq. (5–39) we can write

$$F(r, \ddot{r}, \dot{\theta}) = \ddot{r} - r\dot{\theta}^2 + \frac{\mu}{r^2} = 0 \tag{5–124}$$

Then, using Eq. (5–120) and neglecting the higher-order terms, we obtain

$$F = \ddot{r}_0 - r_0 \dot{\theta}_0^2 + \frac{\mu}{r_0^2} + \delta\ddot{r} - \left(\dot{\theta}_0^2 + \frac{2\mu}{r_0^3}\right) \delta r - 2r_0 \dot{\theta}_0 \, \delta\dot{\theta} = 0 \tag{5–125}$$

where $r = r_0 + \delta r$, $\dot{\theta} = \dot{\theta}_0 + \delta\dot{\theta}$, and so on. But Eq. (5–124) applies at all points along the reference orbit and therefore

$$\ddot{r}_0 - r_0 \dot{\theta}_0^2 + \frac{\mu}{r_0^2} = 0 \tag{5–126}$$

Also, $\ddot{r}_0 = 0$ for this case of a circular orbit; hence

$$r_0 \dot{\theta}_0^2 = \frac{\mu}{r_0^2} \tag{5–127}$$

So, from Eqs. (5–125) and (5–127), we obtain

$$\delta\ddot{r} - 3\dot{\theta}_0^2 \, \delta r - 2r_0 \dot{\theta}_0 \, \delta\dot{\theta} = 0 \tag{5–128}$$

In a similar fashion, we find from Eq. (5–40) that the second equation of motion is

$$r\ddot{\theta} + 2\dot{r}\dot{\theta} = 0$$

and the corresponding perturbation equation is

$$r_0 \, \delta\ddot{\theta} + \ddot{\theta}_0 \, \delta r + 2\dot{r}_0 \, \delta\dot{\theta} + 2\dot{\theta}_0 \, \delta\dot{r} = 0$$

or, noting that $\ddot{\theta}_0$ and $\dot{r}_0$ are zero, we obtain

$$r_0 \, \delta\ddot{\theta} + 2\dot{\theta}_0 \, \delta\dot{r} = 0 \tag{5–129}$$

We wish to solve Eqs. (5–128) and (5–129) for the case where there is a perturbation in the initial velocity. In other words, let us assume that the initial conditions are

$$\delta r(0) = 0, \quad \delta \dot{r}(0) = 0, \quad \delta \dot{\theta}(0) = \frac{\Delta v}{r_0}$$

Then, remembering that

$$r = r_0 + \delta r \qquad\qquad \dot{\theta} = \dot{\theta}_0 + \delta \dot{\theta}$$

$$\dot{r} = \frac{d}{dt}(\delta r) = \delta \dot{r} \qquad\qquad \ddot{\theta} = \frac{d}{dt}(\delta \dot{\theta}) = \delta \ddot{\theta}$$

we integrate Eq. (5–129) with respect to time, obtaining

$$r_0 \, \delta \dot{\theta} + 2 \dot{\theta}_0 \, \delta r = \Delta v \tag{5–130}$$

where the integration constant is evaluated from the initial conditions. Solving for $r_0 \, \delta \dot{\theta}$ from Eq. (5–130), we substitute into Eq. (5–128) and obtain

$$\delta \ddot{r} + \dot{\theta}_0^2 \, \delta r = 2 \dot{\theta}_0 \, \Delta v \tag{5–131}$$

This has the familiar form of the harmonic oscillator equation with a step input. Using a procedure similar to that of Example 3–6, the solution is found to be

$$\delta r = \frac{2 \, \Delta v}{\dot{\theta}_0} (1 - \cos \dot{\theta}_0 t) \tag{5–132}$$

The reference and perturbed orbits are shown in Fig. 5–11.

Now let us compare this result with that obtained previously in Eq. (5–122). There we found that the perturbation in the semimajor axis is

$$\delta a = \left(\frac{\partial a}{\partial v} \right)_0 \delta v = \frac{2 a^2 v}{\mu} \Delta v = \frac{2 r_0^3 \dot{\theta}_0}{\mu} \Delta v$$

or, substituting for μ from Eq. (5–127), we obtain

$$\delta a = \frac{2 \Delta v}{\dot{\theta}_0}$$

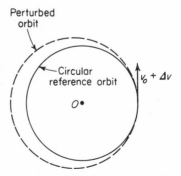

But, from Eq. (5–132) and Fig. 5–11, we see that δa is just half of the maximum amplitude of δr, namely, $2 \, \Delta v / \dot{\theta}_0$. Hence the two methods give identical results for this calculation.

The change in the period can be obtained by substituting into Eq. (5–130) the value of δr found in Eq. (5–132). The result is

$$\delta \dot{\theta} = \frac{-\Delta v}{r_0} (3 - 4 \cos \dot{\theta}_0 t) \tag{5–133}$$

Fig. 5-11. The change in an orbit due to a small increase in speed.

The period T can be found from

$$2\pi = \int_0^T (\dot{\theta}_0 + \delta\dot{\theta})\, dt = \left(\dot{\theta}_0 - \frac{3\Delta v}{r_0}\right)(T_0 + \delta T)$$

$$+ \int_0^T \frac{4\,\Delta v}{r_0} \cos \dot{\theta}_0 t\, dt$$

But

$$\dot{\theta}_0 T_0 = 2\pi$$

Also, the integral involving the cosine is of order $\Delta v\,\delta T$ because the interval of integration deviates by δT from corresponding to exactly one cycle of the cosine function. Therefore, if we omit all terms of order higher than 1 in the small quantities, we obtain

$$-\frac{3\,\Delta v}{r_0}T_0 + \dot{\theta}_0\,\delta T = 0$$

or

$$\delta T = \frac{3\,\Delta v}{r_0 \dot{\theta}_0} T_0 \tag{5-134}$$

We see that this agrees with the result of Eq. (5–123) for this case of a circular reference orbit where $a = r_0$, $v = r_0\dot{\theta}_0$, and $\mu = r_0^3\dot{\theta}_0^2$.

It is interesting to note that the solution for δr given in Eq. (5–132) has the form of a small undamped sinusoidal oscillation. Hence the deviation of the perturbed orbit from the reference orbit remains small. If an infinitesimal disturbance results in an infinitesimal change in the orbit, then the orbit is regarded as *stable*. Notice from Eq. (5–133) that the angle θ drifts slowly relative to the reference value, but this is due to a small change in orbital frequency and is not considered to be an instability.

Example 5–4. A spherical pendulum of length a undergoes conical motion with $\theta = \theta_0$ (Fig. 5–12). Find the frequency of small deviations in θ from this reference value.

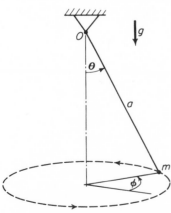

Fig. 5-12. A spherical pendulum.

Let us choose ϕ and θ as coordinates, where ϕ is the usual spherical coordinate and θ is measured from the downward vertical. The differential equation for the θ motion can be obtained by setting the acceleration in the θ direction equal to $-g \sin\theta$, in accordance with Newton's law of motion. Hence we can write

$$a\ddot{\theta} - a\dot{\phi}^2 \sin\theta \cos\theta = -g \sin\theta \tag{5-135}$$

The ϕ equation is a statement of the

conservation of angular momentum about the vertical axis through the point of support.

$$\frac{d}{dt}(ma^2 \sin^2\theta \,\dot\phi) = 0$$

or

$$\sin^2\theta \,\dot\phi = \sin^2\theta_0 \,\dot\phi_0 \tag{5-136}$$

In the reference condition, $\ddot\theta = 0$; therefore we see from Eq. (5-135) that

$$\dot\phi_0^2 = \frac{g}{a \cos\theta_0} \tag{5-137}$$

Noting that the equations of motion apply to the reference and the perturbed conditions, we use Eqs. (5-121) and (5-135), setting $\delta F = 0$, and obtain the following differential equation:

$$a\,\delta\ddot\theta + [a\dot\phi_0^2(\sin^2\theta_0 - \cos^2\theta_0) + g\cos\theta_0]\,\delta\theta \\ - 2a\dot\phi_0 \sin\theta_0 \cos\theta_0 \,\delta\dot\phi = 0 \tag{5-138}$$

Similarly, we obtain from Eq. (5-136) that

$$2\cos\theta_0 \,\dot\phi_0 \,\delta\theta + \sin\theta_0 \,\delta\dot\phi = 0 \tag{5-139}$$

Eliminating $\delta\dot\phi$ between Eqs. (5-138) and (5-139), we obtain

$$a\,\delta\ddot\theta + [a\dot\phi_0^2(\sin^2\theta_0 + 3\cos^2\theta_0) + g\cos\theta_0]\,\delta\theta = 0 \tag{5-140}$$

Finally, substituting the value of $\dot\phi_0^2$ from Eq. (5-137) into Eq. (5-140), we see that

$$\delta\ddot\theta + \frac{g}{a}(4\cos\theta_0 + \sin\theta_0 \tan\theta_0)\,\delta\theta = 0 \tag{5-141}$$

which is the differential equation corresponding to harmonic motion. The angular frequency ω of the small oscillations of $\delta\theta$ is given by

$$\omega = \left[\frac{g}{a}(4\cos\theta_0 + \sin\theta_0 \tan\theta_0)\right]^{1/2} \tag{5-142}$$

It is interesting to consider the case where $\theta_0 \ll 1$. Using series representations of the trigonometric functions, we obtain from Eq. (5-137) the approximate result that

$$\dot\phi_0 \cong \sqrt{\frac{g}{a}}\left(1 + \frac{\theta_0^2}{4}\right) \tag{5-143}$$

where terms of order higher than θ_0^2 have been neglected. In a similar fashion, we see that Eq. (5-142) can be approximated by

$$\omega \cong \sqrt{\frac{g}{a}(4 - \theta_0^2)} = 2\sqrt{\frac{g}{a}}\left(1 - \frac{\theta_0^2}{8}\right) \tag{5-144}$$

Hence there are about two cycles of the motion in $\delta\theta$ for every revolution

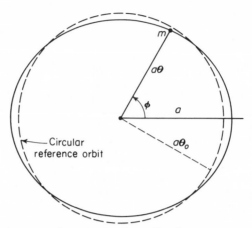

Fig. 5-13. An approximately circular orbit of a spherical pendulum for the case where $\theta_0 \ll 1$.

in the nearly circular orbit. The shape of the resulting orbit is shown in Fig. 5–13.

Now, it can be seen that the time required for two cycles of the motion in $\delta\theta$ is

$$t_1 = \frac{4\pi}{\omega} \cong 2\pi \sqrt{\frac{a}{g}} \left(1 + \frac{\theta_0^2}{8} \right)$$

But, in this time, ϕ moves through an angle

$$\phi_1 = \dot\phi_0 t_1 \cong 2\pi \left(1 + \frac{3\theta_0^2}{8} \right)$$

In other words, ϕ moves through slightly more than one revolution, causing the major axis of the nearly elliptical orbit to *precess* slowly in the direction of the rotation.

5–8.   EXAMPLES

Example 5–5. A particle approaches the earth from a great distance with an initial velocity $v_\infty = v_c$ which is directed along a line missing the center of the earth by a distance b (Fig. 5–14). Solve for the value of b such that the actual orbit is tangent to the earth's surface. What is the eccentricity of this orbit? (Assume a fixed earth.)

First, we note that the velocity is not zero at infinity; therefore the total energy is positive, corresponding to a hyperbolic orbit. Designating the velocity at perigee by v_p, we can equate the initial kinetic energy to the

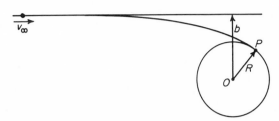

Fig. 5-14. An orbit tangent to the surface of the earth, corresponding to an effective collision radius b.

total energy at perigee. Using Eq. (5-105), we have

$$\frac{1}{2} v_\infty^2 = \frac{1}{2} v_p^2 - \frac{\mu}{R} = \frac{1}{2} v_p^2 - v_c^2 \qquad (5\text{-}145)$$

But, using conservation of angular momentum about the earth's center, we see that

$$b v_\infty = R v_p \qquad (5\text{-}146)$$

From Eqs. (5-145) and (5-146), we obtain

$$\frac{1}{2} v_\infty^2 = \frac{1}{2} \frac{b^2}{R^2} v_\infty^2 - v_c^2$$

or

$$b = R \sqrt{1 + 2 \left(\frac{v_c}{v_\infty}\right)^2} \qquad (5\text{-}147)$$

In this particular case, $v_\infty = v_c$, and therefore

$$b = \sqrt{3}\, R$$

The eccentricity of the orbit is found by evaluating Eq. (5-113) at perigee. From Eq. (5-146), we note that $v_p = \sqrt{3}\, v_c$; hence we find that

$$\epsilon = \sqrt{1 - \left(\frac{r_p}{R}\right)^2 \left(\frac{v_p}{v_c}\right)^2 \left[2\frac{R}{r_p} - \left(\frac{v_p}{v_c}\right)^2\right]} = 2$$

It can be seen that the particle moves initially along the asymptote, but at perigee it moves in a direction perpendicular to the major axis. Thus its velocity vector rotates through an angle $\pi/2 - \phi = \pi/2 - \cos^{-1}(1/\epsilon) = \pi/6$ during its approach.

    The distance b given by Eq. (5-147) is known as the *effective collision radius*, that is, it is the radius of a gravitationless sphere which would be equally effective in intercepting particles having the given velocity v_∞. This concept can be useful in calculating the accuracy required to hit the moon or one of the planets. As a first approximation, we consider the center of the attracting body to be translating uniformly and hence it is fixed in an inertial

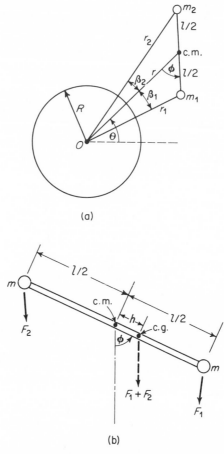

(a)

(b)

Fig. 5-15. A dumbbell satellite, showing the difference between the center of mass and the center of gravity.

frame. Then v_∞ is taken as the relative velocity of the particle at the time when it comes within effective gravitational range of the attracting body. It can be seen that the effective collision radius is larger than the actual radius, the difference decreasing with increasing v_∞.

Example 5–6. Two particles, each of mass m, are connected by a massless rod of length l. The center of mass of this system is in a circular orbit of radius r about the earth—Fig. 5–15 (a). Calculate the frequency of small oscillations of the coordinate ϕ in the plane of the orbit. Assume that $l \ll r$ and $|\phi| \ll 1$.

Our general plan of attack on this problem is to choose the center of mass as the reference point for applying the equation $M_c = \dot{H}_c$. But first let us consider the angular momentum H about the center of the earth. We recall that the total angular momentum is the sum of that due to the motion of the center of mass plus the angular momentum due to motion relative to the center of mass. Thus from Eq. (4–51), we see that

$$H = 2mr^2\dot{\theta} + \frac{ml^2}{2}(\dot{\theta} + \dot{\phi}) \tag{5–148}$$

where we note that $(\dot{\theta} + \dot{\phi})$ is the absolute rotation rate of the system. The gravitational field is a central force field, and therefore it exerts no moment about the attracting center O. Hence H is constant. Now let us differentiate Eq. (5–148) with respect to time and solve for $\ddot{\theta}$, obtaining

$$\ddot{\theta} = -\frac{l^2}{4r^2 + l^2}\ddot{\phi} \tag{5–149}$$

which we note is of second order in the small ratio l/r. Hence we see that $|\ddot{\theta}| \ll |\ddot{\phi}|$.

Now consider the external forces acting on m_1 and m_2. From Eqs. (5–37) and (5–103), we see that the force on m_1 is of magnitude

$$F_1 = \frac{mg_0 R^2}{r_1^2} = \frac{mg_0 R^2}{r^2 + (l^2/4) - rl\cos\phi} \tag{5–150}$$

Similarly,

$$F_2 = \frac{mg_0 R^2}{r_2^2} = \frac{mg_0 R^2}{r^2 + (l^2/4) + rl\cos\phi} \tag{5–151}$$

where r_1 and r_2 are evaluated by using the cosine law. These forces are not quite equal in magnitude and direction because the gravitational field is nonuniform. Hence it is possible for a gravitational moment to exist relative to the center of mass. To show this, we note that

$$M_c = \frac{1}{2}l[F_2\sin(\phi - \beta_2) - F_1\sin(\phi + \beta_1)] \tag{5–152}$$

and, applying the sine law, we obtain

$$\sin(\phi + \beta_1) = \frac{r}{r_1}\sin\phi \tag{5–153}$$

$$\sin(\phi - \beta_2) = \frac{r}{r_2}\sin\phi \tag{5–154}$$

Therefore we find that

$$M_c = \frac{mg_0 R^2 lr}{2}\left(\frac{1}{r_2^3} - \frac{1}{r_1^3}\right)\sin\phi \tag{5–155}$$

Next we expand r_1^{-3} and r_2^{-3} in powers of l/r, where r_1 and r_2 are again evaluated from the cosine law. Assuming $\cos\phi \cong 1$, we obtain the approximations:

$$r_1^{-3} \cong \frac{1}{r^3}\left(1 + \frac{3l}{2r}\right)$$

$$r_2^{-3} \cong \frac{1}{r^3}\left(1 - \frac{3l}{2r}\right)$$

Then, assuming $\sin \phi \cong \phi$, we find that

$$M_c \cong -\frac{3mg_0 R^2 l^2}{2r^3}\phi \tag{5-156}$$

But, in obtaining Eq. (5-148), we saw that the angular momentum about the center of mass is

$$H_c = \frac{ml^2}{2}(\dot{\theta} + \dot{\phi})$$

Therefore, if we set $M_c = \dot{H}_c$ and neglect $\ddot{\theta}$ relative to $\ddot{\phi}$, we obtain

$$-\frac{3mg_0 R^2 l^2}{2r^3}\phi = \frac{ml^2}{2}\ddot{\phi}$$

or

$$\ddot{\phi} + \frac{3g_0 R^2}{r^3}\phi = 0 \tag{5-157}$$

The frequency of the small oscillations in ϕ is

$$\omega = \sqrt{\frac{3g_0 R^2}{r^3}} \tag{5-158}$$

The angular rate ω_0 of the satellite in its circular orbit around the earth is found from Eq. (5-106).

$$\omega_0 = \dot{\theta} = \frac{v}{r} = \sqrt{\frac{g_0 R^2}{r^3}}$$

Hence we obtain the interesting result that

$$\omega = \sqrt{3}\,\omega_0$$

that is, the orbital period is longer than the period of the oscillation in ϕ by a factor $\sqrt{3}$.

This example provides a good illustration of the difference between the *center of mass* and the *center of gravity*. We recall from Eq. (4-6) that the position of the center of mass is a function of just the positions and masses of the particles. On the other hand, the location of the center of gravity of a system depends not only upon its configuration, but also upon the nature of the external gravitational field acting upon the system. The center of gravity is a point through which the total gravitational force is assumed to act. This resultant gravitational force has the same moment about any reference point as the actual system of gravitational forces. In order to specify the exact location of the center of gravity along the line of action of the resultant gravitational force, one can think of giving the system an infinitesimal rigid-body rotation. Then the center of gravity is that point about which one can make an arbitrary infinitesimal rotation of the system without changing the line of action of the resultant gravitational force.

In this particular case, the center of gravity is at a distance h from the center of mass—Fig. 5-15(b)—such that the correct gravitational moment is obtained if the total force is considered to act through that point. To calculate h, we note first that ϕ is assumed to be small and therefore we find from Eqs. (5-153) and (5-154) that

$$\beta_1 + \beta_2 \cong r \left(\frac{1}{r_1} - \frac{1}{r_2} \right) \phi \cong \frac{l}{r} \phi$$

Also, $\cos (\beta_1 + \beta_2) \cong 1$, and the vector sum of the gravitational forces acting on the system is of magnitude $F_1 + F_2$. Furthermore, this resultant force must be directed toward the center of the earth, since the original force system was *concurrent* at this point, that is, each individual force passed through point O (see Sec. 7-10).

So, equating the moment of $F_1 + F_2$ about the center of mass with that obtained previously in Eq. (5-156), we find that

$$-(F_1 + F_2)h\phi = -\frac{3mg_0 R^2 l^2}{2r^3} \phi$$

But

$$F_1 + F_2 \cong \frac{2mg_0 R^2}{r^2}$$

and therefore

$$h = \frac{3l^2}{4r} \tag{5-159}$$

indicating that h is smaller than the particle separation l by an order of magnitude. Thus the assumption implicit in the moment equation, namely, that the line of action of $F_1 + F_2$ is parallel to the position vector r, is valid.

It should be pointed out that the center of mass and the center of gravity coincide for the case of a uniform gravitational field. On the other hand, the center of gravity does not exist for the more general case of a gravitational field which arises from many sources. This is so because a single resultant force can be obtained only for a concurrent force system or the special case of parallel forces. A general set of forces requires at least a force plus a couple as an adequate representation.

Example 5-7. Three particles of equal mass m are initially located at the corners of an equilateral triangle with a mutual separation c. They each have initial velocities $v_0 = (Gm/c)^{1/2}$ in the direction shown in Fig. 5-16. Assuming that the only forces acting are due to gravity, solve for $r_{\min}$ and $r_{\max}$ in the ensuing motion. What is the period?

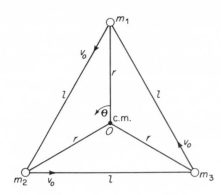

Fig. 5-16. Three mutually-attracting particles in symmetrical motion.

It is apparent that because of the symmetry of the initial configuration and velocities, the particles will always be located at the corners of an equilateral triangle. The separation l will change, however, and consequently the rotation rate $\dot{\theta}$ will vary in a manner such that the angular momentum is conserved about the center of mass.

First let us calculate the total energy of the system. From Eq. (5–10), noting that $l = \sqrt{3}\, r$, we find that the potential energy is

$$V = -\frac{3Gm^2}{l} = -\frac{\sqrt{3}\,Gm^2}{r} \qquad (5\text{–}160)$$

The total kinetic energy is

$$T = \frac{3}{2}mv^2 = \frac{3}{2}m\,(\dot{r}^2 + r^2\dot{\theta}^2) \qquad (5\text{–}161)$$

No external forces act on the system; therefore the total energy is conserved. So, evaluating the total energy from initial conditions, we find that in general,

$$\frac{3}{2}m(\dot{r}^2 + r^2\dot{\theta}^2) - \sqrt{3}\,\frac{Gm^2}{r} = \frac{3}{2}mv_0^2 - 3\frac{Gm^2}{c} \qquad (5\text{–}162)$$

But the total angular momentum about the center of mass is constant. Therefore,

$$H = 3mr^2\dot{\theta} = \frac{\sqrt{3}}{2}mv_0c \qquad (5\text{–}163)$$

where the term on the right is determined from initial conditions. Solving for $\dot{\theta}$, we obtain

$$\dot{\theta} = \frac{cv_0}{2\sqrt{3}\,r^2} \qquad (5\text{–}164)$$

When r is a maximum or a minimum, $\dot{r} = 0$. So, at this time, we obtain from Eqs. (5–162) and (5–164) that

$$mv_0^2\left(\frac{c^2}{8r^2} - \frac{3}{2}\right) = Gm^2\left(\frac{\sqrt{3}}{r} - \frac{3}{c}\right)$$

or, substituting the value of v_0 and rearranging,

$$r^2 - \frac{2c}{\sqrt{3}}r + \frac{c^2}{12} = 0$$

Solving for the roots, we obtain

$$r_{min} = c\left(\frac{1}{\sqrt{3}} - \frac{1}{2}\right) = 0.077c$$

$$r_{max} = c\left(\frac{1}{\sqrt{3}} + \frac{1}{2}\right) = 1.077c$$

Now let us consider the forces acting on the particles. Using Eqs. (4–25) and (5–160), we find that the resultant force on each particle is in the radial direction. Since a change in r results in work being done on all three particles,

$$F_r = -\frac{1}{3}\frac{\partial V}{\partial r} = -\frac{Gm^2}{\sqrt{3}\,r^2} \tag{5–165}$$

Thus each particle is attracted by a force which varies inversely as the square of the distance from the center of mass. Hence we find from Eqs. (5–37) and (5–165) that the effective gravitational coefficient is

$$\mu = Gm/\sqrt{3}$$

Each particle moves in an elliptical orbit with the semimajor axis equal to

$$a = \frac{r_{min} + r_{max}}{2} = \frac{c}{\sqrt{3}}$$

From Eq. (5–79), we find that the period is

$$T = 2\pi\sqrt{\frac{a^3}{\mu}} = 2\pi\sqrt{\frac{c^3}{3Gm}}$$

REFERENCES

1. Baker, R. M., Jr., and M. W. Makemson, *An Introduction to Astrodynamics.* New York: Academic Press, 1960.
2. Blanco, V. M., and S. W. McCuskey, *Basic Physics of the Solar System.* Reading, Mass.: Addison-Wesley Publishing Company, 1961.
3. Ehricke, K. A., *Space Flight*, Vol. 1. Princeton, N.J.: D. Van Nostrand Co., Inc., 1960.
4. Moulton, F. R., *An Introduction to Celestial Mechanics*, 2nd rev. ed. New York: The Macmillan Company, 1914.
5. Synge, J. L., and B. A. Griffith, *Principles of Mechanics*, 3rd ed. New York: McGraw-Hill, Inc., 1959.
6. Thomson, W. T., *Introduction to Space Dynamics.* New York: John Wiley & Sons, Inc., 1961.

PROBLEMS

5–1. Assuming that the orbit of the earth about the sun is circular with a radius of 93×10^6 miles, calculate the absolute velocity at the earth's surface necessary to escape from the solar system. Use an inertial coordinate system fixed at the center of the sun and neglect the gravitational attraction of all other planets except the earth.

5–2. Suppose a straight shaft is bored through the center of the earth. What is the oscillation period of a ball of mass m which is dropped into the shaft from the surface, assuming a spherical nonrotating earth? Calculate the fractional change in the period if one now assumes a frictionless shaft whose ends are at a latitude angle λ on a rotating earth.

5–3. Consider a particle which is moving in an elliptical orbit about a fixed focus due to an inverse-square law of attraction. Find the points in the orbit at which (a) the radial velocity $\dot{r}$ is maximum; (b) the speed is equal to that of a particle in a circular orbit at the same distance.

5–4. A satellite weighing 1000 lbs is in a nearly circular orbit at an altitude of 100 miles above the surface of the earth. If the satellite is losing altitude at the rate of 0.5 miles per day at this time, what is the mean drag force? Show how this drag force actually permits the tangential speed of the satellite to increase.

5–5. What is the longest possible time required for a satellite to travel from the surface of the earth to the moon's orbit which is assumed to be circular with a radius of 239,000 miles? Neglect the attraction of the moon.

5–6. What is the minimum velocity relative to the moon that a missile must have in order to hit the earth? Assume a lunar orbit of 239,000 miles radius which is centered at the earth, and neglect the gravitational attraction of the moon.

5–7. A satellite of mass m is in a circular orbit of radius $2R$ about the earth, where R is the earth's radius. Then a thrust of constant magnitude $F = mg_0/8$ is applied to the satellite, always being directed toward the center of the earth. Assuming that the mass of the satellite is constant, find its minimum distance from the earth and the maximum velocity in the resulting motion.

5–8. Two satellites, each of mass m, are in coplanar circular orbits about the earth. The inner orbit is of radius $3R$ and the outer orbit is of radius $4R$, where R is the radius of the earth. A massless wire of length R connects the two satellites and they move together about the earth while remaining in the given orbits. Find the common period and the force in the wire.

5–9. A satellite consists of two equal spherical masses connected by a rigid rod of negligible mass. Their centers are separated by a distance $l \ll R$. The satellite moves in a circular orbit of radius $5R$ about the earth and the rod is always aligned with the radial direction. If the mass of each sphere is $m = 1$ slug, find the separation l of their centers which results in the force in the rod being zero.

5–10. Consider the sensitivity of the eccentricity of an elliptical orbit to errors in the burnout velocity. For a given r and γ, solve for $(\partial \epsilon / \partial v)$ in terms of the burnout conditions and the geometrical constants a and ϵ.

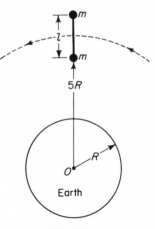

Fig. P5-9

5–11. A particle moves in an elliptical orbit under the influence of an inverse-square attraction to a fixed point. Show that the hodograph of its motion is a circle of radius μ/h with the origin at a distance $\epsilon\mu/h$ from the center of the circle.

5–12. A particle of mass m is attracted to a fixed point O by a force that varies directly with its distance r, that is, $F_r = -kr$. It undergoes orbital motion according to the initial conditions $r(0) = l$, $\dot{r}(0) = l\sqrt{k/m}$, $\theta(0) = 0$, $\dot{\theta}(0) = \sqrt{k/m}$. Find: (a) the differential equation for r, using time as the independent variable; (b) the values of $r_{\max}$ and $r_{\min}$; (c) the period of the orbit.

5–13. A particle is attracted to a fixed point according to the law $F_r = -K/r^n$, where K is a positive constant. Use perturbation theory to show that a stable circular orbit is possible for all cases in which $n < 3$, that is, an infinitesimal disturbance will result in an infinitesimal change in the shape of the orbit.

5–14. A satellite is in a circular orbit at a height R above the earth's surface, where R is the radius of the earth. A missile is fired from the earth with a burnout velocity of 27,000 ft/sec and intercepts the satellite "head-on." Assuming that the missile has a short burning time and that all motion takes place in the same plane about a nonrotating earth, find: (a) the required flight path angle γ of the missile at burnout; (b) the time at which the missile must be fired compared to the appearance of the satellite over the horizon.

5–15. Suppose a rocket is fired into a hyperbolic orbit at the earth's surface and it has a burnout velocity $v(0) = 2v_c$. Solve for the initial flight path angle γ such that its final direction of motion is 30° above the horizon. What is the eccentricity of the orbit? Assume a nonrotating earth.

5–16. A satellite of mass m is initially in a circular orbit of radius $2R$ about the earth. Then a constant thrust of magnitude $10^{-4} mg_0$ is applied in the $\mathbf{e}_\theta$ direction for exactly one orbital revolution. Solve for fractional change in the speed and radial distance of the satellite between the beginning and the end of the thrust interval.

5–17. A satellite is in a circular orbit which passes over the earth's poles and has a period of three hours. An observer P views the satellite from the north pole while another observer E views it from a point on the equator. Suppose that P and E each measure the motion of the satellite when it is midway from the north pole to the equator and is located in the plane of the observers and the earth's

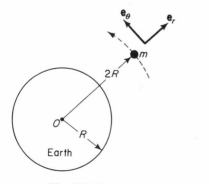

Fig. P5-16

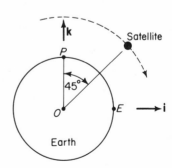

Fig. P5-17

center O. Evaluate the velocity and acceleration of the satellite relative to each observer for the following cases: (a) the observers' reference frames are fixed in the earth and rotate with it; (b) the reference frames do not rotate but translate with each observer.

5-18. Suppose you are inside a windowless box that is moving in *one* of the following ways: (a) translating with uniform velocity in a gravity-free region; (b) translating about the earth in a circular orbit above the atmosphere; (c) falling freely in a uniform gravitational field of magnitude g_0. What mechanical experiments, if any, can you perform inside the box to determine your actual motion?

6

LAGRANGE'S EQUATIONS

Thus far in the study of the dynamics of particles, we have depended upon a rather direct application of Newton's laws of motion. If we are given the motions of a set of particles, we can deduce the forces acting. Conversely, given the forces, we can solve for the motions, at least in theory.

As one analyzes more complicated systems, the direct application of Newton's laws of motion becomes increasingly difficult. The principal reasons are that the equations are vectorial in nature and the forces and accelerations are often difficult to determine. This is particularly true of any forces required to maintain constraints on the motions of particles in the system. Also, each problem seems to require its own particular insights and there are no general procedures for obtaining the differential equations of motion.

In this chapter, we shall present the Lagrangian approach to the formulation of the equations of motion. It will be shown that this method circumvents to some extent the difficulties found in the direct application of Newton's laws of motion to complicated systems. Furthermore, the use of Lagrange's equations presents the equations of motion in a standard convenient form.

6-1. DEGREES OF FREEDOM

An important concept in the description of a dynamical system is that of *degrees of freedom*. The number of degrees of freedom is equal to the number of coordinates which are used to specify the configuration of the system minus the number of independent equations of constraint. For example, if one chooses n coordinates to define the configuration of a system, and if there are m independent equations of constraint relating these coordinates, then there are $(n - m)$ degrees of freedom. Quite frequently, it is possible to find a set of *independent* coordinates which describe the configuration and which can vary freely without violating the constraints. In this case there are as many degrees of freedom as there are coordinates.

Consider the example of a particle which moves on the surface of a fixed sphere of radius R. If its position is specified by the cartesian coordinates (x, y, z), we see immediately that these coordinates are not independent. In fact, they are connected by the equation of constraint

$$(x - x_0)^2 + (y - y_0)^2 + (z - z_0)^2 = R^2$$

where the center of the sphere is located at (x_0, y_0, z_0). Thus there are three coordinates and one equation of constraint, resulting in two degrees of freedom.

It is interesting to note that if we had chosen the origin of the coordinate system at the center of the sphere and had used the ordinary spherical coordinates θ and ϕ, then no additional coordinates would have been needed. The requirement that the motion be confined to the surface of the sphere implies that $r = R$ and therefore r is not a variable. Hence the coordinates θ and ϕ are independent, conform to the constraints of the problem, and represent in an effective manner the two degrees of freedom of the system.

As another example consider two particles m_1 and m_2 that are connected by a rigid, massless rod of length l, as shown in Fig. 6–1. The free motion of this system in space can be described in terms of the cartesian coordinates (x_1, y_1, z_1) of m_1 and the corresponding coordinates (x_2, y_2, z_2) of m_2. The rod causes the separation between the particles to be a constant distance l, and therefore the equation of constraint is

$$(x_1 - x_2)^2 + (y_1 - y_2)^2 + (z_1 - z_2)^2 = l^2$$

In this case there are six coordinates ($n = 6$) and one equation of constraint ($m = 1$). So we see that there are five degrees of freedom, since $n - m = 5$.

On the other hand, the configuration of this system could have been specified by giving the position (x_0, y_0, z_0) of a certain point on the rod,

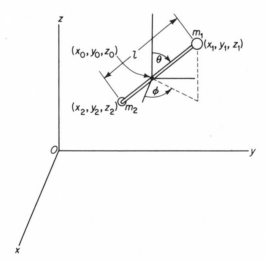

Fig. 6-1. Two particles connected by a rigid rod.

such as the center of mass, and then using the spherical coordinates θ and ϕ to indicate the orientation of the rod in space. This would require a total of five coordinates corresponding to the five degrees of freedom. Since each of these coordinates can vary freely without violating any constraints, there are no equations of constraint.

It is important to realize that the number of degrees of freedom is a characteristic of the system itself and does not depend upon the particular set of coordinates used to describe its configuration. In other words, whereas the choice of coordinates influences n and m, the number $(n - m)$ is fixed for a given system.

6-2. GENERALIZED COORDINATES

We have seen that the configuration of a given system may be expressed in terms of various sets of coordinates. Hence no specific set of coordinates is uniquely suited to the analysis of a given mechanical system. Many coordinate systems are possible; in fact, there is an infinite number. Furthermore, the number of coordinates used in any one system can vary widely, so long as there are at least as many coordinates as there are degrees of freedom. But, in any case, the number of coordinates is equal to the number of degrees of freedom plus the number of independent equations of constraint.

Now suppose that we express the configuration of a system of particles by giving the cartesian coordinates of each of the particles. In this case, $3N$ coordinates would be required for N particles. At any instant, the values of these coordinates could be expressed as a set of $3N$ numbers. On the other hand, if the positions of the particles were expressed in terms of spherical coordinates, then another set of $3N$ numbers would be used to express the configuration of the same system at the given time. The process of obtaining one set of numbers from the other is known as a *coordinate transformation*. Now we can conceive of certain transformations which might result in a set of numbers that do not have a discernible geometrical significance. Nevertheless, they would specify the configuration of the system, and thus could be considered as coordinates in a more general sense. Any set of numbers which serve to specify the configuration of a system are examples of *generalized coordinates*. For systems in motion, these numbers vary with time and are treated as algebraic variables. Note that the term *generalized coordinates* can refer to any of the commonly used coordinate systems, but it can also refer to any of an infinite variety of other sets of parameters which serve to specify the configuration of a system.

As we proceed to a discussion of Lagrange's equations, it will become apparent that in many instances the mathematical analysis of a dynamical system is simplified by choosing a set of *independent* generalized coordinates. In this case, the number of generalized coordinates is equal to the number of degrees of freedom; hence there are no equations of constraint. Furthermore, for a large class of systems, a straightforward procedure exists for obtaining the equations of motion in terms of the generalized coordinates.

Consider now a transformation from a set of k ordinary coordinates to a set of n generalized coordinates. The transformation equations are of the form

$$x_1 = f_1(q_1, q_2, \ldots, q_n, t)$$
$$x_2 = f_2(q_1, q_2, \ldots, q_n, t)$$
$$\cdot \ \cdot \ \cdot$$
$$\cdot \ \cdot \ \cdot \qquad\qquad (6\text{--}1)$$
$$\cdot \ \cdot \ \cdot$$
$$x_k = f_k(q_1, q_2, \ldots, q_n, t)$$

where the x's are ordinary coordinates and the q's are generalized coordinates. Each system of coordinates may have a set of constraint equations associated with it. For example, suppose there are l equations of constraint relating the x's and m equations of constraint relating the q's. If the configuration of the system is completely specified by each set of coordinates, and if the equations of constraint for each set are independent, then, equating the number of degrees of freedom in each case, we find that

$$k - l = n - m \qquad\qquad (6\text{--}2)$$

6–3. CONSTRAINTS

Holonomic Constraints. Now let us take a closer look at the problem of representing and classifying constraints. Consider a system described by n generalized coordinates $q_1, q_2, \ldots, q_n$. Suppose there are m constraint equations of the form

$$\phi_j(q_1, q_2, \ldots, q_n, t) = 0 \quad (j = 1, 2, \ldots, m) \qquad (6\text{--}3)$$

Constraints of this form are known as *holonomic* constraints. In theory, one could use these equations to solve for m coordinates in terms of the $(n - m)$ remaining coordinates and time, thereby retaining only as many generalized coordinates as there are degrees of freedom. Sometimes it is not feasible or even desirable to do this. In these instances one can use an approach such as the Lagrange multiplier method, as we shall see in Sec. 6–7. More frequently, one searches for a set of generalized coordinates which can assume arbitrary values without violating the constraints, thereby permitting a complete description of the configuration of the system without the use of auxiliary equations of constraint. Furthermore, the required number of generalized coordinates in this case is equal to the number of degrees of freedom.

As an example of holonomic constraints, consider the double pendulum of Fig. 6–2. The rods of length l_1 and l_2 are considered to be rigid and massless. Also, the system is pivoted at m_1 and at O such that the motion is confined to a single vertical plane. If we choose the coordinates (x_1, y_1) and (x_2, y_2) to represent the positions of particles m_1 and m_2, respectively, then the constraint equations are of the form

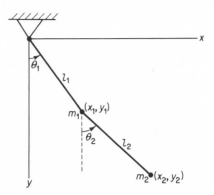

Fig. 6-2. A double pendulum.

$$x_1^2 + y_1^2 = l_1^2$$
$$(x_2 - x_1)^2 + (y_2 - y_1)^2 = l_2^2$$

expressing the fact that the lengths of the rods are constant. These particular holonomic constraints are also classified as *scleronomic* because they do not depend explicitly upon time. Any constraints which are explicit functions of time are known as *rheonomic* constraints. Similarly, a given *system* is classified as rheonomic or scleronomic, depending upon whether or not it contains any constraints which are explicit functions of time, that is, moving constraints.

In this particular example, we have used four coordinates to express the configuration of a system with only two degrees of freedom. But since the constraints are holonomic in nature, it is possible to find a set of independent generalized coordinates such that there are the same number of coordinates as degrees of freedom. For example, the angles θ_1 and θ_2, representing the angles which the rods make with the vertical, could have been used as generalized coordinates. Other choices could also have been made, such as defining θ_2 as the angle of rod l_2 relative to rod l_1.

Notice that the double pendulum constitutes a conservative system. It is usually true that frictionless systems with scleronomic constraints are also conservative.

Now consider an equation of constraint which is of the form

$$f(q_1, q_2, \ldots, q_n, t) \leq 0 \tag{6-4}$$

This sort of a constraint can occur, for example, when a set of particles is contained within a given closed surface. As an illustration, suppose that a free particle is contained within a fixed sphere of radius a which is centered at the origin of a cartesian system. Then, using x, y, and z as the coordinates of the particle, the equation of constraint is

$$x^2 + y^2 + z^2 - a^2 \leq 0$$

So long as the particle is in a position such that the inequality holds, it moves inside the sphere as a free particle with three degrees of freedom. On the other hand, if $x^2 + y^2 + z^2 - a^2 = 0$ at a given time, then the particle is touching the sphere and one must specify the conditions of its motion at this instant. For example, one might give the coefficient of restitution for collisions between the particle and the sphere. It can be seen, therefore, that the analysis proceeds as for a holonomic system during the period between bounces, the coefficient of restitution providing the means of calculating the initial conditions for a given period of the motion from the final conditions of the previous period. On the other hand, if the equality holds for an extended period of time, then the particle slides on the sphere if it moves at all. For this case there are only two degrees of freedom, but again the solution proceeds as for a holonomic system.

Nonholonomic Constraints. Now let us consider constraints which cannot be expressed in the form of Eqs. (6–3) or (6–4) but must be expressed in terms of differentials of the coordinates and possibly time. Constraints of this type are known as *nonholonomic* constraints and are written in the form

$$\sum_{i=1}^{n} a_{ji}\, dq_i + a_{jt}\, dt = 0 \quad (j = 1, 2, \ldots, m) \tag{6–5}$$

where the a's are, in general, functions of the q's and time. Furthermore, if the constraint is nonholonomic, this differential expression is characterized by being *not integrable*. It can be seen that, if it were integrable, the integrated form of the constraint equation would be the same as that given in Eq. (6–3), and we have previously classed constraints of this type as holonomic. As a result of not being able to integrate Eq. (6–5), one cannot eliminate coordinates by using the equations of constraint. Therefore we find that systems containing nonholonomic constraints always require more coordinates for their description than there are degrees of freedom.

As an example of a nonholonomic constraint, consider a vertical disk of radius r which rolls without slipping on the horizontal xy plane, as shown in Fig. 6–3. Let us choose as coordinates the location (x, y) of the point of contact, the angle of rotation ϕ of the disk about a perpendicular axis through its center, and the angle α between the plane of the disk and the yz plane. The requirement of rolling without slipping implies that

$$\begin{aligned} dx - r \sin \alpha \, d\phi &= 0 \\ dy - r \cos \alpha \, d\phi &= 0 \end{aligned} \tag{6–6}$$

since $r\, d\phi$ represents a differential element ds along the path traced by the point of contact, and this infinitesimal line segment ds makes an angle

α with the direction of the y axis. These two equations of constraint are independent. Thus, since there are four coordinates and two equations of constraint, we find that the system has only two degrees of freedom.

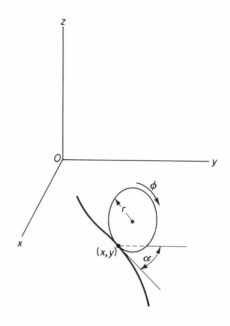

Note particularly that the constraint equations given in Eq. (6–6) are not integrable. In other words, neither of the expressions on the left-hand side is an exact differential of the form

$$dF(x, y, \alpha, \phi) = \frac{\partial F}{\partial x}\, dx + \frac{\partial F}{\partial y}\, dy$$
$$+ \frac{\partial F}{\partial \alpha}\, d\alpha + \frac{\partial F}{\partial \phi}\, d\phi = 0$$

nor can they be made exact through multiplication by an integrating factor. In geometrical terms, we see that if the disk is at a given point (x, y) on the plane and if the tangent

Fig. 6-3. A vertical disk rolling on a horizontal plane.

to its path makes a certain angle α with the y axis, it is nevertheless impossible to specify ϕ because its value depends upon the length of the path followed in reaching that location.

6-4. VIRTUAL WORK

Virtual Displacements and Virtual Work. Consider a system of N particles whose configuration is specified in terms of the cartesian coordinates x_1, $x_2, \ldots, x_{3N}$ where, of course, three coordinates are required to specify the position of each particle. Suppose that the forces $F_1, F_2, \ldots, F_{3N}$ are applied at the corresponding coordinates in the direction of the increasing coordinate in each case. Now let us imagine that, at a given instant, the system is given arbitrary small displacements $\delta x_1, \delta x_2, \ldots, \delta x_{3N}$ of the corresponding coordinates. The work done by the applied forces is

$$\delta W = \sum_{j=1}^{3N} F_j \delta x_j \qquad (6–7)$$

and is known as the *virtual work*. The small displacements are called *virtual displacements* because they are imaginary in the sense that they are assumed to occur without the passage of time, the applied forces remaining constant.

Also, they may not conform to either the kinematic or dynamic restrictions on the motion. Hence the virtual displacement is designated by δx to distinguish it from the corresponding real displacement dx which occurs during the time interval dt. In the usual case, however, the virtual displacements conform to the kinematic constraints which apply at that moment, assuming that any moving constraints are stopped for that instant.

An alternate form of the expression for the virtual work of the applied forces is

$$\delta W = \sum_{i=1}^{N} \mathbf{F}_i \cdot \delta \mathbf{r}_i \qquad (6\text{--}8)$$

where $\mathbf{F}_i$ is the force applied to the particle whose position vector is $\mathbf{r}_i$. This vector formulation serves to emphasize that the concept of virtual work is not associated with a particular coordinate system. Whenever cartesian coordinates are used in the succeeding development, the motivation is convenience rather than necessity.

Now consider the case where the ordinary coordinates $x_1, x_2, \ldots, x_{3N}$ are subject to the holonomic constraints

$$\phi_k(x_1, \ldots, x_{3N}, t) = 0 \quad (k = 1, 2, \ldots, m) \qquad (6\text{--}9)$$

For a virtual displacement consistent with the constraints, the δx's are no longer completely arbitrary but are related by the m equations

$$\frac{\partial \phi_k}{\partial x_1} \delta x_1 + \frac{\partial \phi_k}{\partial x_2} \delta x_2 + \cdots + \frac{\partial \phi_k}{\partial x_{3N}} \delta x_{3N} = 0 \quad (k = 1, 2, \ldots, m)$$

$$(6\text{--}10)$$

which were obtained by setting $d\phi_k = 0$ and replacing the dx's by δx's. Note that we take $\delta t = 0$ because the time remains constant during a virtual displacement. Hence we see that the concept of virtual work applies equally well to scleronomic and rheonomic systems.

In a similar fashion, we find from Eq. (6–5) that in a virtual displacement consistent with one or more nonholonomic constraints, the δx's are restricted by equations of the form

$$a_{j1} \delta x_1 + a_{j2} \delta x_2 + \cdots + a_{j3N} \delta x_{3N} = 0 \qquad (6\text{--}11)$$

Since Eqs. (6–10) and (6–11) are of the same form, we find that nonholonomic constraints also can be considered in using the idea of virtual work. Constraints associated with equations of this form are known as *bilateral* constraints since, for any set of virtual displacements consistent with the constraints, there exists another set which is obtained by reversing each of the original virtual displacements. On the other hand constraints expressed by inequalities, as in Eq. (6–4), are known as *unilateral* constraints.

Constraint Forces. We have been discussing the virtual work of the external forces applied to the system. If the system is subject to constraints, then additional forces are applied to the particles of the system in order to enforce the constraint conditions. Let us consider the work done by some common types of constraint forces as the system undergoes a virtual displacement consistent with the constraints.

First, consider two particles connected by a rigid, massless rod— Fig. 6–4(a). As we have seen previously, the forces transmitted by the rod to the particles must be equal, opposite, and collinear since this is a typical case of mechanical interaction between particles. So if we assume that $\mathbf{R}_1$ is the constraint force on m_1, and $\mathbf{R}_2$ is the constraint force on m_2, then

$$\mathbf{R}_1 = -\mathbf{R}_2 = -R_2\mathbf{e}_r \qquad (6\text{–}12)$$

where $\mathbf{e}_r$ is a unit vector directed from m_1 toward m_2. Assuming virtual displacements $\delta\mathbf{r}_1$ and $\delta\mathbf{r}_2$, we can write the virtual work of the constraint forces in the form of Eq. (6–8), obtaining

$$\delta W = \mathbf{R}_1 \cdot \delta\mathbf{r}_1 + \mathbf{R}_2 \cdot \delta\mathbf{r}_2 \qquad (6\text{–}13)$$

But the displacement components along the rigid rod must be equal; therefore the constraint equation can be written in the form

$$\mathbf{e}_r \cdot \delta\mathbf{r}_1 = \mathbf{e}_r \cdot \delta\mathbf{r}_2 \qquad (6\text{–}14)$$

So from Eqs. (6–12), (6–13), and (6–14), we find that

$$\delta W = (R_2 - R_2)\mathbf{e}_r \cdot \delta\mathbf{r}_2 = 0$$

indicating that the virtual work of the constraint forces is zero.

If we think of a rigid body as consisting of a large number of particles which are rigidly interconnected, we can see that the virtual work done by the constraint forces acting between any two particles is zero if the virtual displacement is consistent with the constraints. Summing over all possible combinations of particle pairs, we conclude that the total virtual work of the internal constraint forces is zero for any rigid body displacement.

Now let us calculate the virtual work of the constraint forces for the case of a body B which slides without friction along a fixed surface S—Fig. 6–4(b). First of all, it is clear that no work can be done on the surface S because its particles cannot move. When we consider the work done by the constraint force $\mathbf{R}_{BS}$ in a virtual displacement of B, we find that $\mathbf{R}_{BS}$ is normal to the surface at that point, whereas any virtual displacement must be tangent to it. Therefore, in accordance with Eq. (6–8), we again find that the virtual work of the constraint forces is zero for a virtual displacement consistent with the constraints. Note that the virtual work would have been zero even if the surface were moving as an explicit function of time. This follows from the assumption that time stands still during a virtual displacement, and therefore the surface would be considered to be fixed at the instant of the virtual displacement.

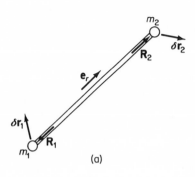

(a)

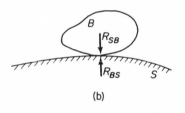

(b)

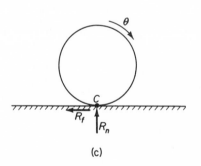

(c)

Fig. 6-4. Examples of workless constraints.

Next consider the work done by the constraint forces in a virtual displacement involving a body which rolls on a fixed surface. As an example, we take the case of a vertical circular disk rolling along a straight path on a horizontal plane, Fig. 6-4(c). Again we find that there can be no work done by forces acting on the fixed surface. Now consider the work of the constraint forces acting on the disk. We see that the force of the surface acting on the disk passes through the instantaneous center C. This force is composed of a frictional component R_f acting tangent to the surface and a normal component R_n. Since, however, we assume no slipping, the particle in the disk at C is instantaneously at rest when the constraint force is applied to it. Therefore the work of the constraint force acting on point C in an infinitesimal displacement $\delta\theta$ must be an infinitesimal of higher order and can be neglected. In fact, if we consider the work on a particle which is at C at the midpoint of the virtual displacement $\delta\theta$, it can be seen that the virtual work due to R_n is zero from symmetry considerations. The virtual work due to R_f can be shown to be of order $(\delta\theta)^3$. Hence the virtual work of the constraint forces is zero for this case of a rolling constraint. Although we have discussed the particular case of a disk rolling on a plane, a similar argument concerning the reactions acting on the instantaneous center would apply to any case of the rolling contact of a body on a fixed surface.

We have shown that for several commonly occurring constraints, the virtual work of the constraint forces is zero for all virtual displacements consistent with the constraints. These are examples of *workless constraints*. More generally, we can define a workless constraint as follows: *A workless constraint is any scleronomic constraint such that the virtual work of the constraint forces acting on the system is zero for any reversible virtual dis-*

placement which is consistent with the constraints. Here we specify that the constraints must be scleronomic because rheonomic constraints, such as moving surfaces, pivots, and so on, will, in general, do work on the system in an actual displacement even though they are considered to be fixed during a virtual displacement. Also, we consider only reversible virtual displacements because, as we shall see later for the case of unilateral constraints, the virtual work of these constraint forces may be non-zero in instances where the virtual displacements cannot be reversed.

It should be realized that workless constraint forces may actually do work on certain particles within the system, and hence may transfer energy from one part of a system to another even though they do no work on the system as a whole. For example, a particle on the rim of a rolling circular disk moves with a changing speed, and therefore a changing kinetic energy. This energy is transferred by means of constraint forces. As the chapter proceeds, we shall find that the concept of workless constraints is important, both in the analysis of the conditions for static equilibrium and in the solution of dynamics problems by energy methods. In either case, it often enables one to avoid calculating the constraint forces, thereby simplifying the analysis.

The Principle of Virtual Work. Now let us consider the conditions required for the static equilibrium of a system of particles. A system is said to be in *static equilibrium* with respect to an inertial frame if all particles of the system are motionless in that frame and if the vector sum of the forces acting on each particle is equal to zero. Hence we see that the condition of static equilibrium implies that each particle has zero velocity and acceleration relative to the inertial reference frame.

Suppose we separate the total force acting on a given particle m_i into a *constraint force* $\mathbf{R}_i$ and an *applied force* $\mathbf{F}_i$ where $\mathbf{R}_i$ is required to enforce the constraints and $\mathbf{F}_i$ includes all other forces. If a system of N particles is in static equilibrium, then for each particle,

$$\mathbf{F}_i + \mathbf{R}_i = 0 \tag{6-15}$$

Therefore the virtual work of all the forces as a result of the virtual displacements $\delta\mathbf{r}_i$ is

$$\sum_{i=1}^{N} (\mathbf{F}_i + \mathbf{R}_i) \cdot \delta\mathbf{r}_i = \sum_{i=1}^{N} \mathbf{F}_i \cdot \delta\mathbf{r}_i + \sum_{i=1}^{N} \mathbf{R}_i \cdot \delta\mathbf{r}_i = 0 \tag{6-16}$$

Now if we assume that the constraints are workless and the $\delta\mathbf{r}_i$ are reversible virtual displacements consistent with those constraints, then the virtual work of the constraint forces is zero.

$$\sum_{i=1}^{N} \mathbf{R}_i \cdot \delta\mathbf{r}_i = 0 \tag{6-17}$$

From Eqs. (6–16) and (6–17), we conclude that

$$\delta W = \sum_{i=1}^{N} \mathbf{F}_i \cdot \delta \mathbf{r}_i = 0 \qquad (6\text{-}18)$$

We have shown that if a system of particles with workless constraints is in static equilibrium, then the virtual work of the *applied forces* is zero for any reversible virtual displacement which is consistent with the constraints.

Now consider a system of particles with workless constraints which is initially motionless but is *not in equilibrium.* Then one or more of the particles must have a net force applied to it, and in accordance with Newton's law, it will tend to move in the direction of that force. Of course, any motion must be compatible with the constraints and so we can always choose a virtual displacement in the direction of the net force at each point. In this case, the virtual work is positive, that is,

$$\sum_{i=1}^{N} \mathbf{F}_i \cdot \delta \mathbf{r}_i + \sum_{i=1}^{N} \mathbf{R}_i \cdot \delta \mathbf{r}_i > 0 \qquad (6\text{-}19)$$

But again the constraints are workless and Eq. (6–17) applies. Therefore, for this system, the virtual work of the applied forces moving through the specified virtual displacements is positive, that is,

$$\delta W = \sum_{i=1}^{N} \mathbf{F}_i \cdot \delta \mathbf{r}_i > 0 \qquad (6\text{-}20)$$

In other words, if the given system is not in equilibrium, then it is always possible to find a set of virtual displacements consistent with the constraints for which the virtual work of the applied forces is positive.

If we consider systems which allow only reversible virtual displacements, then the above results can be summarized in the *principle of virtual work:*

> *The necessary and sufficient condition for the static equilibrium of an initially motionless system which is subject to workless bilateral constraints is that zero virtual work be done by the applied forces in moving through an arbitrary virtual displacement satisfying the constraints.*

The principle of virtual work is of fundamental importance in the study of statics and, if one uses d'Alembert's principle, can be extended to dynamical systems as well. By providing a relatively simple criterion for the equilibrium of a large and important class of systems and by avoiding the necessity of calculating constraint forces in many instances, it simplifies the analysis of a wide variety of problems in mechanics. We shall use virtual work methods frequently in this and succeeding chapters.

Now consider a system of N particles in which all the applied forces are *conservative.* Using cartesian coordinates to define the position of each of the particles, we can write the potential energy function in the form

$$V = V(x_1, x_2, \ldots, x_{3N}) \qquad (6\text{-}21)$$

Using Eq. (4–25), we see that the component of the applied force in the direction of x_i is

$$F_i = -\frac{\partial V}{\partial x_i} \tag{6–22}$$

Hence the virtual work of the applied forces in a virtual displacement which is consistent with the constraints is

$$\delta W = \sum_{i=1}^{3N} F_i \, \delta x_i = -\sum_{i=1}^{3N} \frac{\partial V}{\partial x_i} \, \delta x_i \tag{6–23}$$

But the first variation in the potential energy due to an arbitrary virtual displacement is just

$$\delta V = \sum_{i=1}^{3N} \frac{\partial V}{\partial x_i} \, \delta x_i \tag{6–24}$$

Therefore, using the principle of virtual work, we find that the necessary and sufficient condition for a static equilibrium configuration of a conservative system with bilateral constraints is that

$$\delta V = 0 \tag{6–25}$$

for every virtual displacement consistent with the constraints.

If we express the potential energy in terms of the generalized coordinates $q_1, q_2, \ldots, q_n$, we find that

$$\delta V = \sum_{i=1}^{n} \frac{\partial V}{\partial q_i} \, \delta q_i \tag{6–26}$$

Now assume that a set of *independent generalized coordinates* can be found implying that the constraints are holonomic. Then, since at an equilibrium configuration $\delta V = 0$ for an arbitrary choice of δq's, the coefficients must be zero, that is,

$$\frac{\partial V}{\partial q_1} = 0, \quad \frac{\partial V}{\partial q_2} = 0, \ldots, \frac{\partial V}{\partial q_n} = 0 \tag{6–27}$$

But these are the conditions that the potential energy V have a stationary value. Therefore we conclude that an equilibrium configuration of a conservative system with bilateral holonomic constraints must occur at a position where the potential energy has a stationary value.

Next let us consider briefly the question of the *stability* of this system at a position of static equilibrium. If we expand V in a Taylor series about a reference value V_0, we obtain

$$V = V_0 + \left(\frac{\partial V}{\partial q_1}\right)_0 \delta q_1 + \left(\frac{\partial V}{\partial q_2}\right)_0 \delta q_2 + \cdots + \frac{1}{2}\left(\frac{\partial^2 V}{\partial q_1^2}\right)_0 (\delta q_1)^2 + \cdots$$
$$+ \left(\frac{\partial^2 V}{\partial q_1 \, \partial q_2}\right)_0 \delta q_1 \, \delta q_2 + \cdots \tag{6–28}$$

where the δq's represent infinitesimal displacements from the reference position. Now assume that the equilibrium configuration is taken as the reference position, in which case we can set $V_0 = 0$. But from Eq. (6–27) all the coefficients of linear terms in the δq's are zero. Therefore the potential energy expression consists of terms of at least second order in the small displacements.

Now consider a virtual displacement from the equilibrium position. We have seen that, taking terms of order δq, the potential is zero. So let us consider the higher-order terms. We can write

$$\Delta V = \frac{1}{2} \left(\frac{\partial^2 V}{\partial q_1^2}\right)_0 (\delta q_1)^2 + \left(\frac{\partial^2 V}{\partial q_1\, \partial q_2}\right)_0 \delta q_1\, \delta q_2$$
$$+ \frac{1}{2} \left(\frac{\partial^2 V}{\partial q_2^2}\right)_0 (\delta q_2)^2 + \cdots \qquad (6\text{–}29)$$

where ΔV is the change in the potential energy from its value at equilibrium. Here we use ΔV rather than δV to indicate that higher-order terms are included. If $\Delta V > 0$ for every virtual displacement in which one or more of the δq's is non-zero, then the reference position is one of minimum potential energy corresponding to *stable* equilibrium. On the other hand, if a set of δq's can be found such that $\Delta V < 0$, then the equilibrium position is said to be *unstable*. A third possibility is that $\Delta V > 0$ for every virtual displacement except those for which $\Delta V = 0$. This is the case of *neutral stability*.

Looking at the question of stability from the point of view of conservation of energy we see that if the system is initially motionless at the equilibrium position, then the total energy is zero. Therefore, for an arbitrary infinitesimal displacement,

$$\Delta V + \Delta T = 0 \qquad (6\text{–}30)$$

Since ΔT must equal or exceed zero in order for any displacement to occur, we see that ΔV cannot exceed zero. Hence a system in stable equilibrium cannot move from that position without the addition of energy to the system. But it is quite possible from the viewpoint of energy conservation for motion to occur near a configuration of unstable equilibrium because there are other positions in the immediate vicinity with negative potential energy. The third possibility, neutral stability, implies that other equilibrium positions exist in the neighborhood of the reference configuration, but there is no position in this neighborhood which allows a positive kinetic energy. Note that we have not been concerned with the time required for a given small displacement but rather with the question whether a displacement is possible from an energy standpoint. It turns out that an infinite time is required for a finite displacement from equilibrium in those cases where it is possible.

Taking a more practical view, we might consider the system to be initially in equilibrium but subject to small random disturbances. In this case, a stable system will remain near the reference position; a system with neutral stability will drift slowly but remain essentially in equilibrium; and an unstable system will move away from the reference position with an increasing velocity.

In our discussion of the principle of virtual work, we have assumed that the constraints are workless and bilateral. The idea of virtual work is quite general, however, and is not restricted to such systems. So let us consider the consequences of other assumptions.

First let us assume that damping is associated with the constraints. We note that viscous damping and other frictional forces which are proportional to a positive power of the velocity do not influence the position of static equilibrium. This follows from the fact that all such forces are zero when the system is stationary; hence they do no work in a virtual displacement. Coulomb friction, however, frequently influences the possible configurations of static equilibrium. Suppose we assume that the frictional force at each sliding surface opposes the relative motion and is of maximum magnitude, namely, it equals the coefficient of friction μ times the normal force. Then, considering the friction force as an applied force, we find that an initially motionless system is in static equilibrium if, and only if, the virtual work of the applied forces,

$$\delta W \leq 0 \qquad (6\text{--}31)$$

for all virtual displacements consistent with the constraints. This implies that, in general, a finite additional force is required to cause a system with Coulomb friction to break free of an equilibrium condition and start to move. In contrast, a frictionless system with bilateral constraints requires only an infinitesimal change in an applied force in order to cause the system to move from an equilibrium position. It should be noted that, although the virtual work appears to be easy to calculate, it may be necessary to evaluate the constraint forces before the friction forces can be obtained. Also, it can be seen that, if we consider the actual (rather than the maximum) friction forces to be applied forces, then the principle of virtual work is valid in its usual form given in Eq. (6–18). Either approach, however, often involves calculating constraint forces, thereby losing much of the advantage of the virtual work approach. Furthermore, statically indeterminate forces are more likely with friction present.

Next let us consider the case of a system with unilateral constraints, that is, constraints expressed as an inequality. Also, we assume that the constraints are frictionless and scleronomic. In this case, it is possible for the virtual work of the constraint forces to be positive. Then, since the virtual work of all forces must be zero at a position of static equilibrium, we find that the virtual work of the applied forces may be negative. So, for an initially motionless system with frictionless scleronomic constraints, the necessary

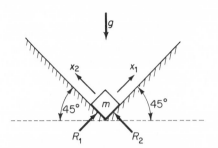

Fig. 6-5. A system with unilateral constraints.

and sufficient condition for static equilibrium is that the virtual work of the applied forces be equal to or less than zero, that is,

$$\delta W \leq 0 \qquad (6\text{–}32)$$

for all virtual displacements consistent with the constraints.

As an example of a unilateral constraint, consider a small cube of mass m which is resting at a corner formed by frictionless, mutually perpendicular planes (Fig. 6–5). Assuming that x_1 and x_2 are measured from the corner, the constraint equations for this system are

$$x_1 \geq 0, \qquad x_2 \geq 0 \qquad (6\text{–}33)$$

The only applied force is due to gravity and it has the components

$$F_1 = F_2 = -\frac{mg}{\sqrt{2}} \qquad (6\text{–}34)$$

Therefore the virtual work of the applied force is

$$\delta W = -\frac{mg}{\sqrt{2}}(\delta x_1 + \delta x_2) \leq 0 \qquad (6\text{–}35)$$

since the δx's must be positive or zero when $x_1 = x_2 = 0$. Hence the system is in static equilibrium in accordance with Eq. (6–32). The same result applies even if the cube is allowed unconstrained motion in the x_3 direction which is perpendicular to the x_1x_2 plane. In this case, F_3 and R_3 are zero, and δx_3 does not appear in the expression for δW.

The *type* of static equilibrium can be determined by the criteria developed previously for the bilateral holonomic case. We find that $\Delta V > 0$ for all possible virtual displacements in the two-dimensional case, corresponding to stable equilibrium. On the other hand, $\Delta V = 0$ if a virtual displacement is made in the x_3 direction. So there is neutral equilibrium in the three-dimensional case. Note that $\partial V/\partial x_1$ and $\partial V/\partial x_2$ are not zero at equilibrium.

It is interesting to calculate the virtual work of the constraint forces for this system. We obtain

$$\delta W_c = R_1 \, \delta x_1 + R_2 \, \delta x_2 \geq 0 \qquad (6\text{–}36)$$

where R_1 and R_2 are assumed to be constant during the virtual displacement. Since the virtual work of all forces is zero, we see that Eqs. (6–35) and (6–36) are compatible.

The principle of virtual work can also be used to calculate the constraint forces R_1 and R_2. The procedure followed is to set the total virtual work of *all* forces equal to zero, that is,

$$\left(R_1 - \frac{mg}{\sqrt{2}}\right)\delta x_1 + \left(R_2 - \frac{mg}{\sqrt{2}}\right)\delta x_2 = 0 \qquad (6\text{-}37)$$

Here we assume that the δx's are *not constrained* and are therefore completely independent. Hence each coefficient in Eq. (6–37) must be zero and we obtain

$$R_1 = R_2 = \frac{mg}{\sqrt{2}}$$

In the discussion of Lagrange multipliers in Sec. 6–7, we shall find another illustration of the general rule that, in order to solve for constraint forces by virtual work methods, the corresponding constraint equation in terms of coordinates is eliminated and is replaced in the analysis by the addition of the constraint forces.

Example 6–1. Two equal masses are connected by a rigid massless bar, as shown in Fig. 6–6. (a) If all surfaces are frictionless, solve for the force F_2 required for static equilibrium of the system. (b) For the case where $F_2 = 0$, what is the minimum friction coefficient μ between m_2 and the floor such that no sliding occurs? The system is initially motionless.

The system of part (a) has workless holonomic constraints if we assume that the masses do not leave the wall or floor. Measuring the displacements x_1 and x_2 from the initial configuration, the equation of constraint states that the length of the bar is constant. Thus,

$$(a - x_1)^2 + (a + x_2)^2 = 2a^2$$

or, equating the total differential of this expression to zero, we find that

$$dx_1 = dx_2 \qquad (6\text{-}38)$$

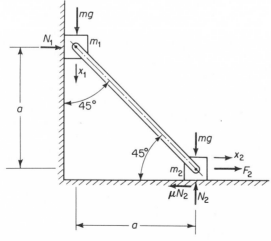

Fig. 6-6. A system which is constrained to move in a vertical plane.

The applied forces are the force F_2 and those due to gravity; all other forces are constraint forces. The gravitational force mg acts on m_2 in a direction perpendicular to δx_2 and hence does no work in a virtual displacement.

Now we apply the principle of virtual work and find that the condition for static equilibrium is that

$$mg\, \delta x_1 + F_2\, \delta x_2 = 0 \qquad (6\text{--}39)$$

But $\delta x_1 = \delta x_2$, since any virtual displacement must be consistent with the constraint of Eq. (6–38). Therefore,

$$F_2 = -mg$$

Consider next the case where there is friction between m_2 and the floor and assume that $F_2 = 0$. We need first to calculate the normal force N_2 in order to find the friction force for incipient slipping. Summing the vertical forces on the system as a whole and equating this sum to zero, we obtain

$$N_2 - 2mg = 0$$

or

$$N_2 = 2mg$$

Hence the frictional force on m_2 is

$$\mu N_2 = 2\mu mg$$

Now we again apply the principle of virtual work, counting the force μN_2 as an applied force. The result is

$$\delta W = mg\, \delta x_1 - 2\mu mg\, \delta x_2 = 0 \qquad (6\text{--}40)$$

But $\delta x_1 = \delta x_2$ and therefore the minimum coefficient of friction for static stability is

$$\mu_{\min} = \frac{1}{2}$$

Note that if μ is larger than this minimum value, then, from Eq. (6–40), we find that the virtual work is negative. Also, if a virtual displacement is taken such that δx_1 and δx_2 are both negative and the direction of the force of friction is reversed, then the virtual work is

$$\delta W = mg\, \delta x_1 + 2\mu mg\, \delta x_2 = (2\mu + 1)mg\, \delta x_1$$

We see that δW is negative for the assumed displacement. Hence we confirm that, by setting $\delta W = 0$ in Eq. (6–40), we are analyzing a virtual displacement in accordance with the actual mode of slipping. In other words, if $\mu = \frac{1}{2}$, the only motion of the system which requires no additional energy input and satisfies the constraints is that motion for which δx_1 and δx_2 are positive.

Example 6–2. Three particles, each of mass m, are located at the pin joints between four massless rods, as shown in Fig. 6–7. The system hangs from support points which have a horizontal separation $2l$. Solve for the angles θ_1 and θ_2 corresponding to a condition of static equilibrium.

It is apparent that in the equilibrium condition, there is symmetry about a vertical line through the central particle; hence we need to consider only the two coordinates θ_1 and θ_2. From the geometry, we can write an equation for the separation of the support points in the form

$$l(2 \sin \theta_1 + 2 \sin \theta_2) = 2l$$

or

$$\sin \theta_1 + \sin \theta_2 = 1 \quad (6\text{--}41)$$

This is an equation of holonomic constraint.

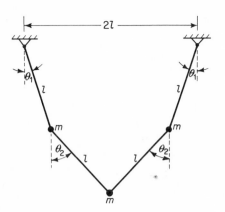

Fig. 6-7. Three particles of mass m which are supported by four massless rods connected by pin joints.

Since this system is conservative, we know from Eq. (6–25) that the static equilibrium configuration occurs at a stationary value of the potential energy. Taking the supports as the reference level for the gravitational potential, we find that

$$V = -mgl(3 \cos \theta_1 + \cos \theta_2) \quad (6\text{--}42)$$

Now, to find a stationary value of this function, we set its total derivative with respect to θ_1 equal to zero, recalling from Eq. (6–41) that θ_2 is a function of θ_1. Thus,

$$\frac{dV}{d\theta_1} = mgl\left(3 \sin \theta_1 + \sin \theta_2 \frac{d\theta_2}{d\theta_1}\right) = 0$$

Evaluating $d\theta_2/d\theta_1$ from Eq. (6–41), we find that

$$3 \sin \theta_1 - \frac{\sin \theta_2 \cos \theta_1}{\cos \theta_2} = 0$$

or

$$3 \tan \theta_1 = \tan \theta_2 \quad (6\text{--}43)$$

Finally, solving for θ_1 and θ_2 from Eqs. (6–41) and (6–43), we obtain

$$\theta_1 = 17.8°$$

$$\theta_2 = 43.9°$$

This is a position of stable equilibrium, corresponding to a minimum value of V. It is interesting to note that another solution is

$$\theta_1 = 162.2°$$

$$\theta_2 = 136.1°$$

which is a position of unstable equilibrium. We can verify this last statement by evaluating $d^2V/d\theta_1^2$. It turns out to be negative, indicating a maximum value of potential energy at this configuration.

D'Alembert's Principle. In Sec. 1–5, d'Alembert's principle was stated for the case of a particle of mass m and absolute acceleration $\mathbf{a}$. Assuming a total external force $\mathbf{F}$, we found that

$$\mathbf{F} - m\mathbf{a} = 0 \tag{6–44}$$

where the term $-m\mathbf{a}$ was considered as an additional force on the particle, namely, an *inertial* force. Hence the result given in Eq. (6–44) can be interpreted as stating that a summation of forces is zero, in the manner of an equation of static equilibrium.

A more general form of d'Alembert's principle uses another criterion of static equilibrium, that is, the method of virtual work, in obtaining the required equations of motion. Suppose we consider a system of N particles which may have workless bilateral constraints. If $\mathbf{F}_i$ is the *applied* force acting on m_i, and if we denote the linear momentum of this particle by

$$\mathbf{p}_i = m_i\mathbf{v}_i \tag{6–45}$$

then we can use the principle of virtual work to obtain another statement of d'Alembert's principle:

$$\sum_{i=1}^{N} (\mathbf{F}_i - \dot{\mathbf{p}}_i) \cdot \delta\mathbf{r}_i = 0 \tag{6–46}$$

where $\delta\mathbf{r}_i$ is a virtual displacement of m_i satisfying the constraints. Thus we use the methods of statics to obtain the equations of motion for a dynamical system. Note that the constraint forces are omitted from the expression for virtual work.

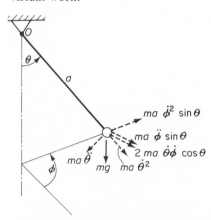

As an example of the use of d'Alembert's principle, let us write the equations of motion for a spherical pendulum of length a. The applied and inertial forces are shown in Fig. 6–8. Assuming arbitrary virtual displacements $\delta\theta$ and $\delta\phi$, we find that

$$ma[(-g\sin\theta - a\ddot{\theta}$$
$$+ a\dot{\phi}^2\sin\theta\cos\theta)\,\delta\theta$$
$$- a\sin\theta(\ddot{\phi}\sin\theta$$
$$+ 2\dot{\theta}\dot{\phi}\cos\theta)\,\delta\phi] = 0$$

Fig. 6-8. The inertial and applied forces acting on a spherical pendulum.

Since $\delta\theta$ and $\delta\phi$ are independent, we can equate each coefficient to zero, obtaining

$$a\ddot{\theta} - a\dot{\phi}^2 \sin\theta \cos\theta = -g \sin\theta$$

and

$$\ddot{\phi} \sin^2\theta + 2\dot{\theta}\dot{\phi} \sin\theta \cos\theta = 0$$

These are seen to be the differential equations of motion. They are essentially the same as Eqs. (5–135) and (5–136) obtained earlier.

6-5. GENERALIZED FORCES

Consider a system of particles whose positions are specified by the cartesian coordinates $x_1, x_2, \ldots, x_k$. If the forces $F_1, F_2, \ldots, F_k$ are applied at the corresponding coordinates, and if their direction is positive in each case, then the virtual work of these forces in an arbitrary virtual displacement consistent with the constraints is

$$\delta W = \sum_{j=1}^{k} F_j \, \delta x_j \qquad (6\text{--}47)$$

in agreement with Eq. (6–7).

Now let us suppose that the ordinary coordinates $x_1, x_2, \ldots, x_k$ are related to the generalized coordinates by equations of the form of Eq. (6–1). Then we can express small virtual displacements of the x's in terms of a set of corresponding virtual displacements of the q's. Differentiating Eq. (6–1) and setting $\delta t = 0$, since we are considering a *virtual* displacement in each case, we obtain

$$\delta x_j = \sum_{i=1}^{n} \frac{\partial x_j}{\partial q_i} \delta q_i \quad (j = 1, 2, \ldots, k) \qquad (6\text{--}48)$$

where we see that the coefficients $\partial x_j / \partial q_i$ are functions of the q's and time. Substituting this expression for δx_j into Eq. (6–47), we obtain

$$\delta W = \sum_{j=1}^{k} \sum_{i=1}^{n} F_j \frac{\partial x_j}{\partial q_i} \delta q_i \qquad (6\text{--}49)$$

Let us change the order of summation. Then we can write Eq. (6–49) in the form

$$\delta W = \sum_{i=1}^{n} Q_i \, \delta q_i \qquad (6\text{--}50)$$

where the *generalized force* Q_i associated with the generalized coordinate q_i is given by

$$Q_i = \sum_{j=1}^{k} F_j \frac{\partial x_j}{\partial q_i} \quad (i = 1, 2, \ldots, n) \qquad (6\text{--}51)$$

Note that the expressions for the virtual work given in Eqs. (6–47) and (6–50) are of the same mathematical form. It can be seen that F_j is the virtual work per unit displacement of x_j when all the other x's are held at zero. In a similar fashion, we see that the generalized force Q_i is equal numerically to the virtual work done per unit displacement of q_i by the forces acting on the system when the other generalized coordinates remain at zero. Usually the generalized coordinates are chosen to be independent; but in case there are equations of constraint relating the generalized coordinates, then these constraints should be ignored in making the virtual displacement.

The dimensions of a generalized force depend upon the dimensions of the corresponding coordinate. In all cases $Q_i \, \delta q_i$ must have the dimensions of work, or $[FL]$. So if q_i corresponds to a linear displacement, the generalized force has the dimensions of force. On the other hand, if q_i is an angle, then Q_i is a moment. We see that the x's are a special case of generalized coordinates. Similarly, an ordinary force F_j is a special case of a generalized force. Hence all results which apply to generalized forces and displacements are valid for ordinary forces and displacements as well.

As an example of a generalized force calculation, consider the system shown in Fig. 6–9. The particles are connected by an elastic string and are constrained such that only transverse motions in a single plane are permitted. Let the ordinary coordinates x_1, x_2, x_3 designate the transverse displacements of the particles. No further constraints are assumed and so there are three degrees of freedom.

Suppose that three independent generalized coordinates q_1, q_2, q_3 are used to describe the same system. In accordance with the form of Eq. (6–1), let us choose transformation equations as follows:

$$x_1 = \frac{3}{4}q_1 + q_2 + \frac{1}{2}q_3$$
$$x_2 = q_1 - q_3 \qquad (6\text{–}52)$$
$$x_3 = \frac{3}{4}q_1 - q_2 + \frac{1}{2}q_3$$

For our present purposes, the coefficients can be chosen in any fashion such that the q's are independent. Now we can calculate the generalized forces directly from Eq. (6–51). Alternatively, if we let $q_1, q_2,$ and q_3 undergo a unit virtual displacement in sequence, we find that the corresponding virtual work expressions are equal to the magnitude of the generalized forces:

$$Q_1 = \frac{3}{4}F_1 + F_2 + \frac{3}{4}F_3$$
$$Q_2 = F_1 - F_3 \qquad (6\text{–}53)$$
$$Q_3 = \frac{1}{2}F_1 - F_2 + \frac{1}{2}F_3$$

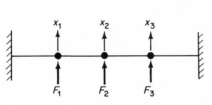

Fig. 6-9. Coordinates and forces associated with the plane transverse motions of a loaded elastic string.

The geometrical significance of each generalized coordinate can be shown by letting it vary separately. For example, if only q_1 is allowed to

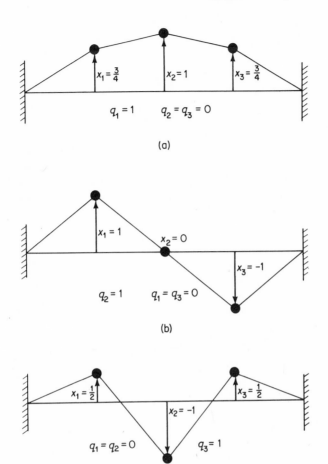

Fig. 6-10. Deflection forms corresponding to generalized coordinates.

vary, we see from Eq. (6–52) that the ordinary coordinates take on values according to the ratio

$$x_1 : x_2 : x_3 = \frac{3}{4} : 1 : \frac{3}{4}$$

A similar procedure can be used to obtain ratios corresponding to q_2 and q_3. Thus each generalized coordinate may have associated with it a deflection form that can be specified in terms of ratios of the ordinary coordinates, as is shown in Fig. 6–10 for this particular example. Nevertheless, the superposition of these deflection forms in the same ratio as the values of the corresponding generalized coordinates will make possible any configuration that could be described in terms of ordinary coordinates. Generalized coordinates of this sort find wide application in the study of linear vibration problems with several degrees of freedom and will be considered in more detail in Chapter 9.

We can use the concept of generalized forces to express the conditions required for static equilibrium. To see this, we note from Eq. (6–50) that if $\delta W = 0$ for an arbitrary virtual displacement of independent q's, then all the generalized forces must be zero. Hence we can use the principle of virtual work to show that the necessary and sufficient condition for the static equilibrium of an initially motionless system which is subject to workless bilateral constraints is that all generalized forces corresponding to the independent generalized coordinates be zero. Constraint forces do not enter directly in this case because the q's are assumed to be independent and hence unconstrained.

6–6. DERIVATION OF LAGRANGE'S EQUATIONS

Let us consider a system of N particles whose positions are given by the cartesian coordinates $x_1, x_2, \ldots, x_{3N}$. For example, we might use (x_1, x_2, x_3) as the coordinates of the first particle, (x_4, x_5, x_6) for the second particle, and so on. Thus we find that the total kinetic energy of the system is

$$T = \frac{1}{2} \sum_{j=1}^{3N} m_j \dot{x}_j^2 \qquad (6\text{–}54)$$

Note that the mass of a given particle appears in three terms, namely, those terms containing the squares of the three cartesian velocity components of that particle.

Now let us obtain an expression for the kinetic energy of the system in terms of generalized coordinates. Using the transformation equations of the form of Eq. (6–1) and differentiating with respect to time, we find that

$$\dot{x}_j = \sum_{i=1}^{n} \frac{\partial x_j}{\partial q_i} \dot{q}_i + \frac{\partial x_j}{\partial t} \qquad (6\text{–}55)$$

where the n generalized coordinates are chosen such that the allowable motions of the system are unchanged. In other words, either coordinate system permits the same number of degrees of freedom, although both the x's and the q's may have constraints associated with them. Hence we see that the $\dot{x}$'s are, in general, functions of the q's, $\dot{q}$'s, and time.

$$\dot{x}_j = g_j(q_1, \ldots, q_n, \dot{q}_1, \ldots, \dot{q}_n, t) \qquad (6\text{–}56)$$

Using Eqs. (6–55) and (6–56), we find that the kinetic energy can be written in the form

$$T = \frac{1}{2} \sum_{j=1}^{3N} m_j \left(\sum_{i=1}^{n} \frac{\partial x_j}{\partial q_i} \dot{q}_i + \frac{\partial x_j}{\partial t} \right)^2 \tag{6-57}$$

where $\partial x_j/\partial q_i$ and $\partial x_j/\partial t$ are functions of the q's and time. Thus we see that the kinetic energy is a function of the q's, $\dot{q}$'s, and time. For the particular case of a *scleronomic* system, $\partial x_j/\partial t$ is zero and we find that the kinetic energy is a homogeneous quadratic function of the $\dot{q}$'s, the coefficients being functions of the q's.

Now let us define the *generalized momentum* p_i associated with the generalized coordinate q_i by the equation

$$p_i = \frac{\partial T}{\partial \dot{q}_i} \tag{6-58}$$

Note that p_i is a scalar quantity. For the case of relatively simple coordinate systems, p_i is just the component of the momentum vector in the direction of the coordinate q_i. As an example, consider a single particle whose position is expressed by cartesian coordinates. Its kinetic energy is

$$T = \frac{1}{2}m(\dot{x}^2 + \dot{y}^2 + \dot{z}^2)$$

The momentum conjugate to the coordinate x is

$$p_x = \frac{\partial T}{\partial \dot{x}} = m\dot{x}$$

which is seen to be the component of the total linear momentum in the direction of the x axis.

Similarly, for the case in which the position of the particle is expressed in terms of spherical coordinates, we find that

$$T = \frac{1}{2}m(\dot{r}^2 + r^2\dot{\theta}^2 + r^2\dot{\phi}^2 \sin^2 \theta)$$

and therefore

$$p_r = \frac{\partial T}{\partial \dot{r}} = m\dot{r}$$

which is the r component of the total linear momentum. Now consider

$$p_\phi = \frac{\partial T}{\partial \dot{\phi}} = mr^2 \sin^2 \theta \, \dot{\phi}$$

This expression can be recognized as the vertical or ϕ component of the total *angular* momentum. Angular momentum rather than linear momentum occurs here because the coordinate ϕ involves angular displacement. Furthermore, we see from the differentiation process that the result must have the dimensions of energy divided by angular velocity.

In the case of a nonorthogonal coordinate system, we find that p_i is not always a momentum *component* as we have defined the term; rather it is the *projection* of the total momentum onto the q_i axis. For even more general coordinates, p_i has no easily expressed physical significance.

Returning now to the system of N particles, we see from Eqs. (6–54) and (6–58) that

$$p_i = \sum_{j=1}^{3N} m_j \dot{x}_j \frac{\partial \dot{x}_j}{\partial \dot{q}_i} \tag{6–59}$$

But from Eq. (6–55) we obtain

$$\frac{\partial \dot{x}_j}{\partial \dot{q}_i} = \frac{\partial x_j}{\partial q_i} \tag{6–60}$$

and therefore

$$p_i = \sum_{j=1}^{3N} m_j \dot{x}_j \frac{\partial x_j}{\partial q_i} \tag{6–61}$$

Now let us find the time rate of change of the generalized momentum.

$$\frac{dp_i}{dt} = \sum_{j=1}^{3N} m_j \ddot{x}_j \frac{\partial x_j}{\partial q_i} + \sum_{j=1}^{3N} m_j \dot{x}_j \frac{d}{dt}\left(\frac{\partial x_j}{\partial q_i}\right)$$

But

$$\frac{d}{dt}\left(\frac{\partial x_j}{\partial q_i}\right) = \sum_{k=1}^{n} \frac{\partial^2 x_j}{\partial q_i\, \partial q_k} \dot{q}_k + \frac{\partial^2 x_j}{\partial q_i\, \partial t}$$

since $\partial x_j / \partial q_i$ is, in general, an explicit function of all the q's and time. Therefore we obtain the result:

$$\frac{dp_i}{dt} = \sum_{j=1}^{3N} \left(m_j \ddot{x}_j \frac{\partial x_j}{\partial q_i} + m_j \dot{x}_j \sum_{k=1}^{n} \frac{\partial^2 x_j}{\partial q_i\, \partial q_k} \dot{q}_k + m_j \dot{x}_j \frac{\partial^2 x_j}{\partial q_i\, \partial t} \right) \tag{6–62}$$

Next, let us use Eq. (6–54) to obtain

$$\frac{\partial T}{\partial q_i} = \sum_{j=1}^{3N} m_j \dot{x}_j \frac{\partial \dot{x}_j}{\partial q_i} \tag{6–63}$$

We can find $\partial \dot{x}_j / \partial q_i$ from Eq. (6–55) in a straightforward manner if we take the precaution of changing the summation index from i to k.

$$\frac{\partial \dot{x}_j}{\partial q_i} = \sum_{k=1}^{n} \frac{\partial^2 x_j}{\partial q_i\, \partial q_k} \dot{q}_k + \frac{\partial^2 x_j}{\partial q_i\, \partial t} \tag{6–64}$$

From Eqs. (6–63) and (6–64), we find that

$$\frac{\partial T}{\partial q_i} = \sum_{j=1}^{3N} m_j \dot{x}_j \left(\sum_{k=1}^{n} \frac{\partial^2 x_j}{\partial q_i\, \partial q_k} \dot{q}_k + \frac{\partial^2 x_j}{\partial q_i\, \partial t} \right) \tag{6–65}$$

Comparing Eqs. (6–62) and (6–65), we see that

$$\frac{dp_i}{dt} = \sum_{j=1}^{3N} m_j \ddot{x}_j \frac{\partial x_j}{\partial q_i} + \frac{\partial T}{\partial q_i} \tag{6–66}$$

Now let us consider the forces acting on a typical particle of the system. The total force acting on the particle in the x_j direction is the sum of the force f_j due to workless constraints and the applied force F_j which includes all other forces. Thus F_j includes externally applied forces, frictional forces, and forces due to moving constraints. Applying Newton's law of motion, we obtain

$$m_j \ddot{x}_j = F_j + f_j \tag{6-67}$$

Now we substitute this expression for $m_j \ddot{x}_j$ into Eq. (6–66) and obtain

$$\frac{dp_i}{dt} = \sum_{j=1}^{3N} F_j \frac{\partial x_j}{\partial q_i} + \sum_{j=1}^{3N} f_j \frac{\partial x_j}{\partial q_i} + \frac{\partial T}{\partial q_i} \tag{6-68}$$

It is apparent from Eq. (6–51) that the first term on the right of the equality is the generalized force Q_i due to the applied forces.

$$Q_i = \sum_{j=1}^{3N} F_j \frac{\partial x_j}{\partial q_i} \tag{6-69}$$

In a similar fashion, the term

$$\sum_{j=1}^{3N} f_j \frac{\partial x_j}{\partial q_i}$$

is a generalized force resulting from the workless constraint forces.

Suppose we calculate the work done by the constraint forces f_j in a virtual displacement. In accordance with Eq. (6–17) we obtain

$$\delta W_c = \sum_{i=1}^{n} \sum_{j=1}^{3N} f_j \frac{\partial x_j}{\partial q_i} \delta q_i = 0 \tag{6-70}$$

for any set of δq's conforming to the constraints of the system, that is, the virtual work of the workless constraints is zero. If the δq's can be chosen independently, then the coefficient of each δq_i in Eq. (6–70) must be zero. So at this point we assume that the q's are unconstrained; hence

$$\sum_{j=1}^{3N} f_j \frac{\partial x_j}{\partial q_i} = 0 \tag{6-71}$$

This assumption that the δq's are independent, coupled with the previous assumption that there are at least as many q's as there are degrees of freedom, results in the following requirement: *There must be exactly as many* q's *as there are degrees of freedom if the generalized forces due to the workless constraints are to vanish.* This also implies that the system must be *holonomic* if the calculation of constraint forces is to be avoided. In case the generalized coordinates are chosen such that they are not independent and hence there are equations of constraint relating the q's, then the generalized constraint forces as given by Eq. (6–71) are not zero, in general, and these forces must be included in the calculation of the Q_i.

So, assuming that the q's are independent, we can use Eqs. (6–69) and (6–71) to simplify Eq. (6–68) to the form

$$\frac{dp_i}{dt} = Q_i + \frac{\partial T}{\partial q_i} \tag{6-72}$$

or, substituting for p_i from Eq. (6–58),

$$\frac{d}{dt}\left(\frac{\partial T}{\partial \dot{q}_i}\right) - \frac{\partial T}{\partial q_i} = Q_i \quad (i = 1, 2, \ldots, n) \tag{6–73}$$

These n equations are known as *Lagrange's equations* and are written here in one of their principal forms.

The physical meaning of Lagrange's equations can best be seen from Eq. (6–72) which states that the time rate of change of the scalar generalized momentum p_i is equal to the generalized force Q_i due to the applied forces plus the term $\partial T/\partial q_i$ which is an inertial generalized force due to motion in the other generalized coordinates. To see this last point more clearly, consider again the example of a particle whose kinetic energy is expressed in terms of spherical coordinates. We find that

$$\frac{\partial T}{\partial r} = mr\dot{\theta}^2 + mr\sin^2\theta\,\dot{\phi}^2$$

The first term on the right is the centrifugal force due to the θ motion, whereas the second term is the r component of the centrifugal force resulting from motion in ϕ.

Another form of Lagrange's equations of motion can be obtained for systems in which all the Q's are derivable from a potential function $V = V(q_1, q_2, \ldots, q_n, t)$ as follows:

$$Q_i = -\frac{\partial V}{\partial q_i} \tag{6–74}$$

We see that this equation applies to systems with time-varying potential functions as well as to conservative systems where V is a function of position only. To clarify this point, note from Eq. (6–50) that the virtual work of the generalized forces in an arbitrary virtual displacement is

$$\delta W = \sum_{i=1}^{n} Q_i\,\delta q_i$$

But time is held fixed during a virtual displacement, and therefore the system can be considered to be conservative in calculating the virtual work. Thus

$$\delta W = -\delta V = -\sum_{i=1}^{n} \frac{\partial V}{\partial q_i}\,\delta q_i \tag{6–75}$$

Since the δq's are assumed to be arbitrary, we can equate coefficients and obtain

$$Q_i = -\frac{\partial V}{\partial q_i}$$

in agreement with Eq. (6–74).

Substituting this expression for Q_i into Eq. (6–73), we find that

$$\frac{d}{dt}\left(\frac{\partial T}{\partial \dot{q}_i}\right) - \frac{\partial T}{\partial q_i} = -\frac{\partial V}{\partial q_i} \tag{6–76}$$

Now let us define the *Lagrangian function L* as follows:

$$L = T - V \tag{6-77}$$

Then, since we have assumed that V is not a function of the $\dot{q}$'s, we can write

$$\frac{d}{dt}\left(\frac{\partial L}{\partial \dot{q}_i}\right) - \frac{\partial L}{\partial q_i} = 0 \quad (i = 1, 2, \ldots, n) \tag{6-78}$$

This is the most common form of Lagrange's equation.

It is interesting to note that this form of Lagrange's equation is applicable to a more general system than was assumed in its derivation. For example, suppose we let

$$L = T - U$$

where $U = U(q_1, \ldots, q_n, \dot{q}_1, \ldots, \dot{q}_n, t)$ is known as a *velocity-dependent potential*. If a function U can be found such that the Q_i are given by

$$\frac{d}{dt}\left(\frac{\partial U}{\partial \dot{q}_i}\right) - \frac{\partial U}{\partial q_i} = Q_i$$

for all generalized coordinates, then it is apparent that Eq. (6-78) applies to this system. An example of such a system is one involving charged particles moving in an electromagnetic field. We shall assume, however, that the Lagrangian function L is given by Eq. (6-77) and will not concern ourselves further with velocity-dependent potentials.

Now let us consider a system in which the generalized forces are not wholly derivable from a potential function. Of course, Eq. (6-73) is always applicable, but it may be more convenient to write Lagrange's equation in the form

$$\frac{d}{dt}\left(\frac{\partial L}{\partial \dot{q}_i}\right) - \frac{\partial L}{\partial q_i} = Q_i' \quad (i = 1, 2, \ldots, n) \tag{6-79}$$

where the Q_i' are those generalized forces not derivable from a potential function. As before, the other forces are obtained from the Lagrangian function L. Examples of typical Q' forces are frictional forces, time-variant forcing functions, and nonholonomic constraint forces (Sec. 6-7).

In this section, we have derived Lagrange's equations from Newton's laws of motion, using the concept of virtual work and expressing the results by means of generalized coordinates and generalized forces. We have seen that a system having n degrees of freedom will, in general, be described by n second-order ordinary differential equations. These equations are equivalent to the equations of motion which would have been obtained by a direct application of Newton's laws; hence they do not contain new and independent physical principles. Nevertheless, the Lagrangian method of obtaining the equations of motion is more systematic and frequently easier to apply than Newton's laws. Only velocities and displacements enter into the Lagrangian function. No accelerations are required, and thereby the need for intricate kinematical calculation is often avoided. Once L is found, the procedures for obtaining the equations of motion are quite straightforward.

Furthermore, these equations tend to have a convenient form, and particularly in the case of linear systems, the equations show a symmetry in the coefficients which might not be apparent in the Newtonian formulation.

It is a remarkable fact that the Lagrange approach enables one to obtain the equations of motion for a broad class of problems from a single scalar function, the Lagrangian function L. The emphasis upon energies rather than forces and accelerations enables one to be concerned primarily with scalar quantities. This analytical approach to mechanics can also be formulated using variational procedures. In more advanced treatments of mechanics, these variational or minimization principles are used as a starting point for writing the equations of motion and, in fact, Lagrange's equations can be derived in this manner.

Example 6-3. Write the differential equations of motion for a particle in a uniform gravitational field using spherical coordinates.

We shall indicate the position of the particle with the usual spherical coordinates r, θ, ϕ, where θ is measured from the vertical. The kinetic and the potential energies are

$$T = \frac{1}{2}m(\dot{r}^2 + r^2\dot{\theta}^2 + r^2\sin^2\theta\,\dot{\phi}^2) \tag{6-80}$$

$$V = mgr\cos\theta \tag{6-81}$$

and therefore the Lagrangian function is

$$L = T - V = \frac{1}{2}m(\dot{r}^2 + r^2\dot{\theta}^2 + r^2\sin^2\theta\,\dot{\phi}^2) - mgr\cos\theta \tag{6-82}$$

Using the standard form of Lagrange's equations given in Eq. (6–78), we obtain

$$\frac{\partial L}{\partial \dot{r}} = m\dot{r}, \qquad \frac{d}{dt}\left(\frac{\partial L}{\partial \dot{r}}\right) = m\ddot{r}$$

Also,

$$\frac{\partial L}{\partial r} = m(r\dot{\theta}^2 + r\sin^2\theta\,\dot{\phi}^2) - mg\cos\theta$$

Therefore the equation of motion associated with the coordinate r is

$$m\ddot{r} - mr\dot{\theta}^2 - mr\dot{\phi}^2\sin^2\theta + mg\cos\theta = 0 \tag{6-83}$$

In a similar manner, the θ equation is found from

$$\frac{\partial L}{\partial \dot{\theta}} = mr^2\dot{\theta}, \qquad \frac{d}{dt}\left(\frac{\partial L}{\partial \dot{\theta}}\right) = mr^2\ddot{\theta} + 2mr\dot{r}\dot{\theta}$$

and

$$\frac{\partial L}{\partial \theta} = mr^2\dot{\phi}^2\sin\theta\cos\theta + mgr\sin\theta$$

resulting in

$$mr^2\ddot{\theta} + 2mr\dot{r}\dot{\theta} - mr^2\dot{\phi}^2 \sin\theta\cos\theta - mgr\sin\theta = 0 \qquad (6\text{–}84)$$

The ϕ equation is obtained as follows:

$$\frac{\partial L}{\partial\dot{\phi}} = mr^2\sin^2\theta\,\dot{\phi}$$

$$\frac{d}{dt}\left(\frac{\partial L}{\partial\dot{\phi}}\right) = mr^2\ddot{\phi}\sin^2\theta + 2mr\dot{r}\dot{\phi}\sin^2\theta + 2mr^2\dot{\theta}\dot{\phi}\sin\theta\cos\theta$$

$$\frac{\partial L}{\partial\phi} = 0$$

Therefore,

$$mr^2\ddot{\phi}\sin^2\theta + 2mr\dot{r}\dot{\phi}\sin^2\theta + 2mr^2\dot{\theta}\dot{\phi}\sin\theta\cos\theta = 0 \qquad (6\text{–}85)$$

Equations (6–83), (6–84), and (6–85) are the differential equations of motion for the particle. If we divide out the common factors in each equation, assuming that neither r nor θ is zero, we obtain

$$\ddot{r} - r\dot{\theta}^2 - r\dot{\phi}^2\sin^2\theta = -g\cos\theta$$

$$r\ddot{\theta} + 2\dot{r}\dot{\theta} - r\dot{\phi}^2\sin\theta\cos\theta = g\sin\theta \qquad (6\text{–}86)$$

$$r\ddot{\phi}\sin\theta + 2\dot{r}\dot{\phi}\sin\theta + 2r\dot{\theta}\dot{\phi}\cos\theta = 0$$

Comparing the left side of each of these equations with the corresponding acceleration component in spherical coordinates, given previously in Eq. (2–33), we see that they are identical.

Example 6–4. Find the differential equations of motion for the spherical pendulum of length l shown in Fig. 6–11.

Let us use spherical coordinates again as we did in Example 6–3. The Lagrangian function L can be obtained by substituting $r = l$ into Eq. (6–82). The result is

$$L = \frac{1}{2}m(l^2\dot{\theta}^2 + l^2\sin^2\theta\,\dot{\phi}^2) - mgl\cos\theta$$

By setting r equal to a constant, we have eliminated it as coordinate, and therefore only two degrees of freedom remain. We can obtain the differential equations for the θ and ϕ motions by setting $r = l$ in Eqs. (6–84) and (6–85).

$$ml^2\ddot{\theta} - ml^2\dot{\phi}^2\sin\theta\cos\theta - mgl\sin\theta = 0$$

$$ml^2\ddot{\phi}\sin^2\theta + 2ml^2\dot{\theta}\dot{\phi}\sin\theta\cos\theta = 0 \qquad (6\text{–}87)$$

The last equation could have been written in the form

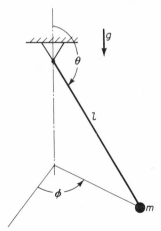

Fig. 6-11. A spherical pendulum.

$$\frac{d}{dt}(ml^2 \sin^2 \theta \, \dot{\phi}) = 0 \qquad (6\text{–}88)$$

which is just a statement of the conservation of angular momentum about a vertical axis through the point of suspension.

Example 6–5. Block m_2 can slide on m_1 which, in turn, can slide on a frictionless, horizontal surface, Fig. 6–12(a). If the absolute position of m_1 is given by x_1, and if the position of m_2 relative to m_1 is given by x_2, solve for the acceleration of m_1. Assume that $\mu < 1$.

First we note that the system has two degrees of freedom. Also, the two coordinates x_1 and x_2 can vary freely without violating the constraints of the system. Hence we can use the coordinates x_1 and x_2 in the Lagrange formulation of the equations of motion. We shall use Eq. (6–79), since the gravitational forces are derivable from a potential function but the frictional forces are not.

It can be seen that the square of the velocity of m_2 is just the sum of the squares of the vertical and horizontal components, or

$$v_2^2 = \frac{\dot{x}_2^2}{2} + \left(\dot{x}_1 - \frac{\dot{x}_2}{\sqrt{2}}\right)^2$$

Hence the kinetic energy is

$$T = \frac{1}{2}m_1 \dot{x}_1^2 + \frac{1}{2}m_2(\dot{x}_1^2 + \dot{x}_2^2 - \sqrt{2}\,\dot{x}_1\dot{x}_2) \qquad (6\text{–}89)$$

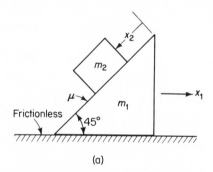

(a)

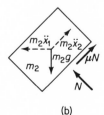

(b)

Fig. 6-12. A system of sliding blocks.

The potential energy is

$$V = -\frac{m_2 g x_2}{\sqrt{2}} \tag{6-90}$$

if we take a reference such that $V = 0$ when $x_2 = 0$. (Note that changing x_1 does not influence the potential energy.) The Lagrangian function is

$$L = T - V = \frac{1}{2} m_1 \dot{x}_1^2 + \frac{1}{2} m_2 (\dot{x}_1^2 + \dot{x}_2^2 - \sqrt{2}\, \dot{x}_1 \dot{x}_2) + \frac{m_2 g x_2}{\sqrt{2}} \tag{6-91}$$

Now we must calculate Q_1' and Q_2'. We do this by considering the virtual work of the frictional forces. First we note that a virtual displacement of x_1 alone does not involve any virtual work because no sliding occurs between blocks and therefore the frictional forces do no work. Hence

$$Q_1' = 0$$

It is apparent, however, that a virtual displacement δx_2 does involve virtual work of the forces of friction. Unfortunately, these forces depend upon the normal constraint force at the sliding surface, and the Lagrangian approach does not indicate the value of this force. So we turn to the free-body diagram shown in Fig. 6–12(b). Since the bodies are in motion at the time of the virtual displacement, we use d'Alembert's principle and include the inertial forces as well as the real external forces. Summing force components normal to the sliding surface, we obtain

$$N + \frac{m_2}{\sqrt{2}} (\ddot{x}_1 - g) = 0$$

or

$$N = \frac{m_2}{\sqrt{2}} (g - \ddot{x}_1)$$

Now we can evaluate the virtual work of the frictional forces.

$$\delta W = -\mu N\, \delta x_2$$

Therefore,

$$Q_2' = -\mu N = -\frac{\mu m_2}{\sqrt{2}} (g - \ddot{x}_1) \tag{6-92}$$

Using Lagrange's equations in the form

$$\frac{d}{dt}\left(\frac{\partial L}{\partial \dot{q}_i}\right) - \frac{\partial L}{\partial q_i} = Q_i'$$

let us first obtain the x_1 equation. We find that

$$\frac{d}{dt}\left(\frac{\partial L}{\partial \dot{x}_1}\right) = m_1 \ddot{x}_1 + m_2 \left(\ddot{x}_1 - \frac{\ddot{x}_2}{\sqrt{2}}\right)$$

$$\frac{\partial L}{\partial x_1} = 0$$

Therefore,

$$(m_1 + m_2)\ddot{x}_1 - \frac{m_2}{\sqrt{2}}\ddot{x}_2 = 0 \qquad (6\text{–}93)$$

This equation states that the rate of change of the horizontal component of the total linear momentum is zero, and therefore the horizontal momentum is conserved. In obtaining the x_2 equation, we see that

$$\frac{d}{dt}\left(\frac{\partial L}{\partial \dot{x}_2}\right) = m_2\ddot{x}_2 - \frac{m_2}{\sqrt{2}}\ddot{x}_1$$

$$\frac{\partial L}{\partial x_2} = \frac{m_2 g}{\sqrt{2}}$$

and therefore the second equation of motion is

$$m_2\ddot{x}_2 - \frac{m_2}{\sqrt{2}}\ddot{x}_1 - \frac{m_2 g}{\sqrt{2}} = -\frac{\mu m_2}{\sqrt{2}}(g - \ddot{x}_1)$$

or

$$m_2\ddot{x}_2 - \frac{m_2}{\sqrt{2}}(1 + \mu)\ddot{x}_1 = \frac{m_2 g}{\sqrt{2}}(1 - \mu) \qquad (6\text{–}94)$$

Now we can use Eqs. (6–93) and (6–94) to solve for $\ddot{x}_1$, the result being

$$\ddot{x}_1 = \frac{m_2(1 - \mu)g}{2m_1 + m_2(1 - \mu)} \qquad (6\text{–}95)$$

Example 6–6. A particle of mass m can slide without friction on the inside of a small tube bent in the form of a circle of radius a. The tube rotates about a vertical diameter at a constant rate of ω rad/sec, as shown in Fig. 6–13. Write the differential equation of motion. If the particle is disturbed slightly from its unstable equilibrium position at $\theta = 0$, find the position of maximum kinetic energy and the position of maximum $\dot{\theta}$, assuming that $\dot{\theta}$ is positive in the initial motion.

Once again we shall use spherical coordinates in describing the configuration. In this case, two holonomic constraints are applied, namely, the workless scleronomic constraint $r = a$ and the working rheonomic constraint $\phi = \omega t + \phi_0$, where ϕ_0 is a constant. Hence we see that there is only one degree of freedom and θ is the corresponding unrestricted coordinate.

From Eqs. (6–80) and (6–81) we see that

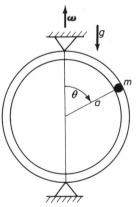

Fig. 6-13. A particle in a whirling tube.

$$T = \frac{1}{2}m(a^2\dot{\theta}^2 + a^2\omega^2 \sin^2\theta) \qquad (6\text{--}96)$$

$$V = mga\cos\theta \qquad (6\text{--}97)$$

and therefore

$$L = T - V = \frac{1}{2}ma^2\dot{\theta}^2 + \frac{1}{2}ma^2\omega^2\sin^2\theta - mga\cos\theta \qquad (6\text{--}98)$$

We use the standard form of Lagrange's equation given in Eq. (6–78) although the system has a working constraint and is not conservative. This follows from the fact that only gravitational forces do work in a virtual displacement $\delta\theta$ and therefore Q'_θ is zero. Note, however, that work is done on the system by means of a non-zero Q'_ϕ. Evaluating the required terms, we obtain

$$\frac{\partial L}{\partial \dot{\theta}} = ma^2\dot{\theta}, \qquad \frac{d}{dt}\left(\frac{\partial L}{\partial \dot{\theta}}\right) = ma^2\ddot{\theta}$$

$$\frac{\partial L}{\partial \theta} = ma^2\omega^2 \sin\theta\cos\theta + mga\sin\theta$$

and therefore the differential equation of motion is

$$ma^2\ddot{\theta} - ma^2\omega^2\sin\theta\cos\theta - mga\sin\theta = 0 \qquad (6\text{--}99)$$

In order to find the value of θ corresponding to the maximum value of the kinetic energy, we need to express T as a function of θ alone, instead of its being expressed as a function of both $\dot{\theta}$ and θ. We can find an expression for $\dot{\theta}$ as a function of θ by an integration of the equation of motion. Making the substitution

$$\ddot{\theta} = \dot{\theta}\frac{d\dot{\theta}}{d\theta}$$

in Eq. (6–99), we obtain

$$\dot{\theta}\,d\dot{\theta} = \left(\omega^2\sin\theta\cos\theta + \frac{g}{a}\sin\theta\right)d\theta$$

which can be integrated directly to yield

$$\dot{\theta}^2 = \omega^2\sin^2\theta + \frac{2g}{a}(1 - \cos\theta) \qquad (6\text{--}100)$$

where the constant of integration has been evaluated from the initial condition that $\dot{\theta} = 0$ when $\theta = 0$. Hence we obtain from Eqs. (6–96) and (6–100) that

$$T = ma^2\omega^2\sin^2\theta + mga(1 - \cos\theta) \qquad (6\text{--}101)$$

To find the maximum value of T, we set $dT/d\theta = 0$ with the result:

$$\frac{dT}{d\theta} = 2ma^2\omega^2\sin\theta\cos\theta + mga\sin\theta = 0$$

or

$$\theta = 0, \pi, \cos^{-1}\left(-\frac{g}{2a\omega^2}\right)$$

The first possibility, $\theta = 0$, clearly corresponds to T_{min} since $T = 0$ at this configuration. T_{max} occurs at $\theta = \pi$ or at $\cos\theta = -g/2a\omega^2$, depending upon whether ω^2 is less than or greater than $g/2a$. Substituting into Eq. (6–101), we see that

$$T_{max} = 2mga \qquad \left(\omega^2 \leq \frac{g}{2a}\right)$$

$$T_{max} = m\left(a^2\omega^2 + ga + \frac{g^2}{4\omega^2}\right) \qquad \left(\omega^2 \geq \frac{g}{2a}\right)$$

The position of maximum $\dot\theta$ is found by differentiating Eq. (6–100) with respect to θ and setting the result equal to zero. We can write

$$\frac{d}{d\theta}(\dot\theta^2) = 2\omega^2 \sin\theta\cos\theta + \frac{2g}{a}\sin\theta = 0$$

since $\dot\theta$ and $\dot\theta^2$ reach a maximum amplitude together. Thus the position of maximum $\dot\theta$ is located at one of the following:

$$\dot\theta = \pi, \cos^{-1}\left(-\frac{g}{a\omega^2}\right)$$

since we found that $\dot\theta = 0$ at $\theta = 0$. A substitution of these values into Eq. (6–100) reveals that

$$\dot\theta_{max} = 2\sqrt{\frac{g}{a}} \qquad \left(\omega^2 \leq \frac{g}{a}\right)$$

$$\dot\theta_{max} = \sqrt{\omega^2 + \frac{2g}{a} + \frac{g^2}{a^2\omega^2}} \qquad \left(\omega^2 \geq \frac{g}{a}\right)$$

Now let us consider this problem from another viewpoint. We have seen that the Lagrangian function L completely determines the equations of motion for systems in which the generalized forces are obtained from a potential function. Had we been given the Lagrangian function of Eq. (6–98) without specifying the system, we might have supposed that the kinetic and potential energies were

$$T' = \frac{1}{2}ma^2\dot\theta^2 \tag{6–102}$$

$$V' = mga\cos\theta - \frac{1}{2}ma^2\omega^2\sin^2\theta \tag{6–103}$$

This corresponds to a conservative system with one degree of freedom, since we note that T' is a quadratic function of $\dot\theta$ and V' is a function of position only. It can be seen that T' is the kinetic energy of the particle, as viewed by an observer rotating with the circular tube. The function V' includes the actual gravitational potential energy plus another term which accounts for the centrifugal force due to the rotation rate ω about the vertical axis. Thus, the coordinate θ moves as though the whirling motion were stopped and the inertial forces due to whirling were considered to originate from a potential function. We note, however, that both T' and V' are fictitious in the sense that the reference frame is noninertial.

Now let us consider the quantity

$$E' = T' + V' \tag{6-104}$$

which is a fictitious total energy. We have seen that this system with one degree of freedom is conservative. So, evaluating the total energy from initial conditions, we find that

$$E' = \frac{1}{2}ma^2\dot{\theta}^2 + mga\cos\theta - \frac{1}{2}ma^2\omega^2\sin^2\theta = mga$$

or

$$\dot{\theta}^2 = \omega^2\sin^2\theta + \frac{2g}{a}(1 - \cos\theta)$$

in agreement with the result obtained previously in Eq. (6–100). Note that we have obtained an expression for $\dot{\theta}$ in this case without the necessity of integrating, or even finding, the differential equation of motion. To find $\dot{\theta}_{max}$ we look for T'_{max}, and this occurs at the same position as V'_{min} since E' is conserved. To find V'_{min} we write

$$\frac{dV'}{d\theta} = 0 = -mga\sin\theta - ma^2\omega^2\sin\theta\cos\theta$$

from which we obtain

$$\theta = 0, \pi, \cos^{-1}\left(-\frac{g}{a\omega^2}\right)$$

If $\omega^2 > g/a$, the first two solutions represent local maxima of V' and the third solution is V'_{min}, as shown in Fig. 6–14. If $\omega^2 \leq g/a$, V'_{min} occurs at $\theta = \pi$. Again the results agree with those obtained previously.

It is interesting to note that V' is symmetric about $\theta = n\pi$, where n is any integer. Therefore, because E' is conserved, the magnitude of $\dot{\theta}$ must show the same symmetry. Thus if $E' > mga$, the sign of $\dot{\theta}$ will not change during the motion. If $-mga < E' < mga$, the solution will consist of an oscillation about $\theta = \pi$. On the other hand, if $E' < -mga$ and $\omega^2 > g/a$, then the solution is an oscillation about $\theta = \cos^{-1}(-g/a\omega^2)$. In all cases, the turning points of the motion occur at positions where $E' = V'$, implying that $\dot{\theta} = 0$.

Example 6-7. The point of support O of a simple pendulum of length l moves with a horizontal displacement

$$x = A\sin\omega t \tag{6-105}$$

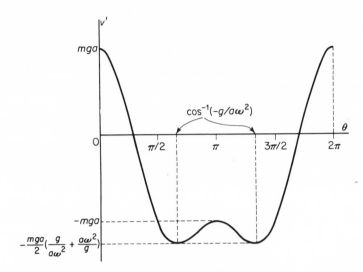

Fig. 6-14. The gravitational plus centrifugal potential energy for the case where $\omega^2 > g/a$.

in the plane of the motion in θ (Fig. 6–15a). Find the differential equation of motion.

This example is identical to that solved previously as Example 4–12. Here we shall use the Lagrangian approach rather than the vectorial approach.

First we note that this problem involves a moving constraint, hence the system is not conservative. Nevertheless, it is apparent that the constraint force acts through the support point O and therefore does not contribute to Q_θ. Thus we can use the standard form of Lagrange's equation.

In order to obtain the kinetic energy we calculate first the absolute velocity v of mass m by the vector addition of the velocities due to $\dot{x}$ and $\dot{\theta}$. In other words, the total velocity is equal to the velocity of O plus the velocity

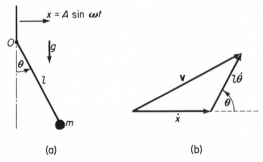

Fig. 6-15. A simple pendulum with a moving support.

of mass m relative to O. Using the cosine law, we see from Fig. 6–15(b) that

$$v^2 = \dot{x}^2 + l^2\dot{\theta}^2 + 2l\dot{x}\dot{\theta}\cos\theta$$

Evaluating $\dot{x}$ from Eq. (6–105), we find that

$$T = \frac{1}{2}mv^2 = \frac{m}{2}(A^2\omega^2\cos^2\omega t + l^2\dot{\theta}^2 + 2Al\omega\dot{\theta}\cos\theta\cos\omega t) \qquad (6\text{–}106)$$

The potential energy is

$$V = -mgl\cos\theta \qquad (6\text{–}107)$$

where the support point is taken as the reference.

Noting again that $L = T - V$, we see that

$$\frac{d}{dt}\left(\frac{\partial L}{\partial\dot{\theta}}\right) = ml^2\ddot{\theta} - mAl\omega^2\cos\theta\sin\omega t - mAl\omega\dot{\theta}\sin\theta\cos\omega t$$

$$\frac{\partial L}{\partial\theta} = -mAl\omega\dot{\theta}\sin\theta\cos\omega t - mgl\sin\theta$$

Therefore, using Eq. (6–78), we find that the equation of motion is

$$ml^2\ddot{\theta} - mAl\omega^2\cos\theta\sin\omega t + mgl\sin\theta = 0 \qquad (6\text{–}108)$$

in agreement with the result of Example 4–12.

A third method of solving this problem is by a direct application of Newton's law of motion. The difficulty of this method depends upon the precise approach; in general, however, the method requires more computation or more insight than the Lagrangian approach. Taking vertical and horizontal forces and accelerations necessitates solving for the force in the supporting wire. On the other hand, if we consider force and acceleration components normal to the wire at any instant, an application of Newton's law results in

$$m(l\ddot{\theta} + \ddot{x}\cos\theta) = -mg\sin\theta$$

or

$$ml\ddot{\theta} - mA\omega^2\cos\theta\sin\omega t = -mg\sin\theta$$

which is equivalent to Eq. (6–108).

6-7. LAGRANGE MULTIPLIERS

In the derivation of Lagrange's equations of motion in Sec. 6–6 we found that, even for a conservative system, there must be no more generalized coordinates used in the analysis than there are degrees of freedom if we are to be able to write the equations of motion in the standard form of Eq. (6–78). Otherwise, generalized force terms representing the constraint forces must appear in the equations. Also, we saw that the nonintegrable nature of the equations of constraint in the case of nonholonomic systems makes it necessary to have more coordinates than degrees of freedom. Therefore it follows that we must include the constraint forces due to non-holonomic constraints when we write the equations of motion for these systems. In this section, we introduce the method of Lagrange multipiers to allow us to solve for the constraint forces. This method will be found to be applicable to holonomic as well as nonholonomic constraints.

Recall from Eq. (6–5) that the general nonholonomic constraint equations are of the form

$$\sum_{i=1}^{n} a_{ji}\, dq_i + a_{jt}\, dt = 0 \qquad (j = 1, 2, \ldots, m)$$

whereas we found in Eq. (6–3) that holonomic constraints are written in the form

$$\phi_j(q_1, q_2, \ldots, q_n, t) = 0 \qquad (j = 1, 2, \ldots, m)$$

Note, however, that a holonomic constraint equation can also be written in the differential form of Eq. (6–5). This can be seen by writing the total differential of ϕ_j, namely,

$$d\phi_j = \frac{\partial \phi_j}{\partial q_1}\, dq_1 + \cdots + \frac{\partial \phi_j}{\partial q_n}\, dq_n + \frac{\partial \phi_j}{\partial t}\, dt = 0$$

and noting that if we let

$$a_{ji} = \frac{\partial \phi_j}{\partial q_i} \quad \text{and} \quad a_{jt} = \frac{\partial \phi_j}{\partial t} \qquad (6\text{–}109)$$

then the form is identical to Eq. (6–5). In general, however, the holonomic constraints would not be written in this form unless one desires to solve for the constraint forces.

To illustrate the method of Lagrange multipliers, consider a system with frictionless constraints which may be either holonomic or nonholonomic. At any instant of time, the variations of the individual generalized coordinates in a virtual displacement must meet the conditions

$$\sum_{i=1}^{n} a_{ji}\, \delta q_i = 0 \qquad (j = 1, 2, \ldots, m) \qquad (6\text{–}110)$$

in accordance with Eq. (6–5), where we recall that δt is zero in a virtual displacement. In other words, the virtual displacement is consistent with the *instantaneous constraints* of the system. We have assumed that the constraints are frictionless; therefore no work is done by the constraint forces in any virtual displacement permitted by Eq. (6–110). So if we let C_i be the *generalized constraint force* corresponding to q_i, then

$$\sum_{i=1}^{n} C_i\, \delta q_i = 0 \qquad (6\text{–}111)$$

for any set of δq's satisfying Eq. (6–110).

Now let us multiply Eq. (6–110) by a factor λ_j known as a *Lagrange multiplier*, obtaining

$$\lambda_j \sum_{i=1}^{n} a_{ji}\, \delta q_i = 0 \qquad (j = 1, 2, \ldots, m) \tag{6–112}$$

where we note that a separate equation is written for each of the m constraints. Next we subtract the sum of the m equations of the form of Eq. (6–112) from Eq. (6–111) and, interchanging the order of summation, we obtain

$$\sum_{i=1}^{n} \left(C_i - \sum_{j=1}^{m} \lambda_j a_{ji} \right) \delta q_i = 0 \tag{6–113}$$

So far the values of the λ's have been considered to be arbitrary, and the δq's have been restricted in accordance with the constraints of Eq. (6–110). But if we choose the λ's such that each of the coefficients of the δq's are zero in Eq. (6–113), then this equation will apply for any values of the δq's, that is, they can be chosen independently. As we shall see, the λ's can be chosen to accomplish this result.

So now let us assume that the λ's have values such that

$$C_i = \sum_{j=1}^{m} \lambda_j a_{ji} \tag{6–114}$$

and, as a result, the q's are considered to be independent. Then, writing Lagrange's equations in the general form of Eq. (6–79), we obtain

$$\frac{d}{dt}\left(\frac{\partial L}{\partial \dot{q}_i}\right) - \frac{\partial L}{\partial q_i} = \sum_{j=1}^{m} \lambda_j a_{ji} \qquad (i = 1, 2, \ldots, n) \tag{6–115}$$

where, from Eq. (6–114), we see that the right-hand side is just the constraint force C_i. What has been accomplished by this procedure is the representation of the constraints in terms of the corresponding constraint forces acting on a system of n independent coordinates. At this point, we have n equations of motion but $(n + m)$ unknowns, namely, the n q's and the m λ's. The required m additional equations are obtained by writing Eq. (6–5) in the form

$$\sum_{i=1}^{n} a_{ji}\dot{q}_i + a_{jt} = 0 \qquad (j = 1, 2, \ldots, m) \tag{6–116}$$

Using Eqs. (6–115) and (6–116), a solution for the λ's as well as the q's can be obtained. The constraint forces vary with time, even for the case of fixed constraints, and therefore the λ's will, in general, be functions of time also.

Hence, considering Eqs. (6–115) and (6–116) together, we have replaced a system having $(n - m)$ unknowns by one with $(n + m)$ unknowns, considering the λ's as additional variables. This may not seem like progress, but quite often this procedure results in simpler equations. Also, the symmetry of the problem is preserved, since there are no preferred coordinates while others are eliminated. In addition, of course, one obtains the constraint forces in the process of obtaining the solution.

The derivation given here has assumed frictionless constraints. In case the constraints also involve dissipative forces, such as sliding friction, then these generalized forces are written as separate terms involving the C_i and possibly other variables of the system.

Example 6-8. Mass m_2 slides relative to m_1 which, in turn, slides relative to a horizontal surface, Fig. 6–16(a). Assuming first that all surfaces are frictionless, write the differential equations of motion for the system and solve for $\ddot{x}_1$ and the constraint force acting between the blocks. Then solve the problem for the case where Coulomb friction is present, again using the Lagrange multiplier method.

The system under consideration is identical with that of Example 6–5. We are asked, however, to solve for the constraint force and therefore we include a new coordinate x_3 which cannot vary freely without violating the constraint in question.

First consider the case where $\mu = 0$. In order to obtain the Lagrangian function L, we must write an expression for the total kinetic energy. Taking horizontal and vertical components of the velocity of m_2, we obtain

$$
T = \frac{1}{2} m_1 \dot{x}_1^2 + \frac{1}{2} m_2 \left[\left(\dot{x}_1 - \frac{\dot{x}_2 + \dot{x}_3}{\sqrt{2}} \right)^2 + \left(\frac{\dot{x}_3 - \dot{x}_2}{\sqrt{2}} \right)^2 \right]
$$
$$
= \frac{1}{2} (m_1 + m_2) \dot{x}_1^2 + \frac{1}{2} m_2 [\dot{x}_2^2 + \dot{x}_3^2 - \sqrt{2} \, (\dot{x}_2 + \dot{x}_3) \dot{x}_1] \quad (6\text{–}117)
$$

where we note that x_3 is considered at this point to be freely variable. The potential energy is

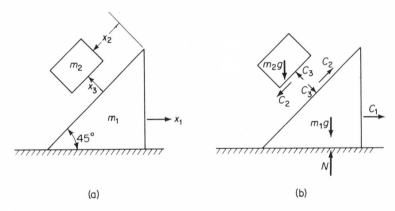

(a)   (b)

Fig. 6-16. The coordinates and forces used in the analysis of the sliding block problem.

$$V = \frac{m_2 g}{\sqrt{2}} (x_3 - x_2) \tag{6-118}$$

and therefore the Lagrangian function is

$$L = \frac{1}{2}(m_1 + m_2)\dot{x}_1^2 + \frac{1}{2}m_2[\dot{x}_2^2 + \dot{x}_3^2 - \sqrt{2}\,(\dot{x}_2 + \dot{x}_3)\dot{x}_1]$$

$$- \frac{m_2 g}{\sqrt{2}}(x_3 - x_2) \tag{6-119}$$

Next we evaluate some of the terms which will be needed in writing the equations of motion.

$$\frac{d}{dt}\left(\frac{\partial L}{\partial \dot{x}_1}\right) = (m_1 + m_2)\ddot{x}_1 - \frac{m_2}{\sqrt{2}}(\ddot{x}_2 + \ddot{x}_3), \qquad \frac{\partial L}{\partial x_1} = 0$$

$$\frac{d}{dt}\left(\frac{\partial L}{\partial \dot{x}_2}\right) = m_2\ddot{x}_2 - \frac{m_2}{\sqrt{2}}\ddot{x}_1, \qquad \frac{\partial L}{\partial x_2} = \frac{m_2 g}{\sqrt{2}}$$

$$\frac{d}{dt}\left(\frac{\partial L}{\partial \dot{x}_3}\right) = m_2\ddot{x}_3 - \frac{m_2}{\sqrt{2}}\ddot{x}_1, \qquad \frac{\partial L}{\partial x_3} = -\frac{m_2 g}{\sqrt{2}}$$

The equation of constraint, written in the form of Eq. (6–116), is

$$\dot{x}_3 = 0 \tag{6-120}$$

and therefore

$$a_{11} = 0$$
$$a_{12} = 0$$
$$a_{13} = 1$$

Note that this constraint is actually holonomic even though we are using the differential form of the constraint equation. Since there is only one equation of constraint, only one Lagrange multiplier is needed and we omit the subscript. From Eq. (6–114), we see that the constraint forces can be written in terms of this Lagrange multiplier as follows:

$$C_1 = 0$$
$$C_2 = 0 \tag{6-121}$$
$$C_3 = \lambda$$

where the positive directions of the generalized constraint forces are shown in Fig. 6–16(b). The differential equations of motion are written in the form of Eq. (6–115).

$$(m_1 + m_2)\ddot{x}_1 - \frac{m_2}{\sqrt{2}}\ddot{x}_2 - \frac{m_2}{\sqrt{2}}\ddot{x}_3 = 0$$

$$-\frac{m_2}{\sqrt{2}}\ddot{x}_1 + m_2\ddot{x}_2 - \frac{m_2 g}{\sqrt{2}} = 0 \tag{6-122}$$

$$-\frac{m_2}{\sqrt{2}}\ddot{x}_1 + m_2\ddot{x}_3 + \frac{m_2 g}{\sqrt{2}} = \lambda$$

Noting from Eq. (6–120) that $\ddot{x}_3 = 0$, we can solve for $\ddot{x}_1$ and λ, obtaining

$$\ddot{x}_1 = \frac{m_2 g}{2m_1 + m_2}$$

and

$$\lambda = C_3 = \frac{\sqrt{2}\, m_1 m_2 g}{2m_1 + m_2}$$

which agrees with the results of Example 6–5 for the case where μ is zero.

Now consider the case in which Coulomb friction exists between the two blocks. We need to include the generalized forces due to friction, so let us write Lagrange's equations in the form

$$\frac{d}{dt}\left(\frac{\partial L}{\partial \dot{q}_i}\right) - \frac{\partial L}{\partial q_i} = \sum_{j=1}^{m} \lambda_j a_{ji} + F_i \qquad (i = 1, 2, \dots, n) \qquad (6\text{–}123)$$

where the F_i are generalized friction forces. In this case, we see that the friction force does work only in the case where m_2 slides relative to m_1. Furthermore, this friction force is equal to the product of μ and the normal force C_3 and opposes the relative motion. So if we assume that $\dot{x}_2 > 0$, we find that

$$F_1 = 0$$
$$F_2 = -\mu C_3 = -\mu\lambda \qquad (6\text{–}124)$$
$$F_3 = 0$$

Using Eq. (6–123), we obtain the differential equations of motion:

$$(m_1 + m_2)\ddot{x}_1 - \frac{m_2}{\sqrt{2}}\ddot{x}_2 - \frac{m_2}{\sqrt{2}}\ddot{x}_3 = 0$$

$$-\frac{m_2}{\sqrt{2}}\ddot{x}_1 + m_2\ddot{x}_2 - \frac{m_2 g}{\sqrt{2}} = -\mu\lambda \qquad (6\text{–}125)$$

$$-\frac{m_2}{\sqrt{2}}\ddot{x}_1 + m_2\ddot{x}_3 + \frac{m_2 g}{\sqrt{2}} = \lambda$$

Again we set $\ddot{x}_3 = 0$ and solve for $\ddot{x}_1$, obtaining

$$\ddot{x}_1 = \frac{m_2 g(1 - \mu)}{2m_1 + m_2(1 - \mu)}$$

in agreement with the previous result of Example 6–5.

Note that we have assumed that Eq. (6–114) applies, even for the case of constraints with friction. With this assumption, the total generalized constraint force corresponding to q_i is the sum of C_i and F_i rather than C_i alone.

Example 6–9. Two wheels are connected by a straight axle of length l. They roll without slipping on the horizontal xy plane. Assume that the total mass is concentrated in two particles of mass m each, which are located at the hub of each wheel, as shown in Fig. 6–17. The coordinates (x, y) give the location of the center of mass; the angle ϕ, measured from a line parallel to the y axis, gives the orientation of the axle. Write the equations of motion for this system and solve for the path followed by the center of mass. Initially, the center of mass is at the origin and moving with velocity v_0 in the direction of the positive x axis. Also, $\dot\phi(0) = \omega$ and $\phi(0) = 0$.

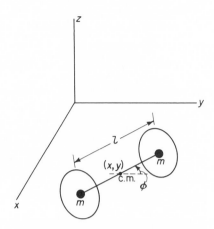

This is a conservative system with a nonholonomic constraint. The wheels roll without slipping and therefore the center of mass must move in a direction perpendicular to

Fig. 6-17. Two wheels, connected by a common axle and rolling on the horizontal xy plane.

the axle. Considering a differential displacement ds along the path followed by the center of mass, we note that the x and y components of ds are

$$dx = \cos\phi\, ds$$
$$dy = \sin\phi\, ds$$

and therefore we obtain

$$\sin\phi\, dx - \cos\phi\, dy = 0$$

which is the differential equation of constraint. If we write the constraint equation in terms of velocities using the form of Eq. (6–116), the result is

$$\sin\phi\, \dot x - \cos\phi\, \dot y = 0 \tag{6–126}$$

The center of mass of the system cannot change its vertical position and therefore the potential energy is constant. Let $V = 0$. The total kinetic energy is found by adding the portion due to the motion of the center of mass and that due to the motion relative to the center of mass. Hence we find that the Lagrangian function is

$$L = T = m(\dot x^2 + \dot y^2) + \tfrac{1}{4}ml^2\dot\phi^2 \tag{6–127}$$

Using Lagrange's equations in the form of Eq. (6–115), we obtain the following differential equations of motion:

$$2m\ddot x = \lambda\sin\phi \tag{6–128}$$
$$2m\ddot y = -\lambda\cos\phi \tag{6–129}$$
$$\tfrac{1}{2}ml^2\ddot\phi = 0 \tag{6–130}$$

Equation (6–130) can be integrated to give the angular velocity

$$\dot{\phi} = \omega \tag{6-131}$$

where the constant of integration is obtained from the initial conditions. Another integration yields

$$\phi = \omega t \tag{6-132}$$

since $\phi(0) = 0$. Now we substitute this expression for ϕ into Eqs. (6-128) and (6-129) and obtain

$$\ddot{x} = \frac{\lambda}{2m} \sin \omega t$$

$$\ddot{y} = -\frac{\lambda}{2m} \cos \omega t$$

From these expressions and Eq. (6-126) we find that λ is constant; hence

$$\dot{x} = -\frac{\lambda}{2m\omega} \cos \omega t + A_1 \tag{6-133}$$

$$\dot{y} = -\frac{\lambda}{2m\omega} \sin \omega t + B_1 \tag{6-134}$$

where A_1 and B_1 are constants. Since $\dot{y}(0) = 0$, we see immediately that $B_1 = 0$. Also, $\dot{x}(0) = v_0$, and therefore

$$A_1 = v_0 + \frac{\lambda}{2m\omega} \tag{6-135}$$

confirming that the Lagrange multiplier λ is constant. We have noted that the system is conservative and the potential energy is constant. It follows that the kinetic energy is constant and, in fact, the translational kinetic energy must also be constant since $\dot{\phi}$ was found to be constant in Eq. (6-131). So we can write

$$\dot{x}^2 + \dot{y}^2 = v_0^2$$

Substituting the values of $\dot{x}$ and $\dot{y}$ from Eqs. (6-133) and (6-134), we obtain

$$\left(\frac{\lambda}{2m\omega}\right)^2 - \frac{\lambda A_1}{m\omega} \cos \omega t + A_1^2 = v_0^2$$

All terms except the cosine term are evidently constant; therefore $A_1 = 0$ since it is clear that λ cannot be zero because this would imply a uniform translational motion in the same direction. So we conclude from Eq. (6-135) that

$$\lambda = -2mv_0\omega \tag{6-136}$$

Now we integrate Eqs. (6-133) and (6-134), noting that $x(0) = y(0) = 0$, and thus we obtain the path of the center of mass:

$$x = \frac{v_0}{\omega} \sin \omega t \tag{6-137}$$

$$y = \frac{v_0}{\omega} (1 - \cos \omega t) \qquad (6\text{-}138)$$

It can be seen from Eqs. (6-137) and (6-138) that

$$x^2 + \left(y - \frac{v_0}{\omega} \right)^2 = \frac{v_0^2}{\omega^2} \qquad (6\text{-}139)$$

indicating that the path is a circle of radius v_0/ω with its center at $x = 0$, $y = v_0/\omega$.

That the path of the center of mass is circular could also have been seen directly from the fact that its velocity v_0 changes direction at a constant angular rate ω. Hence the nonholonomic constraint force is a constant centripetal force, and from Eq. (6-136), its magnitude is seen to be that of the Lagrange multiplier λ.

REFERENCES

1. Banach, S., *Mechanics*, E. J. Scott, trans. Mathematical Monographs, vol. 24, Warsaw, 1951; distributed by Hafner Publishing Co., New York.

2. Corben, H. C., and P. Stehle, *Classical Mechanics*, 2nd ed. New York: John Wiley & Sons, Inc., 1960.

3. Goldstein, H., *Classical Mechanics*. Reading, Mass.: Addison-Wesley Publishing Company, 1950.

4. Lanczos, C., *The Variational Principles of Mechanics*. Toronto: University of Toronto Press, 1949.

5. Whittaker, E. T., *A Treatise on the Analytical Dynamics of Particles and Rigid Bodies*, 4th ed. New York: Dover Publications, Inc., 1944.

PROBLEMS

6-1. A fixed smooth rod makes an angle of 30 degrees with the floor. A small ring of mass m can slide on the rod and supports a string which has one end connected to a point on the floor; the other end is attached to a particle of mass $3m$. Using the principle of virtual work, solve for the angle ψ between the two segments of the string at the static equilibrium position, assuming that the string and the rod lie in the same vertical plane.

6-2. A thin uniform rod of mass m and length $2a$ rests in a smooth hemispherical bowl of radius a. Assuming that the bowl is fixed and its rim is

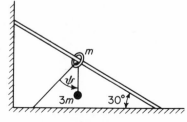

Fig. P6-1

horizontal, find the length of that portion of the rod which remains outside the bowl in the static equilibrium position.

6-3. One end of a thin uniform rod of mass m and length $3l$ rests against a smooth vertical wall. The other end is attached by a string of length l to a fixed point O which is located at a distance $2l$ from the wall. Assuming that the rod and

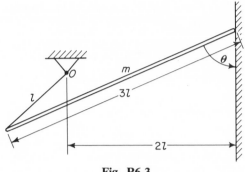

Fig. P6-3

the string remain in the same vertical plane perpendicular to the wall, find the angle θ between the rod and the wall at the position of static equilibrium.

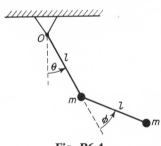

Fig. P6-4

6-4. A double pendulum consists of two massless rods of length l and two particles of mass m which can move in a given vertical plane, as shown. Assuming frictionless joints, and using θ and ϕ as coordinates, obtain the differential equations of motion. What are the linearized equations for small θ and ϕ?

6-5. A particle of mass m is attached by a massless string of length l to a point on the circumference of a horizontal wheel of radius a which is rotating with a constant angular velocity ω about a fixed vertical axis through its center. Write the differential equation of motion for the particle, using as a coordinate the angle θ between the string and a radial line from the center through the attachment point of the string to the wheel. Assume all motion is in the horizontal plane and $-(\pi/2) < \theta < (\pi/2)$.

6-6. The support point O' of a simple pendulum of length l and mass m moves with constant speed in a circular path of radius a centered at the fixed point O. Find the differential equation for θ, assuming that all motion is confined to the vertical plane and $\phi = \omega t$.

6-7. A box of mass m_0 supports a simple pendulum of mass m and length l. A spring of stiffness k forms a horizontal mass-spring system with the box which can slide without friction on a fixed horizontal surface. Find the differential equations of motion of the system, taking x as the translational displacement of the box and θ as the angle of the pendulum from the vertical. What are the linearized equations, assuming that x and θ are small?

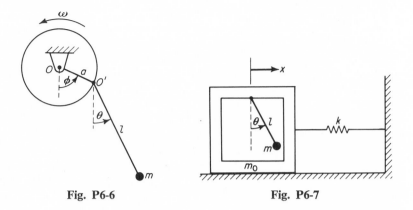

Fig. P6-6   **Fig. P6-7**

6–8. Particles m_1 and m_2, each of mass m, are connected by a massless rod of length l. These particles move on a frictionless horizontal plane, the motion of m_1 being confined to a fixed frictionless circular track of radius R. Denoting the positions of the particles by the angles θ and ϕ, as shown, find the differential equations of motion.

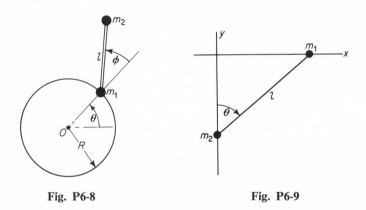

Fig. P6-8   **Fig. P6-9**

6–9. Particles m_1 and m_2 are constrained by frictionless constraints to move in the vertical xy plane such that m_1 remains on the horizontal x axis and m_2 remains on the vertical y axis. The particles are connected by a massless inextensible string of length l. The initial conditions on m_1 are $x(0) = l$, $\dot{x}(0) = 0$, and the corresponding initial conditions on m_2 are $y(0) = 0$, $\dot{y}(0) = 0$. For the case where $m_1 = m_2 = m$, solve for the tension P in the string as a function of the angle θ. What is the period of the motion in θ?

6–10. A massless circular disk of radius r has a particle of mass m embedded at a distance l from the center O. Assuming that the disk can roll without slipping on a horizontal plane, write the differential equation of motion for the system.

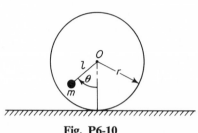

Fig. P6-10

6-11. Given the system of Fig. 6-13, except that the circular tube can rotate freely about the fixed vertical axis, the rotation angle being ϕ. Assuming a moment of inertia $I = ma^2$ for the tube, exclusive of the sliding particle, write the differential equations of motion for the system. Find the maximum value of θ if the initial conditions are $\theta(0) = \pi/2$, $\dot{\theta}(0) = 0$, and $\dot{\phi}(0) = 2\sqrt{g/a}$.

6-12. A particle of mass m can slide without friction in a tube which is rigidly attached at a constant angle $\theta = 60°$ to a vertical shaft which rotates at a constant rate $\omega = \sqrt{2g/r_0}$. If the particle is released with the initial conditions $r(0) = r_0$, $\dot{r}(0) = -\sqrt{gr_0/2}$, find $r_{\min}$. What is the minimum magnitude and sign of the initial radial velocity $\dot{r}(0)$ such that the particle will hit the shaft, that is, reach the position $r = 0$?

6-13. A particle of mass m is constrained to move without friction in the xy plane. The x axis remains horizontal and the y axis is inclined at a constant angle ϕ with the horizontal. In addition, the xy plane translates vertically such that the height of the x axis above a given reference level is $h = A \sin \omega t$. If the particle has the initial conditions $y(0) = y_0$, $\dot{y}(0) = 0$, $x(0) = 0$, $\dot{x}(0) = v_0$, solve for its position (x, y) as a function of time.

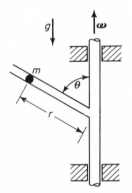

Fig. P6-12

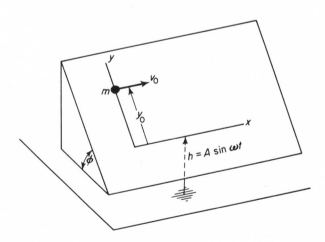

Fig. P6-13

6–14. Consider a system whose configuration is given by the ordinary co-ordinates x_j where $j = 1, 2, \ldots, k$. Suppose a set of impulses $\mathscr{F}_j$ are applied simultaneously at the corresponding x_j. Show that the resulting change in the generalized momentum p_i is given by

$$\Delta p_i = \sum_{j=1}^{k} \mathscr{F}_j \frac{\partial x_j}{\partial q_i}$$

where the right-hand side of the equation can be considered to be the generalized impulse associated with q_i.

6–15. A particle of mass m moves on the smooth inner surface of a fixed inverted circular cone having a vertex angle 2α. (a) Using the angle ϕ about the vertical axis of symmetry and the distance r measured along the cone from the vertex as generalized coordinates, obtain the differential equations of motion. (b) Find $\dot{r}_{\max}$ for the case in which $\alpha = 30°$ and the initial conditions are: $r(0) = a$, $\dot{r}(0) = 0$, $\dot{\phi}(0) = 4\sqrt{g/a}$.

6–16. Consider again the system of problem 6–15. Suppose the particle has a steady circular motion on the inside of the cone at $r = a$. Write the differential equation for small perturbations in r and solve for the circular frequency of this motion.

6–17. A small tube of mass m_0 is bent to form a vertical helix of radius r_0 and pitch angle γ, that is, the tube everywhere makes an angle γ with the horizontal. A particle of mass m slides without friction in the tube. If the helix rotates freely about its vertical axis but does not translate, what are the angular acceleration of the helix and the vertical acceleration of the particle?

6–18. Three masses are connected by two rigid massless rods, each of length l, which are hinged at point B. The system can slide on a smooth horizontal plane. The configuration is given by the cartesian coordinates (x, y) of the center of mass of the system and the angles θ and ϕ, as shown. Write the differential equations of motion for the case where the center of mass remains stationary. If the initial conditions are $\theta(0) = \pi /2$, $\dot{\theta}(0) = -1$ rad/sec, and $\dot{\phi}(0) = 1$ rad/sec, find the minimum value of θ in the ensuing motion. How would the motion of the system differ if the values of $\dot{x}(0)$ and $\dot{y}(0)$ were not zero?

6–19. A particle of mass m is supported by a spring of stiffness k and a dashpot with a damping coefficient c connected in parallel to form a spherical pendulum. (a) Using the spherical co-ordinates (r, θ, ϕ) to express the position of the particle, where θ is measured from the upward vertical, find the differential equations of motion. Assume that the spring is unstressed when $r = l$. (b) Suppose the initial conditions are $r(0) = l$, $\dot{r}(0) = 0$, $\theta(0) = 3\pi/4$, $\dot{\theta}(0) = 0$, $\dot{\phi}(0) = 2\sqrt{g/l}$. What is the motion of the system after a long time, assuming that $|r - l| \ll l$ and neglecting air friction?

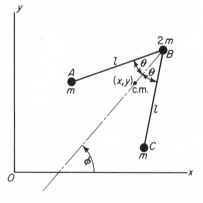

Fig. P6-18

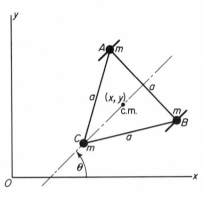

Fig. P6-20

6–20. Three particles, each of mass m, lie in a horizontal plane at the corners of an equilateral triangle. The particles are rigidly connected by massless rods and the whole system can roll without slipping on a horizontal floor. The wheels at A and B revolve about a common axle AB and permit the motion of these two particles in a direction perpendicular to AB. A castered wheel at C allows its motion in any direction consistent with the other constraints. Using the cartesian coordinates (x,y) to specify the location of the mass center, and letting θ be the angle between the x axis and the plane of the wheel at A or B, obtain the differential equations of motion using the Lagrange multiplier method. Assuming the initial conditions $x(0) = y(0) = 0$, $\theta(0) = 0$, $\dot{x}(0) = 0$, $\dot{y}(0) = v_0$; solve for the angular velocity and the constraint force, each as a function of θ. What is the nature of the final motion?

7

BASIC CONCEPTS AND KINEMATICS
OF RIGID BODY MOTION

In the previous chapters, we have developed some of the kinematical tools and dynamical principles that are available for the solution of the problems of particle dynamics. Distributed masses have been avoided and the limited discussions of rigid body motion have been confined to kinematical aspects. In this chapter, we shall further extend the mathematical and kinematical background and then apply some of the principles of particle dynamics to the analysis of rigid body motion in a plane.

7-1. DEGREES OF FREEDOM OF A RIGID BODY

In Sec. 6-1, we found that the number of degrees of freedom associated with a given mechanical system is equal to the number of coordinates used to describe its configuration minus the number of independent equations of constraint. As an example, consider a system of three unconstrained particles. Three cartesian coordinates could be used to describe the position of each particle, making nine independent coordinates in all. Hence this system has nine degrees of freedom.

Now suppose that the three particles are placed at the corners of a triangle whose sides are formed by three rigid rods, but the particles are otherwise free to move. Three independent equations of constraint can be written, giving the constant length of each of the three rods in terms of the coordinates of the particles. So we find that this system has nine coordinates and three independent equations of constraint, resulting in six degrees of freedom. If a fourth particle is rigidly attached to this system, its position can be specified by three additional coordinates; but there are also three additional independent constraint equations specifying the lengths of the three additional rods required to attach the fourth particle rigidly to the original three. So a system of four rigidly connected particles also has six degrees of freedom. It can be shown by similar reasoning that the rigid attachment of further particles does not change the number of degrees of freedom. In the limit, as the number of particles becomes very large, the system of particles can be considered as a rigid body. Hence we conclude that *a rigid body has, in general, six degrees of freedom*. Furthermore, it can be seen that the location and orientation of the rigid body are completely specified by a knowledge of the location of any three noncollinear points in that rigid body. For example, we might think of these points as the locations of any three of the particles that were used in forming the approximation. As we have seen, this group of three particles has six degrees of freedom, as does the rigid body of which they are a part.

In the discussion of the dynamics of a system of particles in Chapter 4, we derived two equations of particular importance. The first was the *translational* equation of motion given by Eq. (4–8):

$$\mathbf{F} = m\ddot{\mathbf{r}}_c \qquad (7\text{–}1)$$

This equation states that the vector sum of the external forces acting on a system is equal to the total mass of the system times the absolute acceleration of the center of mass. The second equation was the *rotational* equation of motion given by Eq. (4–49) or (4–57) which states that

$$\mathbf{M} = \dot{\mathbf{H}} \qquad (7\text{–}2)$$

where the reference point for calculating the applied moment $\mathbf{M}$ and the angular momentum $\mathbf{H}$ is either (1) fixed in an inertial frame or (2) at the center of mass of the system. We saw that these equations apply to any system of particles, including those in which the particles are in motion relative to each other, the only restriction being that the forces between any two particles be equal, opposite, and act along the line connecting the particles. Since we can consider that the rigidity is obtained by massless rods connecting each pair of particles, and the constraint forces consist of axial forces in the rods, these equations are seen to apply to rigid bodies.

Considering the translational equation of motion, we observe that the center of mass moves as though the entire mass m of the rigid body were concentrated in a particle located at this point and all the external forces acted on this particle. It can be seen that the location of the center of mass can be given using three independent coordinates. Hence there are *three translational degrees of freedom*. The remaining three degrees of freedom are associated with the possible motions of the rigid body, assuming that the position of the center of mass is fixed. These motions correspond to changes in the orientation of the body and hence are described by the rotational equation of motion.

We have found that a rigid body possesses a total of six degrees of freedom, three being considered as translational and three as rotational degrees of freedom. In the discussion, we chose the center of mass as the reference point or base point of the body and separated the total motion into (1) translational motion of the center of mass and (2) rotational motion about the center of mass. But let us recall that the number of degrees of freedom of a system is a kinematical property and does not depend upon the location of the center of mass. Therefore any other point in the body could equally well have been chosen as the reference point if kinematics alone governed the choice. The center of mass, however, is normally chosen as the reference point for dynamical reasons. Equations (7–1) and (7–2) which result from this choice allow for dynamically uncoupled (that is, independent) translational and rotational motions; whereas, if an arbitrary reference point were chosen, the rotational equation would contain a term involving the acceleration $\ddot{\mathbf{r}}_p$ of the reference point and the translational equation would contain a term involving the angular acceleration.

7-2. MOMENTS OF INERTIA

Now let us consider some aspects of the rotational motion of a rigid body. The rotational equation of motion for a system of particles involves the rate of change of the angular momentum **H**. So let us calculate **H** for the particular case of a rigid body rotating about a reference point P. Referring to Eq. (4–64), we see that the angular momentum relative to the point P is

$$\mathbf{H}_p = \sum_i \boldsymbol{\rho}_i \times m_i \dot{\boldsymbol{\rho}}_i \tag{7-3}$$

where $\boldsymbol{\rho}_i$ is the position vector of a particle m_i relative to the reference point P and the summation is over all particles. Note that $\dot{\boldsymbol{\rho}}_i$ is the velocity of m_i as viewed by a nonrotating observer translating with P. Assuming that P is fixed in the body, the magnitudes of all the $\boldsymbol{\rho}_i$ are constant. Therefore we see from Eq. (2–85) that

$$\dot{\boldsymbol{\rho}}_i = \boldsymbol{\omega} \times \boldsymbol{\rho}_i \tag{7-4}$$

where $\boldsymbol{\omega}$ is the absolute angular velocity of the body. Hence from Eqs. (7–3) and (7–4) we obtain

$$\mathbf{H}_p = \sum_i m_i \boldsymbol{\rho}_i \times (\boldsymbol{\omega} \times \boldsymbol{\rho}_i) \tag{7-5}$$

For the case of a rigid body with a continuous mass distribution, we let the particle mass m_i be the mass $\rho\, dV$ of a small volume element dV, where the scalar ρ is the mass density at that point. Summing over the whole volume of the rigid body for the limiting case of infinitesimal volume elements, we obtain the integral

$$\mathbf{H}_p = \int_V \rho\boldsymbol{\rho} \times (\boldsymbol{\omega} \times \boldsymbol{\rho})\, dV \tag{7-6}$$

where $\boldsymbol{\rho}$ is the position vector of the volume element dV.

Now let us assume that the reference point is at the origin O of a cartesian coordinate system (Fig. 7–1). Then, considering a volume element at (x, y, z), we can write

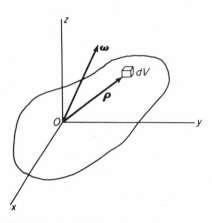

$$\boldsymbol{\rho} = x\mathbf{i} + y\mathbf{j} + z\mathbf{k} \qquad (7\text{-}7)$$

Also, the angular velocity of the body in terms of its cartesian components is

$$\boldsymbol{\omega} = \omega_x\mathbf{i} + \omega_y\mathbf{j} + \omega_z\mathbf{k} \qquad (7\text{-}8)$$

In order to evaluate $\boldsymbol{\omega} \times \boldsymbol{\rho}$ we use the determinant form of the cross product and obtain

$$\boldsymbol{\omega} \times \boldsymbol{\rho} = \begin{vmatrix} \mathbf{i} & \mathbf{j} & \mathbf{k} \\ \omega_x & \omega_y & \omega_z \\ x & y & z \end{vmatrix}$$

Fig. 7-1. A typical volume element in a rotating rigid body.

$$= (z\omega_y - y\omega_z)\mathbf{i} + (x\omega_z - z\omega_x)\mathbf{j}$$
$$+ (y\omega_x - x\omega_y)\mathbf{k}$$

In a similar fashion, we find that

$$\boldsymbol{\rho} \times (\boldsymbol{\omega} \times \boldsymbol{\rho}) = \begin{vmatrix} \mathbf{i} & \mathbf{j} & \mathbf{k} \\ x & y & z \\ (z\omega_y - y\omega_z) & (x\omega_z - z\omega_x) & (y\omega_x - x\omega_y) \end{vmatrix}$$

and, expanding the determinant and collecting terms, we obtain

$$\begin{aligned}\boldsymbol{\rho} \times (\boldsymbol{\omega} \times \boldsymbol{\rho}) = &[(y^2 + z^2)\omega_x - xy\omega_y - xz\omega_z]\mathbf{i} \\ &+ [-yx\omega_x + (x^2 + z^2)\omega_y - yz\omega_z]\mathbf{j} \qquad (7\text{-}9) \\ &+ [-zx\omega_x - zy\omega_y + (x^2 + y^2)\omega_z]\mathbf{k}\end{aligned}$$

Let us define the *moments of inertia* as follows:

$$I_{xx} = \int_V \rho(y^2 + z^2)\, dV$$

$$I_{yy} = \int_V \rho(x^2 + z^2)\, dV \qquad (7\text{-}10)$$

$$I_{zz} = \int_V \rho(x^2 + y^2)\, dV$$

Also, let us define the *products of inertia* to be

$$I_{xy} = I_{yx} = -\int_V \rho xy\, dV$$

$$I_{xz} = I_{zx} = -\int_V \rho xz\, dV \qquad (7\text{-}11)$$

$$I_{yz} = I_{zy} = -\int_V \rho yz\, dV$$

Using these definitions and Eqs. (7-6) and (7-9), we can write the angular momentum in the form

$$\mathbf{H} = (I_{xx}\omega_x + I_{xy}\omega_y + I_{xz}\omega_z)\mathbf{i}$$
$$+ (I_{yx}\omega_x + I_{yy}\omega_y + I_{yz}\omega_z)\mathbf{j} \qquad (7\text{--}12)$$
$$+ (I_{zx}\omega_x + I_{zy}\omega_y + I_{zz}\omega_z)\mathbf{k}$$

or

$$\mathbf{H} = H_x\mathbf{i} + H_y\mathbf{j} + H_z\mathbf{k} \qquad (7\text{--}13)$$

where

$$H_x = I_{xx}\omega_x + I_{xy}\omega_y + I_{xz}\omega_z$$
$$H_y = I_{yx}\omega_x + I_{yy}\omega_y + I_{yz}\omega_z \qquad (7\text{--}14)$$
$$H_z = I_{zx}\omega_x + I_{zy}\omega_y + I_{zz}\omega_z$$

These equations may be written more compactly using summation notation. The angular momentum is

$$\mathbf{H} = \sum_i \sum_j I_{ij}\omega_j\mathbf{e}_i \qquad (7\text{--}15)$$

and the corresponding components are given by

$$H_i = \sum_j I_{ij}\omega_j \qquad (7\text{--}16)$$

where the $\mathbf{e}_i$ are an orthogonal set of unit vectors, namely, $\mathbf{i}$, $\mathbf{j}$, and $\mathbf{k}$ in this case of a cartesian system.

The subscript p has been dropped from the $\mathbf{H}$ vector for convenience, but it is assumed in all cases that the reference point remains at the origin of the xyz system. This reference point is usually chosen to be either fixed in space or at the center of mass in order that the rotational equation will be of the simple form of Eq. (7–2). But, regardless of its location, the expression for $\mathbf{H}$ given in Eq. (7–12) is valid. In the general case of an arbitrary reference point, the rotational equation of motion has the form of Eq. (4–70) which was derived for a general system of particles, and thus applies to this case.

It can be seen that each expression for the moment of inertia given in Eq. (7–10) is just the second moment of the mass distribution with respect to a cartesian axis. For example, I_{xx} is the integral or summation of the mass elements $\rho \, dV$, each multiplied by the square of its distance from the x axis. The effective (or root-mean-square) value of this distance for a certain body is known as its *radius of gyration* with respect to the given axis. Thus we can define the radius of gyration corresponding to I_{jj} by

$$k_j = \sqrt{\frac{I_{jj}}{m}} \qquad (7\text{--}17)$$

where m is the total mass of the rigid body.

Notice that minus signs have been used in the definitions of the products of inertia given in Eq. (7–11). This convention was chosen so that equations such as Eq. (7–12) or (7–16) could be written with positive signs.

We see from Eqs. (7–10) and (7–11) that there are three moments of inertia and three independent products of inertia, making a total of six parameters whose values specify the inertial properties of a rigid body with regard to its rotational motion. This is in contrast to the single inertial parameter, the mass, which is needed to solve for the translational motion. In Sec. 7–3, we shall show how matrix notation can be used to specify the inertial properties of a rigid body in rotational motion.

If we review again the assumptions that have been made in deriving the expression for the angular momentum **H** given by Eq. (7–12), we find that the origin O of the xyz system is located at the reference point for **H**. No assumptions have been made concerning the rotational motion of this co-ordinate system, and in fact, an arbitrary rotation about O can occur. We recall that **ω** is the absolute angular velocity of the rigid body; and the components ω_x, ω_y, and ω_z are the instantaneous projections of **ω** onto the corresponding cartesian axes. Hence the moments and products of inertia, as well as ω_x, ω_y, and ω_z are, in general, functions of time. In order to avoid the difficulties arising from time-varying moments and products of inertia, we often choose an xyz system that is fixed in the rigid body and rotates with it. Such a coordinate system is known as a *body-axis system* and will be discussed further in Chapter 8 when we study the Euler equations of motion and the general rotational equations.

Example 7–1. A thin, uniform rod of mass m, length l lies in the xy plane with its center at the origin. If the rod makes an angle α with the positive x axis, as shown in Fig. 7–2, solve for the moments and products of inertia with respect to this cartesian system. Solve for **H** for the case where the angular velocity is directed along the positive y axis.

Taking η as a position variable along the rod, we can write an expression for an infinitesimal mass element in the form

$$\rho \, dV = \frac{m}{l} \, d\eta$$

Also, for points along the rod,

$$x = \eta \cos \alpha$$
$$y = \eta \sin \alpha$$
$$z = 0$$

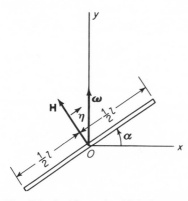

Fig. 7-2. A thin rod which is rotating about the y axis.

Hence, the integrals of Eq. (7–10) can be evaluated to give

$$I_{xx} = \frac{m}{l} \int_{-l/2}^{l/2} \eta^2 \sin^2 \alpha \, d\eta = \frac{ml^2 \sin^2 \alpha}{12}$$

$$I_{yy} = \frac{m}{l} \int_{-l/2}^{l/2} \eta^2 \cos^2 \alpha \, d\eta = \frac{ml^2 \cos^2 \alpha}{12} \qquad (7\text{–}18)$$

$$I_{zz} = \frac{m}{l} \int_{-l/2}^{l/2} \eta^2 \, d\eta = \frac{ml^2}{12}$$

In a similar fashion, the products of inertia are obtained from Eq. (7–11).

$$I_{xy} = -\frac{m}{l} \int_{-l/2}^{l/2} \eta^2 \sin \alpha \cos \alpha \, d\eta = -\frac{ml^2 \sin \alpha \cos \alpha}{12}$$

$$I_{xz} = 0 \qquad (7\text{–}19)$$

$$I_{yz} = 0$$

To solve for **H**, we note first that

$$\boldsymbol{\omega} = \omega \mathbf{j}$$

Then, from Eqs. (7–12), (7–18), and (7–19), we obtain

$$\mathbf{H} = \omega(I_{xy}\mathbf{i} + I_{yy}\mathbf{j} + I_{zy}\mathbf{k})$$

$$= \frac{ml^2 \omega \cos \alpha}{12}(-\sin \alpha \, \mathbf{i} + \cos \alpha \, \mathbf{j})$$

Thus **H** lies in the xy plane, is perpendicular to the rod, and makes an angle α with the direction of $\boldsymbol{\omega}$. In fact, it turns out that the magnitude H is equal to the moment of inertia $ml^2/12$ about a transverse axis through O, multiplied by the component of $\boldsymbol{\omega}$ perpendicular to the rod. This result occurs in this case because the axial component of **H** (that is, the component along the rod) is zero as a consequence of the moment of inertia being zero about this axis.

At this point, we might notice a few properties of the moments and products of inertia. First, we notice that for plane bodies such as this, in which the origin O lies in the plane, those products of inertia involving a coordinate axis which is perpendicular to that plane are zero. Also, the moment of inertia about any perpendicular axis is equal to the sum of the moments of inertia about the other two axes which lie in the plane of the body and pass through the same point. In this particular case,

$$I_{zz} = I_{xx} + I_{yy}$$

regardless of the value of the angle α, as can be seen from Eq. (7–18).

In general, if a three-dimensional body has a plane of symmetry such that the mass distribution on one side of the plane is a mirror image of that on the other side, then all products of inertia involving an axis perpendicular to the plane of symmetry are zero if the other two axes lie in the plane of symmetry. For example, if the xy plane is the plane of symmetry, then I_{xz} and I_{yz} must be zero since, for every mass element at (x, y, z), there exists another at the image point $(x, y, -z)$ whose contribution to these products of inertia is equal and opposite. Furthermore, if one of the coordinate axes is the symmetry axis of a body of revolution, then all products of inertia are zero. This follows from the fact that two of the three coordinate planes are planes of symmetry in this instance.

7–3. MATRIX NOTATION

In Sec. 1–1, we found that scalar and vector quantities can be considered as particular examples of tensors. In the context of this chapter, we might consider some properties of a rigid body moving in space. The mass m is a scalar quantity or zero-order tensor and is expressible as a single, real number. The angular momentum is a vector quantity or first-order tensor which can be expressed in terms of its components H_x, H_y, and H_z; that is, a typical element has a single subscript and the complete set of components is often written as a row or column of numbers. When we consider the rotational inertia of a rigid body, we find that two subscripts are necessary to specify a typical element I_{ij}. The moments and products of inertia are conveniently written as a 3×3 array of numbers, the position in the array being determined by the subscripts. Furthermore, the properties of the moments and products of inertia under a coordinate transformation are such that they qualify as second-order tensors. We shall consider coordinate transformations later in this chapter.

It turns out that the equations of Newtonian mechanics do not involve tensors of higher than the second order; hence the principal parameters in the analysis can be expressed rather concisely using no more complicated notation than two-dimensional arrays of numbers. This circumstance allows many of the equations involving coordinate transformations and rigid-body motion to be written conveniently using matrix notation.

Definition of a Matrix. A *matrix* is a set of numbers (real or complex) which are arranged in rows and columns and which obey certain rules of addition and multiplication. In our use of matrices in the study of dynamics, we shall confine ourselves to *real matrices*, that is, matrices with real numbers as elements, and the discussion of matrix algebra will be based on this assumption.

A *rectangular matrix* of order $m \times n$ has m rows and n columns and is written in the form

$$[a] = \begin{bmatrix} a_{11} & a_{12} & \cdots & a_{1n} \\ a_{21} & a_{22} & \cdots & a_{2n} \\ \cdot & & & \cdot \\ \cdot & & & \cdot \\ \cdot & & & \cdot \\ a_{m1} & a_{m2} & \cdots & a_{mn} \end{bmatrix} \tag{7-20}$$

where we note that a typical element a_{ij} is located at the intersection of the *i*th row and the *j*th column. The actual matrix on the right can be represented in abbreviated form by the symbol $[a]$.

A vector can be expressed as a *row matrix* or as a *column matrix*, that is, $m = 1$ or $n = 1$. For example, consider the angular momentum vector

$$\mathbf{H} = H_x \mathbf{i} + H_y \mathbf{j} + H_z \mathbf{k}$$

Assuming that the directions of the unit vectors are known, the essential information concerning $\mathbf{H}$ can be conveyed by writing the x, y, and z components in a systematic form. Representing $\mathbf{H}$ by a column matrix, we can write

$$\{H\} = \begin{Bmatrix} H_x \\ H_y \\ H_z \end{Bmatrix} \tag{7-21}$$

In general, column matrices will be designated by braces { }, whether or not the individual elements are written out. On the other hand, another vector, such as the force $\mathbf{F}$, might be represented by the matrix

$$\lfloor F \rfloor = \lfloor F_x \quad F_y \quad F_z \rfloor \tag{7-22}$$

where we adopt the half-brackets $\lfloor \ \rfloor$ to designate a row matrix.

A matrix having an equal number of rows and columns is called a *square matrix*. A typical example is the inertia matrix

$$[I] = \begin{bmatrix} I_{xx} & I_{xy} & I_{xz} \\ I_{yx} & I_{yy} & I_{yz} \\ I_{zx} & I_{zy} & I_{zz} \end{bmatrix} \tag{7-23}$$

giving the moments and products of inertia of a rigid body.

A *diagonal matrix* is a square matrix such that the elements $a_{ij} = 0$ for $i \neq j$ and not all the a_{ii} are zero. In other words, the only non-zero elements must lie along the principal diagonal which runs between the upper left and lower right corners.

A *scalar matrix* is a diagonal matrix in which $a_{ii} = a$, that is, all elements along the principal diagonal are equal.

A *unit matrix* is a scalar matrix whose diagonal elements are equal to unity. For example, a 3×3 unit matrix is written as follows:

$$[1] = \begin{bmatrix} 1 & 0 & 0 \\ 0 & 1 & 0 \\ 0 & 0 & 1 \end{bmatrix} \tag{7-24}$$

A *null matrix* or *zero matrix* is a matrix whose elements are all zero. It is designated by $[0]$, $\{0\}$, and so on, depending upon the type of matrix involved.

Equality of Matrices. Two matrices $[a]$ and $[b]$ are equal if $a_{ij} = b_{ij}$ for all i and j, that is, if each element of $[a]$ is equal to the corresponding element of $[b]$. Thus we see that the equality of two matrices requires that they be of the same order.

Transpose of a Matrix. The *transpose* of a matrix is formed by interchanging rows and columns and is designated by the superscript T. Thus if a_{ij} is an element of the matrix $[a]$, then a_{ji} is placed at that location in its transpose $[a]^T$. For a 3×3 matrix we see that

$$[a]^T = \begin{bmatrix} a_{11} & a_{12} & a_{13} \\ a_{21} & a_{22} & a_{23} \\ a_{31} & a_{32} & a_{33} \end{bmatrix}^T = \begin{bmatrix} a_{11} & a_{21} & a_{31} \\ a_{12} & a_{22} & a_{32} \\ a_{13} & a_{23} & a_{33} \end{bmatrix} \tag{7-25}$$

A column or row matrix can be transposed in a similar manner. For example, using a column matrix for $\boldsymbol{\omega}$, the transpose is

$$\{\omega\}^T = \begin{Bmatrix} \omega_x \\ \omega_y \\ \omega_z \end{Bmatrix}^T = \lfloor \omega_x \quad \omega_y \quad \omega_z \rfloor \tag{7-26}$$

It can be seen that the transpose of $[a]^T$ is $[a]$.

A *symmetric matrix* is a square matrix for which $a_{ij} = a_{ji}$, that is, $[a]$ is symmetric if

$$[a] = [a]^T \tag{7-27}$$

A *skew-symmetric* matrix is a square matrix for which $a_{ij} = -a_{ji}$. This requires that the elements a_{ii} on the principal diagonal be zero.

Addition and Subtraction of Matrices. The sum of two matrices $[a]$ and $[b]$ is found by adding corresponding elements. Thus,

$$[c] = [a] + [b] \tag{7-28}$$

if

$$c_{ij} = a_{ij} + b_{ij} \tag{7-29}$$

for all i and j. Of course, this requires that $[a]$ and $[b]$ be of the same order. From Eq. (7-29), we see that

$$b_{ij} = c_{ij} - a_{ij} \tag{7-30}$$

or

$$[b] = [c] - [a] \tag{7-31}$$

Hence the subtraction of matrices involves the subtraction of corresponding elements.

It can be seen that the associative and commutative laws of addition which apply to the scalar elements must apply to matrices as well; that is, the grouping and the order of performing additions and subtractions is immaterial.

Multiplication of Matrices. First consider the multiplication of a matrix $[b]$ by a scalar k. The result is a matrix

$$[a] = k[b] = [b]k$$

where

$$a_{ij} = kb_{ij}$$

Hence multiplication of a matrix by a scalar k results in multiplication of each of its elements by k. Note that multiplication by a scalar is commutative.

The multiplication of two matrices is defined only when they are *conformable*, that is, the number of columns in the first matrix is equal to the number of rows in the second. If the matrices $[a]$ and $[b]$ are of orders $m \times n$ and $n \times q$, respectively, and if

$$c_{ij} = \sum_{k=1}^{n} a_{ik} b_{kj} \tag{7-32}$$

then we define

$$[c] = [a][b] \tag{7-33}$$

where $[c]$ is of order $m \times q$. As an example of matrix multiplication, suppose that

$$[a] = \begin{bmatrix} a_{11} & a_{12} & a_{13} \\ a_{21} & a_{22} & a_{23} \end{bmatrix} \quad \text{and} \quad [b] = \begin{bmatrix} b_{11} & b_{12} \\ b_{21} & b_{22} \\ b_{31} & b_{32} \end{bmatrix}$$

Then

$$[c] = [a][b] = \begin{bmatrix} (a_{11}b_{11} + a_{12}b_{21} + a_{13}b_{31}) & (a_{11}b_{12} + a_{12}b_{22} + a_{13}b_{32}) \\ (a_{21}b_{11} + a_{22}b_{21} + a_{23}b_{31}) & (a_{21}b_{12} + a_{22}b_{22} + a_{23}b_{32}) \end{bmatrix}$$

where we have placed each element of the product within parentheses for clarity. In this example, we say that $[b]$ is *premultiplied* by $[a]$ or that $[a]$ is *postmultiplied* by $[b]$.

It can be seen that matrices which are conformable for a given order of multiplication are not necessarily conformable when multiplied in the opposite order. In the foregoing example, the matrices would be conformable when multiplied in the opposite order only if $m = q$. But even if two given matrices are conformable when multiplied in either order, the product matrices are not equal, in general. In other words, matrix multiplication is *not commutative* in the usual case. For those cases where $[a][b] = [b][a]$, the matrices are said to *commute* or to be *permutable*. A unit matrix or, in fact, any scalar matrix commutes with an arbitrary square matrix of the same order. For example,

$$[a][1] = [1][a] = [a] \tag{7-34}$$

and

$$[a][0] = [0][a] = [0] \tag{7-35}$$

Now let us consider continued products of matrices. It can be seen that

$$\left[[a][b] \right][c] = [a]\left[[b][c] \right] = \sum_k \sum_l a_{ik} b_{kl} c_{lj} \tag{7-36}$$

since the summations may be performed in either order. Hence one can indicate these multiplications without specifying which is to be performed first. Similarly, for the product of more than three matrices, the grouping is unimportant so long as the order is unchanged. Therefore matrix multiplication obeys the *associative law*. Of course, all pairs of adjacent matrices in a continued product must be conformable. The product matrix will have as many rows as the first matrix and as many columns as the last. If $[a]$ is a square matrix, the product of n identical matrices is indicated by $[a]^n$.

From the definition of matrix multiplication given in Eq. (7–32) and the fact that the scalar elements obey the distributive law of multiplication, we see that matrix multiplication is also *distributive:*

$$[a]\left[[b] + [c] \right] = [a][b] + [a][c] \tag{7-37}$$

In summary, it can be seen that all the ordinary laws of algebra apply to matrices, except that matrix multiplication is not commutative.

The physical meaning of matrix multiplication can be seen in the case of two examples associated with rigid-body motion. First consider the product of a row matrix and a column matrix:

$$\lfloor \omega \rfloor \{H\} = \lfloor \omega_x \quad \omega_y \quad \omega_z \rfloor \begin{Bmatrix} H_x \\ H_y \\ H_z \end{Bmatrix} = \omega_x H_x + \omega_y H_y + \omega_z H_z$$

This product is a single number, that is, it is a scalar. Furthermore, it is equal to the scalar product of $\boldsymbol{\omega}$ and $\mathbf{H}$.

$$\lfloor \omega \rfloor \{H\} = \boldsymbol{\omega} \cdot \mathbf{H}$$

Note that the column matrix must be premultiplied by the row matrix if a scalar is to result. The scalar product in this particular case turns out to be equal to twice the rotational kinetic energy, as we shall see in Sec. 7–4.

As another example, consider the set of linear algebraic equations for the components of **H**, as given by Eq. (7–14). The matrix form of this set of equations is

$$\{H\} = [I]\{\omega\} \tag{7-38}$$

where the inertia matrix $[I]$ is given by Eq. (7–23). Here we see that the inertia matrix consists of an orderly tabulation of the coefficients of the components of ω in Eq. (7–14). Performing the indicated operations, we find that a column matrix, when premultiplied by a square matrix, results in another column matrix. In other words, the multiplication by $[I]$ transforms the vector ω into another vector **H** which, in general, has a different magnitude and direction.

The preceding result could also have been obtained by using vector notation and representing the moments and products of inertia in the form of an inertia dyadic and again treating the matrix multiplication as equivalent to the dot product. We shall not, however, use dyadic notation in our treatment of dynamics and the reader is referred to other sources for information on this subject.[1]

Determinant of a Matrix. The determinant of a square matrix $[a]$ is denoted by $|a|$ and is equal to a single number, namely, the value of the determinant formed from the elements of $[a]$ in their given positions. Only square matrices have determinants, since a determinant always applies to a square array of numbers. Of course, all the usual rules for determinants are valid. In particular, the determinant of a matrix is equal to the determinant of its transpose.

The *minor* M_{ij} corresponding to the element a_{ij} is defined to be the determinant obtained by omitting the ith row and the jth column of $|a|$. The *cofactor* of a_{ij} is given by

$$A_{ij} = (-1)^{i+j} M_{ij} \tag{7-39}$$

A determinant can be expanded in terms of the elements of the ith row as follows:

$$|a| = \sum_j a_{ij} A_{ij} \tag{7-40}$$

If the cofactors are taken from any other row, however, then the result is zero.

$$\sum_j a_{ij} A_{kj} = 0 \quad (i \neq k) \tag{7-41}$$

Similarly, if one expands using the elements of the jth column, the result is

$$\sum_i a_{ij} A_{ik} = \begin{cases} 0 & \text{if } j \neq k \\ |a| & \text{if } j = k \end{cases} \tag{7-42}$$

[1]*See* H. Goldstein, *Classical Mechanics* (Reading, Mass.: Addison-Wesley, Publishing Company, 1950), pp. 146*ff.*

Equations (7–40), (7–41), and (7–42) can be summarized by the matrix equation

$$[a][A]^T = [A]^T[a] = |a|[1] \tag{7–43}$$

where $[A]^T$ is the transpose of the matrix of cofactors and is known as the *adjoint matrix* of $[a]$. It can be seen that the right side of the equation is a scalar matrix whose non-zero terms are each equal to the determinant $|a|$.

The Inverse Matrix. The *inverse matrix* of the square matrix $[a]$ is denoted by $[a]^{-1}$ and is defined such that

$$[a][a]^{-1} = [a]^{-1}[a] = [1] \tag{7–44}$$

That this definition is consistent can be shown by premultiplying or post multiplying Eq. (7–43) by $[a]^{-1}$, and using Eq. (7–34), we obtain

$$[a]^{-1}[a][A]^T = [A]^T = |a|[a]^{-1}$$

or, alternatively,

$$[A]^T[a][a]^{-1} = [A]^T = |a|[a]^{-1}$$

In either case, we find that the inverse matrix is given by

$$[a]^{-1} = \frac{[A]^T}{|a|} \tag{7–45}$$

Since a determinant can be evaluated only for a square array of numbers, we see that an inverse matrix exists only for the case of a square matrix. If the determinant $|a|$ is zero, then the matrix is *singular* and again no inverse exists.

The solution of a set of simultaneous linear algebraic equations is a problem of matrix inversion. For example, if

$$\{x\} = [a]\{y\}$$

one can solve for $\{y\}$ by premultiplying both sides of the equation by $[a]^{-1}$, with the result that

$$[a]^{-1}\{x\} = [a]^{-1}[a]\{y\} = \{y\}$$

The requirement that the matrix be square in order that an inverse exist is essentially a requirement that there be as many equations as unknowns when solving a set of simultaneous algebraic equations. The requirement that $[a]$ be nonsingular, that is, $|a| \neq 0$, specifies the independence of the equations.

The Transpose and Inverse of a Product Matrix. It can be seen by expanding the product of two matrices that the transpose of the product is equal to the product of the transposed matrices taken in the reverse order; that is,

$$\Big[[a][b]\Big]^T = [b]^T[a]^T \tag{7–46}$$

A similar rule applies to inverse matrices; that is, the inverse of the product of two matrices is equal to the product of the inverse matrices taken in the reverse order. For example, suppose

$$[a][b] = [c]$$

Then, premultiplying by $[a]^{-1}$ and $[b]^{-1}$ and postmultiplying by $[c]^{-1}$, we see that

$$[b]^{-1}[a]^{-1}[a][b][c]^{-1} = [b]^{-1}[a]^{-1}[c][c]^{-1}$$

But since

$$[a]^{-1}[a] = [b]^{-1}[b] = [c][c]^{-1} = [1]$$

we find that

$$[c]^{-1} = [b]^{-1}[a]^{-1}$$

or

$$\left[[a][b] \right]^{-1} = [b]^{-1}[a]^{-1} \tag{7–47}$$

The results of Eqs. (7–46) and (7–47) can be extended by induction to the product of more than two matrices.

7-4. KINETIC ENERGY

In Sec. 4–2, we obtained an expression for the total kinetic energy of a system of n particles, namely,

$$T = \frac{1}{2}mv_c^2 + \frac{1}{2}\sum_{i=1}^{n} m_i \dot{\rho}_i^2 \tag{7–48}$$

where v_c is the speed of the center of mass of the system and $\dot{\rho}_i$ is the velocity of the ith particle as viewed by a nonrotating observer at the center of mass. For the case of a rigid set of particles rotating with an angular velocity $\boldsymbol{\omega}$, we found in Eq. (7–4) that

$$\dot{\boldsymbol{\rho}}_i = \boldsymbol{\omega} \times \boldsymbol{\rho}_i$$

and hence

$$\dot{\rho}_i^2 = \dot{\boldsymbol{\rho}}_i \cdot \dot{\boldsymbol{\rho}}_i = \dot{\boldsymbol{\rho}}_i \cdot \boldsymbol{\omega} \times \boldsymbol{\rho}_i$$

Now the first term on the right in Eq. (7–48) is called the *translational kinetic energy* and the remaining sum constitutes the *rotational kinetic energy* of this rigid system. So, noting from Eq. (1–30) that a cyclic permutation of the vectors in a scalar triple product does not alter its value, we obtain the following expression for the rotational kinetic energy:

$$T_{\text{rot}} = \frac{1}{2}\sum_{i=1}^{n} m_i \dot{\rho}_i^2 = \frac{1}{2}\sum_{i=1}^{n} \boldsymbol{\omega} \cdot \boldsymbol{\rho}_i \times m_i \dot{\boldsymbol{\rho}}_i \tag{7–49}$$

For the case of a rigid body with a continuous mass distribution, the summation of Eq. (7–49) is replaced by an integral, resulting in

$$T_{\text{rot}} = \frac{1}{2} \int_V \rho \boldsymbol{\omega} \cdot \boldsymbol{\rho} \times \dot{\boldsymbol{\rho}} \, dV \tag{7-50}$$

where $\boldsymbol{\rho}$ is the position vector of the volume element dV relative to the center of mass and ρ is the density. We recall that $\boldsymbol{\omega}$ is not a function of position in a rigid body. Therefore from Eqs. (7–3) and (7–49) or from Eqs. (7–6) and (7–50), we find that

$$T_{\text{rot}} = \frac{1}{2} \boldsymbol{\omega} \cdot \mathbf{H} \tag{7-51}$$

where $\mathbf{H}$ is found with respect to the center of mass. If the rigid body rotates about a fixed point, $\mathbf{H}$ may be calculated with respect to this point, in which case Eq. (7–51) yields the total kinetic energy.

Equation (7–51) can be written conveniently using matrix notation, for example,

$$T_{\text{rot}} = \frac{1}{2} \lfloor \omega \rfloor \{H\} = \frac{1}{2} \{\omega\}^T \{H\} \tag{7-52}$$

Another form is obtained by substituting for $\{H\}$ from Eq. (7–38), obtaining

$$T_{\text{rot}} = \frac{1}{2} \{\omega\}^T [I] \{\omega\} \tag{7-53}$$

or, writing the matrices in detail,

$$T_{\text{rot}} = \frac{1}{2} \begin{Bmatrix} \omega_x \\ \omega_y \\ \omega_z \end{Bmatrix}^T \begin{bmatrix} I_{xx} & I_{xy} & I_{xz} \\ I_{yx} & I_{yy} & I_{yz} \\ I_{zx} & I_{zy} & I_{zz} \end{bmatrix} \begin{Bmatrix} \omega_x \\ \omega_y \\ \omega_z \end{Bmatrix}$$

where the origin of the xyz coordinate system is at the center of mass.

Performing the matrix multiplications, we obtain

$$T_{\text{rot}} = \frac{1}{2} I_{xx} \omega_x^2 + \frac{1}{2} I_{yy} \omega_y^2 + \frac{1}{2} I_{zz} \omega_z^2 + I_{xy} \omega_x \omega_y \\ + I_{xz} \omega_x \omega_z + I_{yz} \omega_y \omega_z \tag{7-54}$$

or

$$T_{\text{rot}} = \frac{1}{2} \sum_i \sum_j I_{ij} \omega_i \omega_j \tag{7-55}$$

It can be seen that if $\boldsymbol{\omega}$ has the same direction as one of the axes at a given moment, then the expression for T_{rot} reduces to a single term. Hence we can write

$$T_{\text{rot}} = \frac{1}{2} I \omega^2 \tag{7-56}$$

where I is the moment of inertia about the instantaneous axis of rotation and ω is the instantaneous angular velocity of the rigid body.

7-5. TRANSLATION OF COORDINATE AXES

The defining equations for the moments and products of inertia, as given by Eqs. (7–10) and (7–11), do not require that the origin of the cartesian coordinate system be taken at the center of mass. So let us calculate the moments and products of inertia for a given body with respect to a set of parallel axes that do not pass through the center of mass. Consider the body shown in Fig. 7–3. The center of mass is located at the origin O' of the primed system and at the point (x_c, y_c, z_c) in the unprimed system. Let us take an infinitesimal volume element dV which is located at (x, y, z) in the unprimed system and at (x', y', z') in the primed system. These coordinates are related by the equations:

$$\begin{aligned} x &= x' + x_c \\ y &= y' + y_c \\ z &= z' + z_c \end{aligned} \qquad (7\text{–}57)$$

The moment of inertia about the x axis can be written in terms of primed coordinates by using Eqs. (7–10) and (7–57).

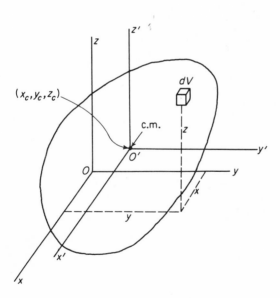

Fig. 7-3. A rigid body, showing the location of parallel coordinate axes and a typical volume element.

$$I_{xx} = \int_V \rho[(y' + y_c)^2 + (z' + z_c)^2]\, dV$$

$$= I_{x'x'} + 2y_c \int_V \rho y'\, dV + 2z_c \int_V \rho z'\, dV + m(y_c^2 + z_c^2) \quad (7\text{--}58)$$

where m is the total mass. Since the origin of the primed coordinate system was chosen to be at the center of mass, we see that

$$\int_V \rho x'\, dV = \int_V \rho y'\, dV = \int_V \rho z'\, dV = 0 \quad (7\text{--}59)$$

and therefore the two integrals on the right side of Eq. (7–58) are zero. A similar procedure can be followed to obtain I_{yy} and I_{zz}; the results are summarized as follows:

$$\begin{aligned} I_{xx} &= I_{x'x'} + m(y_c^2 + z_c^2) \\ I_{yy} &= I_{y'y'} + m(x_c^2 + z_c^2) \\ I_{zz} &= I_{z'z'} + m(x_c^2 + y_c^2) \end{aligned} \quad (7\text{--}60)$$

or, in general,

$$I_{kk} = I_{k'k'} + md^2 \quad (7\text{--}61)$$

where d is the distance between a given unprimed axis and a parallel primed axis passing through the center of mass. This is the *parallel-axis theorem*.

The products of inertia are obtained in a similar manner, using Eqs. (7–11) and (7–57).

$$I_{xy} = -\int_V \rho(x' + x_c)(y' + y_c)\, dV$$

$$= I_{x'y'} - x_c \int_V \rho y'\, dV - y_c \int_V \rho x'\, dV - mx_c y_c$$

Again we see from Eq. (7–59) that the two integrals on the right drop out. The other products of inertia can be calculated in a similar manner and the results written as follows:

$$\begin{aligned} I_{xy} &= I_{x'y'} - mx_c y_c \\ I_{xz} &= I_{x'z'} - mx_c z_c \\ I_{yz} &= I_{y'z'} - my_c z_c \end{aligned} \quad (7\text{--}62)$$

From Eqs. (7–60) and (7–62), it can be seen that a translation of axes away from the center of mass always results in an increase in the moments of inertia. On the other hand, the products of inertia may increase or decrease, depending upon the particular case.

7–6. ROTATION OF COORDINATE AXES

Now consider the case shown in Fig. 7–4 where the origins of the primed and unprimed systems are at the same point O but none of the axes coincide.

An arbitrary vector **r** drawn from O can be expressed in terms of the unprimed coordinates and unit vectors in the form

$$\mathbf{r} = x\mathbf{i} + y\mathbf{j} + z\mathbf{k} \qquad (7\text{-}63)$$

Now let us write expressions for each of these unit vectors in terms of their components in the primed system. Because each unit vector is of unit length and is directed along an unprimed axis, its component in the direction of a given primed axis is just the cosine of the angle between positive portions of the two axes. Hence we can write

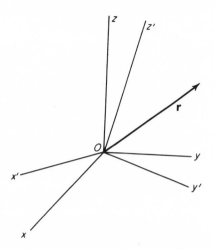

Fig. 7-4. Two cartesian coordinate systems with a common origin and an arbitrary orientation relative to each other.

$$\mathbf{i} = l_{x'x}\mathbf{i}' + l_{y'x}\mathbf{j}' + l_{z'x}\mathbf{k}'$$
$$\mathbf{j} = l_{x'y}\mathbf{i}' + l_{y'y}\mathbf{j}' + l_{z'y}\mathbf{k}' \qquad (7\text{-}64)$$
$$\mathbf{k} = l_{x'z}\mathbf{i}' + l_{y'z}\mathbf{j}' + l_{z'z}\mathbf{k}'$$

where the l's are the cosines of the angles between the axes indicated by the subscripts. In general, these nine cosines have different values. They completely specify the relative orientation of the two coordinate systems and are known as *direction cosines*.

From Eqs. (7-63) and (7-64), we obtain

$$\mathbf{r} = (l_{x'x}x + l_{x'y}y + l_{x'z}z)\mathbf{i}'$$
$$+ (l_{y'x}x + l_{y'y}y + l_{y'z}z)\mathbf{j}' \qquad (7\text{-}65)$$
$$+ (l_{z'x}x + l_{z'y}y + l_{z'z}z)\mathbf{k}'$$

But **r** can be expressed directly in terms of primed components as follows:

$$\mathbf{r} = x'\mathbf{i}' + y'\mathbf{j}' + z'\mathbf{k}' \qquad (7\text{-}66)$$

By comparing Eqs. (7-65) and (7-66), we see that the primed and unprimed coordinates are related by

$$x' = l_{x'x}x + l_{x'y}y + l_{x'z}z$$
$$y' = l_{y'x}x + l_{y'y}y + l_{y'z}z \qquad (7\text{-}67)$$
$$z' = l_{z'x}x + l_{z'y}y + l_{z'z}z$$

or, using matrix notation,

$$\begin{Bmatrix} x' \\ y' \\ z' \end{Bmatrix} = \begin{bmatrix} l_{x'x} & l_{x'y} & l_{x'z} \\ l_{y'x} & l_{y'y} & l_{y'z} \\ l_{z'x} & l_{z'y} & l_{z'z} \end{bmatrix} \begin{Bmatrix} x \\ y \\ z \end{Bmatrix} \qquad (7\text{-}68)$$

Equation (7–68) can be written using a more abbreviated notation as follows:

$$\{r'\} = [l]\{r\} \qquad (7\text{–}69)$$

We have considered the $\{r'\}$ matrix to consist of the components of the vector **r** in the primed system, where the primed system has been rotated relative to the unprimed system. Thus the vector is considered to be fixed and the coordinate system is rotated. On the other hand, the same transformation equations could apply equally well to the case where the coordinate system is fixed and the vector **r** is rotated in the opposite direction, resulting in a different vector **r'**. This approach will be discussed in Sec. 7–11; but for the present, we shall consider other applications of the rotation of coordinate systems.

Now let us suppose that the rigid body of Fig. 7–5 is rotating with an absolute angular velocity **ω**. The common origin for the primed and unprimed systems is chosen at the center of mass. We can calculate the rotational kinetic energy at any given time by using the $[I]$ and $\{\omega\}$ matrices evaluated at that time. Either the primed or unprimed axes may be used, provided that both matrices are evaluated in the same system. Using the unprimed system, for example, we obtain from Eq. (7–53) that

$$T_{\text{rot}} = \frac{1}{2}\{\omega\}^T [I]\{\omega\}$$

and, in terms of the primed quantities,

$$T_{\text{rot}} = \frac{1}{2}\{\omega'\}^T [I']\{\omega'\} \qquad (7\text{–}70)$$

Now the vector **ω** transforms in the same way as the vector **r** of Eq. (7–69). Hence we can write

$$\{\omega'\} = [l]\{\omega\} \qquad (7\text{–}71)$$

or, using Eq. (7–46),

$$\{\omega'\}^T = \{\omega\}^T [l]^T \qquad (7\text{–}72)$$

Therefore, we obtain from Eqs. (7–70), (7–71), and (7–72) that

$$T_{\text{rot}} = \frac{1}{2}\{\omega\}^T [l]^T [I'][l]\{\omega\} \qquad (7\text{–}73)$$

It is apparent that the rotational kinetic energy is independent of the coordinate system used in its calculation since the absolute angular velocity **ω** is used in either case. Thus we can equate the expressions given in Eqs. (7–53) and (7–73), that is,

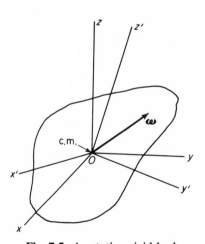

Fig. 7-5. A rotating rigid body.

$$\frac{1}{2}\{\omega\}^T[I]\{\omega\} = \frac{1}{2}\{\omega\}^T[l]^T[I'][l]\{\omega\}$$

for any $\{\omega\}$. Therefore,

$$[I] = [l]^T[I'][l] \qquad (7\text{-}74)$$

This is the transformation equation for the inertia matrix. It can be written in summation notation as follows:

$$I_{ij} = \sum_{m'} \sum_{n'} l_{m'i} l_{n'j} I_{m'n'} \qquad (7\text{-}75)$$

Since each of the indices can assume three different values, we see that Eq. (7–74) or (7–75) actually represents nine equations of nine terms each. A typical equation might be written in the form:

$$\begin{aligned}
I_{xy} = &\, l_{x'x} l_{x'y} I_{x'x'} + l_{y'x} l_{y'y} I_{y'y'} + l_{z'x} l_{z'y} I_{z'z'} \\
&+ (l_{x'x} l_{y'y} + l_{y'x} l_{x'y}) I_{x'y'} + (l_{x'x} l_{z'y} + l_{z'x} l_{x'y}) I_{x'z'} \\
&+ (l_{y'x} l_{z'y} + l_{z'x} l_{y'y}) I_{y'z'}
\end{aligned}$$

where we note that each inertia matrix is symmetric, that is, $I_{x'y'} = I_{y'x'}$, and so on.

It is interesting to apply Eq. (7–74) to the special case of a homogeneous sphere having a unit moment of inertia about any axis through its center. Then, choosing the center as the common origin for the primed and unprimed coordinate systems, we find that

$$[I] = [I'] = [1] \qquad (7\text{-}76)$$

since all products of inertia vanish because of the spherical symmetry of the mass distribution. From Eqs. (7–74) and (7–76), we obtain

$$[l]^T[1][l] = [l]^T[l] = [1] \qquad (7\text{-}77)$$

But, from the definition of an inverse matrix, given in Eq. (7–44), we see that

$$[l][l]^{-1} = [1] \qquad (7\text{-}78)$$

So, postmultiplying Eq. (7–77) by $[l]^{-1}$ and using Eq. (7–78), we obtain

$$[l]^T = [l]^{-1} \qquad (7\text{-}79)$$

Matrices having the property that the inverse matrix is equal to the transposed matrix are known as *orthogonal matrices*. Thus the direction cosine matrix $[l]$ is an orthogonal matrix.

Now let us premultiply Eq. (7–74) by $[l]$ and postmultiply by $[l]^T$. The result is

$$[l][I][l]^T = [l][l]^T[I'][l][l]^T$$

which can be simplified by using Eqs. (7–78) and (7–79) to yield

$$[I'] = [l][I][l]^T \qquad (7\text{-}80)$$

Equations (7–74) and (7–80) constitute the matrix equations for calculating the moments and products of inertia when the coordinate system is changed from the primed to the unprimed system or vice versa, the transformation being an arbitrary rotation about the common origin.

An expansion of Eq. (7–77) results in nine equations, of which six are independent. These equations can be written compactly in the form

$$\sum_{m'} l_{m'i} l_{m'j} = \begin{cases} 1 & \text{if } i = j \\ 0 & \text{if } i \neq j \end{cases}$$

or

$$\sum_{m'} l_{m'i} l_{m'j} = \delta_{ij} \qquad (7\text{--}81)$$

where δ_{ij}, known as the *Kronecker delta*, is a symbol whose value is 1 if the subscripts are equal; otherwise it is 0. Because there are nine direction cosines and six independent equations of constraint of the form of Eq. (7–81), we see once again that there are $9 - 6 = 3$ rotational degrees of freedom for a set of orthogonal coordinate axes or, essentially, for a rigid body since we can consider a set of axes to be embedded in any given rigid body.

7–7. PRINCIPAL AXES

Let us consider again the changes in the moments and products of inertia of a rigid body due to a rotation of coordinate axes. For convenience, we assume that the origin O is fixed, although it need not be taken at the center of mass.

First we note from the definitions of the moments of inertia given in Eq. (7–10) that the moments of inertia cannot be negative. Furthermore,

$$I_{xx} + I_{yy} + I_{zz} = 2 \int_V \rho r^2 \, dV \qquad (7\text{--}82)$$

where the square of the distance from the origin O is given by

$$r^2 = x^2 + y^2 + z^2 \qquad (7\text{--}83)$$

Now the distance r corresponding to any mass element $\rho \, dV$ of the rigid body does not change with a rotation of axes; therefore the sum of the moments of inertia is invariant with respect to a coordinate system rotation. In terms of matrix notation, the sum of the moments of inertia is just the sum of the elements on the principal diagonal of the inertia matrix and is known as the *trace* of that matrix. So the trace of the inertia matrix is unchanged by a coordinate rotation. More generally, it is a theorem of matrix theory that the trace of any square matrix is invariant under an *orthogonal transformation*. An example of an orthogonal transformation is the coordinate rotation of Eq. (7–74).

Considering next the products of inertia, we see that a coordinate rotation can result in changes in the signs of the products of inertia. A 180-degree rotation about the x axis, for example, reverses the signs of I_{xy} and I_{xz}, while I_{yz} is unchanged. This occurs because the directions of the positive y and z axes are reversed. On the other hand, a 90-degree rotation about the x axis reverses the sign of I_{yz}. Furthermore, the products of inertia have no preferred sign, as do the moments of inertia. If we consider a coordinate system which has a random orientation relative to the body, positive and negative products of inertia are equally likely. It can be seen from Eq. (7–75) that the moments and products of inertia vary smoothly with changes in orientation of the coordinate system because the direction cosines vary smoothly. Therefore an orientation can always be found for which a given product of inertia is zero. In fact, it is always possible to find an orientation of the coordinate system relative to a given rigid body such that all products of inertia are zero simultaneously, that is, the inertia matrix is *diagonal*. The three mutually orthogonal coordinate axes are known as *principal axes* in this case, and the corresponding moments of inertia are the *principal moments of inertia*. The three planes formed by the principal axes are called *principal planes*.

Let us assume for the moment that such a set of principal axes has been found for a given rigid body. Then, taking the principal axes as coordinate axes, we see from Eq. (7–14) that if the body is rotating about any axis through the fixed origin, the components of the angular momentum are given by

$$H_x = I_{xx}\omega_x$$
$$H_y = I_{yy}\omega_y \qquad (7\text{-}84)$$
$$H_z = I_{zz}\omega_z$$

since all products of inertia are zero. In particular, if the instantaneous axis of rotation is a principal axis, then $\mathbf{H}$ and $\boldsymbol{\omega}$ are parallel. For example, if $\boldsymbol{\omega}$ is parallel to the x axis, then $\omega_y = \omega_z = 0$, and we see from Eq. (7–84) that $H_y = H_z = 0$; hence $\mathbf{H}$ is also parallel to the x axis. A similar result applies for rotation about the other coordinate axes. So, in general, if $\boldsymbol{\omega}$ lies along one of the principal axes, then

$$\mathbf{H} = I\boldsymbol{\omega} \qquad (7\text{-}85)$$

where I is the moment of inertia about the given principal axis.

Now consider again the more general expressions for the angular momentum components as given in Eq. (7–14), namely,

$$H_x = I_{xx}\omega_x + I_{xy}\omega_y + I_{xz}\omega_z$$
$$H_y = I_{yx}\omega_x + I_{yy}\omega_y + I_{yz}\omega_z$$
$$H_z = I_{zx}\omega_x + I_{zy}\omega_y + I_{zz}\omega_z$$

If the angular velocity $\boldsymbol{\omega}$ is parallel to the x axis, then it is clear that $\mathbf{H}$ cannot be parallel to $\boldsymbol{\omega}$ unless I_{yx} and I_{zx} are zero. But if these two products of inertia are zero, it can be shown from Eq. (7–75) that a certain rotation about the x axis is all that is required to eliminate the third product of inertia. Hence, if $\mathbf{H}$ is parallel to $\boldsymbol{\omega}$, then the x axis is a principal axis. Similar arguments apply if $\mathbf{H}$ and $\boldsymbol{\omega}$ are parallel to another coordinate axis. So we conclude that $\mathbf{H}$ and $\boldsymbol{\omega}$ are parallel only if the rotation is about a principal axis.

In summary, we can state the following: *If a rigid body is rotating about a given reference point O, the angular momentum $\mathbf{H}$ about O is parallel to the angular velocity $\boldsymbol{\omega}$ if, and only if, the rotation is about a principal axis with respect to O.*

Now let us find specific means of calculating the principal moments of inertia and obtaining the directions of the corresponding principal axes. Assuming an arbitrary orientation of the unprimed axes relative to a rotating rigid body, we can write Eq. (7–14) in the form

$$\begin{Bmatrix} H_x \\ H_y \\ H_z \end{Bmatrix} = \begin{bmatrix} I_{xx} & I_{xy} & I_{xz} \\ I_{yx} & I_{yy} & I_{yz} \\ I_{zx} & I_{zy} & I_{zz} \end{bmatrix} \begin{Bmatrix} \omega_x \\ \omega_y \\ \omega_z \end{Bmatrix} \tag{7–86}$$

But we have seen that if $\mathbf{H}$ and $\boldsymbol{\omega}$ are parallel, then the body is rotating about a principal axis. Assuming, then, that $\mathbf{H}$ and $\boldsymbol{\omega}$ are parallel, we can write Eq. (7–85) in matrix form, obtaining

$$\begin{Bmatrix} H_x \\ H_y \\ H_z \end{Bmatrix} = I \begin{Bmatrix} \omega_x \\ \omega_y \\ \omega_z \end{Bmatrix} \tag{7–87}$$

where I is a scalar, namely, the moment of inertia about the instantaneous axis of rotation. This axis is a principal axis in this case; hence I is a principal moment of inertia. Equating the right sides of Eqs. (7–86) and (7–87) and noting that $I = I[1]$, we obtain

$$\begin{bmatrix} (I_{xx} - I) & I_{xy} & I_{xz} \\ I_{yx} & (I_{yy} - I) & I_{yz} \\ I_{zx} & I_{zy} & (I_{zz} - I) \end{bmatrix} \begin{Bmatrix} \omega_x \\ \omega_y \\ \omega_z \end{Bmatrix} = 0 \tag{7–88}$$

We assume that $\boldsymbol{\omega} \neq 0$; therefore the determinant of the coefficient matrix must be zero, that is,

$$\begin{vmatrix} (I_{xx} - I) & I_{xy} & I_{xz} \\ I_{yx} & (I_{yy} - I) & I_{yz} \\ I_{zx} & I_{zy} & (I_{zz} - I) \end{vmatrix} = 0 \tag{7–89}$$

This determinant yields a cubic equation in I known as the *characteristic equation* of the matrix. The three roots of this equation are the three principal moments of inertia.

The principal directions corresponding to the three principal moments of inertia are found by substituting the three roots, one at a time, into Eq. (7–88). It can be seen that one can solve for only the ratios of the components of $\boldsymbol{\omega}$, not their absolute magnitudes. For example, if ω_x, ω_y, and ω_z are each multiplied by the same number, the validity of Eq. (7–88) is unchanged. Only two amplitude ratios of the components of $\boldsymbol{\omega}$ are required; hence only two of the three simultaneous equations given in Eq. (7–88) need to be used. So let us arbitrarily omit the first equation and divide the other two by ω_x, assuming at this point that ω_x is not zero. The result is

$$(I_{yy} - I)\frac{\omega_y}{\omega_x} + I_{yz}\frac{\omega_z}{\omega_x} = -I_{yx}$$

$$I_{zy}\frac{\omega_y}{\omega_x} + (I_{zz} - I)\frac{\omega_z}{\omega_x} = -I_{zx}$$

(7–90)

These equations can be solved to give a set of amplitude ratios for each I, that is, for each principal moment of inertia. The solution in matrix form is as follows:

$$\left\{\begin{matrix}\left(\dfrac{\omega_y}{\omega_x}\right) \\ \left(\dfrac{\omega_z}{\omega_x}\right)\end{matrix}\right\} = \begin{bmatrix}(I_{yy} - I) & I_{yz} \\ I_{zy} & (I_{zz} - I)\end{bmatrix}^{-1}\left\{\begin{matrix}-I_{yx} \\ -I_{zx}\end{matrix}\right\}$$

(7–91)

These ratios indicate the direction of the axis corresponding to the given principal moment of inertia. For example, if we arbitrarily set $\omega_x = 1$, then the values of ω_y and ω_z, together with ω_x, determine the direction of $\boldsymbol{\omega}$, and hence also determine the direction of the corresponding principal axis at the reference point O.

Of course, the amplitude ratios could have been based upon ω_y or ω_z rather than ω_x. Also, another equation could have been omitted rather than the first. In the usual case, these choices make no difference in the result. If one of the coordinate axes is also a principal axis, however, then it may turn out that ω_x, for example, is equal to zero. In this case, a division by ω_x can lead to an erroneous result; hence it would be preferable to divide by ω_y or ω_z, whichever is non-zero for the given principal axis. In any event, the final results should be checked against the complete set of equations given in (7–88).

The problem of finding the principal moments of inertia and the corresponding principal axes is an example of an *eigenvalue problem*. The usual statement of the problem is in the form

$$[I]\{\omega\} - I\{\omega\} = 0$$

(7–92)

which is equivalent to Eq. (7–88). We seek values of the scalar parameter I, called *eigenvalues*, and the corresponding $\{\omega\}$, or *eigenvectors*, such that Eq. (7–92) is satisfied. As we have seen, there are three eigenvalues in this

instance and three eigenvectors, or *modal columns,* as they are sometimes
called. In our study of vibration theory in Chapter 9, we shall consider
another example of an eigenvalue problem. The mathematical treatment
will be similar, but the analysis will not be limited to three eigenvalues.

Consider now two solutions to Eq. (7–92), corresponding to two distinct roots. Designating the eigenvalues by I_1 and I_2, we see that

$$[I]\{\boldsymbol{\omega}_1\} = I_1\{\boldsymbol{\omega}_1\} \qquad (7\text{–}93)$$

and

$$[I]\{\boldsymbol{\omega}_2\} = I_2\{\boldsymbol{\omega}_2\} \qquad (7\text{–}94)$$

where $\{\boldsymbol{\omega}_1\}$ and $\{\boldsymbol{\omega}_2\}$ are the corresponding eigenvectors. It can be shown
that $\boldsymbol{\omega}_1$ and $\boldsymbol{\omega}_2$ are orthogonal and therefore that any two principal axes
are perpendicular if they correspond to distinct roots. To do this, we first
take the transpose of both sides of Eq. (7–94), obtaining

$$\{\boldsymbol{\omega}_2\}^T[I]^T = I_2\{\boldsymbol{\omega}_2\}^T \qquad (7\text{–}95)$$

Now premultiply Eq. (7–93) by $\{\boldsymbol{\omega}_2\}^T$ and postmultiply Eq. (7–95) by
$\{\boldsymbol{\omega}_1\}$. The result is

$$\{\boldsymbol{\omega}_2\}^T[I]\{\boldsymbol{\omega}_1\} = I_1\{\boldsymbol{\omega}_2\}^T\{\boldsymbol{\omega}_1\} \qquad (7\text{–}96)$$

$$\{\boldsymbol{\omega}_2\}^T[I]^T\{\boldsymbol{\omega}_1\} = I_2\{\boldsymbol{\omega}_2\}^T\{\boldsymbol{\omega}_1\} \qquad (7\text{–}97)$$

The inertia matrix is symmetric, that is, $[I] = [I]^T$. Therefore the left sides
of Eqs. (7–96) and (7–97) are equal. So if we subtract Eq. (7–96) from Eq.
(7–97), we obtain

$$(I_2 - I_1)\{\boldsymbol{\omega}_2\}^T\{\boldsymbol{\omega}_1\} = 0 \qquad (7\text{–}98)$$

We assume that the roots I_1 and I_2 are distinct; hence $I_2 - I_1 \neq 0$ and
therefore

$$\{\boldsymbol{\omega}_2\}^T\{\boldsymbol{\omega}_1\} = 0 \qquad (7\text{–}99)$$

or, in vector notation,

$$\boldsymbol{\omega}_2 \cdot \boldsymbol{\omega}_1 = 0 \qquad (7\text{–}100)$$

This result shows the orthogonality of the eigenvectors and indicates that
if all three principal moments of inertia are different, then the corresponding principal axes are mutually orthogonal. If two principal moments of
inertia are equal, then two of the principal axes no longer have definite
directions, although their plane is defined. Nevertheless, it is always possible
to find a set of mutually orthogonal principal axes in this so-called *degenerate case.*

The Ellipsoid of Inertia. A convenient method of representing the rotational inertia characteristics of a rigid body is by means of its ellipsoid of
inertia. As we shall see, the ellipsoid of inertia for a given body and reference
point is essentially a plot of the moment of inertia of the body for all possi-

ble axis orientations through the reference point. This graph in three dimensions turns out to be in the form of an ellipsoidal surface.

In order to see how the inertial characteristics are represented by an ellipsoid, consider a rigid body which is rotating about a fixed point O. From Eqs. (7–53) and (7–56), we see that the kinetic energy is

$$T = \frac{1}{2}\{\omega\}^T[I]\{\omega\} = \frac{1}{2}I\omega^2 \qquad (7\text{–}101)$$

where the scalar I is the moment of inertia about the instantaneous axis of rotation. Now let us define a vector $\boldsymbol{\rho}$ having the same direction as $\boldsymbol{\omega}$ such that

$$\boldsymbol{\rho} = \frac{1}{\omega\sqrt{I}}\,\boldsymbol{\omega} \qquad (7\text{–}102)$$

and let the column matrix $\{\rho\}$ indicate the components of $\boldsymbol{\rho}$ in an xyz system fixed in the body. Then from Eqs. (7–101) and (7–102), we see that

$$\{\rho\}^T[I]\{\rho\} = 1 \qquad (7\text{–}103)$$

If we consider $\boldsymbol{\rho}$ to be drawn from the origin O to a point (x, y, z), then we can express Eq. (7–103) in the form

$$I_{xx}x^2 + I_{yy}y^2 + I_{zz}z^2 + 2I_{xy}xy + 2I_{xz}xz + 2I_{yz}yz = 1 \qquad (7\text{–}104)$$

which is the equation of an ellipsoidal surface centered at O and known as the *ellipsoid of inertia*. From Eq. (7–102) we see that $\boldsymbol{\rho}$ has a magnitude

$$\rho = \frac{1}{\sqrt{I}} = \frac{1}{k_0\sqrt{m}} \qquad (7\text{–}105)$$

where k_0 is the radius of gyration of the body about the given axis through the reference point O, and m is the total mass. Hence the length of a straight line drawn from the center O to a point on the surface of the ellipsoid of inertia is inversely proportional to the radius of gyration of the body about that line (Fig. 7–6).

It is clear that the radius of gyration of a given rigid body depends upon the location of the axis relative to the body, but not upon the position of the body in space. Thus it can be seen that the ellipsoid of inertia is fixed in the body and rotates with it. Furthermore, for a given origin or reference point, the size of the ellipsoid of inertia and its position relative to the rigid body are independent of the coordinate system used in its description. So let

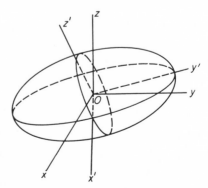

Fig. 7-6. An ellipsoid of inertia.

us write the equation for the inertia ellipsoid of Eq. (7–104) in terms of the primed coordinate system of Fig. 7–6 whose axes are assumed to be the principal axes of the ellipsoid. In this coordinate system, we know from analytic geometry that the equation of the ellipsoid takes the following simple form:

$$I_1 x'^2 + I_2 y'^2 + I_3 z'^2 = 1 \qquad (7–106)$$

But this equation is of the same form as Eq. (7–104) for the case where the *xyz* axes are the principal axes of the rigid body and all products of inertia vanish. Therefore I_1, I_2, and I_3 are the principal moments of inertia of the rigid body and, furthermore, the principal axes of the body coincide with those of the ellipsoid of inertia. From Eq. (7–105) we see that the lengths of the principal semiaxes of the ellipsoid of inertia are

$$\rho_1 = \frac{1}{\sqrt{I_1}}, \qquad \rho_2 = \frac{1}{\sqrt{I_2}}, \qquad \rho_3 = \frac{1}{\sqrt{I_3}} \qquad (7–107)$$

We note that the principal axes are normal to the ellipsoid of inertia at their point of intersection, indicating stationary values of the corresponding moments of inertia. Two of the principal moments of inertia correspond to the largest and smallest of all possible moments of inertia of the body relative to the given reference point. The third principal moment of inertia also corresponds to a stationary value, but it is usually neither a minimum nor a maximum value and corresponds to a so-called saddle point. From the parallel-axis theorem, Eq. (7–61), we see that the minimum moment of inertia about the center of mass is also the smallest possible moment of inertia for the given body with respect to any reference point.

If two rigid bodies have equal masses and identical ellipsoids of inertia about their mass centers, then they are *equimomental* or dynamically equivalent, that is, they will show the same translational and rotational response to the same applied forces. Of course, it is possible for two bodies which are quite different in shape to have the same mass and principal moments of inertia. For example, a homogeneous cube of edge length *a* and a homogeneous sphere of radius *r* are equimomental if they have equal masses and if $a = \sqrt{2.4}\, r$. Hence no detailed information concerning the shape of a body can be obtained from a knowledge of its ellipsoid of inertia.

Suppose we consider next the important case in which the rigid body has a symmetrical distribution of mass and we are interested in the moments and products of inertia with respect to an origin located at the center of mass. First take the case of symmetry about a plane; that is, the mass distribution on one side of the plane of symmetry is the mirror image of that on the other. This case includes all plane bodies and also is approximately true for aircraft, ships, automobiles, and a variety of other objects. It can be seen that the ellipsoid of inertia at O must have the same plane of symmetry as the body, and therefore this is also a principal plane containing two of the principal axes. Hence the third principal axis is perpendicular to the plane of symmetry. If we let the z axis, for example, be perpendicular to the plane of symmetry, then I_{xz} and I_{yz} are zero because of symmetry considerations, regardless of the orientation of the x and y axes in the symmetry plane.

Next let us consider the case in which there are two orthogonal planes of symmetry. Using the reasoning given previously, two principal axes at the center of mass must each be perpendicular to a plane of symmetry; hence the third must lie along their line of intersection. Thus we find that the directions of the principal axes can be determined by inspection.

An interesting special case occurs when we consider a body of revolution. We find that the inertia ellipsoid is an ellipsoid of revolution whose symmetry axis coincides with that of the body. Any two orthogonal planes which intersect along the axis of symmetry can serve as symmetry planes. Therefore, two of the principal axes are not uniquely defined, but the third is the axis of symmetry. Of course, two of the semiaxes of the ellipsoid of inertia are equal in this instance, as are the corresponding moments of inertia.

It is important to note that, even if the inertia ellipsoid is a figure of revolution, the corresponding rigid body need not be. All that is required is that two of the principal moments of inertia be equal. For example, three particles of equal mass rigidly connected at the corners of an equilateral triangle constitute a rigid body whose ellipsoid of inertia is axially symmetric. Furthermore, if a certain rotation ϕ of a body about some axis through the mass center results in no change in the mass distribution in space, and if $0 < |\phi| < \pi$, then that axis is a symmetry axis of the inertia ellipsoid and thus a principal axis. This can be seen from the fact that a rotation $0 < |\phi| < \pi$ of the inertia ellipsoid about a principal axis will always change its appearance in space unless the inertia ellipsoid is a figure of revolution and the given principal axis is its axis of symmetry. This situation occurs for bodies with threefold, fourfold, or higher symmetry. Typical examples are a three-bladed propeller and a right prism of uniform density whose cross section is a regular polygon.

Finally, if the body has three orthogonal planes of symmetry, then the ellipsoid of inertia at the center of mass is a sphere and any axis through the center is a principal axis. Examples are a homogeneous cube or sphere. Using reasoning similar to that of the foregoing paragraph, however, we note that a spherical ellipsoid of inertia does not imply three planes of symmetry. For example, a homogeneous regular tetrahedron has a spherical inertia ellipsoid about its mass center.

Example 7-2. Find the principal moments of inertia about the center of mass of a homogeneous cube of mass m and edge length a. Compare this result with the principal moments of inertia, about the center, of a thin cubical shell of mass m and edge length a whose walls are of uniform thickness.

We note that a homogeneous cube has three orthogonal planes of symmetry. Therefore, its ellipsoid of inertia at the center of mass is spherical, and all axes through the center of mass are principal axes. So let us choose axes parallel to the edges, as shown in Fig. 7–7(a). In this case the moment of inertia about the x axis is found from Eq. (7–10). It is

$$I_{xx} = \int_{-a/2}^{a/2} \int_{-a/2}^{a/2} \int_{-a/2}^{a/2} \rho(y^2 + z^2)\, dx\, dy\, dz$$

where ρ is the uniform density of the cube. Evaluating this integral we obtain

$$I_{xx} = \frac{1}{6} \rho a^5 \tag{7–108}$$

We see that the total mass is $m = \rho a^3$ and also that $I_{xx} = I_{yy} = I_{zz}$. Therefore,

$$I_{xx} = I_{yy} = I_{zz} = \frac{ma^2}{6} \tag{7–109}$$

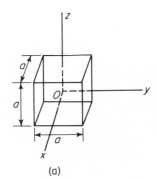

Now suppose that the edge length of the cube increases slightly and the density ρ remains unchanged. The change in the moment of inertia I_{xx} is found by differentiating Eq. (7–108), obtaining

$$dI_{xx} = \frac{dI_{xx}}{da}\, da = \frac{5}{6} \rho a^4\, da \tag{7–110}$$

Since the total moment of inertia of a body about a given axis is simply the summation of the contributions of the individual mass elements, it is clear that the additional moment of inertia dI_{xx} must be due to the layer of thickness $\frac{1}{2}\, da$ which has been added to the original cube. But this added material forms a cubical shell of edge length $a + da$, or approximately a. Also, the mass of the shell is six times the mass of one face, or

$$m = 6\rho a^2 \frac{da}{2} = 3\rho a^2\, da \tag{7–111}$$

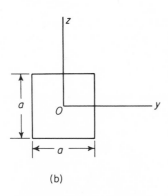

Fig. 7-7 (a) A homogeneous cube and (b) a thin square plate with the origin O at the center of mass.

So if we now let I_{xx}, I_{yy}, and I_{zz} designate the moments of inertia of the *shell*, we replace dI_{xx} in Eq. (7–10) by I_{xx} and use Eq. (7–111) to obtain

$$I_{xx} = I_{yy} = I_{zz} = \frac{5}{18} ma^2 \qquad (7\text{--}112)$$

The method of differentiation which we have used in this example is quite often convenient in obtaining the moments of inertia of shells. Care should be taken, however, in the choice of a proper reference point. Here we chose the center of mass and a closed shell resulted from an incremental change in edge length. If a corner had been chosen as the reference point, the process would have applied to three mutually orthogonal square plates touching the three farthest faces of the cube.

It is interesting to use the results of Eq. (7–109) to calculate the moments of inertia of a thin square plate, such as that of Fig. 7–7(b). If we imagine the particles of the cube to be squeezed together in a direction parallel to the x axis and toward the yz plane, there will be no change in their distances from the x axis. Therefore, in the limiting case of a thin square plate of mass m, we find that

$$I_{xx} = \frac{1}{6} ma^2 \qquad (7\text{--}113)$$

Setting $x = 0$ in Eq. (7–10) for this case where all the mass is in the yz plane, we see that

$$I_{xx} = I_{yy} + I_{zz} \qquad (7\text{--}114)$$

But $I_{yy} = I_{zz}$ and therefore

$$I_{yy} = I_{zz} = \frac{1}{12} ma^2 \qquad (7\text{--}115)$$

We can use these results to check the value of the moment of inertia which we obtained previously for a cubical shell. The moments of inertia of the faces parallel to the yz plane are obtained from Eq. (7–113), whereas the moments of inertia of the other four faces are obtained from Eq. (7–115) with an axis translation of $a/2$, this being the distance between the center of the cube and the center of the face. Hence we obtain

$$I_{xx} = \frac{1}{6} \left[2\left(\frac{ma^2}{6}\right) + 4\left(\frac{ma^2}{12} + \frac{ma^2}{4}\right) \right] = \frac{5}{18} ma^2$$

in agreement with the earlier result of Eq. (7–112). The factor $\frac{1}{6}$ arises because one sixth of the total mass is associated with each face.

Example 7–3. What are the principal moments of inertia at the center of mass of a thin circular disk of mass m and radius r? Find a point P such that any axis through this point is a principal axis.

Because of the axial symmetry of the circular disk, we see that the perpendicular axis passing through the center of mass is a principal axis. Let us calculate the moment of inertia I_{zz} by integrating the contributions of differential elements in the form of rings of radial thickness dr, as shown in Fig. 7–8. The mass of a ring is proportional to its surface area, that is,

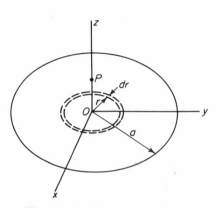

Fig. 7-8. A thin circular disk.

$$dm = \frac{2\pi r m\, dr}{\pi a^2}$$

Therefore,

$$I_{zz} = \int_0^a r^2\, dm = \frac{2m}{a^2} \int_0^a r^3\, dr$$

$$= \frac{1}{2} ma^2 \tag{7-116}$$

As we found in Example 7–2, the moment of inertia of a plane body about a perpendicular axis is equal to the sum of the moments of inertia about orthogonal axes in the plane, assuming a common reference point. So in this case,

$$I_{xx} = I_{yy} = \frac{1}{2} I_{zz} = \frac{1}{4} ma^2 \tag{7-117}$$

Now we come to the problem of finding a point P such that any axis through that point is a principal axis. If the point P meets these conditions, then the line OP must be a principal axis at O, as well as at P. This can be seen by noting that, if a body rotates about a given line through the center of mass and if $\boldsymbol{\omega}$ and $\mathbf{H}$ are parallel for one reference point on the line, then they will be parallel for any reference point on the line, since $\mathbf{H}$ is the same for all such reference points.

So we assume that OP is a principal axis at O; hence it must lie in the disk or be perpendicular to it. That the latter is the actual case follows from the fact that I_{zz} is greater than I_{xx} or I_{yy}; thus a certain translation of the origin from O to P will increase these smaller moments of inertia such that all three are equal. Referring to the parallel-axis theorem of Eq. (7–61), we find that if

$$OP = \frac{a}{2}$$

then the moment of inertia about any axis through P is $ma^2/2$. Note that P can be on either side of the disk but must lie on the z axis.

In general, a point P meeting these conditions is known as a *spherical point* because the corresponding inertia ellipsoid is spherical. Not all bodies have spherical points. In fact, to have a spherical point, two of the principal moments of inertia at the center of mass must be equal to each other and also be less than, or equal to, the remaining principal moment of inertia.

Example 7–4. Given a rigid body with the inertia matrix

$$[I] = \begin{bmatrix} 150 & 0 & -100 \\ 0 & 250 & 0 \\ -100 & 0 & 300 \end{bmatrix} \text{lb sec}^2 \text{ ft}$$

where the reference point is at the center of mass. Solve for the principal moments of inertia and find a coordinate transformation which diagonalizes the inertia matrix.

The characteristic equation of the inertia matrix can be written in determinant form by using Eq. (7–89). We obtain

$$\begin{vmatrix} (150 - I) & 0 & -100 \\ 0 & (250 - I) & 0 \\ -100 & 0 & (300 - I) \end{vmatrix} = 0$$

Expanding the determinant about the central row or column, we obtain

$$(250 - I)[(150 - I)(300 - I) - 10^4] = 0$$

or

$$(250 - I)(I^2 - 450I + 3.5 \times 10^4) = 0$$

which yields the following principal moments of inertia:

$$I_1 = 100 \text{ lb sec}^2 \text{ ft}$$
$$I_2 = 250 \text{ lb sec}^2 \text{ ft}$$
$$I_3 = 350 \text{ lb sec}^2 \text{ ft}$$

In order to obtain the directions of the principal axes, let us first write Eq. (7–88) in detail, using the numerical values of this example. Thus,

$$(150 - I)\omega_x - 100\omega_z = 0$$
$$(250 - I)\omega_y = 0$$
$$-100\omega_x + (300 - I)\omega_z = 0$$

where we recall that ω_x, ω_y, and ω_z are the components of $\boldsymbol{\omega}$ such that **H** and $\boldsymbol{\omega}$ are parallel. At the outset, it is apparent from the second equation that $\omega_y = 0$ unless $I = 250 \text{ lb sec}^2 \text{ ft}$.

First consider the root $I_1 = 100 \text{ lb sec}^2 \text{ ft}$. Of course, $\omega_y = 0$. Also, using either the first or third equation, we find that $\omega_z/\omega_x = \frac{1}{2}$. Thus we see that the principal axis lies in the xz plane and makes an angle with the x axis equal to

$$\alpha_1 = \tan^{-1}\left(\frac{1}{2}\right)$$

Next consider the root $I_3 = 350 \text{ lb sec}^2 \text{ ft}$. Again $\omega_y = 0$, but in this case $\omega_z/\omega_x = -2$, as can be seen from either the first or the third equation.

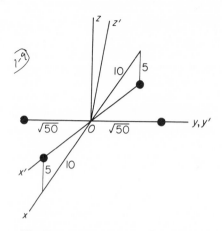

Fig. 7-9. A rigid body consisting of four equal particles.

So the corresponding principal axis lies in the xz plane and makes an angle

$$\alpha_3 = \tan^{-1}(-2)$$

with the x axis.

The third principal axis, corresponding to the root $I_2 = 250$ lb sec^2 ft, lies along the y axis since the other two are in the xz plane. It is interesting to note that in this case the first and third equations become

$$-100\omega_x - 100\omega_z = 0$$

$$-100\omega_x + 50\omega_z = 0$$

yielding $\omega_x = \omega_z = 0$. If, however, we had divided these equations by ω_x and solved for ω_z/ω_x, we would have obtained incorrect results.

In order to visualize the required rotation more clearly, let us suppose that the original inertia matrix was obtained for a system composed of four particles, each of unit mass and located at $(0, \sqrt{50}, 0)$, $(0, -\sqrt{50}, 0)$, $(10, 0, 5)$, and $(-10, 0, -5)$, as shown in Fig. 7–9. Using the primed coordinate system to indicate the principal axes, we can calculate the principal directions from the ratios of the components of $\boldsymbol{\omega}$ just obtained. The primed axes have arbitrarily been labeled such that a right-handed system is formed and a minimum rotation is required relative to the unprimed system. It can be seen that the actual rotation is of magnitude $\alpha_1 = 26.6°$ and occurs about the y axis in a negative sense. Noting that $\cos \alpha_1 = 2/\sqrt{5}$, we find that the direction cosine matrix is

$$[l] = \begin{bmatrix} \dfrac{2}{\sqrt{5}} & 0 & \dfrac{1}{\sqrt{5}} \\ 0 & 1 & 0 \\ \dfrac{-1}{\sqrt{5}} & 0 & \dfrac{2}{\sqrt{5}} \end{bmatrix}$$

We can obtain a check on the preceding results by calculating $[I']$ using Eq. (7–80). Performing the indicated matrix multiplications, we find that

$$[I'] = [l][I][l]^T = \begin{bmatrix} 100 & 0 & 0 \\ 0 & 250 & 0 \\ 0 & 0 & 350 \end{bmatrix} \text{ lb sec}^2 \text{ ft}$$

in agreement with principal moments of inertia obtained previously.

On the basis of the inertia matrix $[I]$ and our previous discussions of inertia ellipsoids, we could have seen immediately that the y axis is a principal axis. This follows from the fact that I_{xy} and I_{yz} are zero; hence the ellipsoid of inertia is symmetric about the xz plane and the y axis is a principal axis. Also, of course, the initial value of I_{yy} turned out to be identical with the value of one of the principal moments of inertia. Note, however, that the intermediate root was involved, rather than the largest or smallest; so the equality of moments of inertia did not, of itself, require that the corresponding axis be a principal axis.

Now let us consider in more general terms the problem of finding the principal axes of a system with only one non-zero product of inertia. Suppose the inertia matrix is of the form

$$[I] = \begin{bmatrix} I_{xx} & 0 & I_{xz} \\ 0 & I_{yy} & 0 \\ I_{zx} & 0 & I_{zz} \end{bmatrix} \qquad (7\text{-}118)$$

A positive rotation of axes through an angle α about the y axis results in a direction cosine matrix

$$[l] = \begin{bmatrix} \cos\alpha & 0 & -\sin\alpha \\ 0 & 1 & 0 \\ \sin\alpha & 0 & \cos\alpha \end{bmatrix} \qquad (7\text{-}119)$$

as can be seen from Fig. 7–10. Performing the transformation

$$[I'] = [l][I][l]^T$$

in accordance with Eq. (7–80) yields the following results:

$I_{x'x'} = I_{xx} \cos^2 \alpha + I_{zz} \sin^2 \alpha - I_{xz} \sin 2\alpha$

$I_{y'y'} = I_{yy}$

$I_{z'z'} = I_{xx} \sin^2 \alpha + I_{zz} \cos^2 \alpha$
$\qquad + I_{xz} \sin 2\alpha \qquad (7\text{-}120)$

$I_{x'z'} = \frac{1}{2}(I_{xx} - I_{zz}) \sin 2\alpha$
$\qquad + I_{xz} \cos 2\alpha$

and, in addition, $I_{x'y'} = I_{y'z'} = 0$. So the primed system becomes a set of principal axes if $I_{x'z'} = 0$; that is, if α is chosen such that

$$\tan 2\alpha = \frac{2I_{xz}}{I_{zz} - I_{xx}} \qquad (7\text{-}121)$$

Knowing the rotation angle α, we can find $I_{x'x'}$ and $I_{z'z'}$ from Eq. (7–120) or from an equivalent pair of equations, namely,

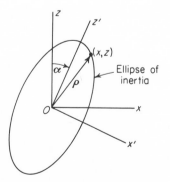

Fig. 7-10. Principal axes of rigid body with a plane of symmetry.

$$I_{x'x'} = \frac{I_{xx} + I_{zz}}{2} + \frac{I_{xx} - I_{zz}}{2} \cos 2\alpha - I_{xz} \sin 2\alpha$$

$$I_{z'z'} = \frac{I_{xx} + I_{zz}}{2} - \frac{I_{xx} - I_{zz}}{2} \cos 2\alpha + I_{xz} \sin 2\alpha$$

(7–122)

Eqs. (7–121) and (7–122) can be represented graphically using a diagram known as *Mohr's circle*.[2]

We have seen that, if one of the coordinate planes is also a plane of symmetry, then at least two of the three products of inertia are zero. In this case, the axis which is perpendicular to the plane of symmetry is a principal axis. Considering only coordinate rotations about this axis, we can think in terms of a two-dimensional *ellipse of inertia* which is formed by the intersection of the inertia ellipsoid and the plane of symmetry. For example, if we again assume that the xz plane is a plane of symmetry, we find by setting $y = 0$ in Eq. (7–104) that the equation of the ellipse of inertia is

$$I_{xx}x^2 + I_{zz}z^2 + 2I_{xz}xz = 1 \qquad (7\text{--}123)$$

Now let α be the angle between the z axis and the radial vector ρ which terminates at (x, z). Then we see that

$$x = \rho \sin \alpha$$
$$z = \rho \cos \alpha$$

and therefore Eq. (7–123) becomes

$$I_{xx} \sin^2 \alpha + I_{zz} \cos^2 \alpha + 2I_{xz} \sin \alpha \cos \alpha = \frac{1}{\rho^2} = I \qquad (7\text{--}124)$$

where we recall from Eq. (7–105) that the moment of inertia about a given axis is equal to $1/\rho^2$.

To obtain the principal moments of inertia, which are also the extreme values of I, we differentiate Eq. (7–124) with respect to α and set the result equal to zero. Thus,

$$\frac{dI}{d\alpha} = (I_{xx} - I_{zz}) \sin \alpha \cos \alpha + 2I_{xz} (\cos^2 \alpha - \sin^2 \alpha) = 0$$

or

$$\tan 2\alpha = \frac{2I_{xz}}{I_{zz} - I_{xx}}$$

in agreement with Eq. (7–121). Two values of 2α which meet this condition differ by π radians. Hence the corresponding values of α differ by $\pi/2$ radians, indicating that the principal axes are orthogonal. The values of the principal moments of inertia are obtained by substituting the two values

[2]Cf. H. Yeh and J. I. Abrams, *Principles of Mechanics of Solids and Fluids—Particle and Rigid-body Mechanics* (New York: McGraw Hill Book Company, Inc., 1960), sec. 11–9.

of α into Eq. (7–124). A little trigonometric manipulation will show that this is equivalent to using the expressions of Eq. (7–122).

7-8. DISPLACEMENTS OF A RIGID BODY

Euler's Theorem. In Sec. 7–1, we saw that a rigid body has six degrees of freedom; three may be considered to be translational and three rotational. The translational aspects of the motion can be specified by the motion of an arbitrary base point which is fixed in the body. For the case of pure translational motion, the orientation of the body in space does not change and all points of the body move along parallel paths with equal velocities. On the other hand, the rotational degrees of freedom are associated with changes in *orientation* of the body. So if we fix the base point, the remaining degrees of freedom can be considered as rotational, and no movement is possible without a change in orientation.

Now let us consider the possible displacements of a rigid body with a point fixed. It can be shown that if this body undergoes a displacement, no matter how complex it is outwardly, the final positions of all points along one line through the fixed point will be the same as their original positions in space. Hence the body may be displaced from its initial position to its final position by a rotation about a single axis. This fact can be expressed as *Euler's theorem:*

The most general displacement of a rigid body with one point fixed is equivalent to a single rotation about some axis through that point.

As a proof of Euler's theorem, consider a rigid body with a fixed point O. We have seen that the position of a rigid body is determined if the positions are given for three noncollinear points which are fixed in the body. For convenience let us choose point O and two other points, A and B, which are at equal distances from O. Then for any displacement of the rigid body, A and B must move on the same spherical surface. Let the final positions of A and B be A' and B', respectively, as shown in Fig. 7–11. Then the great circle arcs AB and $A'B'$ are equal, since the points are fixed in the body. Now

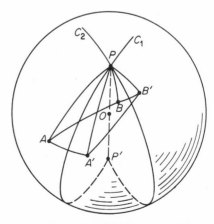

Fig. 7-11. Graphical construction showing the displacement of points A and B on a sphere.

draw the great circle C_1 which is the locus of points on the sphere which are equidistant from A and A'. In a similar fashion, draw the great circle C_2, all points of which are equidistant from B and B'. C_1 and C_2 either intersect at opposite ends of a diameter or else they coincide. Let us assume first that C_1 and C_2 intersect at two points P and P' which lie at the ends of a diameter through O. The great circle arcs PA and PA' are equal; similarly, the arcs PB and PB' are equal. Therefore a rotation about the axis PP' will move B to B'. Also, a rotation about the same axis will move A to A'. Now we shall show that the *same* rotation about PP' will move B to B' and A to A'. We have seen that the great circle arcs AB, BP, and PA are equal to $A'B'$, $B'P$, and PA', respectively. Therefore the spherical triangles ABP and $A'B'P$ are either congruent or symmetrical. The latter possibility is ruled out, however, since it requires that C_1 and C_2 coincide. Hence the angles APB and $A'PB'$ are equal and it follows that the angles APA' and BPB' are also equal. Thus the theorem is proved for the case where C_1 and C_2 intersect.

Now assume that C_1 and C_2 coincide, forming a common great circle C. In this case, the points P and P' defining the axis of rotation are located at the common intersections of C with the two great circles which include the arcs AB and $A'B'$. The spherical triangles ABP and $A'B'P$ degenerate into two great circle arcs and it is apparent that the same angular displacement moves A to A' and B to B'.

It is interesting to consider the effect of two successive rotations of coordinate axes. For example, suppose that the cartesian components of a vector $\mathbf{r}$ in the primed and unprimed coordinate systems are related by an equation similar to Eq. (7–69), namely,

$$\{r'\} = [l_1]\{r\} \tag{7–125}$$

Also, a coordinate transformation from the primed to the double-primed system is of the form

$$\{r''\} = [l_2]\{r'\} \tag{7–126}$$

It can be seen that

$$\{r''\} = [l]\{r\} \tag{7–127}$$

where

$$[l] = [l_2][l_1] \tag{7–128}$$

Now the necessary conditions for a 3×3 transformation matrix to correspond to a real rotation are that it be orthogonal and have a determinant equal to $+1$. The direction cosine matrices $[l_1]$ and $[l_2]$ each meet these conditions, and the question arises whether their product $[l]$ is also a rotation matrix. This question can be decided by noting from Eqs. (7–46) and (7–128) that

$$[l][l]^T = [l_2][l_1][l_1]^T[l_2]^T = [1] \tag{7–129}$$

Therefore $[l]^T = [l]^{-1}$ and thus $[l]$ is orthogonal. Furthermore, the determinant of the product of two square matrices is equal to the product of their determinants. Hence

$$|l| = |l_2| \cdot |l_1| = +1 \qquad (7\text{–}130)$$

Thus the transformation of Eq. (7–127) corresponds to a single rotation about an axis through the common origin O.

By a repeated application of this result, any number of successive rotations about axes through O can be reduced to a single rotation about an axis through O. This is another statement of Euler's theorem.

Chasles' Theorem. In studying the possible displacements of a rigid body with one point fixed, we have been concerned primarily with changes in the *orientation* of a rigid body. If we allow translational motion also, then in the general case there is no point in the body which does not undergo a change of position. Nevertheless, we can think of a general displacement of a rigid body as the superposition of (1) the translational displacement of an arbitrary base point fixed in the body plus (2) a rotational displacement about an axis through that base point. Since the translational displacement does not change the orientation of the body, and since the rotational displacement does not change the location of the base point, it follows that the translational and rotational portions of the total displacement can be performed in either order, or even simultaneously. Hence we have *Chasles' theorem:*

The most general displacement of a rigid body is equivalent to a translation of some point in the body plus a rotation about an axis through that point.

It is apparent that the rotational portion of a general displacement is independent of the choice of the base point. In other words, the direction of the axis of rotation and the amplitude of the angular displacement are the same for all possible base points. The translational portion of the displacement, however, will vary with the choice of the base point because the rotation of the body results in a change of the relative positions of the various possible base points.

It is interesting to note from Euler's theorem that the most general displacement of a rigid body with a point fixed is equivalent to a plane displacement since a rotation about a fixed axis is a plane displacement. Conversely, it can be shown that a general plane displacement of a rigid body is equivalent to a single rotation about a fixed axis perpendicular to the plane of the displacement. As an example, consider the plane displacement of the representative lamina shown in Fig. 7–12. Suppose that an arbitrary point A moves to A' and the lamina undergoes a rotation ϕ. The *center of rotation* C is a point fixed with respect to space and the lamina such that the lamina moves from its initial position to its final position while rotating through an angle ϕ about that point. It is clear that CA and CA' are of

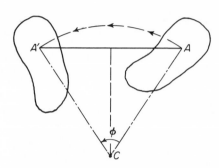

Fig. 7-12. The plane displacement of a representative lamina in a rigid body.

equal length. Hence C must lie on the perpendicular bisector of AA'. Its precise location is such that the angle between CA and CA' is ϕ. Note that the point C may lie inside or outside the lamina, depending upon the displacement considered. In the limiting case of a plane translation, the center of rotation C moves to infinity along a line perpendicular to AA', and the angle ϕ is zero.

We have seen that a plane displacement of a rigid body is equivalent to a single rotation about an axis which is perpendicular to the plane of the displacement and passes through the center of rotation C. Hence it is the necessity for a component of translation *parallel* to the axis of rotation that distinguishes a general displacement from a plane displacement. So we can consider a general displacement as the superposition of a plane displacement and a translation parallel to the axis of rotation. If we think of the translation and rotation as occurring simultaneously, then each point in the body follows a helical path. This most general displacement of a rigid body is known as a *screw displacement*. It can be seen that both the translational and rotational portions of the motion are unique. This is in contrast to the general case of Chasles' theorem which allows a given general displacement to be broken into a certain rotation and one of many possible translational displacements, depending upon the choice of the base point. Note, however, that only one of these possible translations is parallel to the axis of rotation.

7-9. AXIS AND ANGLE OF ROTATION

In Sec. 7-6, we expressed a rotation of coordinate axes in terms of a direction cosine matrix $[l]$. In particular, we found that the components of a vector **r** in the primed and unprimed coordinate systems are related by

$$\begin{Bmatrix} x' \\ y' \\ z' \end{Bmatrix} = \begin{bmatrix} l_{x'x} & l_{x'y} & l_{x'z} \\ l_{y'x} & l_{y'y} & l_{y'z} \\ l_{z'x} & l_{z'y} & l_{z'z} \end{bmatrix} \begin{Bmatrix} x \\ y \\ z \end{Bmatrix}$$

where the vector **r** begins at the origin and ends at (x, y, z) in the unprimed system and at (x', y', z') in the primed system. Now suppose that **r** lies along the axis about which the coordinate system must rotate in moving from the unprimed to the primed axes, or vice versa. It is apparent that the coordinates of any point along this axis will not be changed by the rotation. In other words,

$$\mathbf{r} = \mathbf{r}'$$

or

$$\begin{Bmatrix} x \\ y \\ z \end{Bmatrix} = \begin{bmatrix} l_{x'x} & l_{x'y} & l_{x'z} \\ l_{y'x} & l_{y'y} & l_{y'z} \\ l_{z'x} & l_{z'y} & l_{z'z} \end{bmatrix} \begin{Bmatrix} x \\ y \\ z \end{Bmatrix} \tag{7–131}$$

Hence

$$\begin{bmatrix} (l_{x'x} - 1) & l_{x'y} & l_{x'z} \\ l_{y'x} & (l_{y'y} - 1) & l_{y'z} \\ l_{z'x} & l_{z'y} & (l_{z'z} - 1) \end{bmatrix} \begin{Bmatrix} x \\ y \\ z \end{Bmatrix} = 0 \tag{7–132}$$

Comparing Eq. (7–131) or (7–132) with the general form of the eigenvalue problem given in Eq. (7–92), it can be seen that $+1$ is an eigenvalue corresponding to this real rotation.

For a given $[l]$ matrix, the corresponding axis of rotation can be obtained by solving for the ratios of the x, y, and z components of $\mathbf{r}$ from Eq. (7–132). Noting that Eqs. (7–88) and (7–132) have the same form, we see that the same procedure can be used here to establish the axis of rotation as was used previously in finding a principal axis. Hence we can write a set of equations with a form similar to that of Eq. (7–90), namely,

$$(l_{y'y} - 1) \frac{y}{x} + l_{y'z} \frac{z}{x} = -l_{y'x}$$

$$l_{z'y} \frac{y}{x} + (l_{z'z} - 1) \frac{z}{x} = -l_{z'x} \tag{7–133}$$

This pair of equations can be solved for the ratios y/x and z/x unless x is zero. In case $x = 0$, that is, if the axis of rotation lies in the yz plane, one can solve for the ratio z/y from one of the equations given in Eq. (7–132).

Now comes the problem of finding the angle of rotation corresponding to a given direction cosine matrix. Consider first the particular case where a rotation Φ occurs in a positive sense about the z axis, as shown in Fig. 7–13. Designating the rotation matrix by $[\Phi]$, it can be seen from the figure that

$$[\Phi] = \begin{bmatrix} \cos \Phi & \sin \Phi & 0 \\ -\sin \Phi & \cos \Phi & 0 \\ 0 & 0 & 1 \end{bmatrix} \tag{7–134}$$

Evaluating the trace of this matrix, we obtain

$$\operatorname{tr} [\Phi] = 1 + 2 \cos \Phi \tag{7–135}$$

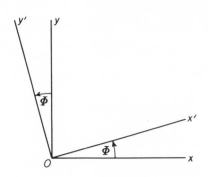

Fig. 7-13. A rotation of the primed coordinate system about the z axis.

The more general case of a rotation about an arbitrary axis can be analyzed by considering the primed and unprimed systems to be originally coincident and then performing three rotations as follows: (1) Rotate the primed system about some axis relative to the unprimed system such that the z' axis becomes the desired rotation axis (Fig. 7–14). (2) Rotate both systems as a rigid body through an angle Φ about the z' axis. (3) Perform the inverse of the first rotation such that the primed system coincides with the new position of the unprimed system. The end result of this sequence of three rotations is a rotation Φ of both systems about the middle position of the z' axis because, in the sense of their relative orientation, the first and last rotations cancel. Using rotation matrices, we can express the final components of a given vector $\mathbf{r}$ in terms of the corresponding initial values by the equation

$$\{r'\} = [l]^T [\Phi][l]\{r\} \qquad (7\text{–}136)$$

where the matrices $[l]$ and $[l]^T$ represent the first and third rotations, respectively, in the foregoing sequence. Now let us recall from Sec. 7–7 that the trace of a square matrix is invariant under an orthogonal transformation. So if we let

$$[\Phi'] = [l]^T[\Phi][l] \qquad (7\text{–}137)$$

then

$$\operatorname{tr}[\Phi'] = 1 + 2\cos\Phi \qquad (7\text{–}138)$$

Hence the angle of rotation Φ can be found by evaluating the trace of the rotation matrix, even for the case of a general rotation.

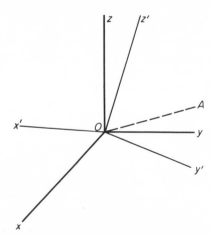

Fig. 7-14. Coordinate systems used for a general rotation. OA is an axis (fixed in both systems) about which the rotations of the primed system relative to the unprimed system are performed.

7–10. REDUCTION OF FORCES

Equipollent Systems. We have seen that the translational and rotational equations of motion, namely,

$$\mathbf{F} = m\ddot{\mathbf{r}}_c$$

$$\mathbf{M} = \dot{\mathbf{H}}$$

can be used to obtain the motion of a rigid body which is acted upon by a given system of external forces. We note that the translational motion of the center of mass, for given initial conditions, is determined by the vector sum of the external forces, regardless of their point of application. On the other hand, the rotational motion depends upon the total moment of the external forces about the center of mass, and therefore is dependent upon the location of the forces. For many analyses, it is convenient to replace the actual force system by an *equipollent* system; that is, a system such that (1) the total force is equal to that of the original system and (2) the total moment about any point is the same as for the original system. It can be seen from

the two equations of motion that the separate application of two equipollent force systems to a given rigid body results in equal dynamic responses. It should be noted, however, that elastic bodies may deform quite differently, depending upon the exact point of application of the forces. Hence equipollent force systems do not produce the same effects upon elastic bodies.

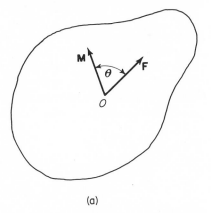

(a)

If we limit ourselves, then, to a consideration of the motions of rigid bodies, we find that it is often convenient to replace the given force system by an equipollent system consisting of the total force **F** acting at the center of mass and a couple whose moment **M** is computed with respect to the center of mass—Fig. 7–15(a). By a *couple* we mean a pair of equal and opposite forces of magnitude P whose lines of action are separated by a distance r—Fig. 7–15(b). It can be seen that the moment of a given couple is the same with respect to any reference point and its magnitude is

$$M = Pr \qquad (7\text{–}139)$$

The direction is perpendicular to the plane of the two forces in accordance

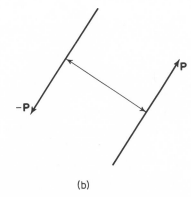

(b)

Fig. 7-15. (a) A force and moment which are equipollent to a general force system acting on a rigid body. (b) A couple.

with the right-hand rule; for example, **M** is directed out of the page in Fig. 7–15(b). For this system of forces, the resultant force **F** has been assumed to pass through the center of mass O. Even though the couple **M** need not have any particular location, it is the usual practice to consider it to be acting at the reference point O.

Now suppose that the reference point in Fig. 7–15(a) had been chosen at some point O' that is not at the center of mass. In this case, we could consider an equipollent system consisting of the same force **F** applied at O' plus a couple **M'**. If O' were chosen along the line of action of **F** acting through O, then **M'** = **M**, since the moment of **F** about any point is unchanged if its line of action is unchanged.

If, however, O' were displaced a distance d in a direction perpendicular to this line of action (Fig. 7–16), then **M'** would differ from **M** by a couple of magnitude Fd whose direction is perpendicular to **F**. This can be seen by considering the equal and opposite forces **F** and $-$**F** to be applied at O'. The added forces are equipollent to zero and will not change the total force or the total moment about any point. But now we can think of a force **F** as being applied at O' and consider the remaining forces **F** and $-$**F** as constituting a couple of magnitude Fd whose direction is normal to **F** and out of the page in Fig. 7–16. Hence by a proper displacement of the reference point, it is possible to cancel the component of **M** that is normal to **F**. The remaining couple **M'** is then parallel to **F** and the required perpendicular displacement is

$$d = \frac{M \sin \theta}{F} \qquad (7\text{--}140)$$

where θ is the angle between **M** and **F**. The resulting equipollent force system consisting of a force **F** and a parallel couple **M'** is known as a *wrench*.

It should be noted that, although a given system of forces can be reduced to a force and a couple in many ways, there is only one equipollent wrench for the system. In other words, there is only one line of action for the total force **F** such that the total moment of the force system about a reference point O' on this line is parallel to **F**. We see that this line of action of the wrench is determined solely from geometrical considerations and will not, in general, pass through the center of mass.

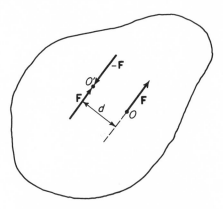

Fig. 7-16. A displacement of the reference point for a force system acting on a rigid body.

We have shown that any force system can be reduced to a force plus a couple, as in Fig. 7–15. Furthermore, the effect of a couple upon the motion of a rigid body does not depend upon its location, but only upon its magnitude and direction. So if we consider the couple to consist of two equal and opposite forces of magnitude P and separation r in accordance with Eq. (7–139), then it can be seen that one of these forces can have a line of action passing through the reference point O (Fig. 7–17). It is clear that the forces P and F acting at O can be replaced by a single force at O equal to their vector sum. The forces $(P + F)$ and $-P$ do not lie in the same plane, in general, and are known as

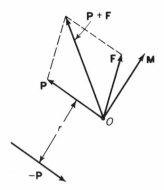

Fig. 7-17. A system of two skew forces which is equipollent to a general force system.

skew forces. Hence we conclude that a general system of forces is equipollent to two skew forces. The skew forces are not unique since the magnitude of P is arbitrary and its direction is restricted only in that it must lie in a plane perpendicular to M.

Parallel Forces. Now let us consider the particular case where all the external forces acting on a rigid body are parallel to each other. Suppose that the forces $F_1, F_2, \ldots, F_n$ are applied at points whose position vectors relative to the reference point O are given by $r_1, r_2, \ldots, r_n$, respectively, as shown in Fig. 7–18. It can be seen that the total force is

$$F = \sum_{i=1}^{n} F_i \qquad (7\text{--}141)$$

and the total moment about O is

$$M = \sum_{i=1}^{n} r_i \times F_i \qquad (7\text{--}142)$$

Hence the given force system is equipollent to a force F applied at O plus a couple M whose location is arbitrary.

We note that M must lie in a plane which is perpendicular to the applied forces since the moment of each force about O lies in this plane. So if the line of action of F is translated in the proper direction through a distance $d = M/F$ in accordance with Eq. (7–140), then, assuming $F \neq 0$, the couple is completely canceled and the force system is equipollent to a single force F. Thus we can find a reference point C such that the system of applied forces is equipollent to a force F acting through this point. If the position of C relative to O is given by r_c and the point of application of F_i relative to C is given by ρ_i, then we see from Fig. 7–18 that

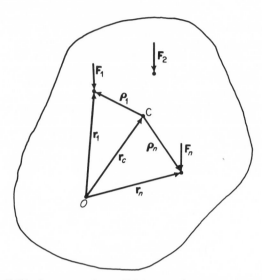

Fig. 7-18. A system of parallel forces acting on a rigid body.

$$\mathbf{r}_i = \mathbf{r}_c + \boldsymbol{\rho}_i \tag{7-143}$$

and from Eqs. (7–141), (7–142), and (7–143), we find that the moment relative to O is

$$\mathbf{M} = \mathbf{r}_c \times \mathbf{F} + \sum_{i=1}^{n} \boldsymbol{\rho}_i \times \mathbf{F}_i \tag{7-144}$$

But, in accordance with the assumption that the moment of the applied forces about C is zero, we obtain

$$\sum_{i=1}^{n} \boldsymbol{\rho}_i \times \mathbf{F}_i = 0 \tag{7-145}$$

Therefore, equating the expressions for the moment given in Eqs. (7–142) and (7–144), we see that

$$\mathbf{r}_c \times \mathbf{F} = \sum_{i=1}^{n} \mathbf{r}_i \times \mathbf{F}_i \tag{7-146}$$

Now let us recall that the $\mathbf{F}_i$ are parallel and use the notation

$$\mathbf{F}_i = F_i \mathbf{e}_f$$
$$\mathbf{F} = F \mathbf{e}_f \tag{7-147}$$

where the unit vector $\mathbf{e}_f$ points in the positive direction of the applied forces. From Eqs. (7–146) and (7–147), we find that

$$\mathbf{r}_c \times F \mathbf{e}_f = \sum_{i=1}^{n} \mathbf{r}_i \times F_i \mathbf{e}_f \tag{7-148}$$

It can be seen that if

$$\mathbf{r}_c = \frac{1}{F} \sum_{i=1}^{n} F_i \mathbf{r}_i \qquad (7\text{-}149)$$

then Eq. (7-148) holds for *any* direction of $\mathbf{e}_f$. Hence the *resultant force* $\mathbf{F}$ passes through C regardless of the direction of the parallel applied forces.

The reference point C whose position is defined by Eq. (7-149) is known as the *center* of the force system. For the particular case of a group of n particles in a uniform gravitational field, the point C is at the center of mass, as may be seen by comparing this result with Eq. (4-7).

Returning now to Eqs. (7-141) and (7-142), it can be seen that if $\mathbf{F} = 0$, then the system of parallel forces is equipollent to a single couple $\mathbf{M}$ whose location is unspecified. Summarizing, we conclude that a system of parallel forces acting on a rigid body may be replaced by an equipollent system which consists of either (1) a force $\mathbf{F}$ acting through the center C or (2) a couple $\mathbf{M}$. The particular instance where both $\mathbf{F}$ and $\mathbf{M}$ are zero can be regarded as a special case of either.

Concurrent Forces. We have found that an external force acting on a rigid body can be considered as a sliding vector; that is, the point of application of the force can be moved along its line of action without influencing its effect upon the motion of the body. Previously, we discussed the reduction to a simpler equipollent system for the general case and also for the case where the lines of action of the applied forces are parallel. Now let us consider a force system in which the lines of action of the applied forces all pass through the same point O (Fig. 7-19). Sets of forces which are all directed through a single point are known as *concurrent forces*. Taking O as the reference point, we see that the moments of all the forces are zero. Hence we conclude that a set of n concurrent forces is equipollent to a single force $\mathbf{F}$ passing through the common point O, where

$$\mathbf{F} = \sum_{i=1}^{n} \mathbf{F}_i$$

An example of a set of concurrent forces is the set of gravitational forces acting on the particles of an extended satellite in orbit about an attracting spherical body. We saw in Example 5-6 that the resultant gravitational force acts along a line which passes through the center of gravity of the satellite and the center of the attracting body.

Next consider a set of forces confined to a plane. In the usual case, any pair of forces can be considered to be concurrent. Hence they can be replaced by a single force equal to

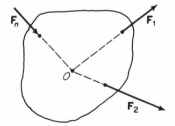

Fig. 7-19. A set of concurrent forces acting on a rigid body.

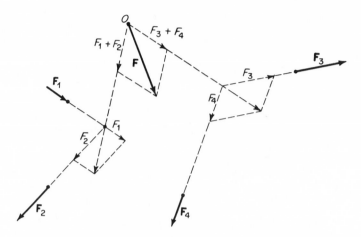

Fig. 7-20. A plane set of forces and the resultant force **F**.

their sum, acting through the point of concurrency. By repeating this procedure we can arrive at the force

$$\mathbf{F} = \sum_{i=1}^{n} \mathbf{F}_i$$

which is equipollent to the plane system of forces (Fig. 7–20). If $\mathbf{F} = 0$,
then the system of forces is equipollent to a couple $\mathbf{M}$ whose direction is
perpendicular to the plane of the forces, or perhaps both $\mathbf{F}$ and $\mathbf{M}$ are zero.

To summarize, we have found that for parallel forces, concurrent forces,
or forces acting in a plane, the simplest force system to which the actual
system is reducible is a force $\mathbf{F}$ with a given line of action, or a couple $\mathbf{M}$,
but not both. In attempting to discover the common characteristic of these
systems which allows their reduction to a single force or couple, we notice
that, in each instance, the equipollent system for an *arbitrary* reference
point consists of a force $\mathbf{F}$ and a couple $\mathbf{M}$ which are *orthogonal*. So we see
that we can apply some results of our earlier discussion of the reduction
of a general force system to a wrench. There we found that, for $F \neq 0$, it
is always possible by a proper choice of reference point to eliminate the
component of $\mathbf{M}$ that is normal to $\mathbf{F}$.

7–11. INFINITESIMAL ROTATIONS

In Sec. 7–6, we saw that an arbitrary rotation of a cartesian set of coordinate axes can be accomplished by premultiplying a column matrix
specifying the position of a typical point by a 3×3 direction cosine matrix
specifying the rotation. Further rotations can be accomplished by further

premultiplications. Thus, if a rotation $[\Phi_1]$ is followed by a second rotation $[\Phi_2]$, the final orientation is identical with that resulting from a single rotation

$$[\Phi] = [\Phi_2][\Phi_1]$$

We recall, however, that matrix multiplication is not commutative in general. Therefore we might expect that it is important to specify the order in which two or more rotations are performed. That this is actually the case can be seen from the following illustration: Suppose that a rectangular parallelepiped is initially in the position shown in Fig. 7–21(a). It is given successive 90-degree rotations about the x and y axes, assuming that the axes are fixed in the body. The resulting orientation is given in Fig. 7–21 (b). On the other hand, if the same rotations are performed in the opposite order, the resulting final orientation is given in Fig. 7–21(c). It is apparent that the orientations given in Figs. 7–21(b) and 7–21(c) are quite different. Furthermore, a similar result will occur if the axes are assumed to be fixed in space rather than in the body.

In studying the kinematics of a particle, we have represented displacements by vectors. We know, for example, that the translational displacements **A** and **B** when taken in either order are together equivalent to a single displacement

$$\mathbf{C} = \mathbf{A} + \mathbf{B} = \mathbf{B} + \mathbf{A}$$

This is in accord with the commutative nature of vector addition. One is tempted to consider rotational displacements in a similar fashion, the direction of the vector being in the direction of the axis of rotation and the magnitude being proportional to the angle of rotation. Thus, by analogy to the foregoing case of translational displacements, one might think that a single rotation $\Phi = \Phi_1 + \Phi_2$ is equivalent to

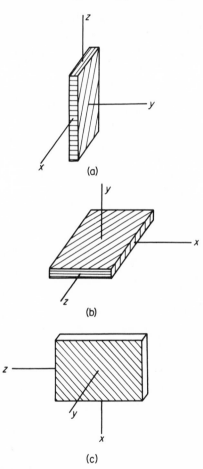

Fig. 7-21. The positions of a block: (a) initially, (b) after 90° rotations about the x and y axes, and (c) after 90° rotations about the y and x axes starting from the initial position.

the rotations Φ_1 and Φ_2 taken in sequence. But we have seen that rigid-body rotations are not commutative, in general, whereas vector addition is commutative. Therefore, a general rigid-body rotation is not a vector quantity.

On the other hand, we have assumed that the angular rotation rate $\boldsymbol{\omega}$ is a vector quantity, and we have represented it in this way many times previously. Also, we note from the definition of $\boldsymbol{\omega}$ given in Eq. (2–3), namely,

$$\boldsymbol{\omega} = \lim_{\Delta t \to 0} \frac{\Delta \boldsymbol{\theta}}{\Delta t}$$

that if $\boldsymbol{\omega}$ is a vector, then the infinitesimal rotation $\Delta \boldsymbol{\theta}$ must also be a vector since Δt is a scalar. Hence we find that an infinitesimal rotation is a vector quantity, whereas a general finite rotation is not.

To study this point further, let us write the rotation matrix corresponding to an infinitesimal rotation ϵ_z about the z axis. Using Eq. (7–134), we obtain

$$[\epsilon_z] = \begin{bmatrix} \cos \epsilon_z & \sin \epsilon_z & 0 \\ -\sin \epsilon_z & \cos \epsilon_z & 0 \\ 0 & 0 & 1 \end{bmatrix} \cong \begin{bmatrix} 1 & \epsilon_z & 0 \\ -\epsilon_z & 1 & 0 \\ 0 & 0 & 1 \end{bmatrix} \tag{7–150}$$

Also, we find from Eq. (7–119) that an infinitesimal rotation ϵ_y about the y axis is represented by

$$[\epsilon_y] = \begin{bmatrix} \cos \epsilon_y & 0 & -\sin \epsilon_y \\ 0 & 1 & 0 \\ \sin \epsilon_y & 0 & \cos \epsilon_y \end{bmatrix} \cong \begin{bmatrix} 1 & 0 & -\epsilon_y \\ 0 & 1 & 0 \\ \epsilon_y & 0 & 1 \end{bmatrix} \tag{7–151}$$

To find the rotation matrix corresponding to an infinitesimal rotation ϵ_z followed by an infinitesimal rotation ϵ_y, we premultiply $[\epsilon_z]$ by $[\epsilon_y]$, obtaining

$$[\epsilon_y][\epsilon_z] = \begin{bmatrix} 1 & 0 & -\epsilon_y \\ 0 & 1 & 0 \\ \epsilon_y & 0 & 1 \end{bmatrix} \begin{bmatrix} 1 & \epsilon_z & 0 \\ -\epsilon_z & 1 & 0 \\ 0 & 0 & 1 \end{bmatrix} = \begin{bmatrix} 1 & \epsilon_z & -\epsilon_y \\ -\epsilon_z & 1 & 0 \\ \epsilon_y & 0 & 1 \end{bmatrix} \tag{7–152}$$

By a similar process, we can show that an infinitesimal rotation about all three axes results in the rotation matrix

$$\begin{bmatrix} 1 & \epsilon_z & -\epsilon_y \\ -\epsilon_z & 1 & \epsilon_x \\ \epsilon_y & -\epsilon_x & 1 \end{bmatrix} = [1] + [\epsilon] \tag{7–153}$$

where $[1]$ is a 3×3 unit matrix and $[\epsilon]$ is the following skew-symmetric matrix:

$$[\epsilon] = \begin{bmatrix} 0 & \epsilon_z & -\epsilon_y \\ -\epsilon_z & 0 & \epsilon_x \\ \epsilon_y & -\epsilon_x & 0 \end{bmatrix} \tag{7–154}$$

It can be seen by evaluating the matrix product $[\epsilon_z]\,[\epsilon_y]$ that the result is identical with that of Eq. (7–152), indicating that the order of the infinitesimal rotations is immaterial. Furthermore, if we neglect the product of infinitesimal quantities, we find that

$$\Big[[1] + [\epsilon_1]\Big]\Big[[1] + [\epsilon_2]\Big] = [1] + [\epsilon_1] + [\epsilon_2] \qquad (7\text{–}155)$$

where $[\epsilon_1]$ and $[\epsilon_2]$ are skew-symmetric matrices representing two arbitrary infinitesimal rotations. Since matrix addition is commutative, we see that

$$[\epsilon_1] + [\epsilon_2] = [\epsilon_2] + [\epsilon_1]$$

and therefore the same result is obtained if the arbitrary infinitesimal rotations are performed in either order. Hence we see once more that infinitesimal rotations are vectorial in nature.

Now let us apply an infinitesimal rotation to a vector **r** whose length is assumed to be constant. Using the matrix notation of Eq. (7–69), we see that

$$\{r'\} = \Big[[1] + [\epsilon]\Big]\{r\} \qquad (7\text{–}156)$$

or

$$\{r'\} - \{r\} = [\epsilon]\{r\} \qquad (7\text{–}157)$$

In the past, we have considered the rotation matrix to refer to a rotation of coordinate axes. The vector **r** was assumed to remain fixed in space, and any changes in the components of **r** were assumed to be due entirely to the coordinate rotation. On the other hand, we could consider the coordinate system to be fixed in space and let **r** rotate in the opposite direction. We shall take the latter viewpoint which is illustrated in Fig. 7–22 for the case of an infinitesimal rotation of **r** about the z axis. Assuming that the rotation of Eq. (7–157) takes place in a time Δt, we can divide this equation by Δt and take the limit as Δt approaches zero. The left side is just $\dot{r}$ expressed as a column matrix. So we can write

$$\{\dot{r}\} = [\omega]\{r\} \qquad (7\text{–}158)$$

where

$$[\omega] = \lim_{\Delta t \to 0} \frac{1}{\Delta t}\,[-\epsilon] \qquad (7\text{–}159)$$

The minus sign is introduced in the definition of $[\omega]$ in order that its elements will refer to the components of the rotation rate of **r** rather than to the rotation rate of the coordinate system. For example, a positive ω_z refers to a counterclockwise rotation of **r** in Fig. 7–22.

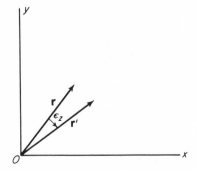

Fig. 7-22. An infinitesimal rotation of a vector **r** about the z axis.

Referring back to Eq. (2–5), we recall that

$$\dot{\mathbf{r}} = \boldsymbol{\omega} \times \mathbf{r} \tag{7–160}$$

for the case where the amplitude of **r** is constant. Now we note from Eqs. (7–158) and (7–160) that these are two forms of the same statement. Hence we find that the premultiplication of the column matrix $\{r\}$ by the skew-symmetric matrix $[\omega]$ corresponds to taking the vector cross product in the order $\boldsymbol{\omega} \times \mathbf{r}$. As an example, consider the matrix product

$$[\omega]\{r\} = \begin{bmatrix} 0 & -\omega_z & \omega_y \\ \omega_z & 0 & -\omega_x \\ -\omega_y & \omega_x & 0 \end{bmatrix} \begin{Bmatrix} x \\ y \\ z \end{Bmatrix} = \begin{Bmatrix} (-\omega_z y + \omega_y z) \\ (\omega_z x - \omega_x z) \\ (-\omega_y x + \omega_x y) \end{Bmatrix} \tag{7–161}$$

Comparing this result with the determinant form of the vector cross product, namely,

$$\boldsymbol{\omega} \times \mathbf{r} = \begin{vmatrix} \mathbf{i} & \mathbf{j} & \mathbf{k} \\ \omega_x & \omega_y & \omega_z \\ x & y & z \end{vmatrix}$$

we see that the corresponding components are identical.

7–12. EULERIAN ANGLES

In Sec. 7–1, we saw that a rigid body has six degrees of freedom, three being translational and three rotational. Also, we saw that if the center of mass is chosen as a reference point, then one can analyze separately the translational and rotational aspects of the motion for a given set of applied forces. Now let us consider the rotational motion only, and indicate some of the preliminary questions that arise in the analysis.

First, it seems quite natural that we should use Lagrange's equations in setting up the equations of motion. Since there are three rotational degrees of freedom, one would prefer to use only three generalized coordinates. Hence the question arises concerning a proper choice of these coordinates. The use of direction cosines as generalized coordinates is not attractive because there are nine direction cosines and six equations of constraint relating them. The resulting equations of motion would be quite complicated.

Another possibility is to write the rotational kinetic energy in terms of the cartesian components ω_x, ω_y, and ω_z, as in Eq. (7–55). Unfortunately, however, one cannot find a set of generalized coordinates corresponding to ω_x, ω_y, and ω_z. In other words, one cannot find a set of three parameters which specify the orientation of the body and also have ω_x, ω_y, and ω_z as time derivatives.

TABLE **7-1**. A COMPARISON OF EULER ANGLE SYSTEMS
(Corresponding Quantities)

Order of Rotation and Corresponding Axes

I	II	III
ψ about z	ψ about z	ϕ about z
θ about y	θ about x	θ about y
ϕ about x	ϕ about z	ψ about z
ψ	$-\psi$	$-\phi$
θ	$\pi/2 - \theta$	$\pi/2 - \theta$
ϕ	ϕ	ψ
X	$-Y$	X
Y	$-X$	$-Y$
Z	$-Z$	$-Z$
x	z	z
y	$-x$	$-y$
z	$-y$	x

Thus we are led to search for a new set of three coordinates by which the orientation of a rigid body can be specified. A set of coordinates of this sort are the Eulerian angles. Several different types of Euler angle systems are in common use (Table 7–1). We shall adopt the system which has been widely employed in aeronautical engineering and, more recently, in the analysis of missiles and other space vehicles.

The Euler angles can be visualized in terms of a coordinate system fixed in the body whose orientation is being described. For example, in the case of an airplane, the xz plane is commonly taken to be the plane of symmetry. The positive x axis is directed forward out of the nose, whereas the positive z axis is directed downward through the bottom of the fuselage. The positive y axis is normal to the other two axes and proceeds roughly along the right wing. In the context of this example, the names given the Euler angles are meaningful. They are the *heading angle* ψ, the *attitude angle* θ, and the *bank angle* ϕ, as shown in Fig. 7–23.

Now let us consider the definitions of these Eulerian angles which give the orientation of the xyz coordinate system relative to the fixed XYZ system. If the two systems are initially coincident, a series of three rotations about the body axes, performed in the proper sequence, is sufficient to allow the xyz system to reach any orientation. The three rotations are:

1.   A positive rotation ψ about the Z axis, resulting in the primed system.

2.   A positive rotation θ about the y' axis, resulting in the double-primed system.

3.   A positive rotation ϕ about the x'' axis, resulting in the final unprimed system.

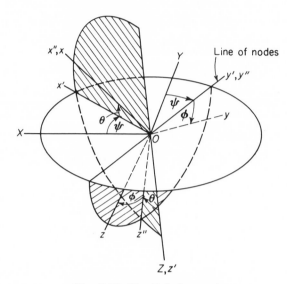

Fig. 7-23. The Eulerian angles.

Another method of defining the Euler angles can be seen directly from Fig. 7–23. The *heading angle* ψ is measured in the horizontal plane and is the angle between the X and x' axes. For example, if $\psi = \pi/2$ and we make the usual assumption that the X axis points due north, then the x' axis coincides with the Y axis and points due east. Using the aircraft example, ψ indicates the direction of the projection of the x axis on the horizontal plane, that is, the heading of the aircraft.

The *attitude angle* θ is measured in the vertical plane and is the angle between the x and x' axes. In other words, it is the angle of the x axis above the horizontal plane. The attitude angle is usually considered to lie in the range $-\pi/2 \leq \theta \leq \pi/2$. The axis about which the θ rotation occurs (y' axis) is known as the *line of nodes*.

The *bank angle* ϕ is the angle between the y and y'' axes and is measured in the plane through O which is perpendicular to the x axis. For example, when $\phi = \pm \pi/2$, the xy plane is vertical, regardless of the value of θ.

We shall be using Euler angles to describe the orientation of general rigid bodies, and will not refer specifically to aircraft rotations. Usually, the xyz system is a system of principal axes at the center of mass or at a fixed point. For the important case of bodies with axial symmetry, the x axis is chosen as the axis of symmetry and two of the three principal moments of inertia are equal.

The Euler angle definitions which we have given are most often used for specifying the orientation of bodies on or near the surface of the earth.

The X axis is assumed to point due north, the Y axis points due east, and the Z axis points directly downward. Nevertheless, we can also use the Euler angles to describe the rotational motion of bodies away from the earth in free space, provided that we make a suitable definition of what is meant by the "vertical" direction. The X and Y axes must be nonrotating in inertial space but their precise direction is usually unimportant, as will be explained in Chapter 8.

Now let us consider the matrix equations which indicate the individual rotations that are performed in going from the original XYZ system to the final xyz system. Referring again to the definitions of ψ, θ, and ϕ, we obtain the following equations:

$$\begin{Bmatrix} x' \\ y' \\ z' \end{Bmatrix} = \begin{bmatrix} \cos\psi & \sin\psi & 0 \\ -\sin\psi & \cos\psi & 0 \\ 0 & 0 & 1 \end{bmatrix} \begin{Bmatrix} X \\ Y \\ Z \end{Bmatrix} \tag{7-162}$$

$$\begin{Bmatrix} x'' \\ y'' \\ z'' \end{Bmatrix} = \begin{bmatrix} \cos\theta & 0 & -\sin\theta \\ 0 & 1 & 0 \\ \sin\theta & 0 & \cos\theta \end{bmatrix} \begin{Bmatrix} x' \\ y' \\ z' \end{Bmatrix} \tag{7-163}$$

$$\begin{Bmatrix} x \\ y \\ z \end{Bmatrix} = \begin{bmatrix} 1 & 0 & 0 \\ 0 & \cos\phi & \sin\phi \\ 0 & -\sin\phi & \cos\phi \end{bmatrix} \begin{Bmatrix} x'' \\ y'' \\ z'' \end{Bmatrix} \tag{7-164}$$

Writing the preceding equations in more abbreviated form, we obtain

$$\{r'\} = [\psi]\{R\}$$
$$\{r''\} = [\theta]\{r'\}$$
$$\{r\} = [\phi]\{r''\}$$

or

$$\{r\} = [\phi][\theta][\psi]\{R\} \tag{7-165}$$

where $\{R\}$ represents the components of a vector $\mathbf{r}$ in the original fixed frame and $\{r\}$ gives the components of the same vector in the final xyz system. Performing the indicated matrix multiplications, we obtain the following result:

$$\begin{Bmatrix} x \\ y \\ z \end{Bmatrix} = \begin{bmatrix} \cos\psi\cos\theta & \sin\psi\cos\theta & -\sin\theta \\ \begin{array}{l}(-\sin\psi\cos\phi \\ +\cos\psi\sin\theta\sin\phi)\end{array} & \begin{array}{l}(\cos\psi\cos\phi \\ +\sin\psi\sin\theta\sin\phi)\end{array} & \cos\theta\sin\phi \\ \begin{array}{l}(\sin\psi\sin\phi \\ +\cos\psi\sin\theta\cos\phi)\end{array} & \begin{array}{l}(-\cos\psi\sin\phi \\ +\sin\psi\sin\theta\cos\phi)\end{array} & \cos\theta\cos\phi \end{bmatrix} \begin{Bmatrix} X \\ Y \\ Z \end{Bmatrix}$$

$$\tag{7-166}$$

The individual rotation matrices as well as the final product matrix are all direction cosine matrices; therefore each is an orthogonal matrix. Hence the inverse transformations are obtained by merely transposing the corresponding matrices.

Assuming that the Euler angles are limited to the ranges

$$0 \leq \psi < 2\pi, \qquad -\frac{\pi}{2} \leq \theta \leq \frac{\pi}{2}, \qquad 0 \leq \phi < 2\pi$$

it can be seen that any possible orientation of the body can be attained by performing the proper rotations in the given order. If the orientation is such that the x axis is vertical, however, then no unique set of values for ψ and ϕ can be found. Thus, if $\theta = \pm \pi/2$, the angles ψ and ϕ are undefined. Nevertheless, it can be shown from Eq. (7–166) that the angle $(\psi - \phi)$ is well-defined if $\theta = \pi/2$; it represents the heading angle of the z axis, that is, the angle between the X and z axes. In a similar fashion, the angle $(\psi + \phi)$ is well defined for $\theta = -\pi/2$ even though the individual angles are not.

The situation in which $\theta = \pm \pi/2$ corresponds to a condition of gyroscope suspensions which is known as *gimbal lock*. We have seen that rotations corresponding to the ψ and ϕ coordinates are not distinct for this case and therefore only two rotational degrees of freedom are represented. The missing degree of freedom corresponds to rotation about the x' axis. In other words, during the condition of gimbal lock an angular velocity component parallel to the x' axis cannot be represented in terms of Euler angle rates.

Earlier it was noted that several systems of Euler angles are in common usage. In order to facilitate the conversion of results from one system to another, we present Table 7–1, showing corresponding quantities for three common systems. The system we are using is designated as type I. The other systems start with the z axis pointing vertically upward before the first rotation. Thus θ is the angle between the vertical and the final position of the z axis.

7–13. EXAMPLES OF RIGID BODY MOTION IN A PLANE

We have discussed many of the principles of dynamics as they apply to a particle or to a system of particles. Now we shall use these principles to solve a few examples concerning rigid bodies which have distributed masses and move in a plane. More complex problems of rigid-body dynamics will be considered in Chapter 8.

Example 7–5. A uniform circular cylinder of mass m and radius a rolls without slipping on a plane inclined at an angle α with the horizontal. Solve for the angular acceleration (Fig. 7–24).

First Method: Taking the center of mass O as the reference point, we can use the rotational equation

$$\mathbf{M} = \dot{\mathbf{H}}$$

It can be seen that for pure rolling motion, the angular velocity $\boldsymbol{\omega}$ must be parallel to the symmetry axis of the cylinder at all times. Furthermore, we note from the free-body diagram—Fig. 7–24(b) that the applied moment $\mathbf{M}$ is parallel to $\boldsymbol{\omega}$. Since the symmetry axis is also a principal axis, it is apparent from Eq. (7–12) that $\mathbf{H}$ and $\dot{\mathbf{H}}$ are parallel. Hence the vector equation shown above can be replaced by the scalar equation

$$M = I\ddot{\theta} \qquad (7\text{–}167)$$

where the reference point is taken at the center O, and where the moment of inertia is

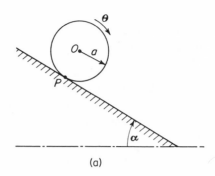

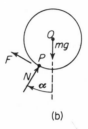

Fig. 7-24. A uniform circular cylinder rolling down an inclined plane.

$$I = \frac{1}{2}ma^2 \qquad (7\text{–}168)$$

in agreement with the value found for the circular disk in Eq. (7–116).

It can be seen from the free-body diagram that the moment of the external forces about O is

$$M = Fa \qquad (7\text{–}169)$$

where M is taken positive in the clockwise sense. From Eqs. (7–167), (7–168), and (7–169), we see that

$$F = \frac{1}{2}ma\ddot{\theta} \qquad (7\text{–}170)$$

Another equation involving F and $\ddot{\theta}$ can be obtained by writing Newton's equation of motion for the force components parallel to the inclined plane. Noting that the linear acceleration of the center of mass is $a\ddot{\theta}$ and is directed down the plane, we find that

$$mg \sin \alpha - F = ma\ddot{\theta} \qquad (7\text{–}171)$$

Eliminating F between Eqs. (7–170) and (7–171), we obtain

$$mg \sin \alpha = \frac{3}{2}ma\ddot{\theta}$$

or

$$\ddot{\theta} = \frac{2g \sin \alpha}{3a} \qquad (7\text{--}172)$$

Second Method: Now let us choose a reference point P which is *fixed in the inclined plane* at the instantaneous center of rotation. We recall that the angular momentum about a point P is equal to the angular momentum about the mass center plus the angular momentum associated with the velocity of the mass center relative to P. Thus we have

$$H_p = I\dot{\theta} + mva = I_p\dot{\theta} \qquad (7\text{--}173)$$

where I_p is the moment of inertia about the instantaneous center.

In this example, we find that $I = \frac{1}{2}ma^2$ and $v = a\dot{\theta}$, yielding

$$I_p = \frac{1}{2}ma^2 + ma^2 = \frac{3}{2}ma^2$$

in agreement with the value of the moment of inertia obtained from Eq. (7–61) for an axis translation. Also, the forces N and F pass through P, with the result that the entire external moment is due to the gravitational force mg.

$$M_p = mga \sin \alpha \qquad (7\text{--}174)$$

Finally, noting that I_p is constant, we can write

$$M_p = \dot{H}_p = I_p\ddot{\theta} \qquad (7\text{--}175)$$

and solving for the angular acceleration $\ddot{\theta}$, we obtain

$$\ddot{\theta} = \frac{2g \sin \alpha}{3a}$$

in agreement with our previous result. It is clear from the symmetry of the cylinder and the uniformity of the incline that the same result would apply for later instants when other points in the plane would successively become instantaneous centers of rotation.

Third Method: Let us again choose the instantaneous center P as the reference point but consider it to be *fixed in the cylinder* rather than in the plane. At this instant the acceleration of P is purely centripetal. It is of magnitude $a\dot{\theta}^2$ and is directed toward the center O, as may be seen by adding the absolute acceleration of O and the acceleration of P relative to O. Since the reference point is accelerating, let us refer to the general rotational equation of (4–70), namely,

$$\mathbf{M}_p - \boldsymbol{\rho}_c \times m\ddot{\mathbf{r}}_p = \dot{\mathbf{H}}_p$$

We recall that $\boldsymbol{\rho}_c$ is the position vector of the center of mass relative to P. Therefore $\boldsymbol{\rho}_c$ and $\ddot{\mathbf{r}}_p$ are parallel, and it follows that

$$\boldsymbol{\rho}_c \times m\ddot{\mathbf{r}}_p = 0$$

The remaining terms have the same values as in the previous case. Hence Eqs. (7–173), (7–174), and (7–175) are valid once again, and we calculate the same value of $\ddot{\theta}$.

We have emphasized previously that the rotational equation of motion, $\mathbf{M} = \dot{\mathbf{H}}$, applies when the reference point is taken at the center of mass or at a fixed point. It is important to remember that by "fixed" we mean that *both the velocity and acceleration are zero relative to some inertial frame.* For example, the term $\boldsymbol{\rho}_c \times m\ddot{\mathbf{r}}_p$ would not, in general, be zero if the center of mass were located at some point away from the geometrical center, even though the velocity of P were instantaneously zero.

Fourth Method: As a final approach, let us consider a reference point P which is the contact point and moves in a straight line down the plane. It is clear that the acceleration of P is of magnitude $a\ddot{\theta}$ and is directed along its path. Hence we find that

$$\boldsymbol{\rho}_c \times m\ddot{\mathbf{r}}_p = ma^2 \ddot{\theta} \mathbf{e}_\theta \tag{7-176}$$

where $\mathbf{e}_\theta$ is a unit vector directed into the page.

The velocity of the center O relative to P is zero because the two points move along parallel paths with equal velocities. Therefore, the angular momentum of the cylinder relative to P is the same as its angular momentum relative to O. Thus we find that

$$\dot{H}_p = \frac{1}{2} ma^2 \ddot{\theta} \tag{7-177}$$

where clockwise rotations are taken as positive. As before, the applied moment is

$$M_p = mga \sin \alpha$$

Substituting into the general rotational equation and solving for $\ddot{\theta}$, we obtain the scalar equation

$$\ddot{\theta} = \frac{2g \sin \alpha}{3a}$$

in agreement with our previous results.

Other choices could have been made of the reference point. Of course, a proper application of the general rotational equation would have given the same expression for $\ddot{\theta}$ in each instance. Nevertheless, a suitable choice of a reference point will often aid considerably in understanding a problem and in finding its solution. It is therefore quite important to gain a facility in this aspect of problem formulation.

Example 7-6. A rigid body of mass m and moment of inertia I about the mass center C is supported at a fixed point O, forming a compound pendulum (Fig. 7-25). Assuming that the points O and C are separated by a distance a, solve for the circular frequency ω_n of small oscillations.

First let us calculate the moment of inertia about the fixed point O. Using the parallel-axis theorem of Eq. (7-61), we find that

$$I_0 = I + ma^2 \tag{7-178}$$

It can be seen that the external moment about O is

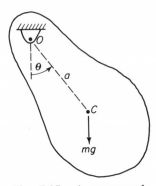

Fig. 7-25. A compound pendulum.

$$M = -mga \sin \theta \qquad (7\text{-}179)$$

since the resultant gravitational force is the only external force not passing through O.

For this case of plane motion, we can write the scalar equation

$$M = I_0 \ddot{\theta} \qquad (7\text{-}180)$$

where both M and θ are taken positive counterclockwise. Thus we obtain from Eqs. (7-178), (7-179), and (7-180) that

$$I_0 \ddot{\theta} + mga \sin \theta = 0 \qquad (7\text{-}181)$$

For the case of small motions, we can approximate $\sin \theta$ by θ, and Eq. (7-181) reduces to

$$I_0 \ddot{\theta} + mga\theta = 0 \qquad (7\text{-}182)$$

This is the equation of simple harmonic motion which was solved in Examples 3-3 and 3-6, and again in Sec. 3-7 for a more general case. The solution is of the form

$$\theta = A \sin \left(\sqrt{\frac{mga}{I_0}} \, t + \alpha \right) \qquad (7\text{-}183)$$

where A and α are constants which are evaluated from the initial conditions. Hence the natural frequency of small oscillations is

$$\omega_n = \sqrt{\frac{mga}{I_0}} \qquad (7\text{-}184)$$

If we let k_c be the radius of gyration about the center of mass, we find that

$$I_0 = m(k_c^2 + a^2)$$

and it follows that

$$\omega_n = \sqrt{\frac{ga}{k_c^2 + a^2}} \qquad (7\text{-}185)$$

Comparing this expression with that for the simple pendulum given in Eq. (3-223), we see that the length l of a simple pendulum with the same period is

$$l = \frac{k_c^2 + a^2}{a}$$

It is interesting to note that l is a minimum for a given body when

$$\frac{dl}{da} = -\frac{k_c^2}{a^2} + 1 = 0$$

or

$$a = k_c$$

So if a given body is supported at a distance k_c from the center of mass, its period will be the minimum possible for small oscillations of the body in that plane.

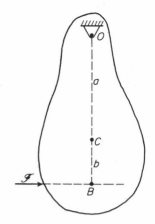

Example 7-7. A rigid body can rotate about a fixed axis through the point O which is located at a distance a from the center of mass C (Fig. 7–26). The body is struck an impulsive blow of magnitude $\mathscr{F}$ which is in the plane of the motion and is perpendicular to the line OC. Find the distance b between the center of mass and the line of action of the impulse such that the body exerts no impulsive reaction on the support at O.

The assumption of no impulsive reaction at O implies that the initial motion of the body is due entirely to the impulse $\mathscr{F}$. From

Fig. 7-26. A transverse impulse applied to a rigid body.

the equation of linear impulse and momentum, the initial velocity of the center of mass is

$$v_c = \frac{\mathscr{F}}{m} \tag{7-186}$$

The angular impulse about C due to $\mathscr{F}$ is

$$\mathscr{M} = b\mathscr{F} \tag{7-187}$$

Therefore, from the equation of angular impulse and momentum, we find that the initial angular velocity is

$$\dot{\theta} = \frac{b\mathscr{F}}{I_c} \tag{7-188}$$

But we have assumed that the support point O is fixed; hence we know that

$$v_c = a\dot{\theta} \tag{7-189}$$

From Eqs. (7–186), (7–188), and (7–189), we obtain that

$$ma = \frac{I_c}{b}$$

or

$$ab = k_c^2 \tag{7-190}$$

where k_c is the radius of gyration of the body about C. Knowing a and k_c, we can solve for the distance b.

The point B is known as the *center of percussion* for the support point O. Conversely, it can be seen from Eq. (7–190) that point O is the center of percussion for the case where the body rotates about an axis through B.

This follows from the fact that a and b can be interchanged in Eq. (7–190) without affecting the equality.

Note that an impulsive reaction must occur at the support point O for all cases in which the impulse $\mathscr{F}$ is applied in any direction other than perpendicular to the line OC. We see that the components of velocity along the line OC must be the same for both O and C. Hence if O is to remain initially motionless under the action of $\mathscr{F}$ alone, then the initial velocity of C cannot have a component along OC, thereby requiring the direction of $\mathscr{F}$ to be perpendicular to OC.

Another interesting result is that if the body is suspended as a compound pendulum, its period for small oscillations is the same whether the fixed axis passes through O or B. This statement can be verified by using the expression for the natural frequency of a compound pendulum which was obtained in Example 7–6, namely,

$$\omega_n = \sqrt{\frac{mga}{I_0}}$$

In order for the frequencies to be equal in the two cases, it is apparent that the moments of inertia about the support points must be proportional to their corresponding distances from the center of mass. Hence we require that

$$\frac{k_c^2 + a^2}{a} = \frac{k_c^2 + b^2}{b}$$

or

$$k_c^2(a - b) = ab(a - b)$$

But we have seen that $k_c^2 = ab$ and therefore the two frequencies are equal.

Example 7–8. A thin uniform bar of length l and mass m is displaced slightly from its vertical equilibrium position at $\theta = 0$. It slides along a frictionless wall and floor due to gravity—Fig. 7–27(a). Assuming that the bar remains in a plane perpendicular to both the wall and the floor, solve for the angle θ at which the bar leaves the wall.

Let us obtain first the differential equation of motion by using Lagrange's equation. In calculating the total kinetic energy of the bar, we add the translational and rotational portions, obtaining

$$T = \frac{1}{2} m v_c^2 + \frac{1}{2} I_c \dot{\theta}^2 \qquad (7\text{–}191)$$

where v_c is the velocity of the center of mass and I_c is the moment of inertia of the bar about a transverse axis through its center of mass.

It can be seen from the geometry—Fig. 7–27(b)—that the center of mass C moves in a circular arc of radius $l/2$. Therefore we obtain

$$v_c = \frac{1}{2} l \dot{\theta} \qquad (7\text{-}192)$$

Also, the moment of inertia about C is

$$I_c = \frac{1}{12} m l^2 \qquad (7\text{-}193)$$

(See Appendix.) Evaluating the total kinetic energy from Eqs. (7-191), (7-192), and (7-193), we obtain

$$T = \frac{1}{8} m l^2 \dot{\theta}^2 + \frac{1}{24} m l^2 \dot{\theta}^2$$

$$= \frac{1}{6} m l^2 \dot{\theta}^2 \qquad (7\text{-}194)$$

The potential energy V is entirely gravitational and is calculated using the floor as the reference level. Thus we see that

$$V = \frac{1}{2} m g l \cos \theta \qquad (7\text{-}195)$$

As long as the bar remains in contact with both the floor and the wall, the system has only one degree of freedom because its configuration can be expressed completely in terms of a single coordinate θ. We note that the system is conservative during this interval; hence we can write Lagrange's equation in the form

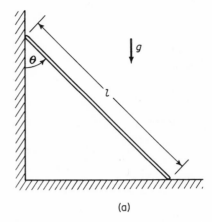

(a)

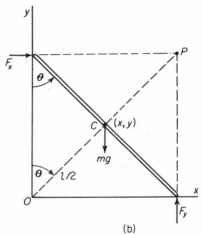

(b)

Fig. 7-27. A thin uniform bar sliding on a smooth wall and floor.

$$\frac{d}{dt}\left(\frac{\partial L}{\partial \dot{\theta}}\right) - \frac{\partial L}{\partial \theta} = 0 \qquad (7\text{-}196)$$

where we recall that $L = T - V$. Evaluating the terms, we find that

$$\frac{d}{dt}\left(\frac{\partial L}{\partial \dot{\theta}}\right) = \frac{1}{3} m l^2 \ddot{\theta}, \qquad \frac{\partial L}{\partial \theta} = \frac{1}{2} m g l \sin \theta$$

Hence the differential equation of motion is

$$\frac{1}{3} m l^2 \ddot{\theta} - \frac{1}{2} m g l \sin \theta = 0$$

or

$$\ddot{\theta} = \frac{3g}{2l} \sin \theta \qquad (7\text{-}197)$$

This equation may be integrated with respect to θ by making the substitution

$$\ddot{\theta} = \dot{\theta} \frac{d\dot{\theta}}{d\theta}$$

Performing the integration we obtain

$$\int \dot{\theta} \, d\dot{\theta} = \frac{3g}{2l} \int \sin \theta \, d\theta$$

or

$$\frac{1}{2} \dot{\theta}^2 = -\frac{3g}{2l} \cos \theta + C$$

where C is the integration constant. From the initial conditions $\dot{\theta}(0) = 0$ and $\theta(0) = 0$, we find that

$$C = \frac{3g}{2l}$$

and therefore

$$\dot{\theta}^2 = \frac{3g}{l} (1 - \cos \theta) \qquad (7\text{-}198)$$

Although we obtained this result by integration of the differential equation of motion, note that it could have been obtained directly from the principle of conservation of energy.

$$T + V = E$$

or

$$\frac{1}{6} ml^2 \dot{\theta}^2 + \frac{1}{2} mgl \cos \theta = \frac{1}{2} mgl$$

from which we again find that

$$\dot{\theta}^2 = \frac{3g}{l} (1 - \cos \theta)$$

We can obtain Eq. (7-197) by differentiation with respect to time.

Now let us consider the conditions which apply at the instant the bar leaves the wall. It is apparent that the wall cannot exert a pull on the bar and therefore F_x—Fig. 7-27(b)—must be positive or zero. On the other hand, our analysis thus far has assumed that the center of mass C follows a circular arc, and from Eq. (7-198), we see that its speed increases with θ. So, at least by the time that $\theta = \pi/2$, it is clear that F_x would have to be negative in order to balance the centrifugal force and keep C moving along its circular path. Thus we conclude that the bar leaves the wall when F_x first tends to become negative, based on the assumption of a circular path for C.

We can find F_x by solving for the horizontal acceleration $\ddot{x}$ of the center of mass and applying Newton's law of motion. Taking the horizontal components of the tangential and centrifugal accelerations, we obtain

$$\ddot{x} = \frac{1}{2}\, l\ddot{\theta}\cos\theta - \frac{1}{2}\, l\dot{\theta}^2\sin\theta$$

from which we find that

$$F_x = m\ddot{x} = \frac{ml}{2}\,(\ddot{\theta}\cos\theta - \dot{\theta}^2\sin\theta) \tag{7–199}$$

This result can be expressed in terms of a single variable, θ, by substituting for $\ddot{\theta}$ and $\dot{\theta}^2$ from Eqs. (7–197) and (7–198), respectively. Hence we see that

$$F_x = \frac{3}{4}\, mg\,[\sin\theta\cos\theta - 2\sin\theta\,(1 - \cos\theta)]$$
$$= \frac{3}{4}\, mg\sin\theta\,(3\cos\theta - 2) \tag{7–200}$$

The force F_x is initially zero at $\theta = 0$, but does not tend to become negative until θ increases to the value

$$\theta = \cos^{-1}\!\left(\frac{2}{3}\right) = 48.2°$$

At this point, the bar leaves the wall. The center of mass henceforth translates uniformly to the right.

Another approach in finding $\ddot{\theta}$ as a function of θ is to use the general rotational equation

$$\mathbf{M}_p - \boldsymbol{\rho}_c \times m\ddot{\mathbf{r}}_p = \dot{\mathbf{H}}_p$$

where we choose the reference point at the instantaneous center P, assuming that P is in a lamina fixed with respect to the bar. This choice eliminates the need of solving for F_x or F_y since the moment of each force about P is zero. We see that the total angular momentum relative to P is found by considering the bar to rotate as a rigid body about P which instantaneously has zero velocity. Thus we obtain

$$\mathbf{H}_p = \left(\frac{ml^2}{12} + \frac{ml^2}{4}\right)\dot{\theta}\mathbf{k} = \frac{1}{3}\, ml^2\dot{\theta}\mathbf{k} \tag{7–201}$$

where we note that $\frac{1}{3}ml^2$ is the moment of inertia of the bar about the instantaneous axis of rotation through P, and the vector $\mathbf{H}_p$ is positive when directed out of the page. Differentiating with respect to time, we see that

$$\dot{\mathbf{H}}_p = \frac{1}{3}\, ml^2\ddot{\theta}\mathbf{k} \tag{7–202}$$

Next we note that $\boldsymbol{\rho}_c$ is directed from P to C. We have considered P to move in a lamina which rotates with the bar. From Example 2–3 we see that it

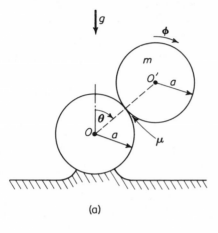

(a)

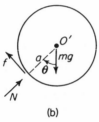

(b)

Fig. 7-28. A sphere of mass m and radius a rolling on a similar fixed sphere.

is instantaneously reversing its velocity at a cusp of its hypocycloidal path; therefore $\ddot{\mathbf{r}}_p$ is also directed from P to C. Thus we find that

$$\boldsymbol{\rho}_c \times m\ddot{\mathbf{r}}_p = 0 \qquad (7\text{-}203)$$

Also,

$$\mathbf{M}_p = \frac{1}{2}\,mgl \sin \theta \, \mathbf{k} \qquad (7\text{-}204)$$

Substituting these expressions into the general rotational equation and equating components, we find that

$$\frac{1}{2}\,mgl \sin \theta = \frac{1}{3}\,ml^2 \ddot{\theta}$$

or

$$\ddot{\theta} = \frac{3g}{2l} \sin \theta$$

in agreement with our previous result.

Example 7-9. A solid uniform sphere of mass m and radius a is placed on top of a fixed sphere of the same radius—Fig. 7-28(a). It is displaced slightly from the equilibrium position at $\theta = 0$ and begins to roll. Assuming a coefficient of sliding friction μ, find the value of θ at which sliding begins.

Let us first obtain the differential equation of motion, assuming that no slipping occurs and, furthermore, that the spheres remain in contact. Once again we shall use the Lagrange formulation. Choosing the reference level for the gravitational potential energy at the center of the fixed sphere, we find that

$$V = 2mga \cos \theta \qquad (7\text{-}205)$$

The kinetic energy is the sum of the translational and rotational portions, namely,

$$T = \frac{1}{2}\,m(2a\dot{\theta})^2 + \frac{1}{2}\,I\dot{\phi}^2 \qquad (7\text{-}206)$$

where the moment of inertia about the center of mass is

$$I = \frac{2}{5}\,ma^2 \qquad (7\text{-}207)$$

Although we have used both θ and ϕ as coordinates, they are not independent as long as no slipping occurs. An equation of holonomic constraint can be found by noting that the velocity of the center O' can be written in terms of either $\dot{\theta}$ alone or $\dot{\phi}$ alone. The point O' moves along a circle of radius $2a$ at an angular rate $\dot{\theta}$. But also we see that the sphere is rotating with an absolute angular velocity $\dot{\phi}$ about an instantaneous center which is the point of contact. So we obtain

$$2a\dot{\theta} = a\dot{\phi}$$

or

$$\dot{\phi} = 2\dot{\theta} \tag{7-208}$$

Substituting from Eqs. (7–207) and (7–208) into Eq. (7–206), we find that the kinetic energy is

$$T = \frac{14}{5} ma^2 \dot{\theta}^2 \tag{7-209}$$

Lagrange's equation for this system is

$$\frac{d}{dt}\left(\frac{\partial L}{\partial \dot{\theta}}\right) - \frac{\partial L}{\partial \theta} = 0$$

and, evaluating the Lagrangian function $L = T - V$, we obtain

$$\frac{d}{dt}\left(\frac{\partial L}{\partial \dot{\theta}}\right) = \frac{28}{5} ma^2 \ddot{\theta}, \qquad \frac{\partial L}{\partial \theta} = 2mga \sin \theta$$

Thus the equation of motion is

$$\frac{28}{5} ma^2 \ddot{\theta} - 2mga \sin \theta = 0$$

or

$$\ddot{\theta} = \frac{5g}{14a} \sin \theta \tag{7-210}$$

Once again we use the substitution $\ddot{\theta} = \dot{\theta}\, d\dot{\theta}/d\theta$, and integrating both sides, we obtain

$$\dot{\theta}^2 = \frac{5g}{7a}(1 - \cos \theta) \tag{7-211}$$

where we have evaluated the constant of integration from the condition that $\dot{\theta} = 0$ when $\theta = 0$.

In order to state the necessary conditions for slipping to begin, let us consider the free-body diagram of Fig. 7–28(b) showing the forces acting on the sphere. We can calculate the normal force N and the frictional force f which result under the assumption of no slipping. Using Newton's law of motion relating the radial components of force and acceleration, we obtain

$$mg \cos \theta - N = 2ma\dot{\theta}^2$$

and substituting for $\dot{\theta}^2$ from Eq. (7–211), this reduces to

$$N = mg\left[\cos\theta - \frac{10}{7}(1 - \cos\theta)\right]$$

$$= \frac{mg}{7}(17\cos\theta - 10) \tag{7-212}$$

The force f can be calculated from the equation

$$M = I\ddot{\phi}$$

where the reference point is at the center of mass of the moving sphere. Thus,

$$af = \frac{2}{5}ma^2\ddot{\phi}$$

or

$$f = \frac{4}{5}ma\ddot{\theta} \tag{7-213}$$

where we note from Eq. (7–208) that $\ddot{\phi} = 2\ddot{\theta}$. The sphere will roll without slipping if $f < \mu N$, that is, the condition of incipient slipping is

$$f = \mu N \tag{7-214}$$

So at the position where slipping begins, we see from Eqs. (7–212), (7–213), and (7–214) that

$$\ddot{\theta} = \frac{5\mu g}{28a}(17\cos\theta - 10) \tag{7-215}$$

Equating the expressions for $\ddot{\theta}$ from Eqs. (7–210) and (7–215), we obtain

$$\frac{5g}{14a}\sin\theta = \frac{5\mu g}{28a}(17\cos\theta - 10)$$

or

$$\sin\theta = \frac{1}{2}\mu(17\cos\theta - 10) \tag{7-216}$$

This transcendental equation can be solved for the value of θ corresponding to a given coefficient of friction μ. For example, if $\mu \ll 1$, we find that slipping begins at $\theta = \sin^{-1}(7\mu/2)$. For $\mu = 0.5$, the value of θ turns out to be $\theta = 41.8°$. In the limiting case of a perfectly rough surface, we set $\mu = \infty$ and find that $\theta = \cos^{-1}(10/17) = 54.0°$.

Example 7–10. A uniform circular cylinder of mass m and radius a is given an initial angular velocity ω_0 and no initial translational velocity. It is placed in contact with a plane inclined at an angle α to the horizontal—Fig. 7–29(a). If there is a coefficient of friction μ for sliding between the cylinder and the plane, find (a) the distance the cylinder moves before sliding stops; (b) the maximum distance it travels up the plane. Assume that $\mu > \tan\alpha$.

A free-body diagram of the forces acting on the cylinder is shown in Fig.

7–29(b), assuming that sliding occurs. Choosing the center of mass O as the reference point, we can use the equation of angular impulse and momentum to calculate the time t required to decrease the angular velocity to a value ω. The angular impulse is

$$\mathcal{M} = -\mu Nat \qquad (7\text{–}217)$$

where a clockwise rotation is considered to be positive. Equating $\mathcal{M}$ to the change in the angular momentum about O, we obtain

$$I(\omega - \omega_0) = -\mu Nat \qquad (7\text{–}218)$$

We recall from Example 7–5 that the moment of inertia about the center is

$$I = \frac{1}{2} ma^2$$

Also, since the component of acceleration normal to the plane is zero, the forces in this direction must sum to zero, resulting in

$$N = mg \cos \alpha$$

Hence we find that the time required to reach an angular velocity ω is

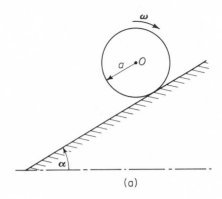

(a)

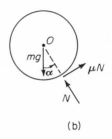

(b)

Fig. 7-29. The forces acting on a uniform circular cylinder while sliding and rotating relative to an inclined plane.

$$t = \frac{a(\omega_0 - \omega)}{2\mu g \cos \alpha} \qquad (7\text{–}219)$$

We can use the equation of linear impulse and momentum to calculate the velocity of the center O in a direction parallel to the plane. Since the translational velocity is zero at $t = 0$, we obtain

$$mg(\mu \cos \alpha - \sin \alpha)t = mv \qquad (7\text{–}220)$$

where the positive direction is up the plane. At the moment sliding stops and pure rolling begins, we see that $v = a\omega$. Therefore, from Eqs. (7–219) and (7–220), we obtain

$$\frac{ag(\mu \cos \alpha - \sin \alpha)(\omega_0 - \omega)}{2\mu g \cos \alpha} = a\omega$$

or

$$\omega = \frac{(\mu \cos \alpha - \sin \alpha)\omega_0}{3\mu \cos \alpha - \sin \alpha} \tag{7-221}$$

From Eqs. (7-219) and (7-221), the time at which sliding stops is found to be

$$t = \frac{a\omega_0}{g(3\mu \cos \alpha - \sin \alpha)} \tag{7-222}$$

The forces acting on the cylinder remain constant during the sliding and therefore the acceleration is constant. Hence the distance traveled while sliding is

$$d_1 = \frac{1}{2} vt = \frac{a^2 \omega_0^2 (\mu \cos \alpha - \sin \alpha)}{2g(3\mu \cos \alpha - \sin \alpha)^2} \tag{7-223}$$

To find the distance traveled by the cylinder from the point where sliding stops until its velocity is zero, we use the principle of conservation of energy. Calling this distance d_2, we equate the loss of kinetic energy during the interval to the gain in potential energy. Thus we can write

$$\frac{1}{2} mv^2 + \frac{1}{2} I\omega^2 = mgd_2 \sin \alpha$$

Substituting for I, v, and ω from the foregoing equations and solving for d_2, we obtain

$$d_2 = \frac{3a^2 \omega_0^2 (\mu \cos \alpha - \sin \alpha)^2}{4g \sin \alpha (3\mu \cos \alpha - \sin \alpha)^2} \tag{7-224}$$

The maximum distance traveled up the plane is

$$d = d_1 + d_2 = \frac{a^2 \omega_0^2 (\mu \cos \alpha - \sin \alpha)}{4g \sin \alpha (3\mu \cos \alpha - \sin \alpha)} \tag{7-225}$$

We have assumed that $\mu > \tan \alpha$. This corresponds to the case where ω is positive and the cylinder is moving up the plane at the time sliding stops. If $\frac{1}{3} \tan \alpha < \mu < \tan \alpha$, Eq. (7-223) remains valid, but since the cylinder moves down the plane throughout its motion, ω is negative at the time sliding stops. On the other hand, if $\mu < \frac{1}{3} \tan \alpha$, sliding never ceases because the angular acceleration multiplied by the radius a is less than the linear acceleration down the plane. If we extrapolate backward in time from the initial condition of the system, we find that the negative time obtained from Eq. (7-222) for this case corresponds to the time when sliding began and the value of d_1 obtained from Eq. (7-223) indicates its position. For all cases where $\mu \leq \tan \alpha$, the initial position of the cylinder is its highest location on the plane.

This example illustrates some important ideas concerning the principle of work and kinetic energy as it applies to rigid bodies. In particular, the use of energy methods in solving this problem requires a clear understanding of the work of the frictional forces. We recall from Sec. 3–2 that the work done by a force **F** acting on a particle as it moves from point A to point B is

$$W = \int_A^B \mathbf{F} \cdot d\mathbf{r}$$

For the case of forces whose point of application moves relative to an extended body, however, the question arises as to whether $d\mathbf{r}$ refers to an infinitesimal displacement of the force, a point on the body, or perhaps neither. In examining this question, it is helpful to write the work integral in the form

$$W = \int_{t_A}^{t_B} \mathbf{F} \cdot \mathbf{v} \, dt \qquad (7\text{--}226)$$

To calculate the work done *on* a given system by the force $\mathbf{F}$, we interpret $\mathbf{v}$ as the velocity (relative to an inertial frame) of the point in the body to which the force is instantaneously being applied. Note that it is *not* the velocity of the point of application of the force, that is, what might be loosely called the velocity of the point of the force vector.

Similarly, to calculate the work done *by* a system exerting a force, we see that $\mathbf{v}$ is the absolute velocity of the point which is instantaneously exerting the force. We note, for example, that if an automobile accelerates along a straight and level road, the force exerted by the road on the tires has a component in the direction of the acceleration. The particles in the road have no motion, however, and therefore the road does no work. The increasing kinetic energy of the automobile arises from the work done by an internal source, namely, the engine.

It is apparent that if two bodies slide relative to each other, then the work done *by* one body may not be equal to the work done *on* the other. This occurs because the velocities of the particles at the contact surface may be quite different for the two bodies. For the case of this example, we see that work is done by the slipping cylinder against friction; but no work is done on the inclined plane since it does not move. In general, if body A slides relative to body B, then the sum of the work done *by* A and the work done *by* B is equal to the energy W_F lost in friction, assuming that no other moving bodies or working forces are involved.

$$W_A + W_B = W_F$$

For example, if A does work on B by means of frictional forces, the work done by A may exceed the mechanical energy $(-W_B)$ received by B. The difference is the energy lost in friction. Although W_A and W_B depend upon which inertial frame is chosen as a reference for these computations, the friction loss W_F is a function of the *relative* motion of the two bodies and hence is the same for any inertial reference frame.

Now let us use these methods in a further analysis of this example in which a cylinder is rolling and sliding on an inclined plane. First let us equate the initial total energy to the total energy at the moment when sliding stops plus the energy lost due to friction. Taking the reference for potential energy at the initial position of the center O, we find that the initial value of the total energy is just the initial kinetic energy, namely,

$$E_i = \frac{1}{4} ma^2 \omega_0^2 \qquad (7\text{-}227)$$

The final value of the total energy is

$$E_f = \frac{3}{4} ma^2 \omega^2 + mgd_1 \sin \alpha \qquad (7\text{-}228)$$

where we note that the translational kinetic energy is twice the rotational kinetic energy for this case of pure rolling of a uniform circular cylinder. The energy lost in friction is equal to the frictional force multiplied by the *relative* displacement of the sliding surfaces.

$$W_f = \mu mg \cos \alpha \, (a\theta - d_1) \qquad (7\text{-}229)$$

where θ is the absolute rotational displacement of the cylinder during the period of sliding. Thus we can write

$$E_i = E_f + W_F$$

or

$$\frac{1}{4} ma^2 (\omega_0^2 - 3\omega^2) = \mu mga\theta \cos \alpha - mgd_1 (\mu \cos \alpha - \sin \alpha) \quad (7\text{-}230)$$

So far, we have lumped together the rotational and translational aspects of the motion in equating energies. A separation can be made, however, by noting that the friction force μN acting on the cylinder at the point of contact—Fig. 7–29(b)—is equipollent to a parallel force acting at the center O plus a couple of magnitude

$$M = \mu Na = \mu mga \cos \alpha$$

applied to the cylinder in a counterclockwise sense. It can be seen that the couple does no work in a translational displacement, whereas the force applied at O does no work in a rotation about O.

Now let us calculate the work of a couple M moving through a displacement θ. For the more general case where the moment $\mathbf{M}$ and the angular velocity $\boldsymbol{\omega}$ of the body have arbitrary directions, we can write the rotational counterpart of Eq. (7–226), namely,

$$W = \int_{t_A}^{t_B} \mathbf{M} \cdot \boldsymbol{\omega} \, dt \qquad (7\text{-}231)$$

For the problem at hand, however, the work done by the cylinder against friction due to a rotational displacement θ is

$$W_\theta = M\theta = \mu mga\theta \cos \alpha$$

But the work W_θ is equal to the loss of rotational kinetic energy during the same interval, or

$$\frac{1}{4}ma^2(\omega_0^2 - \omega^2) = \mu mga\theta \cos \alpha \qquad (7\text{--}232)$$

This is the equation of work and kinetic energy for the rotational motion. Subtracting Eq. (7–232) from Eq. (7–230), we obtain

$$\frac{1}{2}ma^2\omega^2 = mgd_1\,(\mu \cos \alpha - \sin \alpha) \qquad (7\text{--}233)$$

This is the equation of translational work and kinetic energy. It equates the increase in translational kinetic energy to the work done by the force μN and the gravitational force, where both forces are applied at the center O. Of course, this equation could have been written directly.

At this point, we have three equations of work and kinetic energy, but only two are independent. There are three unknowns, however, namely, d_1, ω, and θ. As a third independent equation, we can use Eq. (7–221) which we obtained earlier using impulse and momentum methods. From Eqs. (7–221) and (7–233), we find that the sliding distance is

$$d_1 = \frac{a^2\omega_0^2(\mu \cos \alpha - \sin \alpha)}{2g(3\mu \cos \alpha - \sin \alpha)^2}$$

in agreement with our earlier results. Also, from Eqs. (7–221) and (7–232), we obtain that the total rotation of the cylinder during sliding is

$$\theta = \frac{a\omega_0^2(2\mu \cos \alpha - \sin \alpha)}{g(3\mu \cos \alpha - \sin \alpha)^2} \qquad (7\text{--}234)$$

The total energy dissipated in friction is found by substituting these expressions for d_1 and θ into Eq. (7–229). The result is

$$W_F = \frac{\mu ma^2\omega_0^2 \cos \alpha}{2(3\mu \cos \alpha - \sin \alpha)} \qquad (7\text{--}235)$$

Note that, if $\alpha = 0$, the energy loss in friction is independent of the coefficient of friction μ for $\mu > 0$.

REFERENCES

1. Becker, R. A., *Introduction to Theoretical Mechanics*. New York: McGraw-Hill Book Company, 1954.

2. Frazer, R. A., W. J. Duncan, and A. R. Collar, *Elementary Matrices*. New York: Cambridge University Press, 1960.

3. Goldstein, H., *Classical Mechanics*. Reading, Mass.: Addison-Wesley Publishing Company, 1950.

4. Halfman, R. L., *Dynamics—Particles, Rigid Bodies and Systems*. Reading, Mass.: Addison-Wesley Publishing Company, 1962.

5. Hildebrand, F. B., *Methods of Applied Mathematics*. Englewood Cliffs, N.J.: Prentice-Hall, 1952.

6. Synge, J. L., and B. A. Griffith, *Principles of Mechanics*, 3rd ed. New York: McGraw-Hill Book Company, 1959.

PROBLEMS

7-1. Show that a thin plate of mass m and arbitrary shape is dynamically equivalent to a set of three rigidly-connected particles, each of mass $m/3$, which are placed at the following (x,y) positions in a body-axis system at the mass center: $(0, \sqrt{2I_{xx}/m})$, $(\sqrt{3I_{yy}/2m}, -\sqrt{I_{xx}/2m})$, and $(-\sqrt{3I_{yy}/2m}, -\sqrt{I_{xx}/2m})$.

7-2. Show that a uniform thin triangular plate of mass m is dynamically equivalent to a set of three rigidly-connected particles, each of mass $m/3$, which are placed at the midpoints of the three edges.

7-3. What is the minimum number of equal particles, connected by rigid massless rods, which are necessary to duplicate the inertial properties of a given rigid body? Using a principal axis system at the mass center, solve for an equivalent set of particle positions as a function of the total mass m and the moments of inertia I_{xx}, I_{yy}, I_{zz}.

7-4. Three particles, each of mass m, are located at a unit distance from the origin along the x, y, and z axes, respectively. Determine the inertia matrix in the (x, y, z) system. Find the orientation of three principal axes passing through the center of mass and evaluate the corresponding inertia matrix.

7-5. A certain rigid body has an inertia matrix

$$[I] = \begin{bmatrix} 2 & 0 & 0 \\ 0 & 4 & 0 \\ 0 & 0 & 4 \end{bmatrix} \text{lb sec}^2 \text{ ft}$$

Then a rotation of coordinate axes is performed such that $\Phi = \pi/2$ and the rotation occurs in a positive sense about an axis through the origin in the direction $(\mathbf{i} + \mathbf{j})$. Find the rotation matrix $[l]$ and the resulting inertia matrix.

7-6. A rigid body consists of a particle of unit mass at each of the four corners of a regular tetrahedron of unit edge length. Find the principal moments of inertia about the center of mass.

7-7. At a given instant the angular velocity of a rigid body is $\boldsymbol{\omega} = \omega_x \mathbf{i} + \omega_y \mathbf{j}$ and its angular acceleration is $\dot{\boldsymbol{\omega}} = \alpha_x \mathbf{i}$, where $(\mathbf{i}, \mathbf{j}, \mathbf{k})$ are a set of cartesian unit vectors fixed in the body and ω_x, ω_y, and α_x are nonzero. Assuming that the origin of this body-axis system is fixed, find the location (x, y, z) of a point in the body whose absolute acceleration is $\mathbf{a} = a_x \mathbf{i} + a_y \mathbf{j} + a_z \mathbf{k}$.

7-8. Show that, at a given instant, it is impossible for only one point of a general rigid body to have zero velocity; but there is only one point, in general, which has zero acceleration.

7-9. A certain rigid body has the following inertia matrix:

$$[I] = \begin{bmatrix} 450 & -60 & 100 \\ -60 & 500 & 7 \\ 100 & 7 & 550 \end{bmatrix} \text{lb sec}^2 \text{ ft}$$

Solve for the principal moments of inertia at the same reference point. Find the rotation matrix such that a minimum rotation angle Φ is required. Evaluate Φ.

7–10. Find the moment of inertia of a thin uniform spherical shell of mass m, radius r, about a diameter.

7–11. Find the principal moments of inertia at the center of a thin ellipsoidal shell of mass m. The shell is bounded by two ellipsoidal surfaces of the same shape which have a common center. The semimajor axes of the shell are a, b, and c, respectively.

7–12. Consider a rigid body of mass $m = 4$ lb sec^2/ft whose inertia matrix at the center of mass is

$$[I'] = \begin{bmatrix} 150 & 0 & -100 \\ 0 & 250 & 0 \\ -100 & 0 & 300 \end{bmatrix} \text{lb sec}^2 \text{ ft}$$

as in Example 7–4. Solve for a translation of axes to an unprimed system such that a principal axis system results. Discuss the possibility of obtaining principal axes by translation in the general case.

7–13. A solid homogeneous sphere of mass m_0 and radius a rolls on a triangular prism of mass m which can slide on a frictionless floor. Assuming that the system is initially motionless, solve for the velocity of the prism as a function of time.

7–14. A horizontal turntable with a moment of inertia I is free to rotate without friction about a fixed vertical axis through its center O. With the turntable initially at rest, a dog of mass m is placed at A, and seeing food at B, it walks (relative to the turntable) along the chord to B. Find the angle θ through which the turntable rotates, assuming that the moment of inertia of the dog about its own mass center is negligible.

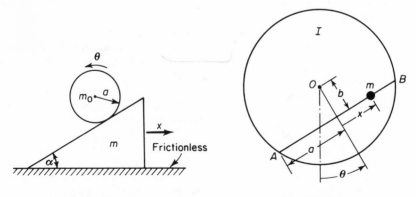

Fig. P7-13   Fig. P7-14

7–15. A solid uniform cube of mass m and edge length a slides along a horizontal plane in a direction normal to its leading edge. What is the minimum value of the coefficient of sliding friction μ which is necessary to cause the cube to tip over, assuming that its initial velocity v_0 is sufficiently large? What changes, if any, occur in the results if the cube slides directly up or down a plane inclined at an angle α to the horizontal?

7–16. A solid cylinder of mass m and radius r rests on top of a thin sheet of paper which, in turn, lies on a horizontal plane. Suddenly the paper is pulled to the right, out from under the cylinder. The coefficient of friction between all surfaces is μ. Assuming that an interval Δt is taken to remove the paper, and sliding of the paper relative to the cylinder occurs throughout this interval, find: (a) the linear and angular velocities of the cylinder just after the paper is removed; (b) the displacement and velocity of the cylinder for very large values of time.

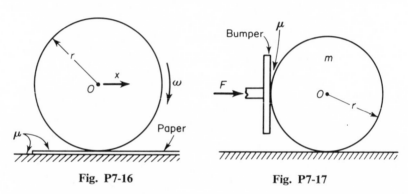

Fig. P7-16   Fig. P7-17

7–17. A truck pushes a solid circular cylinder of mass m and radius r along a level road. It pushes with a force having a horizontal component F. Assuming that the bumper is flat and vertical, calculate the translational acceleration of the cylinder if there is a coefficient of friction μ between the bumper and the cylinder. There is no slipping of the cylinder relative to the road.

7–18. A uniform sphere of mass m and radius r moves with planar motion along

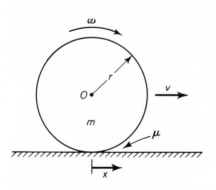

Fig. P7-18

a horizontal surface. Assuming a coefficient of sliding friction μ and initial conditions $x(0) = 0$, $v(0) = v_0$, $\omega(0) = \omega_0$, where $v_0 > r\omega_0$, find: (a) the velocity of the sphere for large t; (b) the displacement x at which sliding stops.

7–19. A thin uniform rod of mass m and length l can rotate freely about a fixed pivot at one end. Initially it hangs motionless. Then a particle of mass m_0 which is traveling horizontally with velocity v_0 hits the rod and sticks to it. Find the distance x from the pivot at which the particle must strike the rod in order that no impulse occurs at the pivot at the instant of impact. What is the angular rotation rate of the rod immediately after impact for this case?

7-20. A thin uniform rod of length l and mass m is pivoted at one end so that it can move in the vertical plane. A particle of mass m can slide without friction on the rod. Write the differential equations of motion using r and θ as coordinates. If the system is released with the initial values $\theta(0) = 0$, $\dot{\theta}(0) = 0$, $r(0) = \frac{1}{2}l$, and $\dot{r}(0) = 0$, find the initial value of the force of the pivot O acting on the rod.

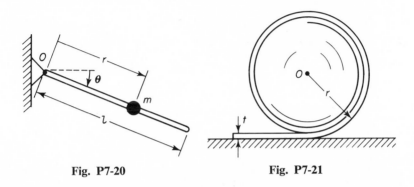

Fig. P7-20   Fig. P7-21

7-21. A flexible rug of uniform thickness t and total mass m is initially rolled up to form a cylinder of radius $r(0) = R$. It is released with zero velocity and proceeds to unroll without slipping on a horizontal floor. Assuming that $t \ll R$, find the linear velocity of the center of the rolled portion of the rug at the moment when its radius r has been reduced to $\frac{1}{2}R$. Where does the energy go at the end of the unrolling process?

7-22. A solid homogeneous sphere of mass m and radius r is dropped such that it strikes a horizontal floor with a vertical velocity v_0. Initially it is spinning with an angular velocity $\omega_0 = v_0/r$ about a horizontal axis. Assuming that a coefficient of friction $\mu = 0.1$ exists between the sphere and the floor, and the coefficient of restitution is $e = 0.5$, solve for the final translational velocity of the sphere. Which stops first, the bouncing or the slipping?

7-23. A uniform disk of mass m and radius a slides on a smooth horizontal circular area of radius $6a$. It is surrounded by a wall at its perimeter such that a coefficient of friction $\mu = 0.01$ and a coefficient of restitution $e = 0.9$ exist for contact between the disk and the wall. Initially the disk is not rotating but is translating with velocity v_0 along a path lying at a perpendicular distance $3a$ from the center O. What is its final motion?

7-24. Two thin uniform rods, each of mass m and length l, are connected in tandem to form a double pendulum. Assuming that the system initially hangs motionless, find the distance x from the pivot O at which a transverse impulse $\mathscr{F}$ should be applied such that the two rods maintain their relative alignment immediately after the impulse is applied.

7-25. Solve for the angular acceleration $\ddot{\theta}$ as a function of θ for the bar of Example 7–8 *after* it leaves the wall. What is the force F_y exerted by the floor at the instant immediately before the bar first becomes horizontal?

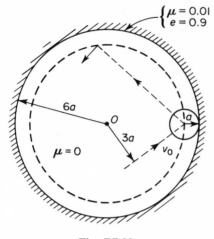

Fig. P7-23   Fig. P7-24

7–26. Write the differential equations for the motion in a vertical plane of a thin rod of mass m and length l which can rotate and slide without friction relative to the edge of a fixed table. Use θ and x as coordinates, where θ is the angle between the rod and the horizontal table surface, and x is the distance between the center of mass of the rod and the contact point at the edge of the table. The center of mass of the rod is assumed to lie beyond the edge.

7–27. A thin uniform rod of mass m and length l is balancing vertically on a horizontal floor. Suddenly the lower end is moved with a constant velocity v_0 along a straight line on the floor. What is the angular velocity of the rod as it strikes the floor?

7–28. A uniform thin rod of mass m and length l is balancing vertically on a smooth horizontal floor. It is disturbed slightly and falls. What is its angular rotation rate when its upper end hits the floor?

7–29. A disk with a moment of inertia $I = \frac{1}{2}mr_0^2$ is free to rotate without friction about a vertical axis through its mass center at O. A particle of mass m can slide without friction in a groove cut in the disk in the form of a spiral $r = r_0 e^{-\theta/2}$, where r is the radial distance from O, and θ is the position angle measured relative to the disk. Initially the disk is motionless and the particle is moving such that $r(0) = r_0$ and $\dot{\theta}(0) = \dot{\theta}_0$. Find $r_{\min}$ for the particle in the ensuing motion.

7–30. Two spinning wheels of radii r_1 and r_2, and corresponding moments of inertia I_1 and I_2, rotate on frictionless parallel shafts. The wheels are pressed together with a normal force N, the shafts being held fixed in the process. Assuming a coefficient of friction μ for sliding between the wheels, find: (a) the final rotation rates of the wheels, and (b) the time required for the sliding to stop.

7–31. Consider again the system of problem 7–30 in which the wheels are pressed together by a force acting between the shafts such that a normal component N occurs at the contact point. In this case, however, no external forces act on the system. Assuming that $m_2/m_1 = r_1/r_2 = 2$, and $I_1 = I_2 = \frac{1}{2}m_1r_1^2$, solve for the

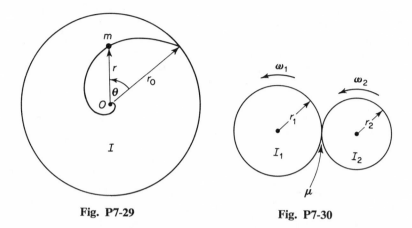

Fig. P7-29   Fig. P7-30

final rotation rates of the wheels and the time required for the slipping to stop.

7-32. A uniform semicircular cylinder of mass m and radius r rests on a horizontal plane. Write the differential equations for small motions about the position of equilibrium for the following cases: (a) the cylinder rolls without slipping, (b) the plane is perfectly smooth. Let θ be the angle of rotation of the cylinder from its equilibrium position.

7-33. A thin uniform rod of mass m and length l is initially balancing vertically on a smooth horizontal surface. A particle of mass m is attached at its upper end. If the system is disturbed slightly and falls, find an expression for its angular velocity $\dot{\theta}$ in terms of θ, where θ is the angle between the rod and the vertical.

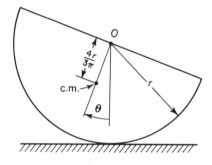

Fig. P7-32

7-34. A thin uniform rod of mass m and length l is balanced vertically on a horizontal surface. An infinitesimal disturbance causes it to fall. Find the coefficient of friction μ if the rod makes an angle $\theta = 30°$ with the vertical at the moment when slipping begins.

7-35. For the system of problem 7-34, find the range in θ over which it is impossible for slipping to begin, regardless of the value of the coefficient of friction.

7-36. A thin uniform rod of mass m and length l moves in the horizontal xy plane. One end is constrained to move without friction along the x axis. Designating the position of the constrained end by x, and the angle between the rod and the positive x axis by θ, find the differential equations of motion for the system. If $\theta(0) = 0$ and $\dot{\theta}(0) = \dot{\theta}_0$, solve for $\dot{\theta}_{max}$.

7-37. For the system of problem 7-36, solve for the angular velocity $\dot{\theta}$ of the rod in terms of θ.

7–38. A despinning device for an axially symmetric satellite consists of two flexible thin ropes, each having a particle of mass m at its free end. These ropes are initially wound around the circular cross-section of the satellite, but are free to unwind in a symmetrical fashion such that each has a straight length l which increases as the unwinding proceeds. There is a moment of inertia I about the axis of symmetry and the initial conditions are $\Omega(0) = \Omega_0$, $l(0) = 0$. Find the rope length l at the moment when the angular velocity Ω becomes zero.

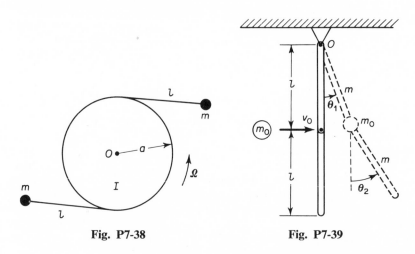

Fig. P7-38   Fig. P7-39

7–39. Two thin uniform bars, each of mass m and length l, are connected by a pin joint to form a double pendulum. Initially the bars are hanging vertically and are motionless. Then a particle of mass m_0, traveling horizontally with velocity v_0, strikes the joint between the bars inelastically. Solve for the angular velocity of each bar immediately after impact.

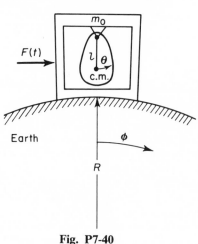

Fig. P7-40

7–40. A horizontal force $F(t)$ is applied to a small box of mass m_0 which slides without friction along a great circle of the earth. A physical pendulum of mass m and radius of gyration k (about the center of mass) is suspended inside the box from a pivot at a distance l from the mass center of the pendulum. Using the angles θ and ϕ to indicate the configuration of the system, where θ is the angle of the pendulum from the local vertical and ϕ gives the position of the box on the earth's surface, write the differential equations of motion, assuming that $\theta \ll 1$ and that l and k are much smaller than R. Solve for the length l which will result in the motion of θ being independent of $F(t)$. For this value of l, solve for the period of the motion in θ. Assume that $R\dot{\phi}^2$ remains much smaller than the acceleration of gravity.

Note: A pendulum having this period is known as a Schuler pendulum.

7–41. Consider the impulse response of a rigid body of mass m having planar motion parallel to a principal plane. Suppose that a linear impulse is applied at point P in the principal plane such that a sudden change $\Delta\mathbf{v}_p$ occurs in the velocity of P. If the vectors $\boldsymbol{\rho}_c$ and $\boldsymbol{\rho}_i$ indicate the positions relative to P of the mass center and a general point P_i, respectively, show that the velocity change of P_i is

$$\Delta\mathbf{v}_i = \Delta\mathbf{v}_p + \frac{m}{I_p}[(\boldsymbol{\rho}_i \cdot \Delta\mathbf{v}_p)\boldsymbol{\rho}_c - (\boldsymbol{\rho}_i \cdot \boldsymbol{\rho}_c)\Delta\mathbf{v}_p]$$

where I_p is the moment of inertia about a transverse axis through P.

7–42. Four thin uniform bars, each of length l and mass m, are joined together by pin joints at their ends to form a rhombus $ABCD$. Initially the system has a square configuration in the vertical plane with point C directly below A. It is dropped through a height h and C strikes the floor inelastically. What is the velocity of point A immediately after the impact?

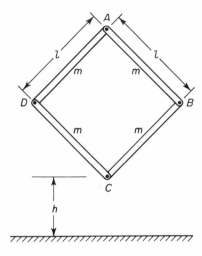

Fig. P7-42

8

DYNAMICS OF A RIGID BODY

Now we turn to a more general consideration of rigid-body dynamics with particular emphasis on rotational motion in three dimensions. A number of methods will be presented. Some are useful in illustrating the physical principles underlying the motion; others are best suited for obtaining solutions to particular types of problems. Sometimes we shall use several approaches to solve an example in order to compare the various methods.

8-1. GENERAL EQUATIONS OF MOTION

In the discussion of the dynamics of a system of particles given in Chapter 4, we obtained the translational equation:

$$\mathbf{F} = m\ddot{\mathbf{r}}_c \tag{8-1}$$

where m is the total mass, $\ddot{\mathbf{r}}_c$ is the acceleration of the center of mass, and $\mathbf{F}$ is the total external force acting on the system. As was shown in Chapter 7, we can consider a rigid body as a particular case of a system of particles; hence this equation of motion applies equally well to rigid bodies.

Also in Chapter 7 we found that the equation of rotational motion, namely,

$$\mathbf{M} = \dot{\mathbf{H}} \tag{8-2}$$

applies in this simple form to a rigid body, provided that the reference point for calculating the external moment $\mathbf{M}$ and the angular momentum $\mathbf{H}$ is taken at the center of mass or at a fixed point. For the case of general motion of a rigid body, one usually chooses the center of mass as the reference point. If the force $\mathbf{F}$ is independent of the rotational motion, and if the moment $\mathbf{M}$ is independent of the motion of the center of mass, then it can be seen that Eqs. (8-1) and (8-2) can be solved separately. A separation of the translational and rotational equations of motion is usually accomplished whenever it is possible. On the other hand, a rigid body with one point fixed has only three degrees of freedom, at most. Since these degrees of freedom correspond to rotational motions, only the rotational equations need be considered for this case.

General Rotational Equations. Now let us use the general expression given in Eq. (8–2) to describe the rotational motion of a rigid body. Suppose that the reference point is taken at the origin O which coincides with the center of mass or a fixed point. Consider first the general case in which the body rotates with an absolute angular velocity $\boldsymbol{\omega}'$ and the xyz system rotates at an absolute rate $\boldsymbol{\omega}$ (Fig. 8–1). Using Eq. (2–90), we can write an expression for the absolute rate of change of **H**. We obtain

$$\dot{\mathbf{H}} = (\dot{\mathbf{H}})_r + \boldsymbol{\omega} \times \mathbf{H} \tag{8–3}$$

where $(\dot{\mathbf{H}})_r$ is the rate of change of the absolute angular momentum with respect to O, as viewed by an observer on the moving xyz system. In other words, if H_x, H_y, and H_z are the instantaneous values of the projections of **H** onto the x, y, and z axes, then

$$(\dot{\mathbf{H}})_r = \dot{H}_x \mathbf{i} + \dot{H}_y \mathbf{j} + \dot{H}_z \mathbf{k} \tag{8–4}$$

where the unit vectors **i**, **j**, and **k** rotate with the xyz system.

Expressions for H_x, H_y, and H_z were given in Eq. (7–14). In terms of the present notation, they are as follows:

$$H_x = I_{xx}\omega'_x + I_{xy}\omega'_y + I_{xz}\omega'_z$$
$$H_y = I_{yx}\omega'_x + I_{yy}\omega'_y + I_{yz}\omega'_z \tag{8–5}$$
$$H_z = I_{zx}\omega'_x + I_{zy}\omega'_y + I_{zz}\omega'_z$$

For the general case in which the angular rotation rate of the coordinate system is arbitrary, it is apparent that the moments and products of inertia vary with time. Hence the evaluation of $\dot{H}_x$, $\dot{H}_y$, and $\dot{H}_z$ involves time derivatives of these moments and products of inertia as well as derivatives of the components of $\boldsymbol{\omega}'$. This makes for rather complicated equations in the general case so we shall make the assumption that the xyz system is fixed in the rigid body and rotates with it. This implies that

$$\boldsymbol{\omega} = \boldsymbol{\omega}' \tag{8–6}$$

Using this *body-axis coordinate system* we see from Eqs. (8–5) and (8–6) that

$$H_x = I_{xx}\omega_x + I_{xy}\omega_y + I_{xz}\omega_z$$
$$H_y = I_{yx}\omega_x + I_{yy}\omega_y + I_{yz}\omega_z \tag{8–7}$$
$$H_z = I_{zx}\omega_x + I_{zy}\omega_y + I_{zz}\omega_z$$

Also,

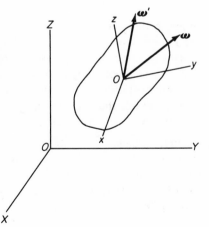

Fig. 8-1. A rotating rigid body.

$$\dot{H}_x = I_{xx}\dot{\omega}_x + I_{xy}\dot{\omega}_y + I_{xz}\dot{\omega}_z$$
$$\dot{H}_y = I_{yx}\dot{\omega}_x + I_{yy}\dot{\omega}_y + I_{yz}\dot{\omega}_z \qquad (8\text{-}8)$$
$$\dot{H}_z = I_{zx}\dot{\omega}_x + I_{zy}\dot{\omega}_y + I_{zz}\dot{\omega}_z$$

Now let us evaluate $\boldsymbol{\omega} \times \mathbf{H}$ using the determinant form of the cross product.

$$\boldsymbol{\omega} \times \mathbf{H} = \begin{vmatrix} \mathbf{i} & \mathbf{j} & \mathbf{k} \\ \omega_x & \omega_y & \omega_z \\ H_x & H_y & H_z \end{vmatrix} \qquad (8\text{-}9)$$

$$= (H_z\omega_y - H_y\omega_z)\mathbf{i} + (H_x\omega_z - H_z\omega_x)\mathbf{j} + (H_y\omega_x - H_x\omega_y)\mathbf{k}$$

From Eq. (8-2), we see that corresponding components of $\mathbf{M}$ and $\dot{\mathbf{H}}$ must be equal. Thus we obtain $\quad H = (H)_r + \omega \times H$

$$M_x = I_{xx}\dot{\omega}_x + I_{xy}\dot{\omega}_y + I_{xz}\dot{\omega}_z + H_z\omega_y - H_y\omega_z \; = \; H_x + (\omega \times H)_x$$
$$M_y = I_{yx}\dot{\omega}_x + I_{yy}\dot{\omega}_y + I_{yz}\dot{\omega}_z + H_x\omega_z - H_z\omega_x \qquad (8\text{-}10)$$
$$M_z = I_{zx}\dot{\omega}_x + I_{zy}\dot{\omega}_y + I_{zz}\dot{\omega}_z + H_y\omega_x - H_x\omega_y$$

These are the general rotational equations of motion in terms of a body-axis coordinate system. They can be given in greater detail by substituting for H_x, H_y, and H_z from Eq. (8-7). Noting that $I_{ij} = I_{ji}$ and collecting similar terms we obtain

$$M_x = I_{xx}\dot{\omega}_x + I_{xy}(\dot{\omega}_y - \omega_x\omega_z) + I_{xz}(\dot{\omega}_z + \omega_x\omega_y)$$
$$+ (I_{zz} - I_{yy})\omega_y\omega_z + I_{yz}(\omega_y^2 - \omega_z^2)$$
$$M_y = I_{xy}(\dot{\omega}_x + \omega_y\omega_z) + I_{yy}\dot{\omega}_y + I_{yz}(\dot{\omega}_z - \omega_x\omega_y)$$
$$+ (I_{xx} - I_{zz})\omega_x\omega_z + I_{xz}(\omega_z^2 - \omega_x^2) \qquad (8\text{-}11)$$
$$M_z = I_{xz}(\dot{\omega}_x - \omega_y\omega_z) + I_{yz}(\dot{\omega}_y + \omega_x\omega_z) + I_{zz}\dot{\omega}_z$$
$$+ (I_{yy} - I_{xx})\omega_x\omega_y + I_{xy}(\omega_x^2 - \omega_y^2)$$

Suppose that the constant values of the moments and products of inertia are known. For a given set of initial conditions, and assuming that $\mathbf{M}$ is a known function of time as well as of the position and velocity of the body; one can, in theory, solve for $\boldsymbol{\omega}$ as a function of time. In general, however, this is a most difficult procedure. On the other hand, if one is given the rotational motion, the calculation of the required external moments is usually a relatively straightforward procedure.

Euler's Equations of Motion. A considerable simplification can be made in the general rotational equations of motion if the *xyz* coordinate axes are chosen such that they are principal axes at the center of mass or at a point fixed in space. With this choice of body axes, all products of inertia vanish and Eq. (8-11) reduces to

$$M_x = I_{xx}\dot{\omega}_x + (I_{zz} - I_{yy})\omega_y\omega_z$$
$$M_y = I_{yy}\dot{\omega}_y + (I_{xx} - I_{zz})\omega_z\omega_x \qquad (8\text{-}12)$$
$$M_z = I_{zz}\dot{\omega}_z + (I_{yy} - I_{xx})\omega_x\omega_y$$

These are known as *Euler's equations of motion*. They are relatively simple and are widely used in solving for the rotational motion of a rigid body. Note, however, that they are nonlinear in general and it may be difficult to solve analytically for **ω** as a function of time. Furthermore, as we have seen previously, the time integrals of ω_x, ω_y, and ω_z do not correspond to any physical angles which might be used to give the orientation of the body. So if one wishes to solve for the orientation, one must transform continuously to another set of coordinates; or perhaps solve the rotational equations in terms of the other coordinates from the beginning. Coordinate transformations are often included as a part of the computation procedure if the equations are solved by means of an electronic computer.

Body-axis Translational Equations. It may be convenient to solve for the translational as well as the rotational motions of a rigid body in terms of a coordinate system fixed in the body, particularly if the applied forces are most easily specified in the body-axis system. This might occur, for example, if the applied forces rotate with the body.

Let us write the translational equation given by Eq. (8–1) in the form

$$\mathbf{F} = m\dot{\mathbf{v}} \qquad (8\text{-}13)$$

where **F** is the total external force acting on the rigid body and **v** is the absolute velocity of the center of mass. Expressing these vectors in terms of their instantaneous body-axis components, we can write

$$\mathbf{F} = F_x\mathbf{i} + F_y\mathbf{j} + F_z\mathbf{k} \qquad (8\text{-}14)$$

and

$$\mathbf{v} = v_x\mathbf{i} + v_y\mathbf{j} + v_z\mathbf{k} \qquad (8\text{-}15)$$

where **i, j, k** are an orthogonal triad of unit vectors which are fixed in the body.

In order to find the absolute acceleration $\dot{\mathbf{v}}$ referred to the body-axis coordinate system, we use Eq. (2–90) and obtain

$$\dot{\mathbf{v}} = (\dot{\mathbf{v}})_r + \boldsymbol{\omega} \times \mathbf{v} \qquad (8\text{-}16)$$

where

$$(\dot{\mathbf{v}})_r = \dot{v}_x\mathbf{i} + \dot{v}_y\mathbf{j} + \dot{v}_z\mathbf{k} \qquad (8\text{-}17)$$

The cross product $\boldsymbol{\omega} \times \mathbf{v}$ is evaluated by using a determinant similar to that of Eq. (8–9). The resulting equation is

$$\boldsymbol{\omega} \times \mathbf{v} = (v_z\omega_y - v_y\omega_z)\mathbf{i} + (v_x\omega_z - v_z\omega_x)\mathbf{j} + (v_y\omega_x - v_x\omega_y)\mathbf{k} \qquad (8\text{-}18)$$

Adding similar components in Eqs. (8–17) and (8–18), we obtain the components of the absolute acceleration of the center of mass. Then, using Eq. (8–13), we can write the following equations of motion:

$$F_x = m(\dot{v}_x + v_z\omega_y - v_y\omega_z)$$
$$F_y = m(\dot{v}_y + v_x\omega_z - v_z\omega_x) \qquad (8\text{–}19)$$
$$F_z = m(\dot{v}_z + v_y\omega_x - v_x\omega_y)$$

The components of ω are usually given or are obtained in the solution of the rotational equations.

Note that even if we solve for v_x, v_y, and v_z as functions of time, we cannot integrate directly to obtain the components of the absolute displacement. This result occurs because the body axes are changing their orientation in space continuously. Hence we must transform the velocity components v_x, v_y, and v_z to a fixed coordinate system if we wish to solve for the absolute displacement.

Example 8–1. Consider a rigid body whose inertia matrix is identical with that of Example 7–4, that is,

$$[I] = \begin{bmatrix} 150 & 0 & -100 \\ 0 & 250 & 0 \\ -100 & 0 & 300 \end{bmatrix} \text{lb sec}^2 \text{ ft}$$

where the reference point is the center of mass. In terms of body-axis components, it rotates with a constant angular velocity such that $\omega_x = 10$ rad/sec and $\omega_y = \omega_z = 0$. Solve for the components of the total external moment applied to the body.

First let us calculate the body-axis components of the angular momentum **H**. From Eq. (8–7) we obtain

$$H_x = I_{xx}\omega_x = 1500 \text{ lb sec ft}$$
$$H_y = 0$$
$$H_z = I_{zx}\omega_x = -1000 \text{ lb sec ft}$$

Now we can use Eq. (8–10) to solve for the components of the applied moment.

$$M_x = 0$$
$$M_y = -H_z\omega_x = 10,000 \text{ lb ft}$$
$$M_z = 0$$

This result could also have been obtained directly from Eq. (8–11).

It can be seen that the applied moment is constant in the body-axis coordinate system; that is, it rotates with the body. This rotating couple is equal and opposite to the inertial couple exerted by the body, and it indicates that the body is *dynamically unbalanced* when rotating about this axis which

is fixed in space as well as in the body. From Fig. 7–9 we note that the inertial couple is due to the centrifugal force of the particles on the x' axis which
are whirling about the x axis.

Now let us analyze the same problem using principal axes. We found in
Example 7–4 that the inertia matrix for principal axes is

$$[I'] = \begin{bmatrix} 100 & 0 & 0 \\ 0 & 250 & 0 \\ 0 & 0 & 350 \end{bmatrix} \text{ lb sec}^2 \text{ ft}$$

where the principal axes are the primed axes of Fig. 7–9. Also, we see that
the components of $\boldsymbol{\omega}$ in the primed system are

$$\{\omega'\} = [l]\{\omega\}$$

where the direction cosine matrix $[l]$ is the same as we found previously in
Example 7–4. Performing the matrix multiplication, we obtain

$$\begin{Bmatrix} \omega_{x'} \\ \omega_{y'} \\ \omega_{z'} \end{Bmatrix} = \frac{1}{\sqrt{5}} \begin{bmatrix} 2 & 0 & 1 \\ 0 & \sqrt{5} & 0 \\ -1 & 0 & 2 \end{bmatrix} \begin{Bmatrix} 10 \\ 0 \\ 0 \end{Bmatrix} \text{ rad/sec}$$

or

$$\omega_{x'} = 4\sqrt{5} \text{ rad/sec}$$
$$\omega_{y'} = 0$$
$$\omega_{z'} = -2\sqrt{5} \text{ rad/sec}$$

Now we can apply the Euler equations given in Eq. (8–12) to this primed
coordinate system. The moments are found to be

$$M_{x'} = 0$$
$$M_{y'} = (I_{x'x'} - I_{z'z'})\omega_{z'}\omega_{x'}$$
$$\quad = 10,000 \text{ lb ft}$$
$$M_{z'} = 0$$

This is in agreement with our previous result since the y and y' axes
coincide.

Example 8–2. A thin uniform
bar of mass m and length l is connected by a pin joint at one end to
a vertical shaft which rotates at a
constant rate Ω as shown in Fig.
8–2. Write the differential equation
of motion for the system in terms
of the angle θ. Assuming $\theta \ll 1$,
find the range of Ω for stability.

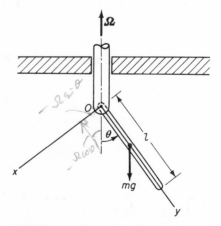

Fig. 8-2. A thin bar which is attached
by a pin joint to a vertical rotating
shaft.

First let us obtain the differential equation of motion using Euler's equation. Choose a body-axis coordinate system such that the y axis lies along the bar and the z axis is the axis of the pin joint. The principal moments of inertia at the fixed point O are

$$I_{xx} = I_{zz} = \frac{1}{3} ml^2, \qquad I_{yy} = 0 \tag{8-20}$$

The components of the absolute angular velocity of the rod are

$$\omega_x = -\Omega \sin \theta$$
$$\omega_y = -\Omega \cos \theta \tag{8-21}$$
$$\omega_z = \dot{\theta}$$

The only applied moment which can be found easily is the gravitational moment

$$M_z = -\frac{1}{2} mgl \sin \theta \tag{8-22}$$

Using the third of Euler's equations given in (8–12) we obtain

$$-\frac{1}{2} mgl \sin \theta = \frac{1}{3} ml^2 \ddot{\theta} - \frac{1}{3} ml^2 \Omega^2 \sin \theta \cos \theta$$

Thus the differential equation of motion is

$$\ddot{\theta} + \left(\frac{3g}{2l} - \Omega^2 \cos \theta \right) \sin \theta = 0 \tag{8-23}$$

Note that, for $\Omega = 0$, this equation is identical in form with that of a simple pendulum of length $2l/3$, as can be seen by comparing it with Eq. (3–220).

Assuming that $\theta \ll 1$ we can replace Eq. (8–23) by

$$\ddot{\theta} + \left(\frac{3g}{2l} - \Omega^2 \right) \theta = 0 \tag{8-24}$$

The natural frequency of small oscillations about $\theta = 0$ is

$$\omega_n = \sqrt{\frac{3g}{2l} - \Omega^2} \tag{8-25}$$

It can be seen that ω_n decreases as Ω increases from zero. The stability limit for motion near $\theta = 0$ is reached when ω_n goes to zero, that is, when

$$\Omega = \pm \sqrt{\frac{3g}{2l}}$$

For $\Omega^2 > 3g/2l$ the coefficient of θ in Eq. (8–24) becomes negative, corresponding to a shift in the form of the transient solution from trigonometric to hyperbolic functions of time. Hence an exponentially increasing term occurs in the solution for θ and it is considered to be unstable in the region near $\theta = 0$.

It is interesting to note that, even for $\Omega^2 > 3g/2l$, one can find a position of *stable* equilibrium for the θ motion. The positions of equilibrium are found by setting $\ddot{\theta} = 0$ in Eq. (8–23). Thus we find that the values

$$\theta = 0, \quad \cos^{-1}\frac{3g}{2l\Omega^2}$$

correspond to positions of equilibrium. For $\Omega^2 > 3g/2l$ we have shown that small motions about $\theta = 0$ are unstable. We can investigate the stability of small motions about $\theta_0 = \cos^{-1}(3g/2l\Omega^2)$ by using the perturbation theory discussed in Sec. 5–7. The equation of motion obtained from Eq. (8–23) is

$$\delta\ddot{\theta} + \Omega^2 \sin^2 \theta_0 \, \delta\theta = 0 \qquad (8\text{–}26)$$

where $\delta\ddot{\theta}$ and $\delta\theta$ are the perturbations from equilibrium. Thus,

$$\theta = \theta_0 + \delta\theta$$
$$\ddot{\theta} = \delta\ddot{\theta}$$

It can be seen that if the system is disturbed slightly from its equilibrium position at $\theta = \theta_0$, the coordinate $\delta\theta$ oscillates sinusoidally with the natural frequency

$$\omega_n = |\,\Omega \sin \theta_0\,| = \left[\Omega^2 - \left(\frac{3g}{2l\Omega}\right)^2\right]^{1/2} \qquad (8\text{–}27)$$

Also, it is apparent from Eq. (8–26) that any small deviation $\delta\theta$ results in a restoring moment tending to reduce $\delta\theta$ to zero. Therefore this is a position of stable equilibrium.

Stability of Rotational Motion about a Principal Axis. Let us consider now the free motion of a rigid body which is rotating about one of its principal axes. Assume that the xyz system forms a set of principal axes at the center of mass and these axes are fixed in the body. We shall consider a general rigid body with unequal principal moments of inertia and assume that the rotation is about the x axis. From Euler's equations given in Eq. (8–12), it is apparent that ω_x remains constant and $\omega_y = \omega_z = 0$, assuming that all applied moments are zero. On the other hand, if the system is given a slight disturbance and then is allowed to continue rotating freely, it will no longer rotate solely about the x axis. Let us analyze the stability of these perturbations in $\boldsymbol{\omega}$. (See Sec. 5–7.) Taking a rotation at the constant rate Ω about the x axis as the reference condition, we can write

$$\omega_x = \Omega + \delta\omega_x$$
$$\omega_y = \delta\omega_y \qquad (8\text{–}28)$$
$$\omega_z = \delta\omega_z$$

Assuming that the perturbations $\delta\omega_x$, $\delta\omega_y$, and $\delta\omega_z$ are much smaller in magnitude than Ω and neglecting the products of these small quantities, we find from the first of Euler's equations of motion that

$$\dot{\omega}_x = 0$$

Since ω_x is constant, we can redefine Ω as the initial value of ω_x, implying that $\delta\omega_x = 0$. The other two Euler equations are

$$I_{yy}\dot{\omega}_y + (I_{xx} - I_{zz})\Omega\omega_z = 0 \qquad (8\text{–}29)$$

$$I_{zz}\dot{\omega}_z + (I_{yy} - I_{xx})\Omega\omega_y = 0 \qquad (8\text{–}30)$$

where we note that both ω_y and ω_z are much smaller than Ω. These are the differential equations for the perturbations in $\boldsymbol{\omega}$.

Now differentiate Eq. (8–29) with respect to time and substitute the value of $\dot{\omega}_z$ from Eq. (8–30). We obtain

$$\ddot{\omega}_y + \frac{(I_{yy} - I_{xx})(I_{zz} - I_{xx})}{I_{yy}I_{zz}} \Omega^2 \omega_y = 0 \qquad (8\text{–}31)$$

The coefficient of the ω_y term is positive if I_{xx} is the *smallest* of the principal moments of inertia. Hence for this case, ω_y will vary sinusoidally with a natural frequency

$$\omega_n = \Omega\sqrt{\frac{(I_{yy} - I_{xx})(I_{zz} - I_{xx})}{I_{yy}I_{zz}}} \qquad (8\text{–}32)$$

It can be seen from Eq. (8–29) that ω_z must vary sinusoidally with the same frequency as ω_y but will have a phase lead of 90 degrees. Also we note that the magnitudes of these oscillations in ω_y and ω_z are constant. Since the magnitude of an arbitrary small disturbance does not grow, the rotational motion about the x axis is said to be *stable*. This result applies regardless of the relative sizes of I_{yy} and I_{zz}.

In a similar fashion, we see that if I_{xx} is the *largest* of the principal moments of inertia, the coefficient of ω_y in Eq. (8–31) is again positive. The solutions for ω_y and ω_z are sinusoidal functions of time and the rotational motion is again considered to be stable. We have shown, then, that the rotation of a rigid body about a principal axis corresponding to either the largest or the smallest moment of inertia is dynamically stable.

Now consider the case in which I_{xx} is the intermediate moment of inertia, that is, its magnitude lies between those of the other two principal moments of inertia. For example, if $I_{yy} < I_{xx} < I_{zz}$, then the coefficient of ω_y in Eq. (8–31) is negative, resulting in a solution for ω_y which has an exponentially increasing term. A similar result occurs if $I_{zz} < I_{xx} < I_{yy}$. Hence a small disturbance in $\boldsymbol{\omega}$ tends to grow if it is applied to a body which is rotating about the principal axis corresponding to the intermediate moment of inertia. So a rotation about this axis is *unstable*.

One can demonstrate the stability of the rotational motion about two of the principal axes, and the instability of motion about the third. This is accomplished by tossing an object, such as a blackboard eraser, into the air so that it spins initially about one of the principal axes. The appearance of the spinning motion about a principal axis of minimum or maximum moment of inertia is relatively smooth. On the other hand, if one attempts to spin the body about the third principal axis, corresponding to the intermediate moment of inertia, the motion is irregular in appearance due to large changes in the location of the axis of rotation relative to the body. The nature of this motion will be further explained in the discussion of the Poinsot construction in Sec. 8–4.

It is interesting to note that if a rigid body is spinning about the principal axis of minimum moment of inertia, it has the largest possible rotational kinetic energy for a given angular momentum about its center of mass. For example, if the rotation is about the x axis, then

$$\omega_x = \frac{H}{I_{xx}}$$

and

$$T_{\text{rot}} = \frac{1}{2} I_{xx} \omega_x^2 = \frac{1}{2} \frac{H^2}{I_{xx}}$$

So if I_{xx} is the minimum moment of inertia, then the kinetic energy T_{rot} is maximum. Conversely, if I_{xx} is the maximum moment of inertia about the center of mass, then T_{rot} is the minimum possible for the given H. This result is important in the analysis of the rotational motion of nearly rigid satellites with internal frictional losses. Because of the presence of internal damping, the kinetic energy will continuously decrease to the minimum possible value consistent with the constant angular momentum, assuming that no external moments are acting. Hence its final condition is a rigid-body rotation about the principal axis of maximum moment of inertia, regardless of its initial axis of rotation. This explains why a satellite which consists of a long narrow rocket spinning about its symmetry axis is unstable in practice and usually begins a tumbling motion within a few hours. On the other hand, an oblate body, such as a circular disk spinning about its axis of symmetry, is stable in the presence of internal damping. In fact, the effect of the damping is to eliminate any initial nutation or wobbling in its motion.

Example 8–3. A uniform sphere of mass m and radius a rolls without slipping on a plane which is inclined at an angle α from the horizontal. Solve for the motion of the sphere, assuming a general set of initial conditions.

Let us express the motion in terms of a fixed cartesian coordinate system which is chosen so that the sphere rolls on the xy plane, as shown in Fig. 8–3. The external forces acting on the sphere are the gravitational force mg acting at the center O and the force $\mathbf{R}$ acting at the contact point P. Consider $\mathbf{R}$ in terms of its two components, namely, a friction force $\mathbf{F}$ which acts in the xy plane and a normal force $\mathbf{N}$.

$$\mathbf{R} = \mathbf{N} + \mathbf{F}$$

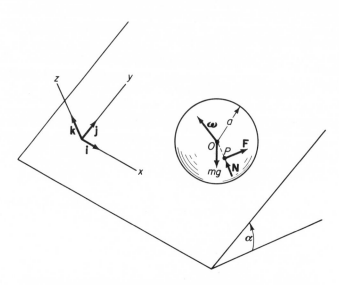

Fig. 8-3. A uniform sphere rolling on an inclined plane.

The center of the sphere moves parallel to the xy plane and therefore the force components normal to this plane must sum to zero. Hence

$$N = mg \cos \alpha$$

We need not consider these force components any further since they do not influence the motion of interest.

With the assumption of no slipping at the contact point P, we can write an expression for the velocity of the center O in terms of the absolute angular velocity $\boldsymbol{\omega}$ of the sphere. We obtain

$$\mathbf{v} = a\boldsymbol{\omega} \times \mathbf{k} \qquad (8\text{--}33)$$

Note that the sphere can spin about an axis normal to the plane at the contact point without slipping. The acceleration of the center is found by differentiating Eq. (8–33).

$$\dot{\mathbf{v}} = a\dot{\boldsymbol{\omega}} \times \mathbf{k} \qquad (8\text{--}34)$$

Using Newton's law of motion, we find that the translational equation is

$$\mathbf{F} - mg \sin \alpha \, \mathbf{j} = ma\dot{\boldsymbol{\omega}} \times \mathbf{k} \qquad (8\text{--}35)$$

Now let us choose the center O as the reference point for the rotational equations. Any axis through O is a principal axis. Consequently the angular momentum is given by

$$\mathbf{H} = I\boldsymbol{\omega} \qquad (8\text{--}36)$$

where, for this case of a uniform sphere,

$$I = \frac{2}{5} ma^2 \tag{8-37}$$

The applied moment about O is due to $\mathbf{F}$ alone and is given by

$$\mathbf{M} = a\mathbf{F} \times \mathbf{k} \tag{8-38}$$

Using the general rotational equation $\mathbf{M} = \dot{\mathbf{H}}$, we obtain

$$a\mathbf{F} \times \mathbf{k} = \frac{2}{5} ma^2 \dot{\boldsymbol{\omega}} \tag{8-39}$$

To solve for the motion of the sphere, substitute the value of $\mathbf{F}$ obtained from Eq. (8–35) into Eq. (8–39). The result is

$$mga \sin \alpha \, \mathbf{i} + ma^2 (\dot{\boldsymbol{\omega}} \times \mathbf{k}) \times \mathbf{k} = \frac{2}{5} ma^2 \dot{\boldsymbol{\omega}}$$

We note that $\mathbf{M}$, and therefore $\dot{\boldsymbol{\omega}}$, each has no z component. Hence

$$(\dot{\boldsymbol{\omega}} \times \mathbf{k}) \times \mathbf{k} = (\dot{\boldsymbol{\omega}} \cdot \mathbf{k})\mathbf{k} - \dot{\boldsymbol{\omega}} = -\dot{\boldsymbol{\omega}}$$

Thus we obtain that

$$mga \sin \alpha \, \mathbf{i} = \frac{7}{5} ma^2 \dot{\boldsymbol{\omega}}$$

or

$$\dot{\boldsymbol{\omega}} = \frac{5g}{7a} \sin \alpha \, \mathbf{i} \tag{8-40}$$

The acceleration of the center O is found by substituting this expression for $\dot{\boldsymbol{\omega}}$ into Eq. (8–34), yielding

$$\dot{\mathbf{v}} = -\frac{5}{7} g \sin \alpha \, \mathbf{j} \tag{8-41}$$

Since the acceleration is constant, we know from our previous experience (Sec. 3–1) that the path of O is a parabola whose major axis is parallel to the y axis. Specifically, if the conditions at $t = 0$ are $x(0) = x_0$, $y(0) = y_0$, $\dot{x}(0) = (\dot{x})_0$, and $\dot{y}(0) = (\dot{y})_0$, we obtain

$$\begin{aligned} x &= x_0 + (\dot{x})_0 t \\ y &= y_0 + (\dot{y})_0 t - \frac{5}{14} gt^2 \sin \alpha \end{aligned} \tag{8-42}$$

Note that the path of the center is identical with that of a particle sliding without friction on a similar inclined plane, but under a uniform gravitational field of magnitude $5g/7$. Hence the rotational motion does not change the general nature of the translational motion for this case of a rolling sphere.

Finally, we can solve for the tangential force $\mathbf{F}$ at the contact point from Eqs. (8–35) and (8–40), the result being

$$\mathbf{F} = \frac{2}{7} mg \sin \alpha \, \mathbf{j} \tag{8-43}$$

Systems of Rigid Bodies. In Chapter 4 we studied the dynamics of a system of particles. Since we have considered a rigid body as a limiting case of a rigidly connected set of particles, we can expect the general dynamical principles for a set of particles to apply equally well to one or more rigid bodies. Nevertheless, it is convenient to use rigid-body parameters, such as the moments and products of inertia, in expressing the angular momentum of such a system. So let us consider briefly the development of expressions for the total angular momentum of a system of rigid bodies and also for its rate of change.

Suppose that a system of $(n + 1)$ rigid bodies consists of a reference body of mass m_0 with its center of mass at P and n other bodies of arbitrary mass. The cartesian $x_0 y_0 z_0$ axes are fixed in the reference body with the origin at P, and rotate with that body at an absolute angular velocity $\boldsymbol{\omega}_0$ as shown in Fig. 8–4. The position of the center of mass of the ith body relative to P is given by the vector $\boldsymbol{\rho}_i$ and its absolute angular velocity is $\boldsymbol{\omega}_i$.

Now let us find an expression for the total angular momentum of this system of rigid bodies with respect to the origin O of the fixed XYZ coordinate system. We found in Sec. 4–5 that the angular momentum of a system of particles with respect to a given point is equal to the sum of (1) the angular momentum due to the translational motion of the center of mass, (2) the angular momentum due to motion relative to the center of mass. Consider-

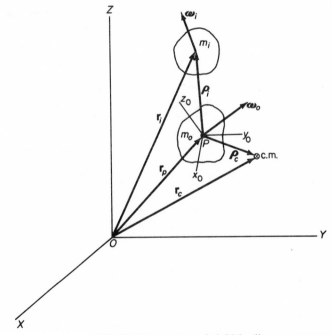

Fig. 8-4. A system of rigid bodies.

ing a rigid body as a system of particles, the relative motion is derived exclusively from its angular velocity. Furthermore, the total angular momentum of the system of rigid bodies is equal to the sum of the angular momenta of the individual bodies. Hence we find that the total angular momentum about O is

$$\mathbf{H} = \mathbf{H}_0 + \sum_{i=1}^{n} \mathbf{H}_i + \mathbf{r}_p \times m_0 \dot{\mathbf{r}}_p + \sum_{i=1}^{n} \mathbf{r}_i \times m_i \dot{\mathbf{r}}_i \qquad (8\text{--}44)$$

where $\mathbf{H}_0$ and the $\mathbf{H}_i$ represent the angular momenta of the bodies about their respective mass centers. From Eq. (7–38), we can write the components of $\mathbf{H}_0$ and $\mathbf{H}_i$ using matrix notation, namely,

$$\{H_0\} = [I_0]\{\omega_0\} \qquad (8\text{--}45)$$

$$\{H_i\} = [I_i]\{\omega_i\} \qquad (8\text{--}46)$$

where $[I_0]$ and $[I_i]$ are the inertia matrices with respect to the mass centers of the individual bodies.

The general rotational equation of motion is obtained by differentiating Eq. (8–44) with respect to time and substituting into Eq. (8–2). Thus,

$$\mathbf{M} = \dot{\mathbf{H}}_0 + \sum_{i=1}^{n} \dot{\mathbf{H}}_i + \mathbf{r}_p \times m_0 \ddot{\mathbf{r}}_p + \sum_{i=1}^{n} \mathbf{r}_i \times m_i \ddot{\mathbf{r}}_i \qquad (8\text{--}47)$$

where $\mathbf{M}$ is the total moment about O of the external forces acting on the system. In evaluating $\dot{\mathbf{H}}_0$ and $\dot{\mathbf{H}}_i$, it is often convenient to use a set of principal axes at the center of mass of each body. The right-hand sides of Euler's equations given in Eq. (8–12) then correspond to the body-axis components of $\dot{\mathbf{H}}$ for the given body with respect to its mass center. Of course, individual rotational equations can be written for each body in many instances.

Another convenient reference point for angular momentum calculations is the center of mass of the system. Referring again to Fig. 8–4, we see that the total angular momentum about the center of mass is

$$\mathbf{H}_c = \mathbf{H}_0 + \sum_{i=1}^{n} \mathbf{H}_i + \boldsymbol{\rho}_c \times m_0 \dot{\boldsymbol{\rho}}_c + \sum_{i=1}^{n} [(\boldsymbol{\rho}_i - \boldsymbol{\rho}_c) \times m_i(\dot{\boldsymbol{\rho}}_i - \dot{\boldsymbol{\rho}}_c)] \quad (8\text{--}48)$$

where we note that $(\boldsymbol{\rho}_i - \boldsymbol{\rho}_c)$ is the position vector of m_i relative to the center of mass. The term $\boldsymbol{\rho}_c \times m_0 \dot{\boldsymbol{\rho}}_c$ represents the angular momentum due to the velocity of the mass center of m_0.

In calculating the location of the center of mass for the system, we can consider the mass of each body to be concentrated at its own mass center. Hence we can use the result for a system of particles given in Eq. (4–63) to obtain

$$\boldsymbol{\rho}_c = \frac{1}{m} \sum_{i=1}^{n} m_i \boldsymbol{\rho}_i \qquad (8\text{--}49)$$

where the total mass is

$$m = m_0 + \sum_{i=1}^{n} m_i \tag{8-50}$$

Now we collect terms in Eq. (8–48) and simplify by using Eqs. (8–49) and (8–50) to yield the following result:

$$\mathbf{H}_c = \mathbf{H}_0 + \sum_{i=1}^{n} \mathbf{H}_i + \sum_{i=1}^{n} \boldsymbol{\rho}_i \times m_i \dot{\boldsymbol{\rho}}_i - \boldsymbol{\rho}_c \times m \dot{\boldsymbol{\rho}}_c \tag{8-51}$$

Next we recall that $\mathbf{M}_c = \dot{\mathbf{H}}_c$, where $\mathbf{M}_c$ is the moment of the external forces about the center of mass. Thus,

$$\mathbf{M}_c = \dot{\mathbf{H}}_0 + \sum_{i=1}^{n} \dot{\mathbf{H}}_i + \sum_{i=1}^{n} \boldsymbol{\rho}_i \times m_i \ddot{\boldsymbol{\rho}}_i - \boldsymbol{\rho}_c \times m \ddot{\boldsymbol{\rho}}_c \tag{8-52}$$

Finally, let us obtain an expression for the angular momentum with respect to the center of mass in terms of the motions of the various bodies relative to the $x_0 y_0 z_0$ system which is fixed in the reference body. The absolute angular velocity $\boldsymbol{\omega}_i$ of the ith body can be written as the sum of $\boldsymbol{\omega}_0$ and the angular velocity relative to the reference body, that is,

$$\boldsymbol{\omega}_i = \boldsymbol{\omega}_0 + \boldsymbol{\Omega}_i \tag{8-53}$$

where we note that $\boldsymbol{\Omega}_i$ is the angular velocity of m_i as viewed by an observer rotating with the $x_0 y_0 z_0$ system. Hence we can rewrite the matrix expression for $\mathbf{H}_i$ as follows:

$$\{H_i\} = [I_i]\{\omega_0\} + [I_i]\{\Omega_i\} \tag{8-54}$$

or, in vector form,

$$\mathbf{H}_i = \mathbf{H}'_i + \mathbf{H}''_i \tag{8-55}$$

where $\mathbf{H}'_i$ and $\mathbf{H}''_i$ are associated with $\boldsymbol{\omega}_0$ and $\boldsymbol{\Omega}_i$, respectively.

Next we remember from Sec. 2–8 that the apparent rate of change of a vector depends upon the angular velocity of the observer. So, using Eq. (2–90), we obtain

$$\dot{\boldsymbol{\rho}}_i = (\dot{\boldsymbol{\rho}}_i)_r + \boldsymbol{\omega}_0 \times \boldsymbol{\rho}_i \tag{8-56}$$

and

$$\dot{\boldsymbol{\rho}}_c = (\dot{\boldsymbol{\rho}}_c)_r + \boldsymbol{\omega}_0 \times \boldsymbol{\rho}_c \tag{8-57}$$

where $(\dot{\boldsymbol{\rho}}_i)_r$ and $(\dot{\boldsymbol{\rho}}_c)_r$ are velocities relative to the $x_0 y_0 z_0$ system. Substituting the expressions of Eqs. (8–55), (8–56), and (8–57) into Eq. (8–51), we find that

$$\mathbf{H}_c = \mathbf{H}_0 + \sum_{i=1}^{n} \mathbf{H}'_i + \sum_{i=1}^{n} \boldsymbol{\rho}_i \times m_i(\boldsymbol{\omega}_0 \times \boldsymbol{\rho}_i) + \sum_{i=1}^{n} \mathbf{H}''_i$$
$$+ \sum_{i=1}^{n} \boldsymbol{\rho}_i \times m_i(\dot{\boldsymbol{\rho}}_i)_r - \boldsymbol{\rho}_c \times m \dot{\boldsymbol{\rho}}_c$$

or

$$\mathbf{H}_c = \mathbf{H}'_p + \mathbf{H}''_p - \boldsymbol{\rho}_c \times m(\dot{\boldsymbol{\rho}}_c)_r - \boldsymbol{\rho}_c \times m(\boldsymbol{\omega}_0 \times \boldsymbol{\rho}_c) \tag{8-58}$$

where

$$\mathbf{H}'_p = \mathbf{H}_0 + \sum_{i=1}^{n} \mathbf{H}'_i + \sum_{i=1}^{n} \boldsymbol{\rho}_i \times m_i(\boldsymbol{\omega}_0 \times \boldsymbol{\rho}_i) \tag{8–59}$$

$$\mathbf{H}''_p = \sum_{i=1}^{n} \mathbf{H}''_i + \sum_{i=1}^{n} \boldsymbol{\rho}_i \times m_i(\dot{\boldsymbol{\rho}}_i)_r \tag{8–60}$$

Note that $\mathbf{H}'_p$ is that portion of the angular momentum with respect to P which is associated with the rotation of the entire system as a rigid body at an angular velocity $\boldsymbol{\omega}_0$. In other words, the system is motionless relative to the $x_0 y_0 z_0$ system and rotates with it. On the other hand, $\mathbf{H}''_p$ represents the angular momentum about P which is due to motions relative to the $x_0 y_0 z_0$ system. The remaining terms account for the change of reference point from P to the center of mass. So we obtain that the total angular momentum about P is

$$\mathbf{H}_p = \mathbf{H}'_p + \mathbf{H}''_p \tag{8–61}$$

where $\mathbf{H}'_p$ and $\mathbf{H}''_p$ are given by Eqs. (8–59) and (8–60). Furthermore, we see that an expression relating the angular momentum about the center of mass to that about P is as follows:

$$\mathbf{H}_c = \mathbf{H}_p - \boldsymbol{\rho}_c \times m\dot{\boldsymbol{\rho}}_c \tag{8–62}$$

in agreement with Eq. (4–65).

Again we can equate the external moment $\mathbf{M}_c$ about the center of mass to the time rate of change of $\mathbf{H}_c$, obtaining

$$\mathbf{M}_c = \dot{\mathbf{H}}_p - \boldsymbol{\rho}_c \times m\ddot{\boldsymbol{\rho}}_c \tag{8–63}$$

or, using Eqs. (2–90) and (8–58),

$$\begin{aligned}
\mathbf{M}_c = (\dot{\mathbf{H}}'_p)_r + (\dot{\mathbf{H}}''_p)_r &- \boldsymbol{\rho}_c \times m(\ddot{\boldsymbol{\rho}}_c)_r + \boldsymbol{\omega}_0 \times \mathbf{H}'_p + \boldsymbol{\omega}_0 \times \mathbf{H}''_p \\
&- \boldsymbol{\rho}_c \times m(\dot{\boldsymbol{\omega}}_0 \times \boldsymbol{\rho}_c) - \boldsymbol{\rho}_c \times m[\boldsymbol{\omega}_0 \times (\boldsymbol{\omega}_0 \times \boldsymbol{\rho}_c)] \\
&\qquad - 2\boldsymbol{\rho}_c \times m[\boldsymbol{\omega}_0 \times (\dot{\boldsymbol{\rho}}_c)_r]
\end{aligned} \tag{8–64}$$

Here we have written the over-all rotational equation for a system of rigid bodies in terms of their motions relative to the $x_0 y_0 z_0$ system (which is a non-inertial rotating frame) and the absolute angular velocity $\boldsymbol{\omega}_0$ of that frame. To solve for the complete motion, one usually must also write equations of motion for the individual bodies.

Example 8–4. The first of a system of two rigid bodies consists of a spherically symmetric body of mass m_0 and moment of inertia I_0. The second body is a thin uniform circular disk of mass m_1 and radius a which is supported on a massless rod connecting the centers of the bodies, as shown in Fig. 8–5. Initially the entire motion of the system consists of a rotation as a single rigid body with an angular velocity $\omega_b \mathbf{e}_b$ about an axis which is perpendicular to the symmetry axis of the system. At $t = 0$, an internal motor applies a couple about this symmetry axis, causing the angular velocity of the disk *relative to the sphere* to be a specified function of time, $\Omega(t)\mathbf{e}_a$, where

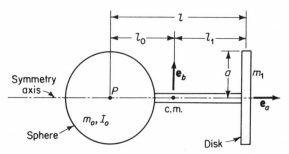

Fig. 8-5. A sphere and a disk connected to form an axially symmetric system.

e_a is a unit vector which is directed along the symmetry axis. Assuming that no external forces or moments act on the system, solve for its rotational motion.

Let us consider first the form of the Euler equations of motion for the case of an axially symmetric rigid body. In general, let us take the x axis as the axis of symmetry and assume that the y and z axes are transverse axes through the center of mass. Also, let $I_{xx} = I_a$ and $I_{yy} = I_{zz} = I_t$, designating the axial and transverse moments of inertia, respectively. Using Eq. (8–12), we can write the equations of motion:

$$M_a = I_a \dot{\omega}_a$$
$$M_y = I_t \dot{\omega}_y - (I_t - I_a)\omega_a \omega_z \qquad (8\text{–}65)$$
$$M_z = I_t \dot{\omega}_z + (I_t - I_a)\omega_a \omega_y$$

where M_a and ω_a are the axial components of the applied moment and the absolute angular velocity, respectively.

Returning now to the problem at hand, we see that the axial component of the absolute angular acceleration is directly proportional to the axial moment applied to the body. Since we are using a principal-axis system, we note that $H_a = I_a \omega_a$ and it follows that

$$M_a = \dot{H}_a \qquad (8\text{–}66)$$

where H_a is the axial component of the total angular momentum of either body about its center of mass. In this instance, we are applying equal and opposite axial moments to the sphere and the disk. Therefore the changes in H_a for the sphere will cancel those for the disk, resulting in a net H_a of zero for the system throughout the motion.

The total angular momentum of the system is constant since there are no external moments. Thus it can be seen that the transverse component of the angular velocity must maintain the same direction and magnitude in space. We note from Eq. (8–65) that the change in the magnitude of the axial component of angular velocity is inversely proportional to the value of

I_a for that body. So we conclude that the angular velocity of the disk is

$$\boldsymbol{\omega}_1 = \omega_b \mathbf{e}_b + \frac{I_0 \Omega}{I_0 + I_a} \mathbf{e}_a \tag{8-67}$$

where the axial moment of inertia of the disk is

$$I_a = \frac{1}{2} m_1 a^2$$

In a similar fashion, the angular velocity of the sphere is

$$\boldsymbol{\omega}_0 = \omega_b \mathbf{e}_b - \frac{I_a \Omega}{I_0 + I_a} \mathbf{e}_a \tag{8-68}$$

It is instructive to use this example to illustrate the meaning of some of the terms introduced in the discussion of systems of rigid bodies. First let us calculate the angular momentum $\mathbf{H}_c$ of the entire system *at* t $= 0$, *noting that* $\Omega(0) = 0$. The sphere is chosen as the reference body and we see that

$$\boldsymbol{\omega}_0 = \boldsymbol{\omega}_1 = \omega_b \mathbf{e}_b \tag{8-69}$$

Hence from Eqs. (8-45) and (8-46), we obtain that the angular momenta of the individual bodies about their respective mass centers are

$$\mathbf{H}_0 = I_0 \omega_b \mathbf{e}_b \tag{8-70}$$

$$\mathbf{H}_1 = I_1 \omega_b \mathbf{e}_b \tag{8-71}$$

where the moment of inertia of the disk about a diameter is

$$I_1 = \frac{1}{4} ma^2$$

Evaluating the remaining two terms of Eq. (8-51) we find that

$$\boldsymbol{\rho}_1 \times m_1 \dot{\boldsymbol{\rho}}_1 = m_1 l^2 \omega_b \mathbf{e}_b \tag{8-72}$$

$$\boldsymbol{\rho}_c \times m \dot{\boldsymbol{\rho}}_c = (m_0 + m_1) l_0^2 \omega_b \mathbf{e}_b \tag{8-73}$$

But from the definition of the mass center for this system we have

$$m_0 l_0 = m_1 l_1 \tag{8-74}$$

and noting that $l = l_0 + l_1$, we obtain

$$\boldsymbol{\rho}_1 \times m_1 \dot{\boldsymbol{\rho}}_1 - \boldsymbol{\rho}_c \times m \dot{\boldsymbol{\rho}}_c = (m_0 l_0^2 + m_1 l_1^2) \omega_b \mathbf{e}_b \tag{8-75}$$

Adding the expressions given in Eqs. (8-70), (8-71), and (8-75), we can evaluate Eq. (8-51), yielding

$$\mathbf{H}_c = (I_0 + I_1 + m_0 l_0^2 + m_1 l_1^2) \omega_b \mathbf{e}_b \tag{8-76}$$

The quantity in parentheses is seen to be the moment of inertia of the system about a transverse axis through the mass center.

Now let us consider the more general case in which $t > 0$ and calculate $\mathbf{H}_p$ in terms of $\boldsymbol{\omega}_0$ and Ω. First we find from Eqs. (8-45) and (8-68) that

$$\mathbf{H}_0 = I_0 \omega_b \mathbf{e}_b - \frac{I_0 I_a}{I_0 + I_a} \Omega \, \mathbf{e}_a \tag{8-77}$$

In a similar fashion, from Eqs. (8–54), (8–55), and (8–68), we obtain

$$\mathbf{H}'_1 = I_1 \omega_b \mathbf{e}_b - \frac{I_a^2 \Omega}{I_0 + I_a} \, \mathbf{e}_a \tag{8-78}$$

$$\mathbf{H}''_1 = I_a \Omega \, \mathbf{e}_a \tag{8-79}$$

Also,

$$\boldsymbol{\rho}_1 \times m_1 (\boldsymbol{\omega}_0 \times \dot{\boldsymbol{\rho}}_1) = m_1 l^2 \omega_b \mathbf{e}_b \tag{8-80}$$

$$\boldsymbol{\rho}_1 \times m_1 (\dot{\boldsymbol{\rho}}_1)_r = 0 \tag{8-81}$$

Adding the terms indicated by Eqs. (8–59) and (8–60), we find that the total angular momentum about P is

$$\mathbf{H}_p = (I_0 + I_1 + m_1 l^2) \omega_b \mathbf{e}_b \tag{8-82}$$

Note that it is constant.

Now we obtain $\mathbf{H}_c$ by using Eq. (8–62). The last term on the right is evaluated from Eq. (8–73), and after some algebraic simplification, we obtain

$$\mathbf{H}_c = (I_0 + I_1 + m_0 l_0^2 + m_1 l_1^2) \omega_b \mathbf{e}_b$$

in agreement with the result of Eq. (8–76).

8–2. EQUATIONS OF MOTION IN TERMS OF EULERIAN ANGLES

We turn now to the use of Lagrange's equations in the formulation of the differential equations of rotational motion for a rigid body. Assuming a coordinate system consisting of a set of *principal axes* at the center of mass, we can use Eq. (7–54) to obtain the following expression for the rotational kinetic energy:

$$T_{\text{rot}} = \frac{1}{2} (I_{xx} \omega_x^2 + I_{yy} \omega_y^2 + I_{zz} \omega_z^2) \tag{8-83}$$

where ω_x, ω_y, and ω_z are the body-axis components of the absolute rotation rate $\boldsymbol{\omega}$ of the body. As we mentioned earlier, however, we cannot substitute this expression for the kinetic energy directly into Lagrange's equations because the variables ω_x, ω_y, and ω_z do not correspond to the time derivatives of any set of coordinates which specify the orientation of the body. Nevertheless, a suitable set of generalized coordinates can be found for the description of the rotational motion. These are the three Eulerian angles which were introduced in Sec. 7–12.

Referring to the definitions of the Eulerian angles, as shown in Fig. 7–23, we see that an arbitrary angular velocity $\boldsymbol{\omega}$ of the xyz system can be expressed in terms of the time derivatives of these angles except for the case where $\theta = \pm\pi/2$. In other words we can write

$$\boldsymbol{\omega} = \dot{\boldsymbol{\psi}} + \dot{\boldsymbol{\theta}} + \dot{\boldsymbol{\phi}} \qquad (8\text{-}84)$$

where $\dot{\boldsymbol{\psi}}$, $\dot{\boldsymbol{\theta}}$, and $\dot{\boldsymbol{\phi}}$ are angular velocity vectors associated with changes in the corresponding Eulerian angles (Fig. 8–6). It is important to realize that $\dot{\boldsymbol{\psi}}$, $\dot{\boldsymbol{\theta}}$, and $\dot{\boldsymbol{\phi}}$ do *not* form a mutually orthogonal vector triad in general. Thus it can be seen that $\dot{\boldsymbol{\theta}}$ is perpendicular to $\dot{\boldsymbol{\psi}}$ and $\dot{\boldsymbol{\phi}}$ but these last two vectors are not perpendicular to each other. We can, however, consider $\dot{\boldsymbol{\psi}}$, $\dot{\boldsymbol{\theta}}$, and $\dot{\boldsymbol{\phi}}$ to be a set of *nonorthogonal components* of $\boldsymbol{\omega}$ since their vector sum is equal to $\boldsymbol{\omega}$. Note particularly that the magnitudes of $\dot{\boldsymbol{\psi}}$ and $\dot{\boldsymbol{\phi}}$ are not equal to the orthogonal projections of $\boldsymbol{\omega}$ on their respective axes.

Now let us obtain expressions for ω_x, ω_y, and ω_z by summing the orthogonal projections of $\dot{\boldsymbol{\psi}}$, $\dot{\boldsymbol{\theta}}$, and $\dot{\boldsymbol{\phi}}$ onto each of the axes of the xyz system. We find that

$$\omega_x = \dot{\phi} - \dot{\psi} \sin \theta$$
$$\omega_y = \dot{\theta} \cos \phi + \dot{\psi} \cos \theta \sin \phi \qquad (8\text{-}85)$$
$$\omega_z = \dot{\psi} \cos \theta \cos \phi - \dot{\theta} \sin \phi$$

as can be seen directly from Fig. 8–6, or by referring to Eq. (7–166) with $\psi = 0$ to obtain the values of the required direction cosines. Conversely, except for the case where $\theta = \pm\pi/2$, we can solve for $\dot{\boldsymbol{\psi}}$, $\dot{\boldsymbol{\theta}}$, and $\dot{\boldsymbol{\phi}}$ in terms of ω_x, ω_y, ω_z, obtaining

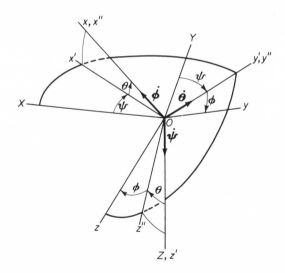

Fig. 8-6. Euler angle components of the angular velocity of the xyz system.

$$\dot{\psi} = (\omega_y \sin \phi + \omega_z \cos \phi) \sec \theta$$

$$\dot{\theta} = \omega_y \cos \phi - \omega_z \sin \phi \qquad (8\text{--}86)$$

$$\dot{\phi} = \omega_x + \dot{\psi} \sin \theta = \omega_x + (\omega_y \sin \phi + \omega_z \cos \phi) \tan \theta$$

The rotational kinetic energy of the body in terms of the Euler angle rates is obtained by substituting into Eq. (8–83) the expressions for ω_x, ω_y, and ω_z from Eq. (8–85), resulting in

$$T_{\text{rot}} = \frac{1}{2} I_{xx}(\dot{\phi} - \dot{\psi} \sin \theta)^2 + \frac{1}{2} I_{yy}(\dot{\theta} \cos \phi + \dot{\psi} \cos \theta \sin \phi)^2$$
$$+ \frac{1}{2} I_{zz}(\dot{\psi} \cos \theta \cos \phi - \dot{\theta} \sin \phi)^2 \qquad (8\text{--}87)$$

We obtain the differential equations for the rotational motion by using Lagrange's equation in the form of Eq. (6–73):

$$\frac{d}{dt}\left(\frac{\partial T}{\partial \dot{q}_i}\right) - \frac{\partial T}{\partial q_i} = Q_i$$

where the generalized force component Q_i is in each case a moment about an axis through the center of mass and involves all external forces, including any arising from a potential function.

Let us calculate first the generalized momenta by using Eq. (6–58). We find that

$$p_\psi = \frac{\partial T}{\partial \dot{\psi}} = -I_{xx}(\dot{\phi} - \dot{\psi} \sin \theta) \sin \theta$$
$$+ I_{yy}(\dot{\theta} \cos \theta \sin \phi \cos \phi + \dot{\psi} \cos^2 \theta \sin^2 \phi)$$
$$+ I_{zz}(\dot{\psi} \cos^2 \theta \cos^2 \phi - \dot{\theta} \cos \theta \sin \phi \cos \phi)$$

$$p_\theta = \frac{\partial T}{\partial \dot{\theta}} = I_{yy}(\dot{\theta} \cos^2 \phi + \dot{\psi} \cos \theta \sin \phi \cos \phi) \qquad (8\text{--}88)$$
$$+ I_{zz}(\dot{\theta} \sin^2 \phi - \dot{\psi} \cos \theta \sin \phi \cos \phi)$$

$$p_\phi = \frac{\partial T}{\partial \dot{\phi}} = I_{xx}(\dot{\phi} - \dot{\psi} \sin \theta)$$

These generalized momenta are scalars equal to the magnitudes of orthogonal projections of the angular momentum vector **H** onto the $\dot{\psi}$, $\dot{\theta}$, and $\dot{\phi}$ axes, respectively. This can be seen by noting that for the case of p_ψ, for example, we have

$$p_\psi = \frac{\partial T}{\partial \dot{\psi}} = I_{xx} \omega_x \frac{\partial \omega_x}{\partial \dot{\psi}} + I_{yy} \omega_y \frac{\partial \omega_y}{\partial \dot{\psi}} + I_{zz} \omega_z \frac{\partial \omega_z}{\partial \dot{\psi}}$$

But the partial derivatives $\partial \omega_x/\partial \dot{\psi}$, $\partial \omega_y/\partial \dot{\psi}$, and $\partial \omega_z/\partial \dot{\psi}$ are just the cosines of the angles between $\dot{\psi}$ and the x, y, and z axes, respectively. Also,

$$H_x = I_{xx} \omega_x, \qquad H_y = I_{yy} \omega_y, \qquad H_z = I_{zz} \omega_z$$

Therefore we find that

$$p_\psi = \frac{1}{\dot\psi}\mathbf{H}\cdot\dot\psi$$

$$p_\theta = \frac{1}{\dot\theta}\mathbf{H}\cdot\dot\theta \tag{8-89}$$

$$p_\phi = \frac{1}{\dot\phi}\mathbf{H}\cdot\dot\phi$$

where the latter two expressions are obtained by a process similar to the first. It follows that because a system of nonorthogonal reference axes is being used, the generalized momenta are not equal to the magnitudes of component momentum vectors whose sum is the total angular momentum $\mathbf{H}$.

Now we can use the expressions given in Eqs. (8–87) and (8–88) in obtaining the differential equations of motion by the Lagrangian method. We have

$$\frac{dp_\psi}{dt} = M_\psi$$

$$\frac{dp_\theta}{dt} - \frac{\partial T}{\partial \theta} = M_\theta \tag{8-90}$$

$$\frac{dp_\phi}{dt} - \frac{\partial T}{\partial \phi} = M_\phi$$

where M_ψ, M_θ, and M_ϕ are the moments of the external forces about the $\dot\psi$, $\dot\theta$, and $\dot\phi$ axes. Note that $\partial T/\partial\psi = 0$.

It can be seen that a substitution of Eqs. (8–87) and (8–88) into Eq. (8–90) results in a rather complicated set of equations in the general case. When we recall that we have made no assumptions concerning the relative magnitudes of the principal moments of inertia, we find that these equations apply to an arbitrary rigid body. In the succeeding sections, however, we shall consider various special cases for which the rotational equations can be obtained in simpler form.

8-3. FREE MOTION OF A RIGID BODY

General Case. As a first simplified case of rotational motion, let us consider the *free motion* of a rigid body. By free motion, we mean that no external forces act on the body, implying that

$$M_\psi = M_\theta = M_\phi = 0 \tag{8-91}$$

It can be seen that the rotational motion is essentially free even if there are forces acting, so long as the total external moment about the center of mass (or a point fixed in space and fixed in the body) remains zero. Hence the following development will apply to these cases as well.

In the past we have chosen $\boldsymbol{\psi}$ to point vertically downward, that is, in the direction of the gravitational force. This direction was assumed to be fixed in inertial space. Now, for this case of free motion in which there is no longer a "vertical," we have the problem of choosing a suitable direction for $\boldsymbol{\psi}$. A convenient choice is to *let* $\boldsymbol{\psi}$ *point in a direction opposite to the angular momentum* $\mathbf{H}$ which, of course, is fixed in inertial space for this case of no external moments. With this assumption, we see that

$$p_\psi = -H \tag{8-92}$$

Also,

$$p_\theta = 0 \tag{8-93}$$

since $\boldsymbol{\theta}$ is always perpendicular to $\boldsymbol{\psi}$, and hence to $\mathbf{H}$.

A set of second-order differential equations in the Eulerian angles can be obtained by substituting the general expressions for p_ψ, p_θ, p_ϕ, and T into the general equations given in Eq. (8–90). Rather than writing out this set of rather lengthy equations, however, and attempting a solution, we can obtain their first integrals more directly in the following manner. First we note from the geometry that

$$\begin{aligned}
H_x &= I_{xx}\omega_x = H \sin \theta \\
H_y &= I_{yy}\omega_y = -H \cos \theta \sin \phi \\
H_z &= I_{zz}\omega_z = -H \cos \theta \cos \phi
\end{aligned} \tag{8-94}$$

Then from Eqs. (8–85) and (8–94), we can write

$$-\dot{\psi} \sin \theta + \dot{\phi} = \frac{H}{I_{xx}} \sin \theta$$

$$\dot{\psi} \cos \theta \sin \phi + \dot{\theta} \cos \phi = -\frac{H}{I_{yy}} \cos \theta \sin \phi \tag{8-95}$$

$$\dot{\psi} \cos \theta \cos \phi - \dot{\theta} \sin \phi = -\frac{H}{I_{zz}} \cos \theta \cos \phi$$

These equations can be solved for $\dot{\psi}$, $\dot{\theta}$, and $\dot{\phi}$ to yield

$$\dot{\psi} = -H \left(\frac{\sin^2 \phi}{I_{yy}} + \frac{\cos^2 \phi}{I_{zz}} \right)$$

$$\dot{\theta} = H \left(\frac{1}{I_{zz}} - \frac{1}{I_{yy}} \right) \cos \theta \sin \phi \cos \phi \tag{8-96}$$

$$\dot{\phi} = H \left(\frac{1}{I_{xx}} - \frac{\sin^2 \phi}{I_{yy}} - \frac{\cos^2 \phi}{I_{zz}} \right) \sin \theta$$

Thus we have obtained the first integrals of the differential equations of motion.

Note that $\dot{\psi}$ cannot be positive, but $\dot{\theta}$ and $\dot{\phi}$ can have either sign, depending upon the values of θ and ϕ and the relative magnitudes of the principal moments of inertia. Additional analytic results can be obtained by integrating the equations given in Eq. (8–96) to yield expressions for ψ, θ, and ϕ, all involving elliptic functions of time. We shall not, however, carry the solution further at this time.[1] Rather, we shall continue a discussion of the qualitative features of the motion when we discuss the Poinsot method in Sec. 8–4.

Axially Symmetric Body. Now let us obtain the general equations of rotational motion for the case of a rigid body which has at least two of its three principal moments of inertia at the mass center equal. From the point of view of its rotational characteristics such a body can be said to be axially symmetric whether its mass distribution is truly symmetrical or not. Let us choose the x axis as the axis of symmetry, thereby implying that I_{yy} and I_{zz} are equal. We shall designate the moments of inertia about the symmetry axis and about a transverse axis through the mass center by I_a and I_t, respectively. Thus

$$I_{xx} = I_a$$
$$I_{yy} = I_{zz} = I_t \tag{8–97}$$

For this case of axial symmetry, we see that the expression for the rotational kinetic energy given by Eq. (8–87) can be simplified to the form

$$T_{\text{rot}} = \tfrac{1}{2} I_a (\dot{\phi} - \dot{\psi} \sin \theta)^2 + \tfrac{1}{2} I_t (\dot{\theta}^2 + \dot{\psi}^2 \cos^2 \theta) \tag{8–98}$$

Similarly, the generalized momenta given in Eq. (8–88) can be expressed as follows:

$$p_\psi = -I_a (\dot{\phi} - \dot{\psi} \sin \theta) \sin \theta + I_t \dot{\psi} \cos^2 \theta$$
$$p_\theta = I_t \dot{\theta} \tag{8–99}$$
$$p_\phi = I_a (\dot{\phi} - \dot{\psi} \sin \theta)$$

Furthermore, it can be shown that p_ψ and p_ϕ are constant. This follows from the fact that the kinetic energy is not an explicit function of ψ or ϕ. Therefore, since the applied moments are zero, we see from Eq. (8–90) that dp_ψ/dt and dp_ϕ/dt are zero.

Let us define the *total spin* Ω by the equation

$$\Omega = \dot{\phi} - \dot{\psi} \sin \theta \tag{8–100}$$

Then we can write Eq. (8–99) in the form

$$p_\psi = -I_a \Omega \sin \theta + I_t \dot{\psi} \cos^2 \theta$$
$$p_\theta = I_t \dot{\theta} \tag{8–101}$$
$$p_\phi = I_a \Omega$$

[1]See J. L. Synge and B. A. Griffith, *Principles of Mechanics*, 3rd ed. (New York: McGraw-Hill Book Company, 1959), pp. 377–80 or E. T. Whittaker, *A Treatise on the Analytical Dynamics of Particle and Rigid Bodies*, 4th ed. (New York: Dover Publications, Inc., 1944) pp. 144–48 for the details of this derivation.

where we see that Ω is constant, since p_ϕ is constant. The total spin Ω is identical with ω_x and represents the amplitude of the orthogonal projection of the absolute rotational velocity $\boldsymbol{\omega}$ onto the axis of symmetry.

So far in this analysis of an axially symmetric body, we have made no assumptions concerning the orientation of the total angular momentum **H** relative to the XYZ system from which the Euler angles are measured. If we now assume once again that **H** *is directed opposite to the* $\dot{\boldsymbol{\psi}}$ *vector*, we see from Eqs. (8–93) and (8–101) that θ is constant. Hence

$$p_\psi = -H$$
$$p_\theta = 0 \qquad\qquad (8\text{–}102)$$
$$p_\phi = H \sin \theta$$

Having found that θ is constant, the nature of the motion can be better understood by solving for $\dot{\psi}$ and $\dot{\phi}$. We see from the middle equation of (8–95) that

$$H = -I_t \dot{\psi} \qquad\qquad (8\text{–}103)$$

indicating that the *precession rate* $\dot{\psi}$ is constant. Solving for $\dot{\psi}$, we obtain

$$\dot{\psi} = -\frac{H}{I_t} = -\frac{I_a \Omega}{I_t \sin \theta} \qquad\qquad (8\text{–}104)$$

The *relative spin* $\dot{\phi}$ is obtained from Eqs. (8–100) and (8–104), the result being

$$\dot{\phi} = \frac{I_a - I_t}{I_a} \dot{\psi} \sin \theta = \frac{I_t - I_a}{I_t} \Omega \qquad\qquad (8\text{–}105)$$

Thus we see that $\dot{\theta}$ is zero, whereas both $\dot{\psi}$ and $\dot{\phi}$ are constant for this case of the free motion of an axially symmetric body.

It is important to distinguish between the total spin Ω and the relative spin $\dot{\phi}$. We have seen that Ω is equal to the axial component ω_x of the absolute angular velocity. On the other hand, $\dot{\phi}$ is independent of any precessional motion the system may have. Note that an increase of 2π in the angle ϕ returns the body to the same position relative to a vertical plane which includes the x axis; that is, the same points of the body once again lie in this plane.

Now let us choose the direction of the positive x axis such that Ω is positive or zero. Then we see from Eq. (8–104) that θ is in the interval $0 \leq \theta \leq \pi/2$. The relative orientations of the various vectors are shown in Fig. 8–7 for a typical case. Notice that the angular velocity $\boldsymbol{\omega}$ can be considered as the sum of an axial component Ω and a transverse vector of magnitude $-\dot{\psi} \cos \theta$. On the other hand, $\boldsymbol{\omega}$ is also the vector sum of $\dot{\boldsymbol{\psi}}$ (whose magnitude is negative in this case so that it is actually directed upward) and $\dot{\boldsymbol{\phi}}$ which is parallel to the symmetry axis. This, of course, is in agreement with Eq. (8–84) for the present case where $\dot{\theta}$ is zero. In a similar fashion, we

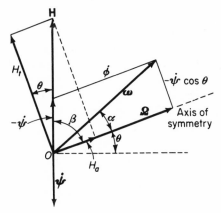

Fig. 8-7. Vector relationships for the free motion of an axially symmetric body. $I_t > I_a$.

see that the total angular momentum **H** has axial and transverse components given by

$$H_a = I_a\Omega = -I_t\dot{\psi}\sin\theta$$
$$H_t = -I_t\dot{\psi}\cos\theta$$
(8–106)

Let us designate the angle between $\boldsymbol{\omega}$ and the axis of symmetry by α and let β be the angle between **H** and the axis of symmetry. We see that

$$\beta = \frac{\pi}{2} - \theta$$
(8–107)

It is apparent from Fig. 8–7 that

$$\tan\alpha = -\frac{\dot{\psi}}{\Omega}\cos\theta$$
(8–108)

and

$$\tan\beta = \frac{H_t}{H_a} = -\frac{I_t\dot{\psi}\cos\theta}{I_a\Omega}$$
(8–109)

Hence

$$\tan\alpha = \frac{I_a}{I_t}\tan\beta$$
(8–110)

Noting that

$$\Omega = \omega\cos\alpha$$
(8–111)

we see from Eqs. (8–107) and (8–108) that

$$\dot{\psi} = -\frac{\sin\alpha}{\sin\beta}\omega$$
(8–112)

Now we substitute for $\sin\beta$ from Eq. (8–110) and obtain

$$\dot{\psi} = -\omega\sqrt{\sin^2 \alpha + [(I_a/I_t) \cos \alpha]^2} \tag{8–113}$$

A geometrical model which illustrates the free motion of an axially symmetric body consists of two right circular cones, one fixed in space and one fixed in the body. The vertex of each cone is located at the mass center of the body and the cones are tangent along a line corresponding to the instantaneous axis of rotation, as shown in Fig. 8–8. We shall consider two cases.

Case 1: $I_t > I_a$. *Direct Precession.* We can represent the rotational motion in terms of a *body cone* which we imagine to be fixed in the body and

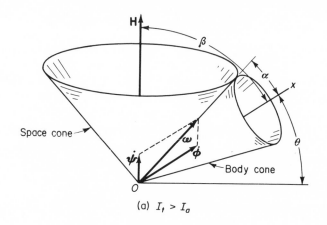

(a) $I_t > I_a$

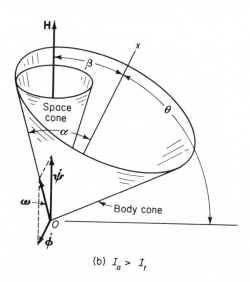

(b) $I_a > I_t$

Fig. 8-8. A geometrical representation of the (a) direct and (b) retrograde precession of a freely rotating body.

rotating with it. For this case of direct precession, the body cone rolls on
the outside surface of another right circular cone which is fixed in space
and is known as the *space cone*. The axis of the body cone coincides with
the symmetry axis of the actual rigid body, whereas the axis of the space cone
points in the direction of the angular momentum **H**. By comparing Figs.
8–7 and 8–8(a) we see the similarity in the vector orientations. Notice that
the total angular velocity **ω** lies along the line of contact between the two
cones. This line is the instantaneous axis of rotation since the body cone
rolls without slipping. Also note that $\beta > \alpha$ and therefore the **ω** vector lies
between the symmetry axis and the **H** vector, all being in the same
plane.

Finally, we see that $\dot{\phi}$ is positive and represents the angular rate with
which the line of contact moves relative to the body cone. If we consider
a circular cross section of the body cone which is rolling on a circular cross
section of the space cone, the condition of rolling requires that the contact
point move an equal distance relative to each cross section during any given
period. Hence the *relative* angular rates of the contact point are inversely
proportional to the corresponding radii, that is,

$$\frac{\dot{\phi}}{-\dot{\psi}} = \frac{\sin(\beta - \alpha)}{\sin \alpha} \tag{8–114}$$

This equation is also valid for Case 2 which follows.

Case 2: $I_a > I_t$. *Retrograde Precession*. This case applies to bodies where
the largest moment of inertia at the mass center is about the axis of symmetry, such as oblate spheroids, for example. Again the body cone rolls on
the fixed space cone, but in this case it is the *inside* surface of the body cone
which rolls on the outside surface of the space cone—see Fig. 8–8(b). From
Eq. (8–110), we see that $\alpha > \beta$; therefore the angular momentum vector
H lies between **ω** and the axis of symmetry. It can be seen from Fig. 8–8(b)
or from Eq. (8–105) that $\dot{\phi}$ is negative. Hence the **ω** vector moves in the
opposite sense relative to the body, as compared to its motion in Case 1.

It is interesting to note that the ratio I_a/I_t lies in the range $0 \le (I_a/I_t)$
≤ 2 for axially symmetric bodies, including the limiting mathematical idealizations of the thin rod and the circular disk. The semivertex angles of the
cones lie in the ranges $0 \le \alpha \le \pi/2$ and $0 \le |\beta-\alpha| < \pi/2$. To illustrate a
few of these limiting cases, consider first a body which is rotating about
its axis of symmetry. Then α and β are both zero and the cones become lines.
Moreover, Ω and $(\dot{\phi}-\dot{\psi})$ are each equal to ω, as can be seen from Eqs. (8–104),
(8–105), and (8–111). On the other hand, if the body is rotating about a
transverse axis, then $\alpha = \beta = \pi/2$. The space cone becomes a vertical
line and the body cone is a vertical plane which revolves about that line.
From Eqs. (8–111), (8–112), and (8–114), it follows that $\dot{\phi} = \Omega = 0$ and
$\dot{\psi} = -\omega$. Finally, for $I_a = 0$ the angle $(\beta-\alpha)$ can come arbitrarily close to

$\pi/2$ as α approaches zero. If $\alpha = 0$ for a finite ω, then $\mathbf{H} = 0$ and the angle $(\beta - \alpha)$ ceases to exist—Fig. 8–8(a).

Example 8–5. An axially symmetric body with $I_t = 10I_a$ is spinning about its vertical axis of symmetry with an angular velocity Ω. Suddenly an angular impulse of magnitude $\mathcal{M} = I_a\Omega$ is applied about a transverse axis through the center of mass. Solve for the rotational motion and find the maximum angular deviation of the symmetry axis from its original position. What is the period of the precessional motion?

The total angular momentum of the body just before the angular impulse is applied is

$$\mathbf{H}_0 = I_a\Omega$$

From the equation of angular impulse and momentum, Eq. (4–76), we know that the change in the angular momentum due to the transverse angular impulse is $\mathcal{M}$. Hence the total angular momentum after the impulse is

$$\mathbf{H} = I_a\Omega + \mathcal{M}$$

as shown in Fig. 8–9. It can be seen that

$$\tan \beta = \frac{\mathcal{M}}{I_a\Omega} = 1$$

The axial component Ω of the total angular velocity $\boldsymbol{\omega}$ is unchanged by the transverse angular impulse, but a transverse component of angular velocity appears which is of magnitude

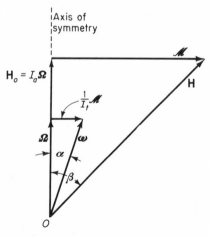

$$\frac{\mathcal{M}}{I_t} = \frac{I_a}{I_t}\,\Omega = 0.1\Omega$$

Hence we find that

$$\tan \alpha = \frac{\mathcal{M}}{I_t\Omega} = 0.1$$

During all further motion of the body, $\mathbf{H}$ remains constant in magnitude and direction, and since $I_t > I_a$, we can use the geometrical model of a body cone rolling on the outside surface of a fixed space cone in order to describe its motion. From Fig. 8–10, we see that the maximum angular deviation of the symmetry axis is

Fig. 8-9. The components of $\mathbf{H}$ and $\boldsymbol{\omega}$ immediately after a transverse angular impulse is applied.

$$2\beta = \frac{\pi}{2}$$

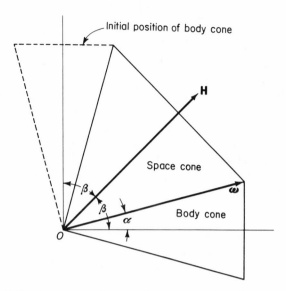

Fig. 8-10. A cross section of the space and body cones at the position of maximum angular deviation.

Noting that $\theta = \beta = \pi/4$, we can solve for the precession rate from Eq. (8–104), obtaining

$$\dot{\psi} = -\frac{I_a \Omega}{I_t \sin \theta} = -\frac{\sqrt{2}}{10} \Omega$$

The time required for one complete precession cycle is

$$T = \frac{2\pi}{|\dot{\psi}|} = \frac{10\sqrt{2}\,\pi}{\Omega}$$

The relative spin $\dot{\phi}$ is obtained from Eq. (8–105).

$$\dot{\phi} = \frac{I_t - I_a}{I_t} \Omega = 0.9\Omega$$

8–4. THE POINSOT METHOD

Now we shall consider another method for analyzing the free motion of a rigid body. This method, stated by Poinsot in 1834, is similar in its geometrical nature to the rolling-cone method and reduces to it for axially symmetric bodies. However, it applies to the free motion of general rigid bodies as well.

Poinsot's Construction. In Sec. 7–7, we saw that the rotational inertia characteristics of a rigid body are conveniently expressed by its *ellipsoid of inertia*. If we choose the center of mass as the origin of a cartesian set of principal axes, the equation of the ellipsoid of inertia is

$$I_1 x^2 + I_2 y^2 + I_3 z^2 = 1 \tag{8–115}$$

where I_1, I_2, and I_3 are the principal moments of inertia at the center of mass. We recall that the distance from the origin O to any point P on the surface of the ellipsoid of inertia is inversely proportional to the square root of the moment of inertia of the body about the axis OP (Fig. 8–11). So if the body is instantaneously rotating with an angular velocity $\boldsymbol{\omega}$ about an axis through O, the direction of rotational motion and the moment of inertia about the instantaneous axis are both given by the vector $\boldsymbol{\rho}$ drawn from O to P. From Eq. (7–102),

$$\boldsymbol{\rho} = \frac{\boldsymbol{\omega}}{\omega \sqrt{I}} \tag{8–116}$$

Now let us evaluate the length of the orthogonal projection of $\boldsymbol{\rho}$ onto the constant vector $\mathbf{H}$. Noting from Eqs. (7–51) and (7–56) that the rotational kinetic energy is

$$T_{\text{rot}} = \frac{1}{2} \boldsymbol{\omega} \cdot \mathbf{H} = \frac{1}{2} I \omega^2$$

we find that the length OS is

$$\frac{\boldsymbol{\rho} \cdot \mathbf{H}}{H} = \frac{\boldsymbol{\omega} \cdot \mathbf{H}}{\omega H \sqrt{I}} = \frac{\sqrt{2 T_{\text{rot}}}}{H}$$

Since both the rotational kinetic energy and the angular momentum are constant, it is apparent that OS is also constant. So we see that the point of

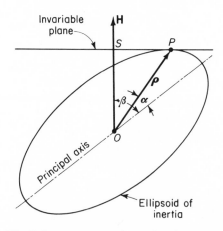

Fig. 8-11. The Poinsot construction, showing a principal cross section of the inertia ellipsoid as it rolls on the invariable plane.

the vector $\boldsymbol{\rho}$ moves in a plane which is perpendicular to the angular momentum vector $\mathbf{H}$ and intersects it at S. This plane is fixed in space and is known as the *invariable plane*.

We have found that the point P at the tip of the $\boldsymbol{\rho}$ vector lies in the invariable plane and also on the surface of the inertia ellipsoid. Now let us calculate the direction of the normal to the ellipsoid of inertia at P. This can be accomplished by finding the gradient of the function

$$F(x, y, z) = I_1 x^2 + I_2 y^2 + I_3 z^2 \tag{8-117}$$

where we recall from Eq. (8–115) that the equation of the ellipsoid of inertia in the principal axis system is

$$F(x, y, z) = 1 \tag{8-118}$$

This procedure can be visualized by noting that, as the right-hand side of Eq. (8–118) is changed slightly, the equation represents other concentric ellipsoids of the same shape. If we consider a series of neighboring ellipsoidal shells, each having a slightly different value of F, it is clear that at a given point on a certain shell, the direction in which F increases most rapidly is the direction of the outward normal, or ∇F. Performing the operations indicated by Eq. (3–87), we find that

$$\nabla F = \frac{\partial F}{\partial x}\mathbf{i} + \frac{\partial F}{\partial y}\mathbf{j} + \frac{\partial F}{\partial z}\mathbf{k}$$
$$= 2I_1 x\mathbf{i} + 2I_2 y\mathbf{j} + 2I_3 z\mathbf{k} \tag{8-119}$$

where $\mathbf{i}, \mathbf{j},$ and $\mathbf{k}$ are the usual cartesian unit vectors which point in the directions of the principal axes in this instance.

Now we write an expression for the angular momentum $\mathbf{H}$ about O, namely,

$$\mathbf{H} = I_1 \omega_x \mathbf{i} + I_2 \omega_y \mathbf{j} + I_3 \omega_z \mathbf{k} \tag{8-120}$$

Comparing Eqs. (8–119) and (8–120), we see that $\mathbf{H}$ and ∇F are parallel if

$$\omega_x : \omega_y : \omega_z = x : y : z \tag{8-121}$$

That this proportionality actually exists is evident from the fact that (x, y, z) are the cartesian components of $\boldsymbol{\rho}$, and $\boldsymbol{\omega}$ and $\boldsymbol{\rho}$ have the same direction. So, having shown that the normal to the inertia ellipsoid at P is parallel to $\mathbf{H}$, we conclude that the invariable plane is *tangent* to the inertia ellipsoid at P. Furthermore, the instantaneous axis of rotation passes through the contact point P at all times. So the ellipsoid of inertia rolls without slipping on the invariable plane, with the point O and the invariable plane remaining fixed in space. Of course, the inertia ellipsoid is fixed in the rigid body, so the rotational motion of the rigid body is also described by the rolling of the inertia ellipsoid on the invariable plane.

It is important to realize that the Poinsot method gives an *exact* representation of the free rotational motion of a rigid body. Also, we see that two bodies having equal inertia ellipsoids will have identical rotational motions, assuming the same initial conditions, even though their masses and exterior shapes might be quite different. Note that the motion usually involves a component of $\boldsymbol{\omega}$ which is normal to the invariable plane. This component causes a pivoting about P in addition to the rolling. If the body is rotating about a principal axis, the motion consists entirely of this pivoting movement.

We recall that the tip of the $\boldsymbol{\rho}$ vector remains in the invariable plane throughout the motion. The path traced by the point P as it moves in this plane is known as the *herpolhode*. On the other hand, P also traces a curve on the inertia ellipsoid, and this curve is known as the *polhode*. Each polhode is a closed curve. Examples of these curves are shown in Fig. 8–12 for a typical ellipsoid, the direction of travel of the tip of the $\boldsymbol{\rho}$ vector being indicated by the arrows.

In the general case, the herpolhode curves are *not closed* even though P moves continuously in the area between the two concentric circles corresponding to the extreme values of $(\beta-\alpha)$. During successive intervals of $\pi/2$ in ϕ, P moves from a point of tangency with one circle to a point of tangency with the other. It can be seen that the form of a herpolhode must repeat at regular intervals, corresponding to repeated circuits of P around the polhode. An interesting characteristic of herpolhode curves is that they are not wavy in the usual sense because they have no inflection points, the center of curvature being always on the same side of the curve as S.

Characteristics of the Motion. Now let us use the Poinsot construction to obtain some of the qualitative features of the free rotational motion of a general rigid body. First we note that if I_1, I_2, and I_3 are all different, the

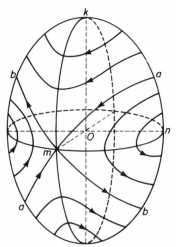

Fig. 8-12. An ellipsoid of inertia showing the polhode curves.

ellipsoid of inertia is not axially symmetric; hence the angle β of Fig. 8–11 oscillates between two limiting values unless the rotation is about a principal axis. This introduces an oscillation in θ or *nutation* to the motion of the given principal axis, in addition to its precession about the **H** vector. Suppose, for example, that

$$I_1 < I_2 < I_3$$

and these principal moments of inertia correspond to the axes Ok, Om, and On, respectively, in Fig. 8–12. Assuming that Ok is the positive x axis and the ellipsoid rolls on a polhode which is near k and encircles it, we see that $\dot{\psi}$, $\dot{\phi}$, and θ are all periodic functions of time, the period being the time required for ϕ to change by π. The precession rate $\dot{\psi}$ is always negative, the relative spin $\dot{\phi}$ is always positive, and θ remains in the range $0 < \theta < \pi/2$.

Consider now the case where On is taken as the x axis and the ellipsoid rolls on a polhode near n. Again $\dot{\psi}$, $\dot{\phi}$, and θ are periodic with two cycles occurring for each revolution of ϕ; but $\dot{\phi}$ is negative in this case.

In general, we notice that the polhodes form closed curves about two of the principal axes, but not about the third. If the motion takes place about a principal axis corresponding to either the smallest or largest moment of inertia, then the motion is stable. This is seen from the fact that the polhodes in the immediate vicinity of k or n are small *ellipses;* hence a small disturbance will not cause a large change in the direction of the axis of rotation relative to the body. On the other hand, we found previously that rotation about the principal axis associated with the intermediate moment of inertia is unstable. This instability is indicated by the *hyperbolic* shape of the polhodes in the vicinity of m. A small disturbance will cause the contact point P to move from m along a polhode near b to a point approximately opposite m, and then to return via a polhode near a to the vicinity of m for further repetitions of the cycle. The time required for this cycle approaches infinity as the polhode approaches m. The polhodes aa and bb also serve to separate the regions where the point P moves about the principal axis of minimum moment of inertia from those in which it moves about the axis of maximum moment of inertia.

Axial Symmetry. Now assume that $I_1 \leq I_2 \leq I_3$ and these principal moments of inertia are about the axes Ok, Om, and On, respectively, of Fig. 8–12. If we let I_2 and I_3 approach each other in magnitude, the polhodes aa and bb become more nearly horizontal. The ellipses in the vicinity of n become elongated, whereas those in the vicinity of k become nearly circular. This implies an increasing stability for rotation about Ok, but a decreasing stability for rotation about On.

Consider the limiting case where $I_2 = I_3$. The axis Ok is an axis of symmetry and the ellipsoid of inertia is an ellipsoid of revolution. The polhodes are all circles lying in planes perpendicular to Ok; hence we see that a rotation about the axis of symmetry is stable. On the other hand, the rotational motion about a transverse principal axis is neutrally stable and a small disturbance will result in a steady drift of the contact point P along a circular polhode passing near m and n.

For this case of axial symmetry in which $I_2 = I_3$, the Poinsot construction indicates that the general motion is that of a prolate spheroid rolling on the invariable plane. Since the polhodes are circles, the vector $\boldsymbol{\rho}$ sweeps out a circular cone in the ellipsoid of inertia. Furthermore, the herpolhodes are circles; therefore $\boldsymbol{\rho}$ also sweeps out a circular cone in space. Thus, comparing Figs. 8–8(a) and 8–11, we see that the Poinsot method and the rolling-cone method (Sec. 8–3) are equivalent for this case.

Now assume that $I_1 = I_2$. The axis On is the axis of symmetry in this instance and the inertia ellipsoid is an oblate spheroid. Again the Poinsot method and the rolling-cone method yield the same results for the motion of the body.

It is interesting to note in these cases of *axial symmetry* that the point P at the tip of the $\boldsymbol{\rho}$ vector moves along the circular polhode with an angular velocity $-\dot{\phi}$. By observing the directions of the arrows in Fig. 8–12, one can see why the sign of this motion relative to the body is dependent upon whether the inertia ellipsoid is prolate or oblate. For the prolate case, the point P moves in a negative sense about the axis Ok; whereas for the oblate case, the motion of P is in a positive sense about On. This provides a further illustration of the change in the sign of $\dot{\phi}$ as the ratio I_t/I_a passes through unity.

Polhode Equations. We have seen that the contact point P traces a polhode on the surface of the inertia ellipsoid as the body rotates in accordance with the Poinsot construction. A second ellipsoid can be described by first writing an expression for the square of the magnitude of the angular momentum, namely,

$$H^2 = I_1^2 \omega_x^2 + I_2^2 \omega_y^2 + I_3^2 \omega_z^2 \qquad (8\text{–}122)$$

Now divide by $2T_{\text{rot}} = I\omega^2$ and again let (x, y, z) represent the components of $\boldsymbol{\rho}$. The resulting equation is

$$I_1^2 x^2 + I_2^2 y^2 + I_3^2 z^2 = \frac{H^2}{2T_{\text{rot}}} = D \qquad (8\text{–}123)$$

where we note that the constant D has the dimensions of a moment of inertia. Equation (8–123) represents an ellipsoid which rotates with the body. Its shape depends upon the values of H and T_{rot} but, in any event, the point P must move on its surface. We see, then, that the polhode is the curve at the intersection of the ellipsoids given by Eqs. (8–115) and (8–123). (In general, two curves are formed, the second being the polhode for a reversed direction of $\boldsymbol{\omega}$.)

Let us continue to assume that $I_1 \leq I_2 \leq I_3$. If the body rotates about the axis Ok of Fig. 8–12, we see from Eqs. (8–116) and (8–123) that $D = I_1$. Similarly, if $\boldsymbol{\omega}$ is directed along the axis On, then $D = I_3$. For other directions of $\boldsymbol{\omega}$, the value of D will lie between these limits:

$$I_1 \leq D \leq I_3$$

Of course, D is constant for any given case of free motion, that is, for any particular polhode.

It can be seen that the ellipsoid of Eq. (8–123) has its principal axes oriented in the same directions as the corresponding axes of the inertia ellipsoid. Its shape is more elongated and exaggerated, however, since it lies outside the inertia ellipsoid near k and inside this ellipsoid in the vicinity of n. The two ellipsoids do not coincide except for the case of spherical symmetry.

The projections of the polhodes on the principal planes can be obtained by eliminating one of the variables between Eqs. (8–115) and (8–123). For example, solving for x^2 from Eq. (8–115), we obtain

$$x^2 = \frac{(1 - I_2 y^2 - I_3 z^2)}{I_1} \tag{8–124}$$

and, substituting into Eq. (8–123), the result is

$$I_2(I_2 - I_1)y^2 + I_3(I_3 - I_1)z^2 = D - I_1 \tag{8–125}$$

This represents the projection of the polhodes onto the yz plane, and for the usual case where $I_1 < I_2 < I_3$, it is the equation of an ellipse. Note that the permissible ranges in y and z are limited by the requirement that x^2 be positive in Eq. (8–124).

Now suppose that we eliminate y between Eqs. (8–115) and (8–123), the result being

$$I_1(I_2 - I_1)x^2 - I_3(I_3 - I_2)z^2 = I_2 - D \tag{8–126}$$

This is the equation of a hyperbola whose major axis is the x axis for the case where $D < I_2$. If $D > I_2$, then the z axis is the major axis of the polhode projection onto the xz plane.

For the third case, z is eliminated and the equation of the polhode projection onto the xy plane is

$$I_1(I_3 - I_1)x^2 + I_2(I_3 - I_2)y^2 = I_3 - D \tag{8–127}$$

Again the projection is an ellipse.

The polhode projections given here provide excellent approximations to the shape of the actual polhode curves in the immediate vicinity of a principal axis. Also, as we saw earlier, the shape of these curves is an indication of the stability of the spinning motion about a principal axis. So let us use Eqs. (8–125), (8–126), and (8–127) in considering the stability of such motion. For a given value of angular momentum, the motion is most stable when the body rotates about a principal axis with circular polhodes surrounding it. This situation occurs when $I_2 = I_3$ for a polhode near k, as seen from Eq. (8–125). Similarly, if $I_1 = I_2$ and the rotation is about On, the motion is relatively stable, as can be seen from Eq. (8–127). As the eccentricity of the polhodes increases, the stability is less because $\boldsymbol{\omega}$ will deviate farther from its nominal axis in the body. An example of neutral stability is provided by setting $I_1 = I_2$ in Eq. (8–125) or (8–126). Here the polhode approximation consists of a pair of lines perpendicular to the z axis. Finally, a hyperbolic polhode in the vicinity of a principal axis indicates unstable rotational motion about that axis which, as we have seen, is always associated with the intermediate moment of inertia.

Limits of the Motion. Next let us consider briefly the limits of the nutational or bobbing motion of a general body which is spinning freely. Referring to Fig. 8–11, we wish to find the extreme values of α and β, where these angles are measured from the principal axis which is encircled by the polhode under consideration. The angle $(\beta_{\max} - \beta_{\min})$ is the amplitude of the nutational motion.

Let us assume that $I_1 < I_2 \leq I_3$ and consider first a polhode about the I_1 axis, that is, the axis Ok of Fig. 8–12. We recall that $\boldsymbol{\rho}$ and $\boldsymbol{\omega}$ point in the same direction and, in fact, their magnitudes are proportional for any given case. So, for a certain body with a given initial value of $\boldsymbol{\omega}$, we can write Eq. (8–115) in the form

$$I_1\omega_1^2 + I_2\omega_2^2 + I_3\omega_3^2 = 2T_{\text{rot}} \tag{8–128}$$

Also, we can express Eq. (8–122) for this case as follows:

$$I_1^2\omega_1^2 + I_2^2\omega_2^2 + I_3^2\omega_3^2 = H^2 \tag{8–129}$$

We see from the Poinsot construction that α reaches a maximum value when $\boldsymbol{\omega}$ lies in the plane determined by the principal axes corresponding to I_1 and I_2, that is, in the plane kOm of Fig. 8–12. Hence, setting $\omega_3 = 0$ and solving for ω_1 from Eqs. (8–128) and (8–129), we obtain

$$\omega_1 = \sqrt{\frac{2I_2T_{\text{rot}} - H^2}{I_1(I_2 - I_1)}} \tag{8–130}$$

In a similar fashion, we find that

$$\omega_2 = \sqrt{\frac{H^2 - 2I_1T_{\text{rot}}}{I_2(I_2 - I_1)}} \tag{8–131}$$

The maximum value of α is found by noting that

$$\tan \alpha_{\max} = \frac{\omega_2}{\omega_1}$$

and substituting the above expressions for ω_1 and ω_2, we obtain

$$\tan \alpha_{\max} = \sqrt{\frac{I_1(H^2 - 2I_1T_{\text{rot}})}{I_2(2I_2T_{\text{rot}} - H^2)}} \tag{8–132}$$

It can be seen from the Poinsot construction that α and β reach corresponding extreme values simultaneously at the moment when both **H** and $\boldsymbol{\omega}$ lie in a principal plane. Hence the maximum value of β can be found by noting that $\omega_3 = 0$ at this time also; thus we obtain

$$\tan \beta_{\max} = \frac{H_2}{H_1} = \frac{I_2 \omega_2}{I_1 \omega_1}$$

or, using Eq. (8–132),

$$\tan \beta_{\max} = \sqrt{\frac{I_2(H^2 - 2I_1 T_{\text{rot}})}{I_1(2I_2 T_{\text{rot}} - H^2)}} \tag{8–133}$$

The minimum values of α and β occur when **H** and $\boldsymbol{\omega}$ lie in the plane kOn; that is, when ϕ differs by $\pi/2$ from the case just considered. Using a procedure similar to the foregoing, we find that the corresponding expressions are identical except for an interchange of the subscripts "2" and "3." The results are

$$\tan \alpha_{\min} = \sqrt{\frac{I_1(H^2 - 2I_1 T_{\text{rot}})}{I_3(2I_3 T_{\text{rot}} - H^2)}} \tag{8–134}$$

and

$$\tan \beta_{\min} = \sqrt{\frac{I_3(H^2 - 2I_1 T_{\text{rot}})}{I_1(2I_3 T_{\text{rot}} - H^2)}} \tag{8–135}$$

In the preceding derivation, we have assumed that the polhode under consideration encircles k. If, on the other hand, it encircles n, then a separate analysis shows that the correct equations are obtained merely by interchanging the "1" and "3" subscripts. The results for this case are

$$\tan \alpha_{\max} = \frac{\omega_2}{\omega_3} = \sqrt{\frac{I_3(2I_3 T_{\text{rot}} - H^2)}{I_2(H^2 - 2I_2 T_{\text{rot}})}} \tag{8–136}$$

$$\tan \beta_{\max} = \sqrt{\frac{I_2(2I_3 T_{\text{rot}} - H^2)}{I_3(H^2 - 2I_2 T_{\text{rot}})}} \tag{8–137}$$

$$\tan \alpha_{\min} = \sqrt{\frac{I_3(2I_3 T_{\text{rot}} - H^2)}{I_1(H^2 - 2I_1 T_{\text{rot}})}} \tag{8–138}$$

$$\tan \beta_{\min} = \sqrt{\frac{I_1(2I_3 T_{\text{rot}} - H^2)}{I_3(H^2 - 2I_1 T_{\text{rot}})}} \tag{8–139}$$

where we assume that $I_1 \leq I_2 < I_3$ and the angles α and β are measured from the On axis.

Solutions can be obtained for ω_1, ω_2, and ω_3 as functions of time. These solutions are somewhat similar to those for the Euler angles in that they involve elliptic functions; however, they will not be presented here.[2]

[2]Cf. W. D. MacMillan, *Dynamics of Rigid Bodies* (New York: Dover Publications, Inc., 1936), pp. 194–96.

Example 8-6. A rigid body with a ratio of principal moments of inertia given by $I_1:I_2:I_3 = 3:4:5$ is undergoing free rotational motion. Initially the direction of $\boldsymbol{\omega}$ is such that $\omega_1:\omega_2:\omega_3 = 3:1:1$. Solve for $(\beta_{max}-\beta_{min})$ and the maximum value of the ratio $-\dot{\psi}/\omega$.

Let us assume reference values of the moment of inertia and angular velocity such that

$$I_1 = 3I_0, \qquad I_2 = 4I_0, \qquad I_3 = 5I_0$$

and

$$\omega_1 = 3\omega_0, \qquad \omega_2 = \omega_3 = \omega_0$$

Then we find from Eqs. (8–128) and (8–129) that

$$2T_{rot} = 36I_0\omega_0^2$$

$$H^2 = 122I_0^2\omega_0^2$$

We notice that $I_1 < I_2 < I_3$ and can solve for the extreme values of β from Eqs. (8–133) and (8–135).

$$\tan \beta_{max} = \sqrt{\frac{I_2(H^2 - 2I_1T_{rot})}{I_1(2I_2T_{rot} - H^2)}} = \sqrt{\frac{56}{66}} = 0.9211$$

$$\tan \beta_{min} = \sqrt{\frac{I_3(H^2 - 2I_1T_{rot})}{I_1(2I_3T_{rot} - H^2)}} = \sqrt{\frac{70}{174}} = 0.6343$$

Hence we obtain

$$\beta_{max} - \beta_{min} = 42.6° - 32.4° = 10.2°$$

This angle is the nutation amplitude of the principal axis which is encircled by the polhode under consideration.

We see that $\dot{\psi}$ is the precession rate of this principal axis about the fixed **H** vector. Looking at Figs. 8–11 and 8–12, consider the Poinsot construction for those instants at which there are extreme values of α and β. We remember that $\boldsymbol{\omega} = \dot{\boldsymbol{\psi}} + \dot{\boldsymbol{\phi}}$. Also, we note from the construction that the $\dot{\boldsymbol{\psi}}$ component is of maximum amplitude and the $\dot{\boldsymbol{\phi}}$ component is of minimum amplitude when $\alpha = \alpha_{max}$. So let us first obtain α_{max} by using Eq. (8–132). The result is

$$\tan \alpha_{max} = \frac{I_1}{I_2} \tan \beta_{max} = 0.6908$$

$$\alpha_{max} = 34.6°$$

Noting that $\boldsymbol{\rho}$ and $\boldsymbol{\omega}$ are related by a constant scalar factor, we see that the component of $\boldsymbol{\omega}$ normal to the given principal axis is

$$\omega \sin \alpha_{max} = -\dot{\psi} \sin \beta_{max}$$

from which we obtain

$$\frac{-\dot{\psi}}{\omega} = \frac{\sin \alpha_{max}}{\sin \beta_{max}} = 0.839$$

This is clearly the maximum value of this ratio, since ω is minimum and $|\dot{\psi}|$ is maximum at this moment.

A check on this ratio can be obtained by using the general solutions for the Euler angle rates given in Eq. (8–96). Noting that $\beta = \pi/2 - \theta$ and taking $\beta = \beta_{\text{max}}$, we obtain

$$\dot{\psi} = -\frac{H}{I_2} = -\frac{\sqrt{122}}{4}\,\omega_0 = -2.761\,\omega_0$$

where we find that $\sin^2 \phi = 1$, $\cos^2 \phi = 0$, and we equate I_{xx}, I_{yy}, I_{zz} with I_1, I_2, I_3, respectively. Also,

$$\dot{\phi} = H \cos \beta \left(\frac{1}{I_1} - \frac{1}{I_2}\right) = 0.6770\,\omega_0$$

and, using Eq. (8–85), we find that

$$\omega = \sqrt{(\dot{\phi} - \dot{\psi}\cos\beta)^2 + \dot{\psi}^2 \sin^2 \beta} = 3.291\,\omega_0$$

Finally,

$$\frac{-\dot{\psi}}{\omega} = 0.839$$

in agreement with our previous result.

8–5.   THE MOTION OF A TOP

In this section, we shall consider the motion of an axially symmetric body, such as a top, which has a fixed point on its axis of symmetry and is acted upon by a uniform force field. The top was chosen because it is a relatively simple example of a body whose forced motion is markedly affected by gyroscopic moments associated with the spin about its axis of symmetry. The results, however, have application in the analysis of other systems, such as gyroscopes and spinning projectiles.

General Equations. Consider the rotational motion of the top shown in Fig. 8–13. It is assumed to spin without friction such that the point O on the axis of symmetry is fixed. The only external moment about O is that due to the constant gravitational force mg acting through its center of mass at C.

Let us analyze the motion of the top by using Lagrange's equations and choosing the Eulerian angles as coordinates. We note that this is an example of forced motion; hence **H** is not fixed in space. So let us use the original Euler angle definition of Sec. 7–12 in which the $\dot{\psi}$ vector is assumed to point vertically downward in the direction of the gravitational force.

If we choose the fixed point O as the reference point, the *total* kinetic energy may be written in terms of the Euler angle rates. Noting that $I_{xx} = I_a$ and $I_{yy} = I_{zz} = I_t$ for this case of symmetry about the x axis, we find from Eq. (8–87) that

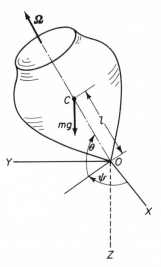

Fig. 8-13. A symmetrical top with a fixed point at O.

$$T = \frac{1}{2} I_a(\dot{\phi} - \dot{\psi} \sin \theta)^2$$
$$+ \frac{1}{2} I_t(\dot{\theta}^2 + \dot{\psi}^2 \cos^2 \theta) \qquad (8\text{--}140)$$

where the moments of inertia are taken about principal axes at O. Assuming a horizontal reference level through O, the gravitational potential energy is

$$V = mgl \sin \theta \qquad (8\text{--}141)$$

where l is the distance of the mass center from the fixed point. Now we can write the Lagrangian function:

$$L = T - V = \frac{1}{2} I_a(\dot{\phi} - \dot{\psi} \sin \theta)^2$$
$$+ \frac{1}{2} I_t(\dot{\theta}^2 + \dot{\psi}^2 \cos^2 \theta) \qquad (8\text{--}142)$$
$$- mgl \sin \theta$$

The generalized momenta are of the same form as we obtained in Eq. (8–101) for the unforced case. We see that

$$p_\psi = \frac{\partial L}{\partial \dot{\psi}} = -I_a \Omega \sin \theta + I_t \dot{\psi} \cos^2 \theta$$

$$p_\theta = \frac{\partial L}{\partial \dot{\theta}} = I_t \dot{\theta} \qquad (8\text{--}143)$$

$$p_\phi = \frac{\partial L}{\partial \dot{\phi}} = I_a \Omega$$

where we recall that the total spin Ω is given by

$$\Omega = \dot{\phi} - \dot{\psi} \sin \theta$$

The standard form of Lagrange's equation, namely,

$$\frac{d}{dt}\left(\frac{\partial L}{\partial \dot{q}_i}\right) - \frac{\partial L}{\partial q_i} = 0$$

is now applied together with Eq. (8–143) to obtain

$$\frac{dp_\psi}{dt} = 0$$
$$\qquad (8\text{--}144)$$
$$\frac{dp_\phi}{dt} = 0$$

from which we see that both p_ψ and p_ϕ are constant. Hence we find that Ω is constant for this case where there is no applied moment about the symmetry axis. Also, the precession rate $\dot\psi$ can be obtained from Eq. (8–143) with the following result:

$$\dot\psi = \frac{p_\psi + I_a\Omega \sin\theta}{I_t \cos^2\theta} \tag{8–145}$$

Now let us use the principle of conservation of energy to obtain an integral of the θ equation of motion. From Eqs. (8–140) and (8–141), we see that the total energy is

$$E = T + V = \frac{1}{2} I_a\Omega^2 + \frac{1}{2} I_t(\dot\theta^2 + \dot\psi^2 \cos^2\theta) + mgl\sin\theta \tag{8–146}$$

where we recall that $\Omega = \dot\phi - \dot\psi\sin\theta$ is constant. It follows that the total energy minus the kinetic energy associated with the total spin Ω is also a constant. Calling this quantity E', we can write

$$E' = E - \frac{1}{2} I_a\Omega^2 = \frac{1}{2} I_t(\dot\theta^2 + \dot\psi^2 \cos^2\theta) + mgl\sin\theta \tag{8–147}$$

Substituting for $\dot\psi$ from Eq. (8–145) and solving for $\dot\theta^2$, we obtain

$$\dot\theta^2 = \frac{2E'}{I_t} - \left(\frac{p_\psi + I_a\Omega\sin\theta}{I_t\cos\theta}\right)^2 - \frac{2mgl}{I_t}\sin\theta \tag{8–148}$$

Note that θ is the only variable on the right-hand side of this equation. Thus we see from Eqs. (8–145) and (8–148) that the precession rate $\dot\psi$ and the nutation rate $\dot\theta$ can be written as functions of θ alone for any given case.

In order to simplify the statement of Eq. (8–148), let us make the substitution

$$u = \sin\theta \tag{8–149}$$

from which it follows that

$$\dot u = \dot\theta \cos\theta \tag{8–150}$$

Also, let us define the constant parameters a, b, c, and e as follows:

$$a = \frac{p_\psi}{I_t}, \qquad b = \frac{I_a\Omega}{I_t}, \qquad c = \frac{2mgl}{I_t}, \qquad e = \frac{2E'}{I_t} \tag{8–151}$$

Now multiply Eq. (8–148) by $\cos^2\theta$ and make the foregoing substitutions. The resulting equation is

$$\dot u^2 = (1 - u^2)(e - cu) - (a + bu)^2 \tag{8–152}$$

If we use the notation

$$f(u) = (1 - u^2)(e - cu) - (a + bu)^2 \tag{8–153}$$

then we can write Eq. (8–152) in the form

$$\dot u^2 = f(u) \tag{8–154}$$

Now let us consider the function $f(u)$ in greater detail. Figure 8–14 shows a plot of $f(u)$ versus u for a typical case. We can assume that the parameter

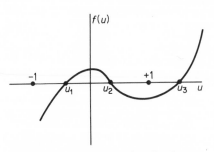

Fig. 8-14. A graph showing the three roots of $f(u)=0$ for a typical case of top motion.

c is positive since we can always consider the distance l from the support point to the center of mass to be positive. Furthermore, we see from Eq. (8–153) that the cubic term predominates for large absolute magnitudes of u. Hence we find that $f(u)$ must be negative for large negative values of u and must be positive for large positive values of u. Now $f(u)$ is a continuous function of u so it must be zero for at least one real value of u. That $f(u)$ is actually zero for three real values of u will now be shown.

It is evident that $\dot{u}^2$ is zero or positive for all physically realizable situations, so $f(u)$ must be zero or positive at some point in the interval $-1 \leq u \leq 1$ which is the range in which u must lie in the actual case. But if we set $u = \pm 1$ in Eq. (8–153) we find that $f(u)$ is zero or negative at these two points since $(a + bu)^2$ must be zero or positive. Therefore there are two roots in the interval $-1 \leq u \leq 1$ and the third root u_3 must lie in the range $u_3 \geq 1$. Summarizing, we can write

$$-1 \leq u_1 \leq u_2 \leq 1 \leq u_3$$

Looking again at Eq. (8–148) or (8–152), we find that the motion in θ stops only when $u = u_1$ or $u = u_2$. It is apparent, then, that u must oscillate between these values and θ will undergo a corresponding oscillation. To analyze the motion of θ in time, let us write Eq. (8–154) in the form

$$\dot{u}^2 = f(u) = c(u - u_1)(u - u_2)(u - u_3) \qquad (8\text{–}155)$$

Now we define

$$w = \sqrt{\frac{u - u_1}{u_2 - u_1}}, \qquad k = \sqrt{\frac{u_2 - u_1}{u_3 - u_1}}, \qquad p = \tfrac{1}{2}\sqrt{c(u_3 - u_1)} \qquad (8\text{–}156)$$

and we see that

$$\dot{w} = \frac{\dot{u}}{2\sqrt{(u - u_1)(u_2 - u_1)}} \qquad (8\text{–}157)$$

Making these substitutions, we can write Eq. (8–155) in the following form:

$$\dot{w}^2 = p^2(1 - w^2)(1 - k^2 w^2) \qquad (8\text{–}158)$$

If we measure the time t from the instant when θ is at its minimum value and $u = u_1$, we find that

$$pt = \int_0^w \frac{dw}{\sqrt{(1 - w^2)(1 - k^2 w)}} = F(\sin^{-1} w, k) \qquad (8\text{–}159)$$

where we recognize the integral as an elliptic integral of the first kind. Conversely, we can solve for w, obtaining

$$w = \text{sn } pt \tag{8-160}$$

where, we recall, the elliptic function, sn, was introduced previously in Sec. 3–9. Now we use the definition of w in Eq. (8–156) to solve for u

$$u = u_1 + (u_2 - u_1) \text{ sn}^2 pt \tag{8-161}$$

Because the sn function is squared, the period of the nutation in θ is just half the period of w. Thus the period of θ or u is

$$T = \frac{2K(k)}{p} = \frac{4K(k)}{\sqrt{c(u_3 - u_1)}} \tag{8-162}$$

where $K(k)$ is the complete elliptic integral of the first kind.

Referring to Eqs. (8–145) and (8–151), we see that the precession rate $\dot{\psi}$ can be expressed as a function of u as follows:

$$\dot{\psi} = \frac{a + bu}{1 - u^2} \tag{8-163}$$

Having solved for u as a function of time, we can also obtain $\dot{\psi}$ as a function of time. Also, from the definition of the total spin Ω given in Eq. (8–100), we see that

$$\dot{\phi} = \Omega + \dot{\psi}u \tag{8-164}$$

where Ω is constant. Hence $\dot{\phi}$ is a known function of time. Note that both $\dot{\psi}$ and $\dot{\phi}$ have the same period as θ.

Path of the Symmetry Axis. Let us consider the path of a point P located on the axis of symmetry at a unit distance from the fixed point O. Taking a horizontal reference through O, we find that $u = \sin \theta$ represents the height of P. The values u_1 and u_2 at which $\dot{u}$ is zero correspond to positions of minimum and maximum height and are known as *turning points*. If we represent the motion of the symmetry axis by the path of P on the surface of a unit sphere, we find that P will remain between two horizontal "latitude" circles given by $\theta = \theta_1$ and $\theta = \theta_2$, corresponding to the extreme values of u.

The path of P in any given case can be classified as one of three general types. In order to simplify this classification, let us define

$$u_0 = -\frac{a}{b} = -\frac{p_\psi}{p_\phi} \tag{8-165}$$

From Eq. (8–163) we see that $\dot{\psi} = 0$ when $u = u_0$ (except if $u_0 = \pm 1$, in which case $\dot{\psi}$ is indeterminate). Hence the motion of the symmetry axis at the instant when $u = u_0$ is such that the path of P is tangent to a vertical circle of "longitude."

Let us consider first the case where $u_0 > u_2$. In other words, u_0 lies outside the possible range of u, and therefore the precessional rate $\dot{\psi}$ does not equal zero at any time during the motion. This case is shown in Fig. 8–15(a), where the arrows indicate the direction of the velocity of P for positive Ω.

Next consider the case in which $u_1 < u_0 < u_2$. As shown in Fig. 8–15(b), we see that $\dot{\psi}$ is zero twice during each nutation cycle at $\theta = \theta_0$, and, for $\theta > \theta_0$, it actually reverses its sign compared to the average value of $\dot{\psi}$. In general, for a certain top with a given total spin Ω, the path traced by P is dependent upon the initial values of θ, $\dot{\theta}$, and $\dot{\psi}$. It can be seen, for example, that the formation of loops as in Fig. 8–15(b) can be accomplished by giving the axis of symmetry an initial precession rate which is opposite in direction to its average value.

Cuspidal Motion. The case in which $u_0 = u_2$ is known as *cuspidal motion* and is illustrated in Fig. 8–15(c). Note that $\dot{u}$ and $\dot{\psi}$ are both zero when $u = u_0$, resulting in a cusp which points upward. This type of motion will occur, for example, if the axis of a spinning top is released with zero initial velocity. The axis begins to fall vertically, causing gyroscopic inertial moments which result in a combination of nutation and precession such that the axis is motionless at a cusp at regular intervals. Incidentally, we see from energy considerations that the axis can be stationary only when P is at its highest point, corresponding to $u = u_2$. In other words, the potential energy must be at a maximum when the kinetic energy is minimum.

To find the limits of the nutation for cuspidal motion, let us set $\dot{u}^2$ or $f(u)$ equal to zero. Consider the following two cases: (1) $\dot{u} = 0$ because $\dot{\theta} = 0$; (2) $\dot{u} = 0$ because $\cos \theta = 0$.

Case 1: $u = u_0$ *when* $\dot{\theta} = 0$. Using the definition of u_0 given in Eq. (8–165), we see from Eq. (8–152) that

$$e = cu_0 \tag{8–166}$$

since $\dot{u} = 0$ at the cusp. We can then write an expression for $f(u)$ in the form

$$f(u) = c(u_0 - u)(1 - u^2) - b^2(u_0 - u)^2 = 0 \tag{8–167}$$

We have already shown that one limit of the nutational motion occurs at $u_2 = u_0$. Dividing out this root, we obtain

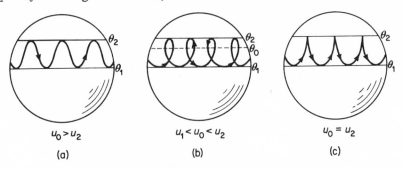

$$u_0 > u_2 \qquad\qquad u_1 < u_0 < u_2 \qquad\qquad u_0 = u_2$$

(a)   (b)   (c)

Fig. 8-15. Possible paths of the axis of symmetry.

$$c(1 - u^2) - b^2(u_0 - u) = 0$$

or

$$u^2 - 2\lambda u + (2\lambda u_0 - 1) = 0 \qquad (8\text{-}168)$$

where λ is a positive constant given by

$$\lambda = \frac{b^2}{2c} = \frac{I_a^2 \Omega^2}{4I_t mgl} \qquad (8\text{-}169)$$

Solving for the roots of Eq. (8-168) and including the one found previously, we can summarize the roots of $f(u) = 0$ for this case as follows:

$$\begin{aligned} u_1 &= \lambda - \sqrt{\lambda^2 - 2\lambda u_0 + 1} \\ u_2 &= u_0 \qquad\qquad\qquad\qquad (8\text{-}170) \\ u_3 &= \lambda + \sqrt{\lambda^2 - 2\lambda u_0 + 1} \end{aligned}$$

where we take the positive square root. For $-1 < u_0 \le 1$, an evaluation of Eq. (8-170) will show that

$$-1 \le u_1 \le u_0 \le 1 \le u_3$$

regardless of the value of λ and in agreement with our earlier results.

The limits of the motion in u or θ are found by evaluating u_0 from Eq. (8-166), λ from Eq. (8-169), and then using Eq. (8-170) to obtain u_1. Note again that u_3 has no physical meaning.

Case 2: $\dot{\theta} \ne 0$ *when* $u = u_0 = \pm 1$. Consider first that $u_0 = 1$. This case applies when the symmetry axis passes through the upper vertical position with a non-zero angular velocity. We see from Eq. (8-165) that

$$a = -b \qquad (8\text{-}171)$$

and, since $\dot{\theta}^2 > 0$ at this moment, we note from Eqs. (8-148) and (8-151) that

$$e > c \qquad (8\text{-}172)$$

So we can express $f(u)$ in the form

$$f(u) = (1 - u^2)(e - cu) - b^2(1 - u)^2 = 0 \qquad (8\text{-}173)$$

and solve for the turning points. Dividing out the known factor $(1-u)$ corresponding to $u_2 = 1$, there remains

$$(1 + u)(e - cu) - b^2(1 - u) = 0$$

which can be written in the form

$$u^2 - \left(\frac{b^2 + e}{c} - 1\right)u + \frac{b^2 - e}{c} = 0 \qquad (8\text{-}174)$$

One of the roots of this equation corresponds to a turning point at u_1. The other root u_3 is greater than 1 and has no physical meaning.

A similar procedure can be followed for the case where $\dot{\theta} \ne 0$ at $u = u_0 =$

-1. In this instance, the symmetry axis passes through the bottom point on the unit sphere with a non-zero velocity. We see from Eq. (8–165) that

$$a = b \tag{8–175}$$

Also, since $\dot{\theta}^2 > 0$ at $u = -1$, we see that

$$e + c > 0 \tag{8–176}$$

So we can write

$$f(u) = (1 - u^2)(e - cu) - b^2(1 + u)^2 = 0 \tag{8–177}$$

Dividing out the known factor $(1 + u)$ corresponding to the root $u_1 = -1$, we obtain

$$u^2 - \left(\frac{b^2 + e}{c} + 1\right) u - \frac{b^2 - e}{c} = 0 \tag{8–178}$$

The roots of this equation are the upper turning point u_2 and the nonphysical root u_3 which again is larger than unity.

Stability of Motion near the Vertical. An example of top motion of particular interest occurs if the axis points vertically upward when $\dot{\theta}$ is zero. This falls under Case 1 of cuspidal motion, and $u_0 = 1$ in this instance. From Eq. (8–170), we see that the roots are

$$u_1, u_2, u_3 = (2\lambda - 1), 1, 1 \tag{8–179}$$

where $(2\lambda–1)$ is designated as u_1 or u_3 according as λ is less than or greater than 1. Now let us consider the motion for the following values of λ: $\lambda < 1$, $\lambda = 1$, and $\lambda > 1$.

Case 1: $\lambda < 1$ *or* $\Omega^2 < 4 I_t mgl / I_a^2$. For this case the cusp occurs at a position of *unstable equilibrium* at $u = 1$. This can be seen from Fig. 8–16 by noting that a small disturbance at this position will cause a nutational motion during which θ decreases until the minimum value corresponding to $u_1 = 2\lambda - 1$ is reached; whereupon a reversal of this motion returns the

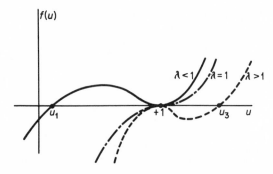

Fig. 8-16. Graphs of $f(u)$ indicating the stability of a vertical top.

axis to the vertical. The double root at $u = u_0 = 1$ is illustrated by the fact that the slope $f'(u)$ is zero at this point, as may be seen from Eq. (8–167).

It is interesting to note that the period of a nutation cycle is infinite. To show this, we refer to Eq. (8–156) and we find that the modulus $k = 1$; hence the complete elliptic integral $K(k)$ is infinite. It follows from Eq. (8–162) that the period is also infinite.

Case 2: $\lambda = 1$ *or* $\Omega^2 = 4I_t mgl/I_a^2$. Here is the borderline case in which there is neutral stability at the vertical. All three roots are equal to unity and thus there is an inflection point of $f(u)$ at $u = 1$.

Case 3: $\lambda > 1$ *or* $\Omega^2 > 4I_t mgl/I_a^2$. The spin is sufficiently large in this case that the vertical position is *stable*, that is, an infinitesimal disturbance will not cause a finite deviation of the symmetry axis from the vertical. Again we find that $f'(1) = 0$, but in this case the third root $u_3 = (2\lambda - 1) > 1$.

A top which is spinning as in Case 3 is known as a *sleeping top*. This name arises because a smooth, axially symmetric top with its axis vertical and $\lambda > 1$ might appear at first glance to be not moving at all, and hence "sleeping."

In an actual case there are small frictional moments which slowly decrease the spin until λ becomes less than unity. At this point a wobble or nutation appears and gradually increases until the body of the top hits the horizontal surface. Incidentally, the ability of an actual top with a large spin to right itself and reach the sleeping condition is due to the action of frictional forces on its rounded point. These forces have been omitted in this analysis.

Nutation Frequency and Amplitude for the Case of Large Spin. Now let us use Eq. (8–170), which was obtained in the analysis of *cuspidal motion*, to find the approximate amplitude and frequency of the nutational motion of a *fast top*, that is, one for which $\lambda \gg 1$. We can write the expression for u_1 in the form

$$u_1 = \lambda - \lambda\sqrt{1 - \frac{2u_0}{\lambda} + \frac{1}{\lambda^2}}$$

or, using the binomial expansion and neglecting terms of order higher than $1/\lambda$, we obtain

$$u_1 \cong \lambda - \lambda\left(1 - \frac{u_0}{\lambda} + \frac{1}{2\lambda^2} - \frac{u_0^2}{2\lambda^2}\right) = u_0 - \frac{1}{2\lambda}(1 - u_0^2) \quad (8\text{–}180)$$

The *amplitude of the nutation* is

$$u_0 - u_1 \cong \frac{1}{2\lambda}(1 - u_0^2) = \frac{2I_t mgl}{I_a^2 \Omega^2}(1 - u_0^2) \quad (8\text{–}181)$$

Note that the nutation amplitude varies inversely with the square of the total spin rate Ω. It becomes zero for $u_0 = 1$, that is, for a vertical top, the spin being much larger than that required for stability.

We see from Eq. (8–170), then, that the roots u_1 and $u_2 = u_0$ have a small separation; but the third root u_3 is approximated by

$$u_3 \cong 2\lambda \tag{8–182}$$

and thus is widely separated from the others.

To find the nutation frequency of a *fast top*, let us refer to Eq. (8–161) which gave the general solution for u in terms of elliptic functions. However, because u_3 is widely separated from u_1 and u_2, we see from Eq. (8–156) that the modulus k is very small; hence we can approximate the elliptic function sn pt by the trigonometric function sin pt. Thus we obtain

$$u = u_1 + (u_0 - u_1) \sin^2 pt \tag{8–183}$$

where the time t is measured from the moment when $u = u_1$. Using Eq. (8–181) and trigonometric identities, we can rewrite this result in the form:

$$u = u_0 - \frac{I_t\, mgl}{I_a^2 \Omega^2} (1 - u_0^2)(1 + \cos 2pt) \tag{8–184}$$

From Eqs. (8–151), (8–156), (8–169), and (8–182), we see that the *circular frequency of the nutation* is

$$2p = \sqrt{c(u_3 - u_1)} \cong \sqrt{2c\lambda} = \frac{I_a \Omega}{I_t} \tag{8–185}$$

An expression for the precession rate $\dot\psi$ can now be obtained, for we note from Eqs. (8–163) and (8–165) that

$$\dot\psi = -b \left(\frac{u_0 - u}{1 - u^2} \right)$$

Substituting for b and $(u_0 - u)$ from Eqs. (8–151) and (8–184), we obtain

$$\dot\psi = -\frac{mgl}{I_a \Omega} \left(\frac{1 - u_0^2}{1 - u^2} \right)(1 + \cos 2pt) \tag{8–186}$$

But we see from Eq. (8–184) that

$$1 - u^2 \cong (1 - u_0^2) + 2u_0(1 - u_0^2) \frac{I_t\, mgl}{I_a^2 \Omega^2} (1 + \cos 2pt)$$

or

$$\frac{1 - u_0^2}{1 - u^2} \cong 1 - \frac{u_0}{2\lambda}(1 + \cos 2pt) \cong 1 \tag{8–187}$$

where we neglect terms of order $1/\lambda$ or higher. So we obtain for this case of a fast top that the *precession rate* is

$$\dot\psi \cong -\frac{mgl}{I_a \Omega}(1 + \cos 2pt) \tag{8–188}$$

the approximation being valid even for cuspidal motion near the vertical, so long as $\lambda \gg 1$.

Precession with No Nutation. Precession with no nutation is indicated by having $u_1 = u_2$, that is, by the presence of a double root of $f(u)$ at a point other than at $u = \pm 1$. The conditions for a double root are that $f(u)$ and $f'(u)$ be zero for the same value of u, as shown in Fig. 8-17. Thus, using Eq. (8-153) we can write

$$f(u) = (1 - u^2)(e - cu) - (a + bu)^2 = 0 \qquad (8\text{-}189)$$

Also,

$$f'(u) = -2u(e - cu) - c(1 - u^2) - 2b(a + bu) = 0 \qquad (8\text{-}190)$$

From Eqs. (8-189) and (8-190), we have

$$e - cu = \frac{(a + bu)^2}{1 - u^2} = \frac{-c(1 - u^2) - 2b(a + bu)}{2u}$$

from which we obtain the following quadratic equation in $(a + bu)$:

$$2u(a + bu)^2 + 2b(1 - u^2)(a + bu) + c(1 - u^2)^2 = 0 \qquad (8\text{-}191)$$

Solving for $(a + bu)$, we obtain

$$a + bu = \frac{(1 - u^2)}{2u}[-b \pm \sqrt{b^2 - 2cu}] \qquad (8\text{-}192)$$

But we recall from Eq. (8-163) that

$$a + bu = \dot{\psi}(1 - u^2) = \dot{\psi} \cos^2 \theta \qquad (8\text{-}193)$$

so we find that the precession rate is

$$\dot{\psi} = \frac{-b}{2u}\left[1 \pm \sqrt{1 - \frac{2cu}{b^2}}\right] \qquad (8\text{-}194)$$

or, in terms of θ,

$$\dot{\psi} = \frac{-b}{2\sin\theta}\left[1 \pm \sqrt{1 - \frac{2c\sin\theta}{b^2}}\right] \qquad (8\text{-}195)$$

It is interesting to note that two steady precession rates are possible, provided that the values of θ and $b = I_a\Omega/I_t$ are such that the square root

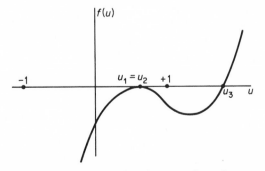

Fig. 8-17. The form of $f(u)$ for the case of steady precession.

in Eq. (8–195) is real. This last requirement is similar to requiring sufficient spin for stability in a vertical top. Thus we see that the condition on the total spin Ω in order that steady precession be possible at a given value of θ is that

$$\Omega^2 > \frac{4I_t\,mgl}{I_a^2}\sin\theta \tag{8–196}$$

Consider now the case in which $\Omega^2 \gg 4I_t mgl \sin\theta/I_a^2$; that is, the spin is large enough so that the second term in the square root of Eq. (8–195) is small compared to unity. Then we can approximate (8–195) by

$$\dot{\psi} = \frac{-b}{2\sin\theta}\left[1 \pm \left(1 - \frac{c\sin\theta}{b^2}\right)\right]$$

from which we obtain the following possible rates of uniform precession:

$$(\dot{\psi})_1 = \frac{-c}{2b} = -\frac{mgl}{I_a\Omega}$$
$$(\dot{\psi})_2 = \frac{-b}{\sin\theta} = -\frac{I_a\Omega}{I_t\sin\theta} \tag{8–197}$$

Note that the slow precession rate $(\dot{\psi})_1$ is independent of θ. This is the precession rate which is usually observed in a fast top or gyroscope and is also equal to the mean value of $\dot{\psi}$ found in Eq. (8–188). On the other hand, the fast precession rate $(\dot{\psi})_2$ is independent of the acceleration of gravity and, in fact, is identical with the free precession rate obtained previously in Eq. (8–104).

Example 8–7. A top with a total spin Ω and velocity v is sliding on a smooth horizontal floor with its symmetry axis vertical (Fig. 8–18). Suddenly, at $t = 0$, the point strikes a crack at O and is prevented from moving further although the angular motion is unhindered. If θ is the angle between the axis of symmetry and the floor, find

  a. The angular velocity $\dot{\theta}(0 +)$ just after the vertex is stopped.

  b. The linear impulse exerted on the top at $t = 0$.

  c. $\theta_{\min}$ in the ensuing motion, assuming that $\Omega = 20v/l = 20\sqrt{g/l}$ and $I_a = \frac{1}{4}ml^2 = I_t/5$, where the moments of inertia are taken about axes through the vertex.

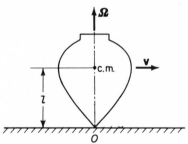

Our approach will be to note that the angular momentum about the vertex O is conserved during the initial instant because the reaction forces of the floor pass through this point. Hence the angular impulse applied at $t = 0$ is zero. Equating expressions for the horizontal com-

Fig. 8-18. A sliding top on a horizontal floor.

ponent of angular momentum before and after the impact at $t = 0$, we obtain

$$H_h = mvl = -I_t\dot\theta(0+) \tag{8-198}$$

where this component is directed into the page. Thus we find that

$$\dot\theta(0+) = -\frac{mvl}{I_t} \tag{8-199}$$

The linear impulse is found by calculating the change in the linear momentum during impact. Noting that

$$v(0+) = -\dot\theta(0+)l$$

we obtain

$$\mathscr{F} = mv(0+) - mv = -mv\left(1 - \frac{ml^2}{I_t}\right) \tag{8-200}$$

where the impulse is positive when directed to the right.

In order to calculate the minimum value of θ, we note first that $\dot\theta \neq 0$ at $u = \sin\theta = 1$. Therefore we have the cuspidal motion which was described previously under Case 2 on page 407. Evaluating the constant parameters b, c, and e from Eq. (8-151), we have

$$b = \frac{I_a\Omega}{I_t} = 4\sqrt{\frac{g}{l}}$$

$$c = \frac{2mgl}{I_t} = \frac{8g}{5l} \tag{8-201}$$

$$e = \frac{2E'}{I_t} = \left(\frac{mvl}{I_t}\right)^2 + \frac{2mgl}{I_t} = \frac{56g}{25l}$$

The turning point is found by substituting these values into Eq. (8-174) with the result:

$$u^2 - 10.4u + 8.6 = 0$$

The roots of this equation are u_1 and u_3, the third root of $f(u) = 0$ being $u_2 = u_0 = 1$. We obtain

$$u_1 = 0.9058$$

$$u_3 = 9.494$$

Thus we find that

$$\theta_{\min} = \sin^{-1}(0.9058) = 64.9°$$

Example 8-8. A uniform circular disk of mass m, radius a, rolls on a horizontal surface in such a manner that its plane is inclined with the vertical at a constant angle θ and its center of mass describes a circular path of radius R—Fig. 8-19(a). Solve for the precession rate $\dot\psi$.

First Method: Although the disk is an axially symmetric body subject to gravitational moments, the geometry and the constraints are different

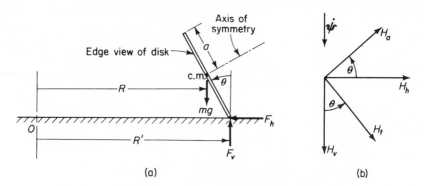

Fig. 8-19. (a) The forces acting on a disk rolling in a circular path.
(b) The components of angular momentum in the vertical plane.

from those encountered in the analysis of a top. So let us start again with the basic rotational equation

$$\mathbf{M} = \dot{\mathbf{H}}$$

where we shall choose the fixed point O as a reference.

Let us use Eulerian angles to describe the orientation of the disk. The precession rate $\dot{\psi}$ can be obtained in terms of the total spin Ω by equating two expressions for the translational velocity of the center of mass.

$$v = R\dot{\psi} = a\Omega$$

from which we have

$$\dot{\psi} = \frac{a}{R}\,\Omega \qquad (8\text{--}202)$$

It can be seen that $\dot{\psi}$ must be constant. A changing $\dot{\psi}$ would imply a changing kinetic energy. But, since θ is constant and no slipping occurs, there are no working forces acting on the disk; hence, by the principle of work and kinetic energy, $\dot{\psi}$ cannot change. From Eq. (8–202), we see that Ω is also constant.

Consider now the axial and transverse components of the angular momentum *about the center of mass*. From Eqs. (8–106) and (8–202), we can write

$$H_a = I_a\Omega = \frac{R}{a}\,I_a\dot{\psi}$$

$$H_t = I_t\dot{\psi}\cos\theta$$

To find the total angular momentum about O, we must add the angular momentum due to the translational velocity of the center of mass about O. Taking horizontal and vertical components of this portion, we obtain

$$H_v = mR^2\dot{\psi}$$

$$H_h = mRa\,\dot{\psi}\cos\theta$$

As the disk travels in its circular path, the system of four component vectors rotates with angular velocity $\dot{\psi}$, as shown in Fig. 8–19(b). Thus we see that the total angular momentum vector $\mathbf{H}$ is of constant magnitude but precesses about a vertical axis at the same rate as the disk.

Now we can evaluate $\dot{\mathbf{H}}$ by noting that

$$\dot{\mathbf{H}} = \dot{\boldsymbol{\psi}} \times \mathbf{H} \tag{8–203}$$

for this case in which the magnitude H is constant. We find that

$$\dot{\mathbf{H}} = \dot{\psi}(H_h + H_a \cos \theta + H_t \sin \theta)\mathbf{e}_\theta \tag{8–204}$$

or, substituting the preceding expressions for H_h, H_a, and H_t, we obtain

$$\dot{\mathbf{H}} = \dot{\psi}^2 \cos \theta \left(mRa + \frac{R}{a} I_a + I_t \sin \theta \right) \mathbf{e}_\theta \tag{8–205}$$

where $\mathbf{e}_\theta$ is a unit vector directed out of the page. The applied moment about O is independent of F_h and is given by

$$\mathbf{M} = mga \sin \theta \, \mathbf{e}_\theta \tag{8–206}$$

where we note that $F_v = mg$ since the mass center has no vertical motion. Equating the expressions on the righthand sides of Eqs. (8–205) and (8–206), we obtain

$$\dot{\psi}^2 \cos \theta \left(mRa + \frac{R}{a} I_a + I_t \sin \theta \right) = mga \sin \theta$$

or

$$\dot{\psi}^2 = \frac{mga \tan \theta}{mRa + \dfrac{R}{a} I_a + I_t \sin \theta} \tag{8–207}$$

For a thin uniform disk, we recall that

$$I_a = \frac{1}{2} ma^2$$

$$I_t = \frac{1}{4} ma^2$$

Substituting these values into Eq. (8–207), we obtain

$$\dot{\psi}^2 = \frac{4g \tan \theta}{6R + a \sin \theta} \tag{8–208}$$

Second Method: Now let us consider this problem from the Lagrangian viewpoint. Using Eq. (8–98) for the rotational kinetic energy and adding the translational kinetic energy, we obtain the total kinetic energy

$$T = \frac{1}{2} m(R^2 \dot{\psi}^2 + a^2 \dot{\theta}^2) + \frac{1}{2} I_a(\dot{\phi} - \dot{\psi} \sin \theta)^2 + \frac{1}{2} I_t(\dot{\theta}^2 + \dot{\psi}^2 \cos^2 \theta) \tag{8–209}$$

In order to use a standard form of Lagrange's equation, the coordinates must be *independent*. But we saw in Eq. (8–202) that the condition of rolling imposes the constraint that

$$\Omega = \dot{\phi} - \dot{\psi} \sin \theta = \frac{R}{a} \dot{\psi}$$

so the coordinates ψ, θ, ϕ are not independent as they stand. If, however, we write the constraint equation in the form

$$\dot{\phi} - \dot{\psi} \sin \theta = \frac{R' - a \sin \theta}{a} \dot{\psi} \tag{8–210}$$

and substitute into the kinetic energy expression of Eq. (8–209), we obtain the result

$$T = \frac{1}{2} \left(m + \frac{I_a}{a^2} \right) (R' - a \sin \theta)^2 \dot{\psi}^2 + \frac{1}{2} (I_t' \dot{\theta}^2 + I_t \dot{\psi}^2 \cos^2 \theta) \tag{8–211}$$

where we have let $I_t' = I_t + ma^2$ and

$$R = R' - a \sin \theta \tag{8–212}$$

in accordance with Fig. 8–19(a). Now ϕ has been eliminated as a generalized coordinate and the remaining coordinates ψ and θ are independent in the kinematic sense. Also, we have introduced a new constant, R', because R is no longer constant if θ is considered to be variable.

It can be seen that the system is conservative and the potential energy is given by

$$V = mga \cos \theta \tag{8–213}$$

The generalized forces are

$$M_\psi = -\frac{\partial V}{\partial \psi} = 0$$

$$M_\theta = -\frac{\partial V}{\partial \theta} = mga \sin \theta \tag{8–214}$$

Let us use Lagrange's equation in the form of Eq. (6–73), namely,

$$\frac{d}{dt} \left(\frac{\partial T}{\partial \dot{q}_i} \right) - \frac{\partial T}{\partial q_i} = Q_i$$

First we obtain the generalized momenta:

$$p_\psi = \frac{\partial T}{\partial \dot{\psi}} = \left[\left(m + \frac{I_a}{a^2} \right) (R' - a \sin \theta)^2 + I_t \cos^2 \theta \right] \dot{\psi}$$

$$p_\theta = \frac{\partial T}{\partial \dot{\theta}} = I_t' \dot{\theta} \tag{8–215}$$

Also, we evaluate the following terms:

$$\frac{\partial T}{\partial \psi} = 0$$

$$\frac{\partial T}{\partial \theta} = -\left[\left(m + \frac{I_a}{a^2}\right)(R' - a \sin \theta)a + I_t \sin \theta\right] \cos \theta \, \dot{\psi}^2$$

(8–216)

From Eqs. (8–214), (8–215), and (8–216), we can now write the ψ equation in the form

$$\frac{dp_\psi}{dt} = 0 \qquad (8\text{–}217)$$

indicating that p_ψ is constant. For this particular case in which θ is constant, it follows from Eq. (8–215) that the precession rate $\dot{\psi}$ is constant.

In a similar fashion we find that the θ equation is

$$I'_t \ddot{\theta} + \left[\left(m + \frac{I_a}{a^2}\right)(R' - a \sin \theta)a + I_t \sin \theta\right] \cos \theta \, \dot{\psi}^2 = mga \sin \theta \quad (8\text{–}218)$$

Setting $\ddot{\theta} = 0$ and solving for $\dot{\psi}^2$, we obtain

$$\dot{\psi}^2 = \frac{mga \tan \theta}{mRa + \dfrac{R}{a} I_a + I_t \sin \theta}$$

in agreement with our earlier result.

8-6. OTHER METHODS FOR AXIALLY SYMMETRIC BODIES

Modified Euler Equations. In our discussion of Euler's equations of motion in Sec. 8–1, we used a body-axis coordinate system with its origin at the center of mass. The primary reason for choosing a system fixed in the body was that it resulted in constant moments of inertia. Furthermore, *principal axes* were chosen in order that the equations be as simple as possible.

In the analysis of axially symmetric bodies, it can be seen that an arbitrary rotation of the body relative to the coordinate system will not cause the moments of inertia to change, provided that the rotation occurs about the axis of symmetry. So let us investigate the form of the rotational equations of motion, using a set of principal axes which are not necessarily fixed in the body, but may rotate relative to the body about the symmetry axis. Again let $\boldsymbol{\omega}'$ be the absolute angular velocity of the *body*, and let $\boldsymbol{\omega}$ be the absolute angular velocity of the *xyz coordinate system*. Choose the origin of the *xyz* system at a fixed point or at the center of mass, and let the axis of symmetry of the body be the *x* axis (Fig. 8–20). Then, using the general equations for the components of angular momentum given in Eq. (8–5), we can write

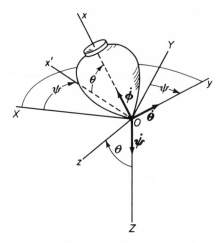

Fig. 8-20. Coordinate systems used in describing the motion of an axially symmetric body.

$$H_x = I_a \omega_x'$$
$$H_y = I_t \omega_y' \qquad (8\text{--}219)$$
$$H_z = I_t \omega_z'$$

Now suppose that the body rotates about the axis of symmetry (*x* axis) with an angular velocity *s* *measured relative to the xyz system.* Then we can write expressions for ω_x', ω_y', and ω_z' in terms of the angular velocity components of the coordinate system as follows:

$$\omega_x' = \omega_x + s$$
$$\omega_y' = \omega_y \qquad (8\text{--}220)$$
$$\omega_z' = \omega_z$$

From Eqs. (8–219) and (8–220), we obtain

$$H_x = I_a(\omega_x + s)$$
$$H_y = I_t \omega_y \qquad (8\text{--}221)$$
$$H_z = I_t \omega_z$$

and by differentiating with respect to time, the rates of change of these scalar components are found to be

$$\dot{H}_x = I_a(\dot{\omega}_x + \dot{s})$$
$$\dot{H}_y = I_t \dot{\omega}_y \qquad (8\text{--}222)$$
$$\dot{H}_z = I_t \dot{\omega}_z$$

Finally, we recall from Eq. (8–3) that

$$\dot{\mathbf{H}} = (\dot{\mathbf{H}})_r + \boldsymbol{\omega} \times \mathbf{H}$$

and, using Eqs. (8–9), (8–221), (8–222), and the general vector equation of motion, $\mathbf{M} = \dot{\mathbf{H}}$, we obtain the *modified Euler equations:*

$$M_x = I_a(\dot{\omega}_x + \dot{s}) = I_a\dot{\Omega}$$
$$M_y = I_t\dot{\omega}_y - (I_t - I_a)\omega_z\omega_x + I_a s\omega_z \qquad (8\text{--}223)$$
$$M_z = I_t\dot{\omega}_z + (I_t - I_a)\omega_x\omega_y - I_a s\omega_y$$

where we note that the total spin Ω is given by

$$\Omega = \omega_x + s \qquad (8\text{--}224)$$

It can be seen that these equations of motion bear a strong resemblance to the Euler equations of Eq. (8–12) for the case of an axially symmetric body and, in fact, are identical for $s = 0$. The added flexibility of being able to choose the spin s *in an arbitrary manner*, however, enables us to obtain more easily the motion of the body in space.

Let us consider the case where we choose s such that the y axis remains in the horizontal plane. We can specify the orientation of the xyz system by using the Euler angles ψ and θ, as shown in Fig. 8–20. In this case, it is evident that the spin s must correspond to the rate of change of the third Euler angle, that is,

$$s = \dot{\phi} \tag{8-225}$$

The rotational motion of the xyz system does not involve ϕ. So if we set ϕ and $\dot{\phi}$ equal to zero in Eq. (8–85), or refer to Fig. 8–20, we see that

$$\omega_x = -\dot{\psi} \sin \theta$$
$$\omega_y = \dot{\theta} \tag{8-226}$$
$$\omega_z = \dot{\psi} \cos \theta$$

Differentiating these equations, we obtain

$$\dot{\omega}_x = -\ddot{\psi} \sin \theta - \dot{\psi}\dot{\theta} \cos \theta$$
$$\dot{\omega}_y = \ddot{\theta} \tag{8-227}$$
$$\dot{\omega}_z = \ddot{\psi} \cos \theta - \dot{\psi}\dot{\theta} \sin \theta$$

and, substituting from Eqs. (8–225), (8–226), and (8–227) into the modified Euler equations of (8–223), we obtain a new set of rotational equations in terms of the Eulerian angles.

$$M_x = I_a \dot{\Omega} = I_a(\ddot{\phi} - \ddot{\psi} \sin \theta - \dot{\psi}\dot{\theta} \cos \theta)$$
$$M_y = I_t(\ddot{\theta} + \dot{\psi}^2 \sin \theta \cos \theta) + I_a\dot{\psi} \cos \theta(\dot{\phi} - \dot{\psi} \sin \theta) \tag{8-228}$$
$$M_z = I_t(\ddot{\psi} \cos \theta - 2\dot{\psi}\dot{\theta} \sin \theta) - I_a\dot{\theta}(\dot{\phi} - \dot{\psi} \sin \theta)$$

Note again that if the axial moment M_x is zero, the total spin $\Omega = \dot{\phi} - \dot{\psi} \sin \theta$ is a constant.

To illustrate the use of Eq. (8–228), let us calculate the steady precession rate of the top of Fig. 8–13 which we analyzed in Sec. 8–5. Because of our particular choice of the xyz coordinate system, the only non-zero component of the applied moment is M_y. This gravitational moment about the point O is given by

$$M_y = -mgl \cos \theta \tag{8-229}$$

Also, we see that $\dot{\theta} = 0$ for the case of steady precession with no nutation. Therefore the second equation of (8–228) becomes

$$I_t\dot{\psi}^2 \sin \theta \cos \theta + I_a\Omega\dot{\psi} \cos \theta = -mgl \cos \theta$$

or

$$\dot{\psi}^2 + \frac{I_a\Omega}{I_t \sin \theta} \dot{\psi} + \frac{mgl}{I_t \sin \theta} = 0 \tag{8-230}$$

Using the definitions of the constants b and c given in Eq. (8–151), we have

$$\dot{\psi}^2 + \frac{b}{\sin\theta}\dot{\psi} + \frac{c}{2\sin\theta} = 0 \qquad (8\text{–}231)$$

from which we obtain

$$\dot{\psi} = -\frac{b}{2\sin\theta}\left[1 \pm \sqrt{1 - \frac{2c\sin\theta}{b^2}}\right]$$

in agreement with the result found previously in Eq. (8–195).

Lagrange's Equations for the General Case. We have used the Lagrangian approach in obtaining the equations for the free motion of an axially symmetric body and for the motion of a top. Now let us consider the differential equations of motion for the more general case of arbitrary external moments. We use Lagrange's equation in the form

$$\frac{d}{dt}\left(\frac{\partial T}{\partial \dot{q}_i}\right) - \frac{\partial T}{\partial q_i} = Q_i$$

and take either a fixed point or the center of mass as the reference point in writing the expression for the rotational kinetic energy given previously in Eq. (8–98):

$$T_{\text{rot}} = \frac{1}{2}I_a(\dot{\phi} - \dot{\psi}\sin\theta)^2 + \frac{1}{2}I_t(\dot{\theta}^2 + \dot{\psi}^2\cos^2\theta)$$

The generalized momenta were given in Eq. (8–99):

$$p_\psi = \frac{\partial T}{\partial \dot{\psi}} = -I_a(\dot{\phi} - \dot{\psi}\sin\theta)\sin\theta + I_t\dot{\psi}\cos^2\theta$$

$$p_\theta = \frac{\partial T}{\partial \dot{\theta}} = I_t\dot{\theta}$$

$$p_\phi = \frac{\partial T}{\partial \dot{\phi}} = I_a(\dot{\phi} - \dot{\psi}\sin\theta)$$

Also, we have

$$\frac{\partial T}{\partial \psi} = 0$$

$$\frac{\partial T}{\partial \theta} = -I_a\dot{\psi}\cos\theta\,(\dot{\phi} - \dot{\psi}\sin\theta) - I_t\dot{\psi}^2\sin\theta\cos\theta \qquad (8\text{–}232)$$

$$\frac{\partial T}{\partial \phi} = 0$$

Performing the required differentiations of the generalized momenta in accordance with the above form of Lagrange's equation, or from Eq. (8–90), we obtain the following equations of motion:

$$I_a[(-\ddot{\phi} + \dot{\psi}\sin\theta + \dot{\psi}\dot{\theta}\cos\theta)\sin\theta - \dot{\theta}(\dot{\phi} - \dot{\psi}\sin\theta)\cos\theta]$$
$$+ I_t(\dot{\psi}\cos^2\theta - 2\dot{\psi}\dot{\theta}\sin\theta\cos\theta) = M_\psi$$
$$I_t(\ddot{\theta} + \dot{\psi}^2\sin\theta\cos\theta) + I_a\dot{\psi}\cos\theta\,(\dot{\phi} - \dot{\psi}\sin\theta) = M_\theta \qquad (8\text{-}233)$$
$$I_a(\ddot{\phi} - \dot{\psi}\sin\theta - \dot{\psi}\dot{\theta}\cos\theta) = M_\phi$$

Comparing these equations with those obtained previously in Eq. (8-228), we find that

$$M_\psi = -M_x \sin\theta + M_z \cos\theta$$
$$M_\theta = M_y \qquad\qquad\qquad (8\text{-}234)$$
$$M_\phi = M_x$$

as may be seen from Fig. 8-20.

The Use of Complex Notation. We have found that it is often important to solve for the motion of the axis of symmetry in space. As we saw in Sec. 8-5, this can be accomplished by solving for the path of a point P located on the axis of symmetry at a unit distance from the reference point O. Evidently this path is traced on the surface of a unit sphere centered at O, as was shown in Fig. 8-15, for example.

Now if the point P remains near the "pole"; that is, if $\theta \cong \pm\pi/2$, then the normal projection of the path onto a plane that is tangent at the pole gives a good approximation to the actual path of P. Suppose that this plane is called the $y'z'$ plane, or w plane, where the y' and z' axes are not rotating in space and the origin is taken at the pole (Fig. 8-21). The motion of P can then be specified conveniently in terms of the *complex variable*

$$w = y' + iz' \qquad\qquad (8\text{-}235)$$

where $i = \sqrt{-1}$.

We shall assume that the reference point O is a fixed point, or at the mass center, and consider the effect of transverse moments on the motion of P. We note first that, under our assumption of small motions, the transverse moment vector remains essentially in the w plane and we can write

$$M_t = M_{y'} + iM_{z'} \qquad (8\text{-}236)$$

where we now consider M_t as a complex variable. In a similar fashion, we can write the transverse component of the angular velocity $\boldsymbol{\omega}$ of the body in the form

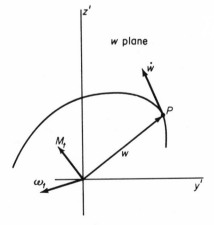

Fig. 8-21. A path traced by P in the complex plane.

$$\omega_t = \omega_{y'} + i\omega_{z'} \tag{8-237}$$

Now let us consider the modified Euler equations of (8–223) for this case. If we choose another primed coordinate system such that the x' axis is the axis of symmetry and set

$$\omega_{x'} = 0 \tag{8-238}$$

where $\omega_{x'}$ is the x' component of the angular velocity of the *coordinate system*, we see that for small motions of the axis of symmetry, the y' z' plane is approximately parallel to the w plane. So, to a first approximation, we can speak of the two planes interchangeably when considering transverse components of various vectors. Now we can write the last two equations of (8–223) in the form

$$
\begin{aligned}
M_{y'} &= I_t \dot{\omega}_{y'} + I_a \Omega \omega_{z'} \\
M_{z'} &= I_t \dot{\omega}_{z'} - I_a \Omega \omega_{y'}
\end{aligned}
\tag{8-239}
$$

where Ω is the total spin of the body and the transverse components $\omega_{y'}$ and $\omega_{z'}$ apply to both the body and the coordinate system. Multiplying the second equation by i, adding it to the first, and using Eq. (8–237), we obtain

$$M_t = M_{y'} + iM_{z'} = I_t \dot{\omega}_t - iI_a \Omega \omega_t \tag{8-240}$$

The last term on the right is a *gyroscopic* term for which the response ω_t is in spatial quadrature to the applied moment.

In order to find the path of P, let us first express ω_t in terms of the velocity $\dot{w}$ as follows:

$$\omega_t = i\dot{w} \tag{8-241}$$

Here we note that $\dot{w}$ is equal to ω_t in magnitude, since P is a unit distance from O. Also, the instantaneous axis of transverse rotation is perpendicular to the velocity of P. From Eqs. (8–240) and (8–241), we obtain

$$M_t = iI_t \ddot{w} + I_a \Omega \dot{w}$$

or

$$\ddot{w} - i\frac{I_a \Omega}{I_t} \dot{w} = -i\frac{M_t}{I_t} \tag{8-242}$$

Now let us apply this equation in the analysis of the *cuspidal motion* of a top near the vertical. Referring to Figs. 8–13 and 8–21, we see that the external moment is given by

$$M_t = imglw \tag{8-243}$$

Note that lw represents the horizontal distance of the mass center from a vertical line through the fixed point O. Thus the equation of motion is

$$\ddot{w} - i\frac{I_a \Omega}{I_t} \dot{w} - \frac{mgl}{I_t} w = 0 \tag{8-244}$$

where Ω is constant since there is no axial moment.

The stability of the top can be investigated by assuming a solution of the form

$$w = Ce^{\lambda t} \tag{8-245}$$

where C is a constant. Substituting Eq. (8–245) into Eq. (8–244), we obtain the characteristic equation

$$\lambda^2 - i\frac{I_a\Omega}{I_t}\lambda - \frac{mgl}{I_t} = 0 \tag{8-246}$$

where the common factor $Ce^{\lambda t}$ has been canceled because it is non-zero for the solutions of interest. The roots of Eq. (8–246) are

$$\lambda_{1,2} = i\frac{I_a\Omega}{2I_t} \pm \sqrt{\frac{mgl}{I_t} - \frac{I_a^2\Omega^2}{4I_t^2}} \tag{8-247}$$

and the general solution is of the form

$$w = C_1 e^{\lambda_1 t} + C_2 e^{\lambda_2 t} \tag{8-248}$$

For the case $\Omega^2 > 4I_t mgl/I_a^2$, we see that both roots are imaginary. Hence any small disturbance (for example, a small initial velocity) will result in a small motion, indicating that the top is *stable*. On the other hand, if $\Omega^2 < 4I_t mgl/I_a^2$, one of the roots will have a positive real part, indicating *unstable* motion. The actual limits of the motion cannot be obtained in this last case because nonlinear effects limit the motion, whereas the analysis of Eq. (8–244) is linear.

As an illustration of cuspidal motion, consider again the sliding top of Example 8–7. In this instance we find that

$$\Omega^2 = 400 \frac{g}{l}$$

and

$$\frac{4I_t mgl}{I_a^2} = 80 \frac{g}{l}$$

so the motion is stable. The characteristic equation (8–246) is found to be

$$\lambda^2 - i4\sqrt{\frac{g}{l}}\lambda - \frac{4g}{5l} = 0$$

which yields the roots

$$\lambda_{1,2} = i\left(2 \pm \frac{4}{\sqrt{5}}\right)\sqrt{\frac{g}{l}}$$

$$= i0.211\sqrt{\frac{g}{l}} \quad , \quad i3.789\sqrt{\frac{g}{l}}$$

If we assume that the initial velocity of the top is in the direction of the positive real axis, we can use Eq. (8–199) to obtain the following initial conditions:

$$w(0) = 0$$

$$\dot{w}(0) = \dot{\theta}(0+) = \frac{4}{5}\sqrt{\frac{g}{l}}$$

Evaluating the constants C_1 and C_2 in Eq. (8–248) from the initial conditions, we find that the general solution is

$$w = i0.224\,(e^{\lambda_1 t} - e^{\lambda_2 t})$$

The initial portion of the solution is shown in Fig. 8–22(a). The maximum deviation of the axis from the vertical occurs when w attains its greatest magnitude. We see that

$$|w|_{\max} = |C_1| + |C_2| = 0.448$$

This result is slightly larger than the comparable result, that is, $\cos 64.9° = 0.424$, which was obtained in Example 8–7. The difference occurs because the axis of symmetry deviates about 25 degrees from the vertical in this case

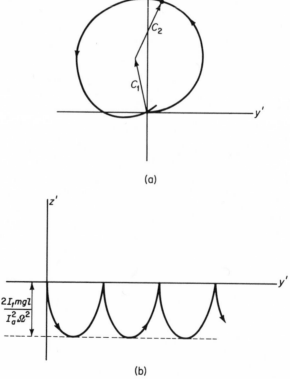

(a)

(b)

Fig. 8-22. Typical paths of the axis of symmetry for (a) a nearly vertical top and (b) a nearly horizontal top.

and the small-motion assumption is only approximately true. The period for a nutation cycle is calculated to be

$$T = \frac{2\pi i}{\lambda_2 - \lambda_1} = 1.76 \sqrt{\frac{l}{g}}$$

We have seen that the point P is confined to the surface of a unit sphere. In the previous examples, the w plane was considered to be tangent to this unit sphere at $\theta = \pi/2$. We chose the point $w = 0$ to correspond to a vertical top. But complex notation is also useful in studying the motion of an axially symmetric body near $\theta = 0$. In this case, the path of the point P is projected normally onto a circular cylinder which is tangent to the unit sphere along the unit circle at $\theta = 0$. We take the imaginary axis parallel to the axis of the cylinder. The real axis is the circle tangent to the unit sphere. Thus it can be seen that if the cylindrical surface is cut vertically and unrolled, the resulting coordinate grid is the usual rectangular complex plane.

To illustrate this application of the complex notation method, consider the cuspidal motion of a top or gyroscope which moves near $\theta = 0$ under the action of gravity. Again we assume that the center of mass is on the axis of symmetry at a distance l from the fixed point O. For small motions, we use the approximation $\cos \theta \cong 1$; hence the transverse moment is

$$M_t = mgl \tag{8-249}$$

Equation (8–242) applies in this case, so we have

$$\ddot{w} - i \frac{I_a \Omega}{I_t} \dot{w} = -i \frac{mgl}{I_t} \tag{8-250}$$

Using standard methods, it can be seen that the solution for $\dot{w}$ is of the form

$$\dot{w} = \frac{mgl}{I_a \Omega} + C e^{i(I_a \Omega/I_t)t} \tag{8-251}$$

If we assume that the motion starts from a cusp, the initial velocity of P is $\dot{w}(0) = 0$. Thus we obtain from Eq. (8–251) that

$$C = -\frac{mgl}{I_a \Omega}$$

and the solution is

$$\dot{w} = \frac{mgl}{I_a \Omega} [1 - e^{i(I_a \Omega/I_t)t}] \tag{8-252}$$

A further integration, assuming the initial condition $w(0) = 0$, yields the complex displacement

$$w = \frac{mgl}{I_a \Omega} \left[t - i \frac{I_t}{I_a \Omega} (1 - e^{i(I_a \Omega/I_t)t}) \right] \tag{8-253}$$

This result is plotted in Fig. 8–22(b). Note that the axis of symmetry follows a cycloidal path whose peak-to-peak amplitude is $2I_t mgl/I_a^2 \Omega^2$. The nutation frequency is $I_a \Omega/I_t$ and the mean precession rate is $mgl/I_a \Omega$. If we assume that the spin is large, that is, $\Omega^2 \gg I_t mgl/I_a^2$, then the nutation amplitude is small and these results are the same as those obtained previously in Eqs. (8–184), (8–185), and (8–186).

In summary, we have seen that when the complex notation method is applicable, problems are often solved quite directly and effectively by this approach. Further examples of its use will be given in the next section.

8–7. EXAMPLES

Example 8–9. An axially symmetric rocket is fired in free space at time $t = 0$. The principal moments of inertia are

$$I_a = 10^4 \text{ lb sec}^2 \text{ ft}$$

$$I_t = 5 \times 10^5 \text{ lb sec}^2 \text{ ft}$$

and are assumed to be constant. Owing to a thrust misalignment, the rocket motor produces a constant moment (about the center of mass) having the components

$$M_z = 2 \times 10^4 \text{ lb ft}$$

$$M_x = M_y = 0$$

where the xyz axes are principal axes fixed in the body and the x axis is the axis of symmetry. Initial conditions are

$$\omega_x(0) = 10 \text{ rad/sec}$$

$$\omega_y(0) = \omega_z(0) = 0$$

Find the motion in space of the axis of symmetry.

We shall use this numerical example to illustrate several of the methods which we have presented for analyzing the rotational motion of an axially symmetric body.

First method: The possibility of using Euler's equations immediately comes to mind because the components of the applied moment are constant in a body-axis system. From Eq. (8–12) or (8–65), we obtain the following set of differential equations:

$$I_a \dot{\omega}_x = 0$$

$$I_t \dot{\omega}_y - (I_t - I_a) \omega_x \omega_z = 0 \qquad (8\text{–}254)$$

$$I_t \dot{\omega}_z + (I_t - I_a) \omega_x \omega_y = M_z$$

From the first equation, we see that ω_x is constant, so we have $\omega_x = 10$

rad/sec for all $t > 0$. Substituting numerical values into the last two equations, we obtain

$$\dot{\omega}_y - 9.8\,\omega_z = 0$$
$$\dot{\omega}_z + 9.8\,\omega_y = 0.04 \qquad\qquad (8\text{--}255)$$

Now we eliminate ω_z between these two equations and obtain the following differential equation in ω_y:

$$\ddot{\omega}_y + (9.8)^2\,\omega_y = 0.392$$

Solving for ω_y by standard methods (see Example 3–6), we have

$$\omega_y = A \cos 9.8t + B \sin 9.8t + 4.08 \times 10^{-3} \qquad (8\text{--}256)$$

and it follows that

$$\dot{\omega}_y = 9.8(-A \sin 9.8t + B \cos 9.8t) \qquad (8\text{--}257)$$

The constants A and B are evaluated from the initial conditions $\omega_y(0) = 0$ and $\dot{\omega}_y(0) = 0$, the latter condition being obtained with the aid of Eq. (8–255). The resulting values are

$$A = -4.08 \times 10^{-3}, \qquad B = 0$$

Consequently the solutions for ω_y and ω_z are found to be

$$\omega_y = 4.08 \times 10^{-3}(1 - \cos 9.8t)$$
$$\omega_z = 4.08 \times 10^{-3} \sin 9.8t \qquad (8\text{--}258)$$

A plot of ω_y versus ω_z is shown in Fig. 8–23(a).

Thus far we have solved for the motion of $\boldsymbol{\omega}$ relative to the body. In order to obtain the motion of the axis of symmetry in space, let us first obtain the components of $\boldsymbol{\omega}$ in terms of the primed coordinate system of Fig. 8–23(b). The primed coordinate system has the same origin as the unprimed system and, in fact, coincides with it at $t = 0$. The primed system, however, *does not rotate* in space. So we find that

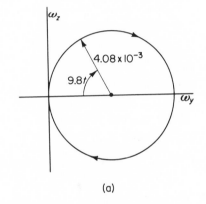

(a)

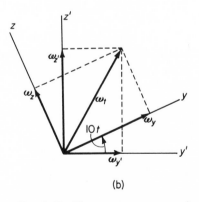

(b)

Fig. 8-23. The transverse angular velocity in the body-axis system and the transformation to nonrotating primed axes.

for small motions of the symmetry axis, the components of the absolute angular velocity $\boldsymbol{\omega}$ of the body, referred to the primed system, are

$$\omega_{x'} = \omega_x$$
$$\omega_{y'} = \omega_y \cos 10t - \omega_z \sin 10t \qquad (8\text{--}259)$$
$$\omega_{z'} = \omega_y \sin 10t + \omega_z \cos 10t$$

and, using Eq. (8–258) and trigonometric identities, we obtain

$$\omega_{x'} = 10$$
$$\omega_{y'} = 4.08 \times 10^{-3}(\cos 10t - \cos 0.2t) \qquad (8\text{--}260)$$
$$\omega_{z'} = 4.08 \times 10^{-3}(\sin 10t - \sin 0.2t)$$

In the nonrotating, primed coordinate system, it is a straightforward process to write expressions for the velocity components of the point P which is located on the positive x axis at a unit distance from the origin. Using y' and z' to specify the position of P relative to the primed coordinate system, it can be seen that

$$\dot{y}' = \omega_{z'} = 4.08 \times 10^{-3}(\sin 10t - \sin 0.2t)$$
$$\dot{z}' = -\omega_{y'} = 4.08 \times 10^{-3}(-\cos 10t + \cos 0.2t) \qquad (8\text{--}261)$$

Integrating these equations with respect to time and using the initial conditions $y'(0) = z'(0) = 0$, we obtain the following path for P:

$$y' = 2.04 \times 10^{-2} \cos 0.2t - 4.08 \times 10^{-4} \cos 10t - 0.02$$
$$z' = 2.04 \times 10^{-2} \sin 0.2t - 4.08 \times 10^{-4} \sin 10t \qquad (8\text{--}262)$$

This path (Fig. 8–24) is an epicycloid formed by a point on a circle of radius 4.08×10^{-4} rolling on the outside of a circle of radius 0.02, centered at $y' = -0.02$, $z' = 0$.

Second method: Now let us solve this example by the complex notation method. We choose the $y'z'$ plane as the complex plane. Then, noting again

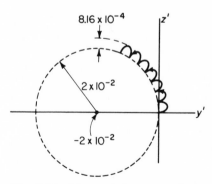

Fig. 8-24. The path in space of a point P on the axis of symmetry.

that the applied moment consists solely of a constant component M_z which rotates with the body, we see that Ω or ω_x remains constant at its initial value of 10 rad/sec. Furthermore, by a coordinate transformation similar to that of Eq. (8–259), we see that

$$M_{y'} = -M_z \sin 10t$$
$$M_{z'} = M_z \cos 10t$$

Consequently, the complex transverse moment is

$$M_t = M_{y'} + iM_{z'} = M_z(-\sin 10t + i\cos 10t)$$

or

$$M_t = i\,2 \times 10^4 e^{i10t} \text{ lb ft}$$

Substituting into the general equation given in (8–242), we obtain the following differential equation in the complex displacement w:

$$\ddot{w} - i\,0.2\dot{w} = 0.04e^{i10t} \qquad (8\text{–}263)$$

Since we were given that $\omega_y(0) = \omega_z(0) = 0$, it is clear that $\omega_t(0) = 0$; hence from Eq. (8–241), we see that $\dot{w}(0) = 0$. Integrating Eq. (8–263) and using this initial condition to evaluate the constant of integration, we obtain,

$$\dot{w} = i\,4.08 \times 10^{-3}\,(e^{i0.2t} - e^{i10t}) \qquad (8\text{–}264)$$

A further integration, assuming that the initial displacement of P is $w(0) = 0$, yields the solution

$$w = -0.02 + 2.04 \times 10^{-2}\,e^{i0.2t} - 4.08 \times 10^{-4}e^{i10t} \qquad (8\text{–}265)$$

Thus the path of the axis of symmetry is identical with that obtained previously in Eq. (8–262). It is shown in Fig. 8–24.

Third method: We have seen that the moment applied to the rocket is constant in the body-axis system, that is, it rotates with the rocket. This suggests a similarity to the case in which the rocket is dynamically unbalanced for a spinning motion about the x axis. So let us find the required product of inertia such that the inertial moment about the z axis due to ω_x is equal to the actual external moment M_z. We recall from d'Alembert's principle that the inertial and applied forces or moments are equal and opposite. Thus we find from Eq. (8–11) that

$$-M_z = I_{xy}\omega_x^2$$

from which we obtain the required product of inertia:

$$I_{xy} = -\frac{M_z}{\omega_x^2} = -200 \text{ lb sec}^2 \text{ ft}$$

where M_z is the external moment.

Of course, if we assume $I_{xy} \neq 0$, then the x axis is no longer a principal axis. So let us find the direction of a new principal axis, x'', which is known to lie in the xy plane because I_{xy} is the only non-zero product of inertia. This direction can be found by requiring that $\boldsymbol{\omega}$ and $\mathbf{H}$ be parallel when the rotation takes place about the x'' axis (Fig. 8–25). For this case,

$$\frac{H_y}{H_x} = \frac{I_{xy}\omega_x + I_{yy}\omega_y}{I_{xx}\omega_x + I_{xy}\omega_y} = \frac{\omega_y}{\omega_x}$$

Now, we recall that each of the principal axes intersects the surface of the ellipsoid of inertia perpendicularly; hence a small rotation of axes will result in a very small change in the moments of inertia. Therefore we can use the approximations $I_{xx} \cong I_a$ and $I_{yy} \cong I_t$ in the preceding equation and obtain

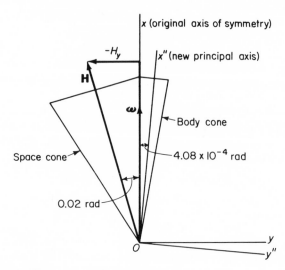

Fig. 8-25. The space and body cones for a rocket with a small thrust misalignment.

$$\frac{\omega_y}{\omega_x} \cong -\frac{I_{xy}}{I_t - I_a} = 4.08 \times 10^{-4}$$

implying that the x and x'' axes are separated by an angle of 4.08×10^{-4} radians.

Having replaced the actual external moment by an inertial moment, we can now consider the *free motion* of a body which is approximately axially symmetric with the x'' axis as its axis of symmetry. We shall use the rolling-cone method of Sec. 8–3, and since $I_t > I_a$, we expect a case of direct precession with the body cone rolling on the outside of the fixed space cone. We see immediately that the semivertex angle of the body cone is 4.08×10^{-4} radians since that is the angle between the x and x'' axes.

To find the semivertex angle of the space cone we calculate the angle between the $\boldsymbol{\omega}$ and $\mathbf{H}$ vectors. In terms of the unprimed coordinate system, we see that

$$\frac{H_y}{H_x} = \frac{I_{xy}\omega_x}{I_{xx}\omega_x} = -0.02$$

Thus the semivertex angle of the space cone is 0.02 radians.

The rotational motion of the rocket in space is now apparent. Remembering that the x axis rotates with the body cone, the motion of a point P at a unit distance from O on the x axis is found to be essentially that of a small circle of radius 4.08×10^{-4} which is rolling on the outside of a fixed circle of radius 0.02. This results in the epicycloidal motion which we obtained previously (Fig. 8–24).

Example 8–10. Rolling disk. In Example 8–8, we considered the steady motion of a disk which is rolling along a circular path. Now let us analyze its rolling motion in the more general case.

A thin circular disk of radius a and mass m rolls on a horizontal plane. Obtain the differential equations of motion. Check the steady precession rate found in Example 8–8. Also discuss the stability of a vertical disk which is rolling in a straight line.

We note that the disk is an axially symmetric body. Let us choose a cartesian coordinate system with its origin O at the center of mass. The x axis is the axis of symmetry; the y axis is in the plane of the disk and remains horizontal, whereas the z axis is in the plane of the disk and passes through the contact point P (Fig. 8–26). The orientation of the xyz system is specified by the Eulerian angles ψ and θ; the rotation of the disk relative to the xyz system is given by the third Euler angle ϕ.

Now let us write an expression for the velocity of the mass center. Using an orthogonal triad of unit vectors $\mathbf{i}, \mathbf{j}, \mathbf{k}$ directed along the x, y, z axes, respectively, we find that the disk is rotating about its contact point with an absolute angular velocity

$$\boldsymbol{\omega}' = (\dot{\phi} - \dot{\psi} \sin \theta)\mathbf{i} + \dot{\theta}\mathbf{j} + \dot{\psi} \cos \theta\,\mathbf{k} \qquad (8\text{–}266)$$

We note that the position vector drawn from P to O is $-a\mathbf{k}$, and assuming that the disk rolls without slipping, we obtain

$$\mathbf{v} = \boldsymbol{\omega}' \times (-a\mathbf{k}) = -a\dot{\theta}\mathbf{i} + a(\dot{\phi} - \dot{\psi} \sin \theta)\mathbf{j} \qquad (8\text{–}267)$$

The acceleration of O is found from

$$\mathbf{a} = (\dot{\mathbf{v}})_r + \boldsymbol{\omega} \times \mathbf{v} \qquad (8\text{–}268)$$

where $\boldsymbol{\omega}$ is the absolute angular velocity of the *xyz coordinate system*. Thus we see that $\boldsymbol{\omega}$ is equal to $\boldsymbol{\omega}'$ for the case where $\dot{\phi} = 0$.

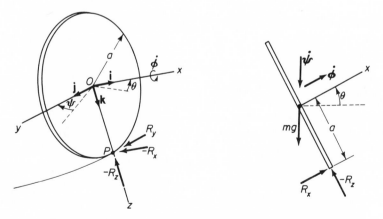

Fig. 8-26. A thin circular disk rolling on a horizontal plane.

$$\boldsymbol{\omega} = -\dot{\psi}\sin\theta\,\mathbf{i} + \dot{\theta}\mathbf{j} + \dot{\psi}\cos\theta\,\mathbf{k} \tag{8-269}$$

From Eq. (8–267), we obtain

$$(\dot{\mathbf{v}})_r = -a\ddot{\theta}\mathbf{i} + a(\ddot{\phi} - \ddot{\psi}\sin\theta - \dot{\psi}\dot{\theta}\cos\theta)\mathbf{j} \tag{8-270}$$

Evaluating the vector product $\boldsymbol{\omega} \times \mathbf{v}$ and adding the result of Eq. (8–270), we obtain the acceleration of O:

$$\mathbf{a} = a_x\mathbf{i} + a_y\mathbf{j} + a_z\mathbf{k} \tag{8-271}$$

where

$$a_x = a[-\ddot{\theta} - \dot{\psi}\cos\theta\,(\dot{\phi} - \dot{\psi}\sin\theta)]$$
$$a_y = a[\ddot{\phi} - \ddot{\psi}\sin\theta - 2\dot{\psi}\dot{\theta}\cos\theta] \tag{8-272}$$
$$a_z = a[\dot{\theta}^2 - \dot{\psi}\sin\theta\,(\dot{\phi} - \dot{\psi}\sin\theta)]$$

The force acting on the disk at the contact point P can be written in the form

$$\mathbf{R} = R_x\mathbf{i} + R_y\mathbf{j} + R_z\mathbf{k} \tag{8-273}$$

Thus, adding the x and z components of the gravitational force, we can write the translational equations as follows:

$$-ma[\ddot{\theta} + \dot{\psi}\cos\theta\,(\dot{\phi} - \dot{\psi}\sin\theta)] = R_x - mg\sin\theta$$
$$ma[\ddot{\phi} - \ddot{\psi}\sin\theta - 2\dot{\psi}\dot{\theta}\cos\theta] = R_y \tag{8-274}$$
$$ma[\dot{\theta}^2 - \dot{\psi}\sin\theta\,(\dot{\phi} - \dot{\psi}\sin\theta)] = R_z + mg\cos\theta$$

The moment of the external forces about the mass center O is

$$\mathbf{M} = -aR_y\mathbf{i} + aR_x\mathbf{j} \tag{8-275}$$

We have chosen the xyz system to move with the disk such that the x axis is the axis of symmetry and the y axis remains horizontal; hence we can obtain the rotational equations by a direct application of the modified Euler equations written in terms of Euler angles. From Eqs. (8–228) and (8–275), we have

$$M_x = -aR_y = I_a(\ddot{\phi} - \ddot{\psi}\sin\theta - \dot{\psi}\dot{\theta}\cos\theta)$$
$$M_y = aR_x = I_t(\ddot{\theta} + \dot{\psi}^2\sin\theta\cos\theta) + I_a\dot{\psi}\cos\theta\,(\dot{\phi} - \dot{\psi}\sin\theta) \tag{8-276}$$
$$M_z = 0 = I_t(\ddot{\psi}\cos\theta - 2\dot{\psi}\dot{\theta}\sin\theta) - I_a\dot{\theta}(\dot{\phi} - \dot{\psi}\sin\theta)$$

Thus we obtain the six equations of motion given in Eqs. (8–274) and (8–276) from which to solve for the time histories of the three Euler angles and the three components of the reaction $\mathbf{R}$. An analytic solution cannot be obtained in the general case, but we shall apply these equations to certain special cases.

First consider again the disk of Example 8–8 which is *rolling steadily with constant* θ such that its mass center moves in a circle of radius R. From the first equation of (8–274) we obtain

$$R_x = mg\sin\theta - ma\dot{\psi}\cos\theta\,(\dot{\phi} - \dot{\psi}\sin\theta)$$

and substituting this expression for R_x into the middle equation of (8–276), we have

$$mga \sin \theta = I_t \dot{\psi}^2 \sin \theta \cos \theta + (ma^2 + I_a)(\dot{\phi} - \dot{\psi} \sin \theta)\dot{\psi} \cos \theta$$

But the condition of rolling implies the kinematic constraint of Eq. (8–202), namely,

$$R\dot{\psi} = a(\dot{\phi} - \dot{\psi} \sin \theta)$$

so we see that

$$mga \sin \theta = I_t \dot{\psi}^2 \sin \theta \cos \theta + \frac{R}{a}(ma^2 + I_a)\dot{\psi}^2 \cos \theta$$

Hence we obtain

$$\dot{\psi}^2 = \frac{mga \tan \theta}{mRa + (R/a)I_a + I_t \sin \theta}$$

in agreement with the previous result of Eq. (8–207).

Now consider the *stability of a vertical disk* rolling in a straight line. The only variables in Eqs. (8–274) and (8–276) which are non-zero in the reference condition are

$$\dot{\phi} = \dot{\phi}_0$$
$$R_z = -mg$$

Let us write perturbation equations for small deviations from this reference condition. In this instance, this is equivalent to writing the translational and rotational equations of (8–274) and (8–276), and then neglecting all terms of order higher than the first in the small quantities. The translational equations become

$$mg\theta - ma(\ddot{\theta} + \dot{\phi}_0\dot{\psi}) = R_x$$
$$ma\ddot{\phi} = R_y \qquad (8\text{–}277)$$
$$R_z + mg = 0$$

The rotational equations are

$$I_t\ddot{\theta} + I_a\dot{\phi}_0\dot{\psi} = aR_x$$
$$I_a\ddot{\phi} = -aR_y \qquad (8\text{–}278)$$
$$I_t\ddot{\psi} - I_a\dot{\phi}_0\dot{\theta} = 0$$

From the middle equations of each set we find that R_y and $\ddot{\phi}$ are zero; hence $\dot{\phi} = \dot{\phi}_0$ throughout the motion. Eliminating R_x between the first equations of each set, we obtain

$$(ma^2 + I_t)\ddot{\theta} - mga\theta + (ma^2 + I_a)\dot{\phi}_0\dot{\psi} = 0 \qquad (8\text{–}279)$$

The last equation of (8–278) can be integrated to yield

$$\dot{\psi} = \frac{I_a}{I_t}\dot{\phi}_0\theta \qquad (8\text{–}280)$$

where we assume that $\dot{\psi}$ and θ are initially zero. Finally, substituting this expression for $\dot{\psi}$ into Eq. (8–279), we have

$$(ma^2 + I_t)\ddot{\theta} + \left[(ma^2 + I_a)\frac{I_a}{I_t}\dot{\phi}_0^2 - mga\right]\theta = 0 \qquad (8\text{–}281)$$

This is recognized as the harmonic equation, whose transient solution consists of sinusoidal oscillations, provided that

$$\dot{\phi}_0^2 > \frac{I_t mga}{I_a (ma^2 + I_a)}$$

This is the condition for stability. If the spin $\dot{\phi}_0$ is smaller in magnitude than this critical value, the solution to the preceding perturbation equation will contain hyperbolic functions of time and hence will be unstable.

Note that if the system is stable, the solution for $\dot{\psi}$ contains terms of the same frequency as the solution for θ.

Example 8–11. A uniform solid sphere of mass m and radius a rolls without slipping on a plane horizontal surface which rotates with constant angular velocity Ω about a vertical axis through the fixed point O (Fig. 8–27). Solve for the motion of the center C of the sphere.

Let $\mathbf{R}$ be the *horizontal component* of the total reaction force passing through the contact point P. It is evident that the vertical component of the force on the sphere at P is equal and opposite to the gravitational force mg because the center C undergoes no vertical acceleration. Furthermore, these forces are collinear and thus do not enter the rotational equations.

The translational equation of motion can be written immediately in the form:

$$m\ddot{\mathbf{r}} = \mathbf{R} \tag{8–282}$$

We see that the rotational equation is relatively simple because the moment of inertia is the same for any axis through the center. Thus we have

$$\mathbf{M} = \dot{\mathbf{H}} = I\dot{\boldsymbol{\omega}} \tag{8–283}$$

where $\boldsymbol{\omega}$ is the absolute angular velocity of the sphere. The external moment about C is

$$\mathbf{M} = -a\mathbf{k} \times \mathbf{R} \tag{8–284}$$

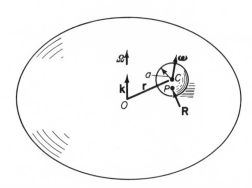

Fig. 8-27. A solid sphere rolling on a rotating horizontal surface.

Combining these equations, we obtain

$$I\dot{\boldsymbol{\omega}} = -ma\mathbf{k} \times \ddot{\mathbf{r}} \tag{8-285}$$

The velocity of the instantaneous contact point P can be found by considering it to be fixed in the sphere and also fixed in the rotating plane. Equating the corresponding expressions for the velocities, we obtain

$$\dot{\mathbf{r}} - a\boldsymbol{\omega} \times \mathbf{k} = \Omega\mathbf{k} \times \mathbf{r} \tag{8-286}$$

where we note that $\mathbf{r} - a\mathbf{k}$ is the position vector of P. Differentiating this equation with respect to time, we have

$$\ddot{\mathbf{r}} - a\dot{\boldsymbol{\omega}} \times \mathbf{k} = \Omega\mathbf{k} \times \dot{\mathbf{r}}$$

or

$$\ddot{\mathbf{r}} = \mathbf{k} \times (\Omega\dot{\mathbf{r}} - a\dot{\boldsymbol{\omega}}) \tag{8-287}$$

Now we substitute the value of $\dot{\boldsymbol{\omega}}$ from Eq. (8–285) into Eq. (8–287) and find that

$$\ddot{\mathbf{r}} = \mathbf{k} \times \left(\Omega\dot{\mathbf{r}} + \frac{ma^2}{I}\mathbf{k} \times \ddot{\mathbf{r}}\right)$$

$$= \Omega\mathbf{k} \times \dot{\mathbf{r}} - \frac{ma^2}{I}\ddot{\mathbf{r}}$$

or, solving for $\ddot{\mathbf{r}}$,

$$\ddot{\mathbf{r}} = \frac{I\Omega}{I + ma^2}\mathbf{k} \times \dot{\mathbf{r}} \tag{8-288}$$

Integrating this equation with respect to time, we obtain

$$\dot{\mathbf{r}} = \frac{I\Omega}{I + ma^2}\mathbf{k} \times (\mathbf{r} - \mathbf{r}_c) \tag{8-289}$$

where $\mathbf{r}_c$ is a constant vector such that Eq. (8–289) conforms to the initial conditions on $\mathbf{r}$ and $\dot{\mathbf{r}}$. We see that $\dot{\mathbf{r}}$ is always perpendicular to $(\mathbf{r} - \mathbf{r}_c)$. Therefore the magnitude $|\mathbf{r} - \mathbf{r}_c|$ is constant. It follows from Eq. (8–289) that the center of the sphere moves in a circular path at a constant speed. For convenience we take $\mathbf{r}_c$ as the position vector of the center of this path, although it may be the position vector of any point on a vertical line through the center of the circular path.

For the particular case of a uniform solid sphere, we have

$$I = \frac{2}{5}ma^2$$

Thus we find from Eq. (8–289) that the center of the sphere moves with a speed

$$v = \frac{2}{7}\Omega\,|\mathbf{r} - \mathbf{r}_c| \tag{8-290}$$

Also, from Eqs. (8–285), (8–288), and (8–289), we obtain the angular acceleration of the sphere.

$$\dot{\boldsymbol{\omega}} = -\frac{5\Omega}{7a}\, \mathbf{k} \times (\mathbf{k} \times \dot{\mathbf{r}}) = \frac{5\Omega}{7a}\,\dot{\mathbf{r}}$$

$$= \frac{10\Omega^2}{49a}\, \mathbf{k} \times (\mathbf{r} - \mathbf{r}_c) \tag{8–291}$$

Note that $\dot{\boldsymbol{\omega}}$ is horizontal, implying that any initial vertical component of $\boldsymbol{\omega}$ remains unchanged. Also note that the period of the circular motion of the center is

$$T = \frac{2\pi\,|\mathbf{r} - \mathbf{r}_c|}{v} = \frac{7\pi}{\Omega} \tag{8–292}$$

so it is independent of the initial conditions.

Example 8–12. Gravitational moments. Find the gravitational moment acting on a general rigid body which is attracted toward a fixed point by an inverse-square field. Apply this result in obtaining the rotational equations for the problem of the dumbbell satellite discussed in Example 5–6.

Let the *xyz* axes be principal axes at the mass center (Fig. 8–28). Suppose $\mathbf{r}$ is the position vector, relative to the attracting center *O*, of a typical mass element *dm*, and $\boldsymbol{\rho}$ is its position vector relative to the center of mass. The force acting on *dm* due to the external gravitational field is

$$d\mathbf{F} = -\frac{\mu}{r^3}\,\mathbf{r}\,dm = -\frac{\mu}{r^3}\,(\mathbf{r}_c + \boldsymbol{\rho})\,dm \tag{8–293}$$

where μ is the gravitational constant per unit mass and $\mathbf{r}_c$ is the position vector of the mass center. Integrating over the body, we find that the total moment about the center of mass due to the gravitational field is

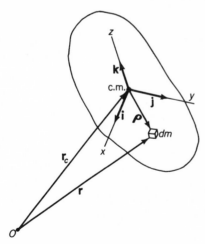

$$\mathbf{M} = -\mu \int_m \frac{\boldsymbol{\rho} \times (\mathbf{r}_c + \boldsymbol{\rho})}{r^3}\,dm \tag{8–294}$$

In the evaluation of this integral, we shall assume that the dimensions of the body are small compared to its distance from the attracting center. Using the law of cosines, we can write an expression for the distance r from *O* to a mass element *dm*. We find that

$$r^2 = r_c^2 + 2\mathbf{r}_c \cdot \boldsymbol{\rho} + \rho^2$$

$$= r_c^2 \left[1 + \frac{2\mathbf{r}_c \cdot \boldsymbol{\rho}}{r_c^2} + \left(\frac{\rho}{r_c}\right)^2 \right] \tag{8–295}$$

Fig. 8-28. A general body which is attracted by an inverse-square gravitational force acting toward *O*.

Now we use the binomial expansion to obtain an approximation for r^{-3}. Neglecting terms of order higher than the first in the small ratio ρ/r_c, we have

$$\frac{1}{r^3} \cong \frac{1}{r_c^3}\left(1 - \frac{3\mathbf{r}_c \cdot \boldsymbol{\rho}}{r_c^2}\right) \tag{8–296}$$

It is convenient to express $\boldsymbol{\rho}$ in terms of its body-axis components as follows:

$$\boldsymbol{\rho} = x\mathbf{i} + y\mathbf{j} + z\mathbf{k} \tag{8–297}$$

Let α, β, γ be the direction cosines of the positive x, y, z axes relative to the positive direction of $\mathbf{r}_c$. Then we see that

$$\mathbf{r}_c \cdot \boldsymbol{\rho} = r_c(\alpha x + \beta y + \gamma z) \tag{8–298}$$

and, from Eq. (8–296),

$$\frac{1}{r^3} = \frac{1}{r_c^3}\left[1 - \frac{3(\alpha x + \beta y + \gamma z)}{r_c}\right] \tag{8–299}$$

Referring to the integral for the gravitational moment $\mathbf{M}$ in Eq. (8–294), we see that

$$\boldsymbol{\rho} \times (\mathbf{r}_c + \boldsymbol{\rho}) = \boldsymbol{\rho} \times \mathbf{r}_c$$
$$= r_c[(\gamma y - \beta z)\mathbf{i} + (\alpha z - \gamma x)\mathbf{j} + (\beta x - \alpha y)\mathbf{k}] \tag{8–300}$$

Hence we obtain

$$\mathbf{M} = -\frac{\mu}{r_c^2}\int_m\left[1 - \frac{3(\alpha x + \beta y + \gamma z)}{r_c}\right]$$
$$[(\gamma y - \beta z)\mathbf{i} + (\alpha z - \gamma x)\mathbf{j} + (\beta x - \alpha y)\mathbf{k}]\,dm \tag{8–301}$$

We note that the direction cosines α, β, γ are constant for this integration. Also, we have chosen the xyz axes to be principal axes at the mass center. Thus we see that

$$\int_m x\,dm = \int_m y\,dm = \int_m z\,dm = 0$$

and

$$\int_m xy\,dm = \int_m xz\,dm = \int_m yz\,dm = 0$$

Furthermore,

$$\int_m (y^2 - z^2)\,dm = I_{zz} - I_{yy}$$

and so on. Therefore Eq. (8–301) reduces to

$$\mathbf{M} = -\frac{3\mu}{r_c^3}[\beta\gamma(I_{yy} - I_{zz})\mathbf{i} + \alpha\gamma(I_{zz} - I_{xx})\mathbf{j} + \alpha\beta(I_{xx} - I_{yy})\mathbf{k}] \tag{8–302}$$

This is the equation for the gravitational moment acting on an arbitrarily oriented body. Note that a rotation about $\mathbf{r}_c$ does not change these body-axis components.

Now consider an axially symmetric satellite which is in a circular orbit about the earth. We shall assume that $I_t > I_a$ and also that the angular deviations of the symmetry axis from the radial direction are small. Let us choose the x axis as the axis of symmetry and require the y axis to remain in the orbital plane. The angle between the position vector $\mathbf{r}_c$ and the xz plane is ψ, whereas θ is the angle between the normal to the orbital plane and the z axis, as shown in Fig. 8–29. In other words, ψ refers to the in-plane motion of the symmetry axis relative to $\mathbf{r}_c$, and θ indicates the out-of-plane motion. Thus we are not using a body-axis system.

For small ψ and θ, we see that the direction cosines are given by

$$\alpha = -1$$
$$\beta = \psi \qquad\qquad (8\text{--}303)$$
$$\gamma = -\theta$$

Furthermore, an arbitrary rotation of the body about its axis of symmetry does not change the gravitational moment or the orientation of the coordinate system. Hence, neglecting second-order terms in small quantities, we find from Eq. (8–302) that the components of the gravitational moment are approximated by

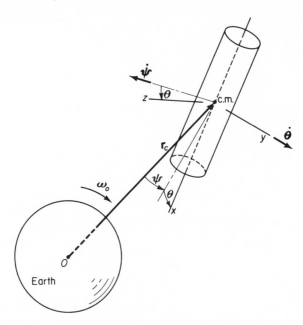

Fig. 8-29. An axially symmetric satellite in a circular orbit about the earth.

$$M_x = 0$$
$$M_y = -3\omega_0^2(I_t - I_a)\theta \qquad (8\text{-}304)$$
$$M_z = -3\omega_0^2(I_t - I_a)\psi$$

where we have taken $I_{xx} = I_a$ and $I_{yy} = I_{zz} = I_t$. Also, we recall that the orbital angular velocity of the satellite is

$$\omega_0 = \sqrt{\frac{\mu}{r_c^3}} = \sqrt{\frac{g_0 R^2}{r_c^3}} \qquad (8\text{-}305)$$

as was shown in Example 5-6.[3]

Now let us substitute the results of Eq. (8-304) into the modified Euler equations given in (8-223) to obtain

$$I_a\dot{\Omega} = 0$$
$$I_t\dot{\omega}_y - I_t\omega_z\omega_x + I_a\Omega\omega_z = -3\omega_0^2(I_t - I_a)\theta \qquad (8\text{-}306)$$
$$I_t\dot{\omega}_z + I_t\omega_x\omega_y - I_a\Omega\omega_y = -3\omega_0^2(I_t - I_a)\psi$$

where we note that the total spin Ω is given by

$$\Omega = \omega_x + s$$

For this case of small θ and ψ, the absolute angular velocity $\boldsymbol{\omega}$ of the xyz coordinate system has the following components:

$$\omega_x = \omega_0\theta$$
$$\omega_y = \dot{\theta} \qquad (8\text{-}307)$$
$$\omega_z = -\omega_0 + \dot{\psi}$$

The rotational motion of the satellite about its axis of symmetry is obtained from the first equation of (8-306) which indicates that the total spin Ω is constant. The equations of motion of the symmetry axis itself can be obtained by substituting from Eq. (8-307) into the last two equations of (8-306), yielding

$$I_t\ddot{\theta} + \omega_0^2(4I_t - 3I_a)\theta + I_a\Omega\dot{\psi} = I_a\Omega\omega_0$$
$$I_t\ddot{\psi} + 3\omega_0^2(I_t - I_a)\psi - I_a\Omega\dot{\theta} = 0 \qquad (8\text{-}308)$$

where we have neglected the products of small quantities.

As an example, let us apply Eq. (8-308) to the particular case of the dumbbell satellite of Example 5-6. In this instance, we set $I_a = 0$ and obtain

$$\ddot{\theta} + 4\omega_0^2\theta = 0$$
$$\ddot{\psi} + 3\omega_0^2\psi = 0 \qquad (8\text{-}309)$$

[3]A consideration of the requirements for the conservation of energy and the conservation of angular momentum will show that slight changes in orbital speed must accompany any angular oscillations of the satellite. We neglect these changes in this analysis.

The motions are independent or uncoupled in this case and are of different frequencies. The second equation, which refers to the in-plane oscillations, is in agreement with the result obtained previously in Eq. (5–157).

REFERENCES

1. Goldstein, H., *Classical Mechanics*. Reading, Mass.: Addison-Wesley Publishing Company, 1950.

2. Housner, G. W., and D. E. Hudson, *Applied Mechanics—Dynamics*, 2nd ed. Princeton, N.J.: D. Van Nostrand Co., Inc., 1959.

3. MacMillan, W. D., *Dynamics of Rigid Bodies*. New York: Dover Publications, Inc., 1936.

4. Synge, J. L., and B. A. Griffith, *Principles of Mechanics*, 3rd ed. New York: McGraw-Hill Book Company, 1959.

5. Thomson, W. T., *Introduction to Space Dynamics*. New York: John Wiley & Sons, Inc., 1961.

6. Webster, A. G., *The Dynamics of Particles and of Rigid, Elastic, and Fluid Bodies*, 2nd ed. New York: Dover Publications, Inc., 1959.

7. Yeh, H., and J. I. Abrams, *Principles of Mechanics of Solids and Fluids—Particle and Rigid-Body Mechanics*. New York: McGraw-Hill Book Company, 1960.

PROBLEMS

8–1. A rigid body is moving freely with a general angular velocity $\boldsymbol{\omega}_0$. Then an angular impulse having components $(\mathcal{M}_x, \mathcal{M}_y, \mathcal{M}_z)$ is applied to the body, where the reference point is at the mass center. Assuming a general body-axis cartesian frame at the center of mass, solve for the change $\Delta\boldsymbol{\omega}$ in the angular velocity at the time of the impulse.

8–2. An axially symmetric rigid body with $I_t = 2I_a$ is spinning in free space such that $\alpha = 45°$, where α is the angle between the angular velocity $\boldsymbol{\omega}$ and the axis of symmetry. Find the magnitude of the least angular impulse which can be applied to the body and result in α being reduced to zero.

8–3. Consider the free rotational motion of an axially symmetric body with $I_a = 2I_t$. What is the largest possible value of the angle between $\boldsymbol{\omega}$ and $\mathbf{H}$?

8–4. A prolate spheroid and an oblate spheroid are each rotating in free space about their respective axes of symmetry. Both have the same angular momentum $\mathbf{H}_0$ and total spin Ω. If equal transverse angular impulses are applied to each, which will undergo the largest angular deviation of its symmetry axis during the resulting motion? Which has the largest precession rate $\dot{\psi}$?

8–5. Consider the free rotation of a rigid body having principal moments of inertia $I_1 < I_2 < I_3$. Under what conditions on H and T_{rot} does the polhode encircle the principal axis of maximum moment of inertia?

8–6. A rigid body is rotating freely such that its angular velocity at a given instant has components $(\omega_x, \omega_y, \omega_z)$, where the xyz axes are principal axes at the mass center. Assuming that $I_{xx} < I_{yy} < I_{zz}$ and $H^2 < 2I_{yy} T_{\text{rot}}$, find the magnitude of the smallest possible angular impulse which will result in a steady free rotation about a principal axis. The time of application of this impulse can be chosen at will.

8–7. Explain how a top with a spin parameter λ much greater than unity is able to right itself to a vertical position. Consider the forces of friction acting on a rounded point.

8–8. A solid circular cylinder of mass m, radius a, and length l is pivoted about a transverse axis through its center of mass. This axis rotates with a constant angular velocity Ω, as shown. Assuming that $l > \sqrt{3}\, a$, find: (a) the frequency of small oscillations of the cylinder about $\theta = \pi/2$; (b) the angular velocity $\dot{\theta}$ when $\theta = \pi/2$ if the cylinder is released from $\theta = 0$ with a very small positive value of $\dot{\theta}$.

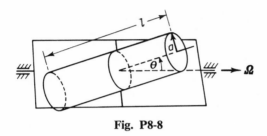

Fig. P8-8

8–9. A thin uniform rod of mass m and length l is rotated at a constant rate ω_x about a transverse horizontal shaft through its center of mass. The shaft is

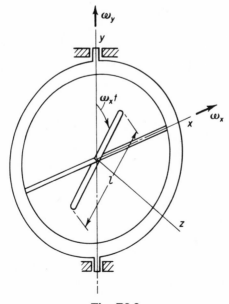

Fig. P8-9

supported by a gimbal which rotates at a constant rate ω_y about a fixed vertical axis. If the x axis rotates with the gimbal and the y axis remains vertical, solve for the driving moments about the x axis and the y axis, respectively. Assume that the rod is vertical initially.

8–10. A rocket is rotated at a constant rate ω about a fixed transverse axis through the center of mass at O. The transverse moment of inertia about O is $I_t = \frac{1}{3}ml^2$, where $m = m_0 - bt$. The rocket is fired at $t = 0$, and mass is ejected at a constant rate b, the exhaust velocity relative to the rocket being v_e. What external moment must be applied to the rocket?

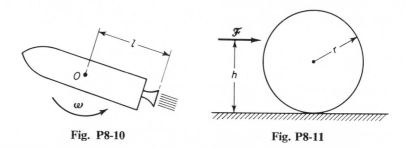

Fig. P8-10   **Fig. P8-11**

8–11. A uniform sphere of mass m and radius r lies motionless on a smooth horizontal plane. Suddenly it is given a horizontal impulse $\mathscr{F}$ whose line of action intersects the vertical line through the mass center. (a) Solve for the height h of the line of action of the impulse such that no slipping occurs in the subsequent motion. (b) Now suppose that the line of action is moved horizontally a distance $r/2$ to the right, as viewed from the rear. What changes occur in the motion of the sphere?

8–12. A uniform spherical bowling ball of mass m and radius r is placed on a horizontal floor so that the initial velocity of its center is $\mathbf{v}(0) = v_0\mathbf{i}$ and its initial angular velocity is $\boldsymbol{\omega}(0) = \omega_0\mathbf{i}$. Assuming a coefficient of friction μ, solve for the velocity $\mathbf{v}$ when the sliding stops. Express the result in terms of a fixed set of cartesian unit vectors with $\mathbf{k}$ pointing vertically upward.

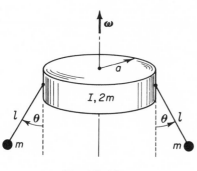

Fig. P8-14

8–13. While observing the free motion of an axially symmetric body, the following characteristics are noted: (1) $\boldsymbol{\omega}$ and $\mathbf{H}$ are separated by $30°$; (2) the precession rate $\dot{\psi}$ is larger in magnitude than ω; and (3) the axis of symmetry changes its direction in space by $90°$ during half of a precession cycle. Solve for the ratio I_a/I_t and evaluate $\dot{\psi}$ and $\dot{\phi}$ in terms of ω.

8–14. A satellite consists of a disk of radius a, mass $2m$, and moment of inertia I about the axis of symmetry. Two particles, each of mass m, are attached by rigid massless rods of length l to opposite points on the circumference. The frictionless pin joints are oriented so that the two rods always lie in a plane which contains the symmetry axis and rotates with the disk. Initially, the disk is spinning about its axis of symmetry at ω_0 rad/sec with $\theta(0) = 0$ and $\dot\theta(0) = 0$. For free motion, solve for the satellite rotation rate ω and also the value of $\dot\theta$ as a function of θ.

8–15. A solid right circular cone of vertex angle 2β and height h rolls without slipping on a horizontal plane. Show that the angular velocity $\dot\psi$ of the line of contact which is required in order for the vertex to lift off the plane is given by

$$\dot\psi^2 = \frac{mgh \sin\beta (1 + 3\sin^2\beta)}{4 I_g \cos^2 \beta}$$

where m is the mass and I_g is the moment of inertia about the line of contact.

8–16. A gyroscope consists of a uniform thin disk of radius a and mass m spinning about one end of a rigid massless rod of length l whose other end is supported at a fixed point O. For the initial conditions $\theta(0) = 0$, $\dot\theta(0) = 0$, $\dot\phi(0) = \Omega$, the rod is found to move through the bottom position at $\theta = -\pi/2$. Solve for the initial precession rate $\dot\psi(0)$ and the magnitude of $\dot\theta$ when $\theta = -\pi/2$.

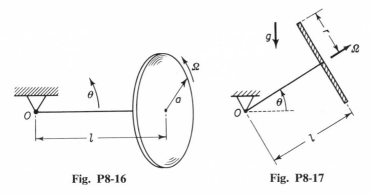

Fig. P8-16   **Fig. P8-17**

8–17. A uniform circular disk of mass m and radius r has a total spin $\Omega = 10\sqrt{g/r}$ about a massless rod of length $l = 2r$ which lies along its symmetry axis. The rod is initially supported without friction at the fixed point O and the disk is precessing uniformly with $\theta = 30°$. Solve for the slower precession rate. Now suppose that the support at O suddenly breaks. Find the precession rate of the free motion and solve for the maximum angular deviation of the axis of symmetry from its position when the support breaks.

8–18. A solid right circular cone whose base diameter $2r$ is equal to its altitude is spinning as a top on a horizontal plane. Assuming that the total spin is $\Omega = 10\sqrt{2g/r}$, and the symmetry axis makes an angle of 45° with the horizontal, find the slower rate of uniform precession for the following cases: (a) the vertex of the cone is fixed; (b) the cone can slide without friction.

8–19. Consider the motion of the sliding top of Example 8–7. Compare the approximate nutation period obtained by using the complex notation method with the exact period found by using elliptic integrals.

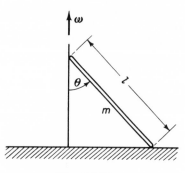

Fig. P8-20

8–20. One end of a thin uniform rod of mass m and length l is constrained to a fixed vertical axis, along which it can slide without friction. The other end slides along a smooth horizontal floor. At $t = 0$, the rod is whirling with an angular velocity ω_0 about the vertical axis. Other initial conditions are $\theta(0) = 45°$ and $\dot{\theta}(0) = 0$. If the rod remains in contact with the floor, find ω and $\dot{\theta}$ as functions of θ. What is the smallest value of ω_0 for which the rod will leave the floor?

8–21. A solid hemisphere of mass m and radius a is at rest in free space when it is struck by an impulse of magnitude $m\sqrt{ga}$ applied tangent to the rim. Find: (a) the initial magnitude and direction of $\boldsymbol{\omega}$; (b) the maximum angular deviation of the symmetry axis from its initial position.

8–22. Suppose that the solid hemisphere of problem 8–21 is resting on a smooth horizontal floor when the given impulse is applied. Solve for the initial velocity of the point P on the rim at which the impulse is applied. Show that the rim will never come in contact with the floor, regardless of the magnitude of the impulse.

8–23. An axially symmetric rocket is spinning in free space with an angular velocity Ω about its axis of symmetry. Its center of mass is motionless. Then a particle of mass m_0 is ejected with an absolute velocity v_0 along a line perpendicular to the symmetry axis and intersecting this axis at a distance l from the mass center of the rocket. The rocket, exclusive of the particle, has a mass m and moments of inertia I_a and I_t about its center of mass. Find: (a) the transverse angular velocity of the rocket immediately after the particle is ejected; (b) the maximum angular deviation of the symmetry axis from its initial direction; (c) the distance traveled by the center of mass of the rocket before the symmetry axis returns to its original orientation.

8–24. An axially symmetric space vehicle with $I_t/I_a = 10$ spins in free space. The angle between the angular momentum vector **H** and the axis of symmetry is $45°$. Then symmetrically placed internal masses are moved slowly toward the axis of symmetry by internal forces. At the end of this process, the rotational kinetic energy is found to be three times its former value, whereas I_a is halved and I_t is 80 percent of its original size. Find the final angle between **H** and the symmetry axis.

8–25. An axially symmetric body with $I_t = 10I_a$ is supported by a massless, frictionless gimbal system in such a manner that its center of mass is fixed. Using Euler angles to specify the orientation of the body, the initial conditions are $\dot{\psi}(0) = \dot{\theta}(0) = 0$, $\dot{\phi}(0) = 100\sqrt{k/I_t}$, and $\theta(0) = 0$. The only applied moment is $M_\theta = k[(\pi/2) - \theta]$, where the spring stiffness k is positive. Assuming that θ remains small, find $\theta_{\max}$ and $\dot{\psi}_{\max}$ in the resulting motion.

8–26. An axially symmetric rocket is spinning in free space about its axis of symmetry with an angular velocity $\omega_r = 10$ rad/sec. Freely suspended at the mass center of the rocket by massless gimbals is a gyro rotor which is rotating at an absolute rate $\omega_g = 10^4$ rad/sec about its own axis of symmetry which is perpendicular to that of the rocket. Suddenly the gimbals are clamped and the rotor motion relative to the rocket is stopped. The gyro rotor has moments of inertia $I_a = 0.1$ lb sec^2 ft and $I_t = 0.05$ lb sec^2 ft. The rocket, exclusive of the rotor, has moments of inertia $I_a = 10^3$ lb sec^2 ft and $I_t = 10^4$ lb sec^2 ft.

Solve for the following: (a) the x,y,z components of the angular impulse applied to the rocket by the gyro at the instant the rotor is stopped; (b) the maximum angular deviation of the axis of symmetry of the rocket from its initial position. Note that approximate axial symmetry is maintained after the gimbals are clamped. (c) Find the time required for a complete precession cycle of the rocket.

8–27. Solve problem 8–26 for the case where the gimbals are suddenly clamped but the rotor is allowed to continue spinning about its own axis of symmetry.

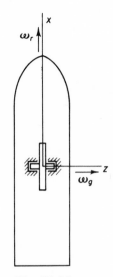

Fig. P8-26

8–28. Find the separation h between the center of mass and the center of gravity of an axially symmetric satellite at a distance r from the earth's center. Assume that the axis of symmetry is oriented in the radial direction and $I_t > I_a$.

8–29. Suppose a satellite with $I_{zz} > I_{yy} > I_{xx}$ is in a circular orbit of radius r_c about the earth. Assuming that the xyz system is a principal axis system at the mass center, write the differential equations describing the small angular motions about a reference condition in which the x axis is directed toward the center of the earth and the z axis is normal to the orbital plane. Let the angles ψ and θ be defined as in Fig. 8–29, and let ϕ be the angle between the y axis and the intersection of the yz and orbital planes, where the positive sense is chosen so that the vector ϕ is directed along the positive x axis.

9

VIBRATION THEORY

The equations of motion that have been obtained in our study of dynamical systems have been, for the most part, *nonlinear* differential equations. Quite frequently one cannot find analytical solutions for these equations and must use approximate methods, and often a computer, in obtaining time histories of the variables.

Now we turn to the study of the motions of mathematically simpler systems which are adequately described by a set of *linear* differential equations with constant coefficients. Some of these systems are assumed to be linear at the outset. Others are actually nonlinear but can be approximated by linear equations for small motions or perturbations in the vicinity of an equilibrium position. These equations are usually much easier to solve than the more complete nonlinear equations and provide an insight into important characteristics of the system.

In addition to limiting ourselves to the study of linear or "linearized" systems, we shall consider only those systems which have a finite number of degrees of freedom. No continuous elastic systems are analyzed, that is, none whose mathematical description involves partial differential equations. For an explanation of these topics the reader is referred to a textbook on vibration theory.[1]

9–1. REVIEW OF SYSTEMS WITH ONE DEGREE OF FREEDOM

In our study of the mass-spring-damper system in Sec. 3–7, we discussed many of the more important aspects of the motion of a linear, time-invariant system with a single degree of freedom. We shall now review some of these results briefly before we consider more complex systems. First, we recall from Eq. (3–161) that the differential equation of motion is of the form

$$m\ddot{x} + c\dot{x} + kx = F(t) \qquad (9\text{–}1)$$

where $F(t)$ is the external force and x is the displacement of the mass, as shown in Fig. 3–17. This equation can also be written in the form

[1]For example, see K. N. Tong, *Theory of Mechanical Vibration* (New York: John Wiley & Sons, Inc., 1960).

$$\ddot{x} + 2\zeta\omega_n\dot{x} + \omega_n^2 x = \frac{1}{m} F(t) \qquad (9\text{-}2)$$

where $\omega_n = \sqrt{k/m}$ is the *undamped natural frequency* and $\zeta = c/2\sqrt{km}$ is the *damping ratio*.

Transient Solution. The solution to the homogeneous equation

$$\ddot{x} + 2\zeta\omega_n\dot{x} + \omega_n^2 x = 0 \qquad (9\text{-}3)$$

is known as the *transient solution*. It has the general form

$$x_t = C_1 e^{\lambda_1 t} + C_2 e^{\lambda_2 t} \qquad (9\text{-}4)$$

where λ_1, λ_2 are the roots of the *characteristic equation*

$$\lambda^2 + 2\zeta\omega_n\lambda + \omega_n^2 = 0 \qquad (9\text{-}5)$$

obtained by substituting a trial solution $x = Ce^{\lambda t}$ into Eq. (9-3). In general,

$$\lambda_{1,2} = -\zeta\omega_n \pm \omega_n\sqrt{\zeta^2 - 1} \qquad (9\text{-}6)$$

These roots must be both real or both imaginary or form a complex conjugate pair, depending upon the value of the damping ratio ζ. Let us summarize the possible solution forms.

Case 1: Underdamped: $0 \le \zeta < 1$. The roots form a complex conjugate pair, resulting in a transient solution of the form

$$x_t = e^{-\zeta\omega_n t}(C_1 \cos \omega_n\sqrt{1 - \zeta^2}\, t + C_2 \sin \omega_n\sqrt{1 - \zeta^2}\, t) \qquad (9\text{-}7)$$

or, equivalently,

$$x_t = Ce^{-\zeta\omega_n t} \cos (\omega_n\sqrt{1 - \zeta^2}\, t + \theta) \qquad (9\text{-}8)$$

where the constants C_1, C_2, or C, θ are evaluated from the conditions at $t = 0$ (or at some other time). Note that these constants are not evaluated until the steady-state solution is known. Also, we see that for the undamped case where $\zeta = 0$, a pair of imaginary roots occur; hence the transient solution is a sine function of circular frequency ω_n.

Case 2. Critically Damped. $\zeta = 1$. The two roots are identical in this case, $\lambda_1 = \lambda_2 = -\omega_n$, and the transient solution is

$$x_t = (C_1 + C_2 t)e^{-\omega_n t} \qquad (9\text{-}9)$$

Case 3: Overdamped: $\zeta > 1$. A pair of negative real roots occurs whose difference increases with increasing damping ratio ζ, as can be seen from Eq. (9-6). The transient solution is

$$x_t = C_1 e^{-(\zeta+\sqrt{\zeta^2-1})\omega_n t} + C_2 e^{-(\zeta-\sqrt{\zeta^2-1})\omega_n t} \qquad (9\text{-}10)$$

Unforced Motion. The arbitrary constants in the foregoing solutions can be evaluated immediately for the case in which the $F(t) = 0$. Let us assume that the initial conditions are $x(0) = x_0$ and $\dot{x}(0) = v_0$.

For the *underdamped* case, we can write the solution in the form

$$x = e^{-\zeta\omega_n t}\left(x_0 \cos \omega_d t + \frac{v_0 + \zeta\omega_n x_0}{\omega_d} \sin \omega_d t\right) \qquad (9\text{-}11)$$

where the *damped natural frequency* is

$$\omega_d = \omega_n\sqrt{1 - \zeta^2} \quad (\zeta < 1) \tag{9-12}$$

In a similar fashion, we find that the unforced solution for the *critically damped* case is

$$x = e^{-\omega_n t}[x_0 + (v_0 + \omega_n x_0)t] \tag{9-13}$$

Finally, the unforced solution for the *overdamped* case is found to be

$$x = \frac{1}{\lambda_1 - \lambda_2}[(v_0 - \lambda_2 x_0)e^{\lambda_1 t} - (v_0 - \lambda_1 x_0)e^{\lambda_2 t}] \tag{9-14}$$

where λ_1 and λ_2 are given by Eq. (9–6).

Steady-State Solution for a Sinusoidal Forcing Function. Now consider the case where the system is driven by the force $F(t) = F_0 \cos \omega t$. We wish to find the steady-state solution, that is, a particular integral satisfying the complete differential equation:

$$m\ddot{x} + c\dot{x} + kx = F_0 \cos \omega t \tag{9-15}$$

Using the procedures of Sec. 3–7, we see that the steady-state solution can be written in the form

$$x_s = A \cos(\omega t + \phi) \tag{9-16}$$

where the constants A and ϕ depend solely upon the system parameters as well as the amplitude and frequency of the forcing function. Substituting Eq. (9–16) into (9–15), it can be shown that

$$A = \frac{F_0/k}{\sqrt{[1 - (\omega^2/\omega_n^2)]^2 + [2\zeta(\omega/\omega_n)]^2}} \tag{9-17}$$

and

$$\phi = \tan^{-1}\left[\frac{-2\zeta(\omega/\omega_n)}{1 - (\omega^2/\omega_n^2)}\right] \tag{9-18}$$

Complex Notation. The preceding expressions for the amplitude A and phase angle ϕ of the steady-state response can be obtained conveniently by considering both the forcing function and the response to be complex variables. Noting that the steady-state response varies sinusoidally with the same frequency as the forcing function, we can let $x = \text{Re}(z)$ and write the differential equation of (9–15) in the form

$$\text{Re}(m\ddot{z} + c\dot{z} + kz) = \text{Re}(F_0 e^{i\omega t})$$

where we remember that m, c, k, and the time t are real. A similar equation applies to the imaginary parts, namely,

$$\text{Im}(m\ddot{z} + c\dot{z} + kz) = \text{Im}(F_0 e^{i\omega t})$$

since this corresponds to a 90-degree shift in both the forcing function and the response. Adding these two equations, we conclude that

$$m\ddot{z} + c\dot{z} + kz = F_0 e^{i\omega t} \qquad (9\text{-}19)$$

Having established this result, it is convenient to revert more closely to the former notation. We *now consider* x *and* A *to be complex* and let

$$x = A e^{i\omega t} \qquad (9\text{-}20)$$

The differential equation of (9-15) is written in the form:

$$m\ddot{x} + c\dot{x} + kx = F_0 e^{i\omega t} \qquad (9\text{-}21)$$

Substituting for x from Eq. (9-20) and canceling the common factor $e^{i\omega t}$, we obtain

$$(k - m\omega^2 + ic\omega)A = F_0$$

or

$$A = \frac{F_0}{k - m\omega^2 + ic\omega} = \frac{F_0/k}{1 - (\omega^2/\omega_n^2) + i2\zeta(\omega/\omega_n)} \qquad (9\text{-}22)$$

If we assume that F_0 is a complex number and plot both F_0 and A on the complex plane (Fig. 9-1), we find that k times the ratio of their magnitudes is

$$\left| \frac{A}{F_0/k} \right| = \frac{1}{\sqrt{[1 - (\omega^2/\omega_n^2)]^2 + [2\zeta(\omega/\omega_n)]^2}} \qquad (9\text{-}23)$$

in agreement with Eq. (9-17). This is just the *amplification factor* of Fig. 3-19. Furthermore, the angle between A and F_0 is the phase angle ϕ of Eq. (9-18) which is plotted versus frequency in Fig. 3-20. So we see that the principal results of the formal solution for the steady-state response x_s are obtained in a rather direct fashion by using complex notation.

A and F_0 are represented in Fig. 9-1 by arrows, which might lead one to believe that complex numbers can be treated as vectors. This is only partially true, however; hence these directed line segments are usually called *phasors* rather than *vectors*. We note that they obey the parallelogram rule for vector

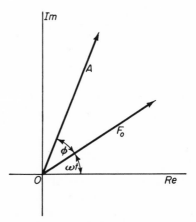

Fig. 9-1. The representation on the complex plane of the relative amplitude and phase of sinusoidally varying quantities.

addition. Also, the time derivative can be found by solving for the velocity of the tip of the arrow, as for a vector drawn from a fixed origin. The product of two phasors, or complex numbers, however, does not correspond to either the dot or the cross product of vector analysis. Rather, we see that their magnitudes are multiplied and the phase angles (measured from the real axis) are added in obtaining the product.

Referring again to Fig. 9–1, we see that the solution for A in Eq. (9–22) actually indicates its amplitude and phase *relative to* F_0. These are constant for any given driving frequency. But if we multiply both sides of Eq. (9–22) by $e^{i\omega t}$, this has the effect of rotating both A and F_0 counterclockwise at a constant angular velocity ω. Now we take the real parts of both sides, corresponding to the projection of the rotating phasors onto the real axis. The result is the actual time variation of x and $F(t) = F_0 \cos \omega t$ in the solution of the original scalar differential equation given in Eq. (9–15).

We have given a brief review of some of the principal results and methods of analysis applicable to linear, second-order systems having a single degree of freedom. In the succeeding sections of this chapter, we shall consider systems having more than one degree of freedom. Many of the methods, however, will be similar to those encountered previously. In particular, we shall show that certain oscillation *modes* can be analyzed in the manner of independent systems, each having a single degree of freedom.

9–2. EQUATIONS OF MOTION

Now let us formulate the equations of motion for a system which has a finite number of degrees of freedom. We shall consider a system of N particles whose configuration is specified by $3N$ ordinary coordinates. But since we allow for workless constraints, the results are also valid for systems which include rigid bodies.

Conditions for Equilibrium. In discussing the principle of virtual work in Sec. 6–4, we stated the necessary and sufficient conditions for the static equilibrium of a system of N particles. Suppose the system configuration is specified by the ordinary coordinates $x_1, x_2, \ldots, x_{3N}$. If F_j is the *applied* force acting in the x_j direction at the corresponding particle, then the necessary and sufficient condition for static equilibrium is that the virtual work of the applied forces is zero, that is,

$$\delta W = \sum_{j=1}^{3N} F_j \, \delta x_j = 0 \qquad (9\text{–}24)$$

for all virtual displacements consistent with the constraints. The constraints are assumed to be workless and bilateral.

Now assume that the applied forces F_j are derivable from a potential function according to the usual relationship

$$F_j = -\frac{\partial V}{\partial x_j} \qquad (9\text{–}25)$$

where V is a function of the x's and possibly time. We see from Eqs. (9–24) and (9–25) that

$$\delta W = -\sum_{j=1}^{3N} \frac{\partial V}{\partial x_j} \delta x_j = 0$$

or

$$\delta V = 0 \qquad (9\text{–}26)$$

where we recall that $\delta t = 0$ in a virtual displacement.

We note that, because of the constraints, the δx's are *not independent* in general. If, however, we assume *holonomic* constraints, it is possible to find a set of n *independent* generalized coordinates $q_1, q_2, \ldots, q_n$ which describe the configuration of the system. Transforming to these coordinates, we see that the condition for static equilibrium is

$$\delta W = \sum_{j=1}^{3N} \sum_{i=1}^{n} F_j \frac{\partial x_j}{\partial q_i} \delta q_i = 0$$

or, using Eq. (9–25),

$$\delta W = -\sum_{i=1}^{n} \frac{\partial V}{\partial q_i} \delta q_i = 0 \qquad (9\text{–}27)$$

in agreement with Eq. (9–26). Since the δq's are assumed to be independent, the virtual work is zero in all cases only if the coefficients of the δq's are zero at the equilibrium condition; that is, if

$$\left(\frac{\partial V}{\partial q_i}\right)_0 = 0 \quad (i = 1, 2, \ldots, n) \qquad (9\text{–}28)$$

Now let us expand the potential function V in a Taylor series about the position of equilibrium in the manner of Eq. (6–28). If we choose to measure the q's from the equilibrium position, we obtain

$$V = V_0 + \sum_{i=1}^{n} \left(\frac{\partial V}{\partial q_i}\right)_0 q_i + \frac{1}{2} \sum_{j=1}^{n} \sum_{i=1}^{n} \left(\frac{\partial^2 V}{\partial q_i \partial q_j}\right)_0 q_i q_j + \cdots \qquad (9\text{–}29)$$

We can arbitrarily set the potential energy at the reference condition equal to zero, that is,

$$V_0 = 0 \qquad (9\text{–}30)$$

Then we see from Eqs. (9–28) and (9–29) that

$$V = \frac{1}{2} \sum_{j=1}^{n} \sum_{i=1}^{n} \left(\frac{\partial^2 V}{\partial q_i \partial q_j}\right)_0 q_i q_j + \cdots$$

or, neglecting terms of order higher than the second in the small displacements,

$$V = \frac{1}{2} \sum_{j=1}^{n} \sum_{i=1}^{n} k_{ij} q_i q_j \qquad (9\text{–}31)$$

where the *stiffness coefficients* are

$$k_{ij} = k_{ji} = \left(\frac{\partial^2 V}{\partial q_i \, \partial q_j}\right)_0 \tag{9–32}$$

Thus we see that the potential energy V is a homogeneous quadratic function of the q's for small motions near a position of equilibrium.

Quadratic Forms. The expression for the potential energy V given in Eq. (9–31), where we now assume that the k_{ij} are all constant, is an example of a quadratic form. If V is positive for all possible values of the q's, except when $q_1 = q_2 = \ldots = q_n = 0$, then V is said to be *positive definite*. This, of course, puts a restriction on the allowable values of the k_{ij}. If we consider the matrix $[k]$, it is clear that the terms on the main diagonal must all be positive; for if one, say k_{ii}, were negative or zero, then V would be negative or zero when q_i is the only non-zero coordinate. This condition, however, is insufficient to ensure the positive definite nature of V. The necessary and sufficient condition that V be positive definite is that[2]

$$k_{11} > 0, \quad \begin{vmatrix} k_{11} & k_{12} \\ k_{21} & k_{22} \end{vmatrix} > 0, \cdots, \quad \begin{vmatrix} k_{11} & k_{12} & \cdots & k_{1n} \\ k_{21} & k_{22} & & \cdot \\ \cdot & & & \cdot \\ \cdot & & & \cdot \\ k_{n1} & \cdot & \cdot & k_{nn} \end{vmatrix} > 0 \tag{9–33}$$

For the system under consideration, we see that the reference equilibrium configuration at $q_1 = q_2 = \ldots = q_n = 0$ is *stable* if V is a positive definite quadratic form.

Now let us suppose that one or more of the principal minors of matrix (k) are zero, but none are negative. Then V is a *positive semidefinite* quadratic form. In other words, the potential energy is nonnegative but may be zero when one or more of the q's is different from zero. This corresponds to a condition of *neutral stability* at $q_1 = q_2 = \ldots = q_n = 0$ and, as we shall see, also implies the existence of one or more pairs of zero roots of the characteristic equation.

Finally, if the principal minors of (k) are of both signs, the quadratic form is *indefinite* and the system is *unstable* at the given equilibrium configuration.

Kinetic Energy. We shall be using the Lagrange formulation of the equations of motion and must first obtain an expression for the kinetic energy of the system. If we again let the x's be the ordinary coordinates specifying the location of the N particles and assume a transformation to generalized coordinates of the form given in Eq. (6–1), namely,

[2]See R. A. Frazer, W. J. Duncan, and A. R. Collar, *Elementary Matrices* (New York: Cambridge University Press, 1960), p. 30.

$$x_1 = f_1(q_1, q_2, \ldots, q_n, t)$$
$$x_2 = f_2(q_1, q_2, \ldots, q_n, t)$$

$$\cdot$$
$$\cdot \qquad\qquad\qquad\qquad\qquad (9\text{--}34)$$
$$\cdot$$

$$x_{3N} = f_{3N}(q_1, q_2, \ldots, q_n, t)$$

then the kinetic energy is

$$T = \frac{1}{2} \sum_{k=1}^{3N} m_k \left(\sum_{i=1}^{n} \frac{\partial x_k}{\partial q_i} \dot{q}_i + \frac{\partial x_k}{\partial t} \right)^2 \qquad (9\text{--}35)$$

as we found previously in Eq. (6–57).

It can be seen that the kinetic energy is not, in general, a homogeneous quadratic function of the $\dot{q}$'s, but also includes terms that are linear in the $\dot{q}$'s as well as other terms that are independent of the $\dot{q}$'s. Furthermore, the coefficients $\partial x_k/\partial q_i$ may be explicit functions of time.

Now let us simplify the analysis by making the following assumptions: (1) the coordinate transformation equations relating the x's and the q's are independent of time; (2) the equilibrium configuration is not a function of time. These assumptions imply that there are no moving constraints. Specifically,

$$\left(\frac{\partial x_k}{\partial t} \right)_0 = 0 \qquad (9\text{--}36)$$

and

$$\left(\frac{\partial x_k}{\partial q_i} \right)_0 = \text{constant} \qquad (9\text{--}37)$$

where the coefficients are evaluated at the equilibrium position. With these assumptions, the kinetic energy can be written in the form

$$T = \frac{1}{2} \sum_{i=1}^{n} \sum_{j=1}^{n} m_{ij} \dot{q}_i \dot{q}_j \qquad (9\text{--}38)$$

where

$$m_{ij} = m_{ji} = \sum_{k=1}^{3N} m_k \left(\frac{\partial x_k}{\partial q_i} \right)_0 \left(\frac{\partial x_k}{\partial q_j} \right)_0 \qquad (9\text{--}39)$$

Thus the *inertia coefficients* m_{ij} are assumed to be constant for the present case of small motions. We can show that the kinetic energy T is a *positive definite quadratic function* of the $\dot{q}$'s. It is evidently positive definite because it is the sum of the positive kinetic energies of the individual particles (assuming that none of the m_k is zero). Its quadratic form is apparent from Eqs. (9–38) and (9–39). Note that the self-inertia coefficients m_{ii} must be positive. Some of the mutual inertia coefficients m_{ij} may be negative, however, so long as none of the principal minors of the $[m]$ matrix is negative. This last requirement is similar to that of Eq. (9–33).

Equations of Motion. Consider a system in which the only forces which do work in an arbitrary virtual displacement are derivable from a potential function. Assuming holonomic bilateral constraints, we can use the standard form of Lagrange's equation:

$$\frac{d}{dt}\left(\frac{\partial L}{\partial \dot{q}_i}\right) - \frac{\partial L}{\partial q_i} = 0 \tag{9--40}$$

where the Lagrangian function L is given by

$$L = T - V$$

First let us discuss briefly the general case in which T and V are functions of time and we desire the equations of motion in the neighborhood of the instantaneous position of equilibrium[3]. Let us use the notation

$$a_{ki} = \left(\frac{\partial x_k}{\partial q_i}\right)_0, \qquad c_k = \left(\frac{\partial x_k}{\partial t}\right)_0 \tag{9--41}$$

The kinetic energy expression of Eq. (9--35) then becomes

$$T = \frac{1}{2}\sum_{k=1}^{3N} m_k \left[\sum_{i=1}^{n} a_{ki}(t)\dot{q}_i + c_k(t)\right]^2 \tag{9--42}$$

Similarly, from Eqs. (9--31) and (9--32), the potential energy is of the form

$$V = \frac{1}{2}\sum_{j=1}^{n}\sum_{i=1}^{n} k_{ij}(t)q_i q_j \tag{9--43}$$

Now we find that

$$\frac{\partial L}{\partial \dot{q}_i} = \sum_{k=1}^{3N} m_k a_{ki}\left(\sum_{j=1}^{n} a_{kj}\dot{q}_j + c_k\right)$$

$$\frac{\partial L}{\partial q_i} = -\sum_{j=1}^{n} k_{ij}q_j$$

and, using Eq. (9--40), we see that the equations of motion are

$$\sum_{j=1}^{n}[m_{ij}(t)\ddot{q}_j + b_{ij}(t)\dot{q}_j + k_{ij}(t)q_j] + g_i(t) = 0 \tag{9--44}$$

where

$$m_{ij} = \sum_{k=1}^{3N} m_k a_{ki} a_{kj}$$

$$b_{ij} = \sum_{k=1}^{3N} m_k \frac{d}{dt}(a_{ki}a_{kj}) = \dot{m}_{ij} \tag{9--45}$$

$$g_i = \sum_{k=1}^{3N} m_k \frac{d}{dt}(a_{ki}c_k)$$

Having obtained the general differential equations for small motions in the vicinity of the instantaneous position of equilibrium, let us now consider

[3]Here we define equilibrium with the aid of the principle of virtual work and note that $\delta t = 0$.

the special case of a *conservative, scleronomic system.* We found the kinetic and potential energies previously; hence we can use Eqs. (9–31) and (9–38) to show that the Lagrangian function is

$$L = \frac{1}{2} \sum_{i=1}^{n} \sum_{j=1}^{n} m_{ij}\dot{q}_i\dot{q}_j - \frac{1}{2} \sum_{i=1}^{n} \sum_{j=1}^{n} k_{ij}q_iq_j \qquad (9\text{--}46)$$

where the m_{ij} and k_{ij} are constants. Now we substitute into Lagrange's equation, Eq. (9–40), and obtain the following equations of motion:

$$\sum_{j=1}^{n} m_{ij}\ddot{q}_j + \sum_{j=1}^{n} k_{ij}q_j = 0 \quad (i = 1, 2, \ldots, n) \qquad (9\text{--}47)$$

In summary, we note that the linearized system is described by a set of n second-order ordinary differential equations with constant coefficients. The kinetic energy T is a positive definite quadratic form in the $\dot{q}$'s. The potential energy V is quadratic in the q's, but is positive definite only if the reference equilibrium position is stable.

9-3. FREE VIBRATIONS OF A CONSERVATIVE SYSTEM

Natural Modes. Let us consider a conservative system whose kinetic and potential energies can be written as follows:

$$T = \frac{1}{2} \sum_{i=1}^{n} \sum_{j=1}^{n} m_{ij}\dot{x}_i\dot{x}_j \qquad (9\text{--}48)$$

$$V = \frac{1}{2} \sum_{i=1}^{n} \sum_{j=1}^{n} k_{ij}x_ix_j \qquad (9\text{--}49)$$

or, in matrix form,

$$T = \frac{1}{2}\{\dot{x}\}^T[m]\{\dot{x}\} \qquad (9\text{--}50)$$

$$V = \frac{1}{2}\{x\}^T[k]\{x\} \qquad (9\text{--}51)$$

where the $[m]$ and $[k]$ matrices are *constant* and *symmetric.* The x's are usually considered to be ordinary coordinates, but could be any set of generalized coordinates which are independent, specify the configuration of the system, and are measured from the equilibrium position.

In accordance with Eq. (9–47), we can write the equations of motion as follows:

$$[m]\{\ddot{x}\} + [k]\{x\} = 0 \qquad (9\text{--}52)$$

Let us assume solutions of the form:

$$x_j = A_jC \cos(\lambda t + \phi) \quad (j = 1, 2, \ldots, n) \qquad (9\text{--}53)$$

where we note that the amplitude is the product of the constants A_j and C. It is evident that C acts as an over-all scale factor for the x's, whereas the A_j indicate their relative magnitudes. If we substitute the trial solutions from Eq. (9–53) into (9–52), we obtain

$$\sum_{j=1}^{n} (-\lambda^2 m_{ij} + k_{ij})A_j C \cos(\lambda t + \phi) = 0 \quad (i = 1, 2, \ldots, n)$$

This set of equations must apply for all values of time; hence the factor $C \cos(\lambda t + \phi)$ cannot be zero except in the trivial case where $C = 0$ and all the x's remain zero. Therefore we conclude that

$$\sum_{j=1}^{n} (-\lambda^2 m_{ij} + k_{ij})A_j = 0 \quad (i = 1, 2, \ldots, n) \tag{9–54}$$

or, in matrix notation,

$$([k] - \lambda^2[m])\{A\} = 0 \tag{9–55}$$

This is a set of linear homogeneous equations in the amplitude coefficients A_j. If the A_j are not to be all zero, then the determinant of their coefficients must vanish; that is,

$$\begin{vmatrix} (k_{11} - \lambda^2 m_{11}) & (k_{12} - \lambda^2 m_{12}) & \cdots & (k_{1n} - \lambda^2 m_{1n}) \\ (k_{21} - \lambda^2 m_{21}) & (k_{22} - \lambda^2 m_{22}) & & \cdot \\ \cdot & & & \cdot \\ \cdot & & & \cdot \\ (k_{n1} - \lambda^2 m_{n1}) & \cdot & \cdot & (k_{nn} - \lambda^2 m_{nn}) \end{vmatrix} = 0 \tag{9–56}$$

The evaluation of this determinant results in an nth-degree algebraic equation in λ^2 which is called the *characteristic equation*. The n roots λ_k^2 are known as *characteristic values* or *eigenvalues*, each being the square of a natural frequency (usually expressed in rad/sec). It can be shown[4] from matrix theory that the roots λ_k^2 are all real and finite if T is positive definite and if both $[m]$ and $[k]$ are real symmetric matrices. Furthermore, the λ_k^2 are positive and the motion occurs about a position of stable equilibrium if, in addition, V is positive definite. If V is positive semidefinite, the system is in neutral equilibrium in the reference configuration and at least one of the λ_k^2 is zero. For the case of motion near a position of unstable equilibrium, at least one of the λ_k^2 is negative real.

Let us summarize the effect of a given root λ_k^2 on the stability of the solution as follows:

Case 1: $\lambda_k^2 > 0$. The solution for each coordinate contains a term of the form $C_k \cos(\lambda_k t + \phi_k)$ in accordance with the assumed solution of Eq. (9–53). This term is stable, but, of course, the over-all stability depends upon the other roots as well.

[4]See Goldstein, *Classical Mechanics* (Reading, Mass.: Addison-Wesley Publishing Company, 1950), p. 323.

Case 2: $\lambda_k^2 = 0$. This results in a repeated zero root λ_k, corresponding to a term in the solution of the form $C_k t + D_k$. This implies that a steady drift can occur in one or more of the coordinates and is characteristic of a neutrally stable system.

Case 3: $\lambda_k^2 < 0$. The corresponding pair of roots are imaginary and they result in terms of the form $C_k \cosh \lambda_k t + D_k \sinh \lambda_k t$, implying an unstable solution.

Note that the symmetry, or lack of it, in the $[m]$ and $[k]$ matrices may depend upon the procedure used in formulating the equations of motion. For example, if we use Newton's laws in obtaining the equations of motion, it may occur that no symmetry is apparent in either the $[m]$ or $[k]$ matrices; yet we are analyzing a linear bilateral system for which symmetry is expected. The loss of symmetry in the equations can occur merely by multiplying one of the equations by a constant, or by adding equations. This would not change the nature of the solution, but nevertheless would change the appearance of the equations. An advantage of the Lagrange formulation of the equations of motion is that the systematic approach preserves the symmetry of the coefficient matrices for those cases where T and V are adequately represented by quadratic functions of the velocities and displacements, respectively.

Returning now to Eq. (9–54) or Eq. (9–55), we find that for each λ_k^2 we can write a set of n simultaneous algebraic equations involving the n amplitude coefficients, A_j. Because the equations are homogeneous, however, there is no unique solution for these amplitude coefficients, but only for the ratios among them. For convenience, we shall solve for the amplitude ratios with respect to A_1. Also, let us arbitrarily eliminate the first equation of (9–55) and divide the others by A_1, resulting in $(n-1)$ equations from which to solve for the $(n-1)$ amplitude ratios. Using matrix notation we can write

$$[(k_{ij} - \lambda^2 m_{ij})] \left\{ \left(\frac{A_j}{A_1} \right) \right\} = \{(\lambda^2 m_{i1} - k_{i1})\} \quad (i, j = 2, \ldots, n) \quad (9\text{–}57)$$

where we have shown a typical element of the matrix in parentheses in each case. The subscripts refer to the original rows and columns in Eq. (9–55).

Now we solve Eq. (9–57) for the amplitude ratios and obtain

$$\left\{ \left(\frac{A_j}{A_1} \right) \right\} = [(k_{ij} - \lambda^2 m_{ij})]^{-1} \{(\lambda^2 m_{i1} - k_{i1})\} \quad (i, j = 2, \ldots, n) \quad (9\text{–}58)$$

where we assume that *the eigenvalues are distinct* and $A_1 \neq 0$, thereby ensuring the existence of the inverse matrix.[5] In case $A_1 = 0$, another coordinate should be chosen as the reference. In fact, one can use any ampli-

---

[5]The *rank* of the matrix must be $(n-1)$ in order that the $(n-1)$ amplitude ratios can be determined directly. (See Frazer, Duncan, and Collar, *Elementary Matrices*, p. 18.)

tude normalization scheme which preserves the correct relative magnitudes of the coordinates.

Note that a set of $(n - 1)$ amplitude ratios can be obtained for each natural frequency λ_k. Each set defines what is known as an *eigenvector* or *modal vector*, since the $(n - 1)$ ratios plus a unit reference amplitude can be considered as the components of a vector in n dimensions. On the other hand, a set of amplitude ratios can be considered to define the *mode shape* associated with a given natural frequency λ_k. Each natural frequency with its corresponding set of amplitude ratios defines a *natural mode of vibration*. The terms *principal mode* and *normal mode* are also used.

For the case of the conservative systems just considered for which the λ_k^2 are real, the amplitude ratios are also real. Therefore we see from Eq. (9–53) that if the initial conditions are such that the system is vibrating in a single mode, all the coordinates are executing sinusoidal motion at the same frequency. The relative phase angle between any two coordinates is either $0°$ or $180°$, depending upon whether the particular amplitude ratio is positive or negative.

The *zero frequency* modes are somewhat different physically in that no elastic deformation occurs. For this reason, they are known as *rigid-body modes*. Both the potential energy and the kinetic energy are constant, implying a uniform translational or rotational motion. The amplitude ratios are calculated in the usual fashion and are perhaps more easily considered as *velocity ratios*.

Comparing the mathematical methods employed here in finding the eigenvalues and the corresponding amplitude ratios with those methods used in Sec. 7–7 in finding the principal moments of inertia and the corresponding principal directions reveals a strong similarity. To illustrate this mathematical similarity, we found in the discussion of principal moments of inertia that

$$[I]\{\omega\} = I\{\omega\}$$

where the scalar, I, is the eigenvalue. In the present vibration problem we have

$$[k]\{A\} = \lambda^2 [m]\{A\}$$

or, assuming that $[m]^{-1}$ exists,

$$[m]^{-1}[k]\{A\} = \lambda^2\{A\}$$

where λ^2 is the eigenvalue. It is evident that the square matrix $[m]^{-1}[k]$ plays a mathematically similar role to the inertia matrix $[I]$. The eigenvectors or *modal columns* are $\{A\}$ and $\{\omega\}$, respectively.

An important difference between the two problems is that the characteristic determinant for the vibration problem has n rows and columns rather than the three shown in Eq. (7–89). Also, we note that the eigenvalue

λ^2 may appear in each element of the determinant rather than being confined to the principal diagonal. Nevertheless, we can think of the eigenvectors of the vibration problem as defining principal directions in an n-dimensional space.

Example 9-1. Calculate the natural frequencies and amplitude ratios for the system of Fig. 9–2, where $m_1 = 4m_2 = 4\,\text{lb sec}^2/\text{in.}$ and $k_1 = k_2 = 1000\,\text{lb/in.}$

In this relatively simple example, we can write the inertia and stiffness matrices directly, or we can use the Lagrangian approach, in which case we find that

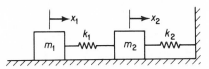

Fig. 9-2. A mass-spring system with two degrees of freedom.

$$m_{ij} = \frac{\partial^2 T}{\partial \dot{q}_i\, \partial \dot{q}_j} \tag{9-59}$$

$$k_{ij} = \frac{\partial^2 V}{\partial q_i \partial q_j} \tag{9-60}$$

In either event, we obtain

$$[m] = \begin{bmatrix} 4 & 0 \\ 0 & 1 \end{bmatrix} \text{lb sec}^2/\text{in.} \qquad [k] = \begin{bmatrix} 1000 & -1000 \\ -1000 & 2000 \end{bmatrix} \text{lb/in.}$$

We note that m_{ij} is numerically equal to the inertial force which *opposes* a positive displacement of x_i and which arises from a unit acceleration of x_j, assuming that all other coordinates remain at zero. In a similar fashion, k_{ij} is numerically equal to the elastic force opposing a positive displacement of x_i, the force arising from a unit displacement of x_j.

The equations of motion are found directly from Eq. (9–52).

$$\begin{bmatrix} 4 & 0 \\ 0 & 1 \end{bmatrix} \begin{Bmatrix} \ddot{x}_1 \\ \ddot{x}_2 \end{Bmatrix} + \begin{bmatrix} 1000 & -1000 \\ -1000 & 2000 \end{bmatrix} \begin{Bmatrix} x_1 \\ x_2 \end{Bmatrix} = 0$$

Assuming solutions of the form

$$x_j = A_j C \cos(\lambda t + \phi)$$

we obtain

$$\begin{bmatrix} (1000 - 4\lambda^2) & -1000 \\ -1000 & (2000 - \lambda^2) \end{bmatrix} \begin{Bmatrix} A_1 \\ A_2 \end{Bmatrix} = 0$$

The characteristic equation is

$$\begin{vmatrix} (1000 - 4\lambda^2) & -1000 \\ -1000 & (2000 - \lambda^2) \end{vmatrix} = 0$$

or

$$4\lambda^4 - 9000\lambda^2 + 10^6 = 0$$

The roots are

$$\lambda_1^2 = 117.2 \text{ rad}^2/\text{sec}^2 \qquad \lambda_2^2 = 2133 \text{ rad}^2/\text{sec}^2$$

$$\lambda_1 = 10.83 \text{ rad/sec} \qquad \lambda_2 = 46.18 \text{ rad/sec}$$

In this simple example, we can solve directly for the amplitude ratio, obtaining

$$\frac{A_2}{A_1} = \frac{1000}{2000 - \lambda^2}$$

For the first mode, we obtain

$$\frac{A_2^{(1)}}{A_1^{(1)}} = \frac{1000}{1883} = 0.531$$

Similarly, for the second mode,

$$\frac{A_2^{(2)}}{A_1^{(2)}} = \frac{1000}{-133} = -7.53$$

where the numbers in parentheses refer to the mode.

The amplitude ratios indicate that the two masses move in phase when oscillating in the first mode. On the other hand, the negative sign indicates that the motion is 180 degrees out of phase in the second mode. It is characteristic of the oscillations of these conservative systems that adjacent masses tend to move in phase for the low-frequency modes and move out of phase for the high-frequency modes. The reasons for this property will become more evident when we discuss Rayleigh's principle in Sec. 9–4. In general, then, we find that, if a conservative linear system is oscillating in a single mode of vibration, we can consider all coordinates to be moving in phase except for a possible sign reversal.

Principal Coordinates. We have seen that a linear conservative system with n degrees of freedom has n eigenvalues or natural frequencies. It is possible for the system to oscillate at a single frequency if the initial conditions are properly established; for example, if all velocities are zero and the coordinates have initial values proportional to the amplitude coefficients in the modal column $\{A\}$ for the given mode. For the general case, however, no single mode matches the initial conditions, and a superposition of modes is required. So, using solutions of the form of Eq. (9–53), we can write

$$x_j = \sum_{k=1}^{n} A_{jk} C_k \cos (\lambda_k t + \phi_k) \tag{9–61}$$

where the subscript k refers to the mode.

Let us define the *principal coordinate* U_k corresponding to the kth mode by the equation

$$U_k = C_k \cos (\lambda_k t + \phi_k) \quad (k = 1, 2, \ldots, n) \tag{9–62}$$

Then from Eqs. (9–61) and (9–62) we obtain

$$x_j = \sum_{k=1}^{n} A_{jk} U_k \qquad (9\text{–}63)$$

or, in matrix notation,

$$\{x\} = [A]\{U\} \qquad (9\text{–}64)$$

Conversely, we can solve for the principal coordinates in terms of ordinary coordinates as follows:

$$\{U\} = [A]^{-1}\{x\} \qquad (9\text{–}65)$$

The *modal matrix* [A] is a square matrix whose columns are the modal columns for the various modes. We assume here that the modal columns are linearly independent, a condition which is assured if the eigenvalues are distinct.

Note that the principal coordinates are generalized coordinates of a particular type. We see from Eq. (9–63) that if only one of the coordinates U_k is non-zero, the x's must be proportional to the corresponding amplitude coefficients for that mode. In this case, Eq. (9–62) indicates that U_k oscillates sinusoidally at the frequency of the given mode. So if only one mode is excited, the entire motion is described by using only one principal coordinate. Of course, a general motion requires all n principal coordinates. Nevertheless, since we are often interested in the transient response within a certain frequency range, the use of coordinates associated specifically with a given frequency makes possible a reduction in the number of coordinates used in the analysis. As we shall see in Sec. 9–5, a similar reduction in the effective number of degrees of freedom can often be accomplished when one analyzes forced vibrations by using principal coordinates.

Orthogonality of the Eigenvectors. In the discussion of principal axes in Sec. 7–7, we found that if the vectors $\boldsymbol{\omega}_1$ and $\boldsymbol{\omega}_2$ are directed along any two of the three principal axes, the orthogonality of these axes can be expressed by Eq. (7–99), namely,

$$\{\omega_2\}^T\{\omega_1\} = 0$$

An equivalent vector statement is $\boldsymbol{\omega}_1 \cdot \boldsymbol{\omega}_2 = 0$, indicating that the eigenvectors corresponding to the principal moments of inertia are mutually orthogonal. Furthermore, the coordinate transformation to a principal-axis system results in the reduction of the general quadratic form for the rotational kinetic energy given in Eq. (7–54) to the simple form:

$$T = \frac{1}{2}(I_1\omega_1^2 + I_2\omega_2^2 + I_3\omega_3^2)$$

In other words, the transformation to principal axes *diagonalizes* the inertia matrix [I], since all off-diagonal terms, corresponding to the products of inertia, are zero.

Returning to vibration theory, we are confronted by *two* quadratic forms, namely, the kinetic energy T and the potential energy V. Nevertheless, certain mathematical similarities between the two eigenvalue problems will be apparent. We shall find that the transformation to principal coordinates diagonalizes *both* the $[m]$ and $[k]$ matrices, thereby simplifying the expressions for T and V. An additional result is that the eigenvectors are shown to be orthogonal, but not in the same geometrical sense as was the case for the principal axes of a rigid body.

Now let us write Eq. (9–55) for the kth natural mode in the form

$$[k]\{A^{(k)}\} = \lambda_k^2[m]\{A^{(k)}\} \tag{9–66}$$

Similarly, for the lth mode we have

$$[k]\{A^{(l)}\} = \lambda_l^2[m]\{A^{(l)}\} \tag{9–67}$$

Premultiply both sides of Eq. (9–66) by $\{A^{(l)}\}^T$ and premultiply both sides of Eq. (9–67) by $\{A^{(k)}\}^T$. The resulting equations are

$$\{A^{(l)}\}^T[k]\{A^{(k)}\} = \lambda_k^2\{A^{(l)}\}^T[m]\{A^{(k)}\} \tag{9–68}$$

$$\{A^{(k)}\}^T[k]\{A^{(l)}\} = \lambda_l^2\{A^{(k)}\}^T[m]\{A^{(l)}\} \tag{9–69}$$

Because of the symmetry of the $[k]$ and $[m]$ matrices, the kth and lth modal columns can be interchanged on both sides of either equation. Performing this operation on Eq. (9–68), we note that the left-hand sides of the two equations are equal. Now subtract Eq. (9–69) from Eq. (9–68), obtaining

$$(\lambda_k^2 - \lambda_l^2)\{A^{(k)}\}^T[m]\{A^{(l)}\} = 0 \tag{9–70}$$

If the eigenvalues λ_k^2 and λ_l^2 are *distinct*, that is, $\lambda_k^2 \neq \lambda_l^2$, then it follows that

$$\{A^{(k)}\}^T[m]\{A^{(l)}\} = 0 \quad (k \neq l) \tag{9–71}$$

This is the orthogonality condition for the kth and lth eigenvectors with respect to the inertia matrix $[m]$. Insight into the physical meaning of Eq. (9–71) can be obtained by noting that a modal column of amplitude coefficients gives the velocity and acceleration ratios as well as the displacement ratios for the given mode. So the orthogonality condition can be interpreted as the dot product of the eigenvector for one mode and the inertial force vector for another mode. The zero value of this dot product indicates that the two vectors are orthogonal in an n-dimensional space. In practical terms, there is no inertial coupling between the principal coordinates corresponding to distinct modes.

It is interesting to note that if one of the modal columns of Eq. (9–71) corresponds to a zero frequency mode, then the orthogonality condition implies that the appropriate momentum component is conserved in the vibration of the elastic mode. If several zero frequency modes exist, then several momentum components will be zero in the remaining modes. For example, the *elastic* modes of a body in free space must have zero linear and angular momentum.

If the natural modes are not distinct, that is, for $k = l$, then the matrix product of Eq. (9–71) is no longer zero. It cannot be zero because it is proportional to the kinetic energy for motion in the given mode, and the kinetic energy has been shown to be positive definite. So let us write

$$\{A^{(k)}\}^T[m]\{A^{(k)}\} = M_{kk} \quad (k = 1, 2, \ldots, n) \tag{9-72}$$

where M_{kk} is a positive constant which is just the inertial coefficient for the kth mode.

We can summarize Eqs. (9–71) and (9–72) as follows:

$$[A]^T[m][A] = [M] \tag{9-73}$$

where we recall that each column of $[A]$ is a modal column. Because of the orthogonality of different eigenvectors, we see that $[M]$ is a *diagonal matrix;* each element along the diagonal is the *generalized mass* or inertial coefficient corresponding to a particular mode. Thus we find that the coordinate transformation to principal coordinates given in Eq. (9–64) has resulted in the diagonalization of the mass matrix, indicating a lack of inertial coupling among the natural modes.

We can show that the transformation to principal coordinates also diagonalizes the stiffness matrix. Let us write Eqs. (9–68) and (9–69) in the form

$$\frac{1}{\lambda_k^2}\{A^{(l)}\}^T[k]\{A^{(k)}\} = \{A^{(l)}\}^T[m]\{A^{(k)}\} \tag{9-74}$$

$$\frac{1}{\lambda_l^2}\{A^{(k)}\}^T[k]\{A^{(l)}\} = \{A^{(k)}\}^T[m]\{A^{(l)}\} \tag{9-75}$$

where we assume that λ_k^2 and λ_l^2 are non-zero.

Again we note that $[k]$ and $[m]$ are symmetric; hence the modal columns for the kth and lth modes can be interchanged. Performing this interchange of matrices in Eq. (9–74) and subtracting Eq. (9–75), we obtain

$$\left(\frac{1}{\lambda_k^2} - \frac{1}{\lambda_l^2}\right)\{A^{(k)}\}^T[k]\{A^{(l)}\} = 0 \tag{9-76}$$

If the modes are distinct, we have $\lambda_k^2 \neq \lambda_l^2$, and it follows that

$$\{A^{(k)}\}^T[k]\{A^{(l)}\} = 0 \tag{9-77}$$

For a zero frequency mode, the same result is obtained directly from Eq. (9–69).

This is the orthogonality condition with respect to the stiffness matrix. The product of $[k]$ and a certain modal column $\{A^{(l)}\}$ gives the elastic force components corresponding to a given displacement of that mode. So Eq. (9–77) states that the scalar product of the eigenvector for a given mode and the elastic force vector for another mode is zero. In other words, the principal modes are not elastically coupled because no work is done by the elastic forces of one mode in moving through the displacements of a second mode.

For the case where $\lambda_k^2 = \lambda_l^2$, we can write

$$\{A^{(k)}\}^T[k]\{A^{(k)}\} = K_{kk} \tag{9-78}$$

where K_{kk} is the *generalized stiffness coefficient* for the kth mode. In contrast with M_{kk} which must be positive, we find that K_{kk} can be positive, zero, or negative according to whether the kth mode is stable, neutrally stable, or unstable. In general,

$$[A]^T[k][A] = [K] \tag{9-79}$$

where $[K]$ is a *diagonal* matrix.

In summary, the transformation to principal coordinates has resulted in the diagonalization of both the inertia and stiffness matrices. This implies that the modes have neither inertial nor elastic coupling and therefore are *independent* for the case of free vibration.

The kinetic energy can be written in terms of principal coordinates by using the familiar form

$$T = \frac{1}{2}\{\dot{U}\}^T[M]\{\dot{U}\} \tag{9-80}$$

or, since $[M]$ is diagonal,

$$T = \frac{1}{2}M_{11}\dot{U}_1^2 + \frac{1}{2}M_{22}\dot{U}_2^2 + \cdots + \frac{1}{2}M_{nn}\dot{U}_n^2 \tag{9-81}$$

Similarly, the potential energy is given by

$$V = \frac{1}{2}\{U\}^T[K]\{U\} \tag{9-82}$$

or

$$V = \frac{1}{2}K_{11}U_1^2 + \frac{1}{2}K_{22}U_2^2 + \cdots + \frac{1}{2}K_{nn}U_n^2 \tag{9-83}$$

Now we can express the Lagrangian function $L = T - V$ in terms of principal coordinates, and from Lagrange's equation, we obtain

$$[M]\{\ddot{U}\} + [K]\{U\} = 0 \tag{9-84}$$

or

$$M_{11}\ddot{U}_1 + K_{11}U_1 = 0$$
$$M_{22}\ddot{U}_2 + K_{22}U_2 = 0$$
$$\vdots \qquad \vdots \tag{9-85}$$
$$M_{nn}\ddot{U}_n + K_{nn}U_n = 0$$

These differential equations of motion indicate that the free vibrations of the entire system can be described in terms of the vibrations of n *independent* undamped second-order systems, each system representing a single mode. We note that the natural frequency of each mode is given by

$$\lambda_k^2 = \frac{K_{kk}}{M_{kk}} \quad (k = 1, 2, \ldots, n) \tag{9-86}$$

Finally, let us use the system of Example 9-1 to illustrate the orthogonality of the natural modes. The $[A]$ matrix was calculated to be

$$[A] = \begin{bmatrix} 1 & 1 \\ 0.531 & -7.53 \end{bmatrix} \text{ in.}$$

where we arbitrarily set the amplitude of the motion in x_1 equal to unity for each mode. Using Eq. (9-73), we obtain

$$[M] = [A]^T[m][A] = \begin{bmatrix} 4.28 & 0 \\ 0 & 60.7 \end{bmatrix} \text{ lb sec}^2 \text{ in.}$$

Similarly, from Eq. (9-79) we have

$$[K] = [A]^T[k][A] = \begin{bmatrix} 0.502 & 0 \\ 0 & 129 \end{bmatrix} \times 10^3 \text{ lb in.}$$

From Eq. (9-86) we obtain $\lambda_1^2 = K_{11}/M_{11} = 117.2 \text{ sec}^{-2}$ and $\lambda_2^2 = K_{22}/M_{22} = 2133 \text{ sec}^{-2}$. Note that each element of the modal matrix $[A]$ has the units of length.

Normal Coordinates. We have seen that the modal columns give the relative amplitudes of the ordinary coordinates when the system is oscillating in a single natural mode. Thus far, we have arbitrarily chosen the amplitude coefficient to be unity for a given coordinate, such as x_1, and have calculated the other amplitude coefficients accordingly. But we could have used other amplitude normalization procedures, such as choosing another coordinate as a reference, or letting the reference amplitude be some other non-zero value. It turns out that the equations of motion in terms of modal coordinates can be expressed in a particularly simple form if we normalize the modal columns in the following manner: Let us choose a new modal column $\{A'^{(k)}\}$ which is related to the former column $\{A^{(k)}\}$ by

$$\{A'^{(k)}\} = \frac{1}{\sqrt{M_{kk}}} \{A^{(k)}\} \tag{9-87}$$

It is evident from Eqs. (9-72) and (9-87) that

$$\{A'^{(k)}\}^T[m]\{A'^{(k)}\} = 1 \quad (k = 1, 2, \ldots, n) \tag{9-88}$$

Similarly, from Eqs. (9-71) and (9-87) we obtain

$$\{A'^{(k)}\}^T[m]\{A'^{(l)}\} = 0 \quad (k \neq l) \tag{9-89}$$

Hence we find that the generalized inertia matrix for the primed system is just the *unit matrix*, that is,

$$[M'] = [A']^T[m][A'] = [1] \tag{9-90}$$

Let us use the term *normal coordinates* to designate the generalized coordinates U'_k associated with the primed system. We can solve for U'_k in terms of U_k by equating the expressions in the primed and unprimed systems for the displacement x_j due to the kth mode. We see that

$$A_{jk}U_k = A'_{jk}U'_k$$

and using Eq. (9–87), we obtain

$$U'_k = \sqrt{M_{kk}}\, U_k \tag{9-91}$$

Now let us calculate the stiffness matrix in the primed coordinate system. For the kth mode we have, using Eq. (9–87),

$$K'_{kk} = \{A'^{(k)}\}^T[k]\{A'^{(k)}\} = \frac{1}{M_{kk}}\{A^{(k)}\}^T[k]\{A^{(k)}\} \tag{9-92}$$

So we see from Eqs. (9–78) and (9–86) that

$$K'_{kk} = \frac{K_{kk}}{M_{kk}} = \lambda_k^2 \quad (k = 1, 2, \ldots, n) \tag{9-93}$$

The off-diagonal terms are again zero and we obtain

$$[K'] = [A']^T[k][A'] = \begin{bmatrix} \lambda_1^2 & 0 & \cdot & \cdot & \cdot \\ 0 & \lambda_2^2 & & & \\ \cdot & & \cdot & & \\ \cdot & & & \cdot & \\ \cdot & & & & \lambda_n^2 \end{bmatrix} \tag{9-94}$$

In terms of normal coordinates, the kinetic and potential energies reduce to particularly simple quadratic forms, namely,

$$T = \frac{1}{2}\dot{U}_1'^2 + \frac{1}{2}\dot{U}_2'^2 + \cdots + \frac{1}{2}\dot{U}_n'^2 \tag{9-95}$$

$$V = \frac{1}{2}\lambda_1^2 U_1'^2 + \frac{1}{2}\lambda_2^2 U_2'^2 + \cdots + \frac{1}{2}\lambda_n^2 U_n'^2 \tag{9-96}$$

Now we can use Lagrange's equation to obtain

$$\ddot{U}'_k + \lambda_k^2 U'_k = 0 \quad (k = 1, 2, \ldots, n) \tag{9-97}$$

or, using matrix notation,

$$\{\ddot{U}'\} + [K'][U'] = 0 \tag{9-98}$$

where

$$[K'] = [M]^{-1}[K] \tag{9-99}$$

Repeated Roots. In the previous discussion of free vibrations, we assumed that the roots are distinct. With this assumption, we found that the eigenvectors are mutually orthogonal with respect to both the $[m]$ and $[k]$ matrices; that is, both matrices are diagonalized by a transformation to principal coordinates.

Now let us consider a so-called *degenerate system* in which the eigenvalues are not all distinct. Specifically, first consider the case of a *double root* such that $\lambda_p^2 = \lambda_{p+1}^2$. If we use Eq. (9–58) to solve for the amplitude ratios corresponding to all the distinct modes, we find that these modes are mutually orthogonal. But if we choose $\lambda^2 = \lambda_p^2$, we find that the inverse matrix $[(k_{ij} - \lambda_p^2 m_{ij})]^{-1}$ does not exist, indicating that the corresponding set of $(n - 1)$ simultaneous algebraic equations is not linearly independent.

For this case of a *simply degenerate system*, we can choose *two* amplitude coefficients arbitrarily when $\lambda^2 = \lambda_p^2$; for example, we might take $A_1^{(p)} = 1$, $A_2^{(p)} = 0$. Then we can solve for the $(n - 2)$ remaining amplitude coefficients from the $(n - 2)$ independent equations contained in (9-57). Furthermore, we find that the modal column $\{A^{(p)}\}$ is orthogonal to all the modal columns corresponding to the distinct modes, since the assumptions for the derivation of Eq. (9–71) are still valid.

The principal problem arises in finding the final modal column $\{A^{(p+1)}\}$ such that it is orthogonal to $\{A^{(p)}\}$. We can, however, again set $A_1^{(p+1)} = 1$. Then we use the $(n - 2)$ independent equations from (9–57) plus the orthogonality condition

$$\{A^{(p)}\}^T [m]\{A^{(p+1)}\} = 0 \qquad\qquad (9–100)$$

Thus we have $(n - 1)$ equations from which to solve for the $(n - 1)$ remaining amplitude coefficients in $\{A^{(p+1)}\}$.

For the general case of an *m-fold degeneracy*, that is, $\lambda_p^2 = \lambda_{p+1}^2 = \cdots = \lambda_{p+m-1}^2$, we note first that the m amplitudes $A_1^{(p)}, A_2^{(p)}, \ldots, A_m^{(p)}$ may be chosen arbitrarily. But now only $(n - m)$ of the equations in (9–57) are independent. Using these equations, we solve for the remaining amplitude coefficients in $\{A^{(p)}\}$. Next we find $\{A^{(p+1)}\}$ by choosing $A_1^{(p+1)}, A_2^{(p+1)}, \ldots,$ $A_{m-1}^{(p+1)}$ arbitrarily and solving for the other $(n - m + 1)$ amplitudes using $(n - m)$ independent equations obtained from (9–57) and the condition that $\{A^{(p+1)}\}$ be orthogonal to $\{A^{(p)}\}$. We continue in this fashion, solving for $\{A^{(p+2)}\}$, $\{A^{(p+3)}\}$, and so on, in order. Each succeeding calculation involves one less arbitrary choice of an amplitude coefficient, but has one more orthogonality condition with the other vectors of the degenerate set. The final eigenvector, $\{A^{(p+m-1)}\}$, is found by setting $A_1^{(p+m-1)} = 1$ and solving for the other amplitudes using the $(n - m)$ equations from (9–57) and the orthogonality conditions with the $(m - 1)$ other vectors of the set.

By this procedure, we obtain a set of n mutually orthogonal eigenvectors, even for the degenerate case. This complete set of modal columns forms a modal matrix $[A]$ which diagonalizes both the $[m]$ and $[k]$ matrices in the usual manner. We should note, however, that the amplitude ratios are no longer unique, as they were for the case of distinct roots, since we found that some of the amplitude ratios can now be chosen arbitrarily. In fact, any linear combination of the modal columns corresponding to a repeated root forms another possible modal column for that root. Columns formed in this way, however, are not necessarily mutually orthogonal.

Initial Conditions. The form of the transient solution for a system having n degrees of freedom was given in Eq. (9–61), namely,

$$x_j = \sum_{k=1}^{n} A_{jk} C_k \cos(\lambda_k t + \phi_k)$$

Assuming that the eigenvalues λ_k^2 and the modal matrix $[A]$ have been found, we must now solve for the n C_k's and the n ϕ_k's from the $2n$ initial conditions. We can use the orthogonality conditions to simplify this process.

Let us assume that the initial displacements and velocities are given. We obtain from Eq. (9–61) that

$$x_j(0) = \sum_{k=1}^{n} A_{jk} C_k \cos \phi_k \tag{9–101}$$

and

$$\dot{x}_j(0) = -\sum_{k=1}^{n} A_{jk} C_k \lambda_k \sin \phi_k \tag{9–102}$$

Now multiply each of these equations by $m_{ij} A_{il}$ and sum over i and j. Because of the orthogonality condition of Eq. (9–71), the right-hand side of each equation will be zero, except when $k = l$. Thus we obtain from Eq. (9–101) that

$$\sum_{i=1}^{n} \sum_{j=1}^{n} x_j(0) m_{ij} A_{il} = \sum_{i=1}^{n} \sum_{j=1}^{n} \sum_{k=1}^{n} m_{ij} A_{il} A_{jk} C_k \cos \phi_k$$

$$= M_{ll} C_l \cos \phi_l$$

or

$$C_l \cos \phi_l = \frac{1}{M_{ll}} \sum_{i=1}^{n} \sum_{j=1}^{n} x_j(0) m_{ij} A_{il} \tag{9–103}$$

In a similar fashion, we obtain from Eq. (9–102) that

$$C_l \sin \phi_l = \frac{-1}{\lambda_l M_{ll}} \sum_{i=1}^{n} \sum_{j=1}^{n} \dot{x}_j(0) m_{ij} A_{il} \tag{9–104}$$

Thus we find that we can use Eqs. (9–103) and (9–104) to solve directly for the C_l and ϕ_l for each mode, rather than being forced to solve $2n$ simultaneous equations.

As an illustration of this method, let us calculate the transient response of the system of Example 9–1 for the following initial conditions:

$$\dot{x}_1(0) = 10 \text{ in./sec}, \qquad x_1(0) = 0$$

$$\dot{x}_2(0) = 0, \qquad\qquad x_2(0) = 0$$

From Eq. (9–61), we see that the transient solution in this case is of the form:

$$x_1 = A_{11}C_1 \cos(\lambda_1 t + \phi_1) + A_{12}C_2 \cos(\lambda_2 t + \phi_2)$$
$$x_2 = A_{21}C_1 \cos(\lambda_1 t + \phi_1) + A_{22}C_2 \cos(\lambda_2 t + \phi_2) \tag{9–105}$$

Thus we need to solve for the four constants C_1, C_2, ϕ_1, and ϕ_2.

Now we evaluate the right-hand sides of Eqs. (9–103) and (9–104), using the numerical values of the masses and amplitude coefficients which we obtained previously. We obtain for the first mode that

$$C_1 \cos \phi_1 = 0$$
$$C_1 \sin \phi_1 = -\frac{(10)(4)(1)}{(10.83)(4.28)} = -0.863$$

from which we obtain

$$C_1 = -0.863, \qquad \phi_1 = \frac{\pi}{2}$$

Similarly, for the second mode we have

$$C_2 \cos \phi_2 = 0$$
$$C_2 \sin \phi_2 = -\frac{(10)(4)(1)}{(46.2)(60.7)} = -0.0143$$

with the result that

$$C_2 = -0.0143, \qquad \phi_2 = \frac{\pi}{2}$$

Therefore, substituting numerical values for the constants in Eq. (9–105), we find that the transient solutions for x_1 and x_2 are

$$x_1 = 0.863 \sin 10.83t + 0.0143 \sin 46.2t$$
$$x_2 = 0.458 \sin 10.83t - 0.108 \sin 46.2t \tag{9–106}$$

Example 9–2. Find the differential equations of motion and the natural frequencies for the system whose equilibrium configuration is shown in Fig. 9–3. Obtain a set of mutually orthogonal modal columns.

Let us obtain the equations of motion by the Lagrangian method. Noting that the square of the velocity of the mass $2m$ is $v^2 = \dot{x}_3^2 + \dot{x}_4^2$, we see that the total kinetic energy of the system is

$$T = m\left(\frac{1}{2}\dot{x}_1^2 + \frac{1}{2}\dot{x}_2^2 + \dot{x}_3^2 + \dot{x}_4^2\right) \tag{9–107}$$

The potential energy is found by adding the energies of the individual springs, where we assume small motions such that the angles remain essentially unchanged. We obtain

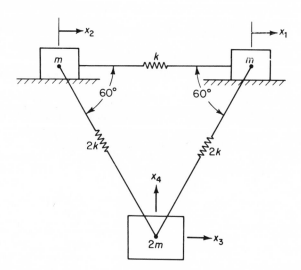

Fig. 9-3. A mass-spring system with four degrees of freedom.

$$V = \frac{1}{2} k(x_1 - x_2)^2 + k\left(\frac{x_1 - x_3}{2} - \frac{\sqrt{3}}{2} x_4\right)^2 + k\left(\frac{x_2 - x_3}{2} + \frac{\sqrt{3}}{2} x_4\right)^2$$

$$= k\left(\frac{3}{4} x_1^2 + \frac{3}{4} x_2^2 + \frac{1}{2} x_3^2 + \frac{3}{2} x_4^2 - x_1 x_2 - \frac{1}{2} x_1 x_3 - \frac{1}{2} x_2 x_3\right.$$

$$\left. - \frac{\sqrt{3}}{2} x_1 x_4 + \frac{\sqrt{3}}{2} x_2 x_4\right) \tag{9-108}$$

Now let $L = T - V$ and substitute into the standard form of Lagrange's equation:

$$\frac{d}{dt}\left(\frac{\partial L}{\partial \dot{q_i}}\right) - \frac{\partial L}{\partial q_i} = 0$$

The resulting equations of motion are

$$m\ddot{x}_1 + k\left(\frac{3}{2} x_1 - x_2 - \frac{1}{2} x_3 - \frac{\sqrt{3}}{2} x_4\right) = 0$$

$$m\ddot{x}_2 + k\left(-x_1 + \frac{3}{2} x_2 - \frac{1}{2} x_3 + \frac{\sqrt{3}}{2} x_4\right) = 0$$

$$2m\ddot{x}_3 + k\left(-\frac{1}{2} x_1 - \frac{1}{2} x_2 + x_3\right) = 0 \tag{9-109}$$

$$2m\ddot{x}_4 + k\left(-\frac{\sqrt{3}}{2} x_1 + \frac{\sqrt{3}}{2} x_2 + 3x_4\right) = 0$$

Assuming solutions of the form

$$x_j = A_j C \cos(\lambda t + \phi)$$

as in Eq. (9–53), we obtain the following equation in terms of the amplitude coefficients:

$$
\begin{bmatrix}
\left(\frac{3}{2}k - m\lambda^2\right) & -k & -\frac{1}{2}k & -\frac{\sqrt{3}}{2}k \\
-k & \left(\frac{3}{2}k - m\lambda^2\right) & -\frac{1}{2}k & \frac{\sqrt{3}}{2}k \\
-\frac{1}{2}k & -\frac{1}{2}k & (k - 2m\lambda^2) & 0 \\
-\frac{\sqrt{3}}{2}k & \frac{\sqrt{3}}{2}k & 0 & (3k - 2m\lambda^2)
\end{bmatrix}
\begin{Bmatrix}
A_1 \\ A_2 \\ A_3 \\ A_4
\end{Bmatrix} = 0
$$

$$(9\text{--}110)$$

The characteristic equation is found by setting the determinant of the square matrix equal to zero. The resulting equation can be written as

$$m^3\lambda^8 - 5m^2 k\lambda^6 + 7mk^2\lambda^4 - 3k^3\lambda^2 = 0$$

or, in factored form,

$$\lambda^2 \left(\lambda^2 - \frac{k}{m}\right)^2 \left(\lambda^2 - \frac{3k}{m}\right) = 0 \qquad (9\text{--}111)$$

Thus we see that this system is simply degenerate and also has a zero frequency mode.

Consider first the zero frequency mode. Let $\lambda^2 = \lambda_1^2 = 0$ in Eq. (9–110) and take $A_1^{(1)} = 1$. Omitting the first equation, we can use Eq. (9–58) to solve for the modal column for the first mode. We find that

$$\{A^{(1)}\}^T = \lfloor 1 \quad 1 \quad 1 \quad 0 \rfloor$$

In a similar fashion, we find for the fourth mode, $\lambda_4^2 = 3k/m$, that

$$\{A^{(4)}\}^T = \left\lfloor 1 \quad -1 \quad 0 \quad -\frac{1}{\sqrt{3}} \right\rfloor$$

Now we consider the modal columns corresponding to the repeated roots $\lambda_2^2 = \lambda_3^2 = k/m$. Let us arbitrarily choose $A_1^{(2)} = 1$ and $A_3^{(2)} = 0$. Omitting the first two equations of (9–110), we have

$$-\frac{1}{2} A_2^{(2)} = \frac{1}{2}$$

$$\frac{\sqrt{3}}{2} A_2^{(2)} + A_4^{(2)} = \frac{\sqrt{3}}{2}$$

from which we find that $A_2^{(2)} = -1$, $A_4^{(2)} = \sqrt{3}$. The corresponding modal column is given by

$$\{A^{(2)}\}^T = \lfloor 1 \quad -1 \quad 0 \quad \sqrt{3} \rfloor$$

For the remaining mode, $\lambda_3^2 = k/m$, we can choose only one amplitude coefficient arbitrarily. Let us take $A_1^{(3)} = 1$. If we again omit the first two equations of (9–110), the remaining two can be written as

$$-\frac{1}{2} A_2^{(3)} - A_3^{(3)} = \frac{1}{2}$$

$$\frac{\sqrt{3}}{2} A_2^{(3)} + A_4^{(3)} = \frac{\sqrt{3}}{2}$$

The required third equation is a statement of the orthogonality of the second and third modes with respect to the mass matrix. Using Eq. (9–71) and the numerical values for $\{A^{(2)}\}$, we have

$$m(A_1^{(3)} - A_2^{(3)} + 2\sqrt{3}\ A_4^{(3)}) = 0$$

Hence we can solve for $\{A^{(3)}\}$ with the following result:

$$\{A^{(3)}\}^T = \lfloor 1 \quad 1 \quad -1 \quad 0 \rfloor$$

These modal columns can be collected to form the modal matrix:

$$[A] = \begin{bmatrix} 1 & 1 & 1 & 1 \\ 1 & -1 & 1 & -1 \\ 1 & 0 & -1 & 0 \\ 0 & \sqrt{3} & 0 & \dfrac{-1}{\sqrt{3}} \end{bmatrix}$$

It is interesting to note that all modes except the first have zero linear momentum in the horizontal direction. This is a result of the orthogonality of the zero frequency mode relative to the others and the consequent lack of inertial coupling.

Finally, we can solve for the generalized inertia and generalized stiffness matrices from Eqs. (9–73) and (9–79). These are

$$[M] = m \begin{bmatrix} 4 & 0 & 0 & 0 \\ 0 & 8 & 0 & 0 \\ 0 & 0 & 4 & 0 \\ 0 & 0 & 0 & \frac{8}{3} \end{bmatrix}, \qquad [K] = k \begin{bmatrix} 0 & 0 & 0 & 0 \\ 0 & 8 & 0 & 0 \\ 0 & 0 & 4 & 0 \\ 0 & 0 & 0 & 8 \end{bmatrix}$$

9–4. THE USE OF SYMMETRY

We have observed that the characteristic equation for a linearized vibratory system having n degrees of freedom is of nth order in the eigenvalue λ^2. The characteristic equation is obtained by equating to zero the determinant of the coefficient matrix for the general modal column $\{A\}$, as shown in Eq. (9–56). Since the effort required to evaluate a determinant increases rapidly with its size, it is apparent that any procedure which allows us to consider fewer degrees of freedom at one time will be most welcome. For systems having a plane of symmetry, we will show that the n required roots may be found by solving two characteristic equations, each obtained by evaluating determinants which have approximately $n/2$ rows and columns.

Let us consider a system having a plane of symmetry; that is, in the reference position of static equilibrium, a plane can be found such that the system is divided into two parts which are mirror images of each other with respect to the plane. If we think of the displacements and forces at the interface between the two portions, we find that the displacements must preserve the *continuity* of the system, whereas the forces acting on the two portions must be *equal and opposite* in accordance with Newton's law of action and reaction. Given these conditions at the plane of symmetry, and assuming small displacements, it can be seen that the motion in any distinct mode must fall into one of two classifications: (1) *symmetric modes* in which the displacement at each point in one half of the system is the mirror image of the displacement of the corresponding point in the other half; (2) *anti-symmetric modes* in which the displacement at each point is the *negative* of the mirror image of the displacement for the corresponding point in the other half.

As an example of the effect of these symmetry assumptions upon the boundary conditions, we note that particles in the plane of symmetry must remain in that plane for the symmetric modes, but must move in a direction perpendicular to the plane of symmetry for the antisymmetric modes. Similarly, we find that only normal forces and stresses occur at the symmetry plane for the case of the symmetric modes, whereas only shear forces and stresses occur at this boundary when there is an antisymmetric vibration.

Note that the symmetry assumptions can be expressed in the form of a set of simple holonomic constraint equations. Usually these equations state that the ratio of a given coordinate to its image coordinate is either plus one or minus one. If, however, the coordinate refers to a point in the plane of symmetry, then this coordinate must be *zero* for either the symmetric or the antisymmetric case. In the usual case, the numbers of symmetric and antisymmetric modes are roughly equal. But if there are more coordinates at the boundary which are associated with one type of symmetry, then there will be that many more modes with the given symmetry.

Example 9-3. Find the natural frequencies and amplitude ratios for the system shown in Fig. 9-4(a). All motion occurs along the same straight line.

$$m_1 = m_4 = 4 \text{ lb sec}^2/\text{in.} \qquad k_1 = k_3 = 1000 \text{ lb/in.}$$
$$m_2 = m_3 = 1 \text{ lb sec}^2/\text{in.} \qquad k_2 = 500 \text{ lb/in.}$$

This system has four degrees of freedom. One could obtain expressions for T and V and use the Lagrangian method to find the equations of motion. An alternative plan for this relatively simple system would be to write the

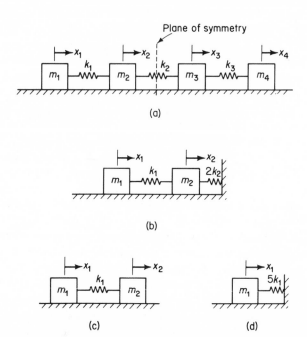

Fig. 9-4. The use of symmetry in breaking a system into smaller parts.

$[m]$ and $[k]$ matrices directly by inspection. In either event, the characteristic equation would be of third order in λ^2 after factoring out the zero root.

Because of the symmetry of the physical system, however, let us use symmetry methods in the solution. First consider the *symmetric modes*. For this case, the coordinates are related according to the equations:

$$x_1 = -x_4, \qquad x_2 = -x_3 \qquad (9\text{--}112)$$

We note that the center of the spring k_2 cannot move, since any movement of this point would be antisymmetric in nature. So let us assume that the center of spring k_2 is fixed and consider just one half of the system—the left-hand side, for example. We recall that if the length of a given spring is halved, its stiffness is doubled. So we find that the system to be analyzed, Fig. 9–4(b), is identical with the system of Example 9–1.

Using the results of Example 9–1, we can immediately give the frequencies and amplitude ratios for the symmetric modes. They are

$$\lambda_2 = 10.83 \text{ rad/sec}, \qquad \{A^{(2)}\}^T = \lfloor 1 \qquad 0.531 \qquad -0.531 \qquad -1 \rfloor$$

$$\lambda_4 = 46.18 \text{ rad/sec}, \qquad \{A^{(4)}\}^T = \lfloor 1 \qquad -7.53 \qquad 7.53 \qquad -1 \rfloor$$

Next, consider the *antisymmetric modes*. For this case we assume that

$$x_1 = x_4, \qquad x_2 = x_3 \qquad (9\text{--}113)$$

One of the antisymmetric modes is clearly the zero frequency mode corresponding to a uniform translation such that $x_1 = x_2 = x_3 = x_4$. But in general, we see that the condition $x_2 = x_3$ implies that no force is exerted by the spring k_2 for any of the antisymmetric modes. Therefore we can omit this spring entirely and consider the system of Fig. 9-4(c).

We recall that the zero frequency mode contains the entire translational momentum for the system; otherwise the remaining modes would not be orthogonal to it. Hence these other modes must have zero translational momentum in the given direction of motion. Also, by symmetry the momenta of the two halves are equal. So we can set the linear momentum of each half equal to zero. For the system of Fig. 9-4(c), we have

$$m_1 \dot{x}_1 + m_2 \dot{x}_2 = 0$$

and, since the displacement ratio is constant and is equal to the velocity ratio, we see that

$$\frac{x_2}{x_1} = -\frac{m_1}{m_2} = -4$$

Now we can reduce this system one step further. Assuming a uniform spring between m_1 and m_2, the foregoing ratio implies that a point which is one-fifth of the distance from x_1 to x_2 must remain motionless. (In other words, the mass center of this small system is stationary.) If we fix this point and again consider the left-hand system, we see that the effective spring constant is $5k_1$. The resulting mass-spring system is shown in Fig. 9-4(d). Evidently the natural frequency of this mode is

$$\lambda_3 = \sqrt{\frac{5k_1}{m_1}} = 35.4 \text{ rad/sec}$$

So we can summarize the antisymmetric modes as follows:

$$\lambda_1 = 0, \qquad \{A^{(1)}\}^T = \lfloor 1 \quad 1 \quad 1 \quad 1 \rfloor$$

$$\lambda_3 = 35.4 \text{ rad/sec}, \qquad \{A^{(3)}\}^T = \lfloor 1 \quad -4 \quad -4 \quad 1 \rfloor$$

The modal matrix is

$$[A] = \begin{bmatrix} 1 & 1 & 1 & 1 \\ 1 & 0.531 & -4 & -7.53 \\ 1 & -0.531 & -4 & 7.53 \\ 1 & -1 & 1 & -1 \end{bmatrix}$$

We have shown that the use of symmetry, plus the knowledge that certain elastic modes have zero momentum, can result in a considerable saving of labor in the solution of vibration problems. Many physical systems of interest, such as ships, aircraft, and missiles, have at least approximate symmetry; and whether the system is lumped or continuous, symmetry methods can be used to simplify the analysis. We have considered the case of a single plane of symmetry. It might be that more than one plane of symmetry exists, thereby allowing further divisions of the system. Also, it is possible that the symmetry exists with respect to a line or a point, rather than with respect to a plane.

Example 9-4. Find the equations of motion, the mode frequencies, and the amplitude ratios for the system shown in Fig. 9–5(a). Also obtain the [M] and [K] matrices corresponding to a convenient set of principal coordinates. Assume that the wires have a constant tensile force P and are massless. Let $m_1 = m_3 = m$ and $m_2 = 3m/4$.

We note that the system is symmetric about m_2; hence we can consider just half of the system at a time. Let us take the left-hand half, as shown in Fig. 9–5(b), and consider first the *symmetric modes*. We see that the plane of symmetry bisects m_2, and because of the even symmetry, the force P acts normal to the boundary, that is, it has no x_2 component.

The kinetic energy T is just

$$T = \frac{1}{2} m\dot{x}_1^2 + \frac{3}{16} m\dot{x}_2^2 = \frac{m}{16}(8\dot{x}_1^2 + 3\dot{x}_2^2) \tag{9–114}$$

The potential energy is found by calculating the work done on the wires for a given small displacement. The tensile force P is constant, so the work done equals P multiplied by the total stretch of the wire. Assuming that V is zero at equilibrium, we can write

$$V = (\sqrt{l^2 + x_1^2} + \sqrt{l^2 + (x_2 - x_1)^2} - 2l)P \tag{9–115}$$

or, expanding the square roots in terms of the small quantities x_1/l and x_2/l, we obtain

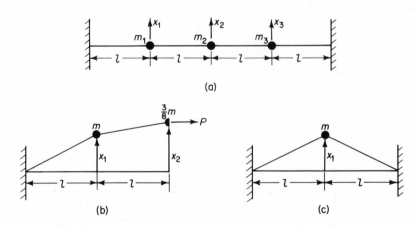

(a)

(b)   (c)

Fig. 9-5. (a) The complete system. (b) The partial systems associated with the symmetric modes and (c) the antisymmetric modes.

$$V = \frac{P}{2l}(2x_1^2 - 2x_1 x_2 + x_2^2) \tag{9-116}$$

where all higher-order terms have been neglected.

Using Lagrange's equations, we find that the equations of motion are

$$m\ddot{x}_1 + \frac{2P}{l}x_1 - \frac{P}{l}x_2 = 0$$

$$\frac{3}{8}m\ddot{x}_2 - \frac{P}{l}x_1 + \frac{P}{l}x_2 = 0 \tag{9-117}$$

An alternate approach is to write the inertia and stiffness matrices directly, the latter by a consideration of the elastic forces acting on each particle. The result, using primes to indicate a partial system, is

$$[m'] = m\begin{bmatrix} 1 & 0 \\ 0 & \frac{3}{8} \end{bmatrix} \tag{9-118}$$

$$[k'] = \frac{P}{l}\begin{bmatrix} 2 & -1 \\ -1 & 1 \end{bmatrix} \tag{9-119}$$

Now the equations of motion can be written by inspection.

Following the usual procedure of assuming solutions of the form

$$x_j = A_j C \cos(\lambda t + \phi)$$

and substituting into the equations of motion, we obtain

$$\left(\frac{2P}{l} - m\lambda^2\right)A_1 - \frac{P}{l}A_2 = 0$$

$$-\frac{P}{l}A_1 + \left(\frac{P}{l} - \frac{3}{8}m\lambda^2\right)A_2 = 0 \tag{9-120}$$

The characteristic equation is

$$\begin{vmatrix} \left(\frac{2P}{l} - m\lambda^2\right) & -\frac{P}{l} \\ -\frac{P}{l} & \left(\frac{P}{l} - \frac{3}{8}m\lambda^2\right) \end{vmatrix} = 0$$

or

$$\lambda^4 - \frac{14P}{3ml}\lambda^2 + \frac{8}{3}\frac{P^2}{m^2 l^2} = 0 \tag{9-121}$$

This equation yields the following eigenvalues for this case of symmetric modes:

$$\lambda_1^2 = \frac{2P}{3ml} \quad \text{or} \quad \lambda_1 = 0.816\sqrt{\frac{P}{ml}}$$

$$\lambda_3^2 = \frac{4P}{ml} \quad \text{or} \quad \lambda_3 = 2\sqrt{\frac{P}{ml}}$$

The corresponding amplitude ratios for the complete system, obtained by using Eq. (9–120) and symmetry, are

$$\{A^{(1)}\}^T = \lfloor 0.75 \quad 1 \quad 0.75 \rfloor$$

$$\{A^{(3)}\}^T = \lfloor 0.50 \quad -1 \quad 0.50 \rfloor$$

where we have arbitrarily set the largest amplitude equal to unity for each mode.

Next we consider the *antisymmetric modes*. It is evident from symmetry considerations that m_2 does not move in this case. Hence the left half of the system is as shown in Fig. 9–5(c). The equation of motion for this simple system is

$$m\ddot{x}_1 + 2\frac{P}{l}x_1 = 0 \tag{9–122}$$

from which we obtain the frequency

$$\lambda_2 = \sqrt{\frac{2P}{ml}}$$

The corresponding modal column for the entire system is

$$\{A^{(2)}\}^T = \lfloor 1 \quad 0 \quad -1 \rfloor$$

Now we can write the complete modal matrix as follows:

$$[A] = \begin{bmatrix} 0.75 & 1 & 0.50 \\ 1 & 0 & -1 \\ 0.75 & -1 & 0.50 \end{bmatrix} \tag{9–123}$$

It is interesting to consider the transformation to principal coordinates, namely,

$$x_1 = 0.75\,U_1 + U_2 + 0.50\,U_3$$

$$x_2 = U_1 - U_3 \tag{9–124}$$

$$x_3 = 0.75\,U_1 - U_2 + 0.50\,U_3$$

If we compare these equations with those of (6–52), we see that they are of identical form, the U's in this case corresponding to the q's of Eq. (6–52). Furthermore, the deflection forms or *mode shapes* for the three natural modes are those shown in Fig. 6–10. Thus we see that the system can perform a free vibration with the particles moving sinusoidally and in phase according to any one of the three possible mode shapes. A general free motion would involve all three modes oscillating at their individual frequencies; hence the motion of any single particle would not be sinusoidal.

Returning now to the problem of finding the generalized inertia and stiffness matrices, let us first write the $[m]$ and $[k]$ matrices for the complete system. Noting that the individual coefficients are additive at the plane of symmetry, we find from Eqs. (9–118) and (9–119) that

$$[m] = m \begin{bmatrix} 1 & 0 & 0 \\ 0 & 0.75 & 0 \\ 0 & 0 & 1 \end{bmatrix} \tag{9-125}$$

$$[k] = \frac{P}{l} \begin{bmatrix} 2 & -1 & 0 \\ -1 & 2 & -1 \\ 0 & -1 & 2 \end{bmatrix} \tag{9-126}$$

Finally, we can use the general equations given in (9–73) and (9–79) to obtain

$$[M] = [A]^T [m][A] = m \begin{bmatrix} 1.875 & 0 & 0 \\ 0 & 2 & 0 \\ 0 & 0 & 1.250 \end{bmatrix} \tag{9-127}$$

$$[K] = [A]^T [k][A] = \frac{P}{l} \begin{bmatrix} 1.25 & 0 & 0 \\ 0 & 4 & 0 \\ 0 & 0 & 5 \end{bmatrix} \tag{9-128}$$

Rayleigh's Principle. Suppose we consider a conservative linear system described by the coordinates $x_1, x_2, \ldots, x_n$. If such a system is oscillating in a single natural mode, we know that the motions of all coordinates are in phase, in the sense that all reach their maximum amplitudes together and all pass through zero at the same instant. Furthermore, if the reference value of potential energy is chosen such that it is zero at the equilibrium position, then the kinetic energy is maximum when potential energy is zero, and vice versa. Since the total energy is constant, we find that the maximum values of the kinetic and potential energy are equal.

Now we assume for convenience that the maximum value of the given principal coordinate is unity. Then, from Eqs. (9–61) and (9–62),

$$(x_j)_{\text{max}} = A_j \tag{9-129}$$

and

$$(\dot{x}_j)_{\text{max}} = A_j \omega \tag{9-130}$$

where ω is the natural circular frequency of the given mode. It follows from Eqs. (9–48) and (9–49) that

$$T_{\text{max}} = \frac{\omega^2}{2} \sum_{i=1}^{n} \sum_{j=1}^{n} m_{ij} A_i A_j \tag{9-131}$$

and

$$V_{\text{max}} = \frac{1}{2} \sum_{i=1}^{n} \sum_{j=1}^{n} k_{ij} A_i A_j \tag{9-132}$$

Thus, by setting T_{max} equal to V_{max}, we obtain

$$\omega^2 = \frac{\omega^2 V_{\text{max}}}{T_{\text{max}}} = \frac{\sum_{i=1}^{n} \sum_{j=1}^{n} k_{ij} A_i A_j}{\sum_{i=1}^{n} \sum_{j=1}^{n} m_{ij} A_i A_j} \tag{9-133}$$

This is *Rayleigh's method* for obtaining mode frequencies.

We see that the preceding equation provides a convenient means of determining the natural frequency of a given mode without finding the equations of motion or solving the characteristic equation, provided that the modal column $\{A\}$ is known. It is apparent, then, that this method can always be applied to systems with a single degree of freedom. But it is also useful for more complicated systems in which the modal column is known from symmetry considerations, or from the requirement of zero momentum, or perhaps is estimated or approximated in some fashion.

As an example, consider again the non-zero antisymmetric mode for the system of Fig. 9–4(a). If we take just that portion shown in Fig. 9–4(c), we find that

$$\omega^2 = \frac{k_1(A_1 - A_2)^2}{m_1 A_1^2 + m_2 A_2^2} \tag{9–134}$$

But we recall that $A_2/A_1 = -4$ in order that the linear momentum be zero. Also, it was given that $m_2/m_1 = 1/4$. So we see that

$$\omega^2 = \frac{k_1(1 + 4)^2}{m_1(1 + 4)} = 1250 \text{ rad}^2/\text{sec}^2$$

in agreement with the value of λ_3^2 found in Example 9–3.

Another important result is *Rayleigh's principle* which states that, if small changes are made in the modal column in the vicinity of its true values for a given mode, the changes in the value of ω^2 computed using Eq. (9–133) will be of higher order in the small quantities. In other words, the *stationary* values of ω^2 correspond to the true eigenvalues. So we find that

$$\frac{\partial \omega_k^2}{\partial A_i} = 0 \quad (i = 1, 2, \ldots, n) \tag{9–135}$$

at an eigenvalue. These equations can be used to solve for the unknown amplitude coefficients.

As an illustration of the use of Rayleigh's principle, consider again the system of Example 9–3. Let us calculate the amplitude ratios for the symmetric modes. Taking the portion of the system that is shown in Fig. 9–4(b), we find that

$$T_{\text{max}} = \frac{\omega^2}{2}(m_1 A_1^2 + m_2 A_2^2) \tag{9–136}$$

and

$$V_{\text{max}} = \frac{1}{2} k_1 (A_1 - A_2)^2 + k_2 A_2^2 \tag{9–137}$$

If we arbitrarily let $A_1 = 1$ and note that $k_1 = 2k_2$, $m_1 = 4m_2$, we find from Eq. (9–133) that

$$\omega^2 = \frac{k_1}{m_2}\left(\frac{2A_2^2 - 2A_2 + 1}{A_2^2 + 4}\right) \tag{9–138}$$

We see that A_2 is the only variable. So we can write

$$\frac{d\omega^2}{dA_2} = 0$$

from which we obtain the following equation:

$$A_2^2 + 7A_2 - 4 = 0$$

The roots are

$$A_2 = \frac{1}{2}(-7 \pm \sqrt{65}) = 0.531, \ -7.53$$

in agreement with our earlier results in Example 9–3.

It is interesting to note that the value of ω^2 obtained by using Eq. (9–133) reaches its minimum value when the amplitude coefficients correspond exactly to those of the fundamental or lowest frequency mode. Hence if one uses an approximate mode shape in calculating ω^2, the result must be larger than the true eigenvalue. Nevertheless, small errors in the assumed mode shape or eigenvector can usually be tolerated because, according to Rayleigh's principle, the resulting error in ω^2 must be of higher order in the small quantities.

In a similar fashion, the largest eigenvalue corresponds to the maximum possible value of ω^2 for any set of A's. Hence any approximation to this mode shape will result in a value of ω^2 that is too small. For the intermediate eigenvalues the error can have either sign. These characteristics of the function ω^2, which is the ratio of homogeneous quadratic forms, remind one of similar properties of the moment of inertia I of a rigid body about an axis near one of the principal axes.

It is apparent from Eq. (9–133) that the higher modes have a larger V_{max} for a given value of T_{max}. Since the elastic members normally connect adjacent masses, this explains why the adjacent masses tend to move in the same direction for the low-frequency modes and in the opposite direction for the high-frequency modes.

Example 9–5. Given the system shown in Fig. 9–6, consisting of two uniform circular disks connected by a spring between corresponding points on the circumference of each. Find the natural frequencies and amplitude ratios for small oscillations about this reference position for the following cases: (a) there is no slipping of the disks on the floor; (b) the floor is perfectly smooth.

Consider first the case of no slipping. The system has only two degrees of freedom and can be described in terms of the coordinates x_1 and x_2. Furthermore, there is symmetry about a vertical plane through the center of the spring, so we can look at just one half of the system at a time.

The antisymmetric mode is characterized by the constraint $x_1 = x_2$. Since this motion involves no change in the length of the spring, it follows that the only antisymmetric mode is a zero frequency mode.

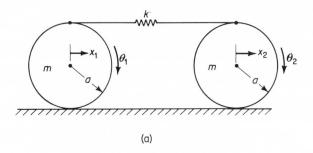

(a)

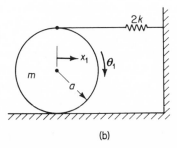

(b)

Fig. 9-6. (a) The complete system consisting of two uniform circular disks and a spring. (b) The partial system for symmetric modes.

Now consider the symmetric mode, for which $x_1 = -x_2$. It can be seen that the center of the spring is fixed; hence the left-hand half of the system can be analyzed separately. For the system shown in Fig. 9–6(b), we obtain

$$T_{\max} = \frac{1}{2} m \dot{x}_1^2 + \frac{1}{2} I \dot{\theta}_1^2$$

$$= \frac{3}{4} m \omega^2 A_{x_1}^2 \qquad (9\text{–}139)$$

where we note that $I = \frac{1}{2} ma^2$. The maximum potential energy is

$$V_{\max} = 4k A_{x_1}^2 \qquad (9\text{–}140)$$

So, using Eq. (9–133), we find that

$$\omega = 4 \sqrt{\frac{k}{3m}}$$

Summarizing the results, we can write

$$\omega_1 = 0, \qquad \omega_2 = 4 \sqrt{\frac{k}{3m}}, \qquad [A] = \begin{bmatrix} 1 & 1 \\ 1 & -1 \end{bmatrix}$$

Now consider the case in which the floor is frictionless. Here the complete system has four degrees of freedom. For the antisymmetric modes we have $x_1 = x_2$ and $\theta_1 = \theta_2$. It is evident, then, that both antisymmetric modes involve no change in the length of the spring. Therefore these modes have zero frequency. For these repeated roots, we see that a possible set of eigenvectors which conform to the constraints and are mutually orthogonal is given by

$$\lfloor A_{x_1}^{(1)} \quad A_{x_2}^{(1)} \quad A_{\theta_1}^{(1)} \quad A_{\theta_2}^{(1)} \rfloor = \lfloor 1 \quad 1 \quad 0 \quad 0 \rfloor$$

and

$$\lfloor A_{x_1}^{(2)} \quad A_{x_2}^{(2)} \quad A_{\theta_1}^{(2)} \quad A_{\theta_2}^{(2)} \rfloor = \lfloor 0 \quad 0 \quad 1 \quad 1 \rfloor$$

For the case of the symmetric modes, we can again consider the partial system of Fig. 9–6(b), but now the floor is assumed to be frictionless. We obtain that

$$T_{\max} = m\omega^2 \left(\frac{1}{2} A_{x_1}^2 + \frac{1}{4} a^2 A_{\theta_1}^2 \right) \tag{9-141}$$

and

$$V_{\max} = k(A_{x_1} + aA_{\theta_1})^2 \tag{9-142}$$

So, from Eq. (9–133) we have

$$\omega^2 = \frac{4k(A_{x_1}^2 + 2aA_{x_1}A_{\theta_1} + a^2 A_{\theta_1}^2)}{m(2A_{x_1}^2 + a^2 A_{\theta_1}^2)} \tag{9-143}$$

Now let us arbitrarily set $A_{x_1} = 1$ and use Rayleigh's principle to find A_{θ_1}. We see that

$$\frac{d\omega^2}{dA_{\theta_1}} = 0$$

Performing the differentiation and simplifying, we obtain

$$a^2 A_{\theta_1}^2 - aA_{\theta_1} - 2 = 0 \tag{9-144}$$

The roots of Eq. (9–144) are

$$A_{\theta_1} = -\frac{1}{a}, \quad \frac{2}{a}$$

Substituting these values of A_{θ_1} into Eq. (9–143), we obtain the corresponding natural frequencies, namely,

$$\omega = 0, \quad \sqrt{\frac{6k}{m}}$$

Using the foregoing results and the symmetry properties, we see that the eigenvectors for the symmetric modes are

$$\lfloor A_{x_1}^{(3)} \quad A_{x_2}^{(3)} \quad A_{\theta_1}^{(3)} \quad A_{\theta_2}^{(3)} \rfloor = \left\lfloor 1 \quad -1 \quad -\frac{1}{a} \quad \frac{1}{a} \right\rfloor$$

$$\lfloor A_{x_1}^{(4)} \quad A_{x_2}^{(4)} \quad A_{\theta_1}^{(4)} \quad A_{\theta_2}^{(4)} \rfloor = \left\lfloor 1 \quad -1 \quad \frac{2}{a} \quad -\frac{2}{a} \right\rfloor$$

We can summarize the mode shapes in the modal matrix:

$$[A] = \begin{bmatrix} 1 & 0 & 1 & 1 \\ 1 & 0 & -1 & -1 \\ 0 & 1 & -\dfrac{1}{a} & \dfrac{2}{a} \\ 0 & 1 & \dfrac{1}{a} & -\dfrac{2}{a} \end{bmatrix}$$

We have seen that three out of four natural modes for the frictionless case are zero frequency modes. It is interesting to consider again the nature of the single non-zero mode. We note that plane motion is involved; hence the disks will rotate about their respective *instantaneous centers* which remain fixed in space for this case of small motion. Since no other horizontal forces are applied, the instantaneous centers must also be located at the *centers of percussion* with respect to the spring attachment points. Using Eq. (7–190), we find that the center of percussion is halfway between the center of the disk and the contact point with the floor. It follows that the ratio of linear and angular accelerations for the disk are such that this point does not move during the vibration.

9–5. FORCED VIBRATIONS OF A CONSERVATIVE SYSTEM

Transfer Functions. Consider a conservative system whose configuration is described in terms of n independent coordinates $x_1, x_2, \ldots, x_n$. Suppose that external forces are applied to the system such that the force $f_j(t)$ corresponds to x_j, that is, the virtual work done by the external forces in an arbitrary virtual displacement is

$$\delta W = \sum_{j=1}^{n} f_j \, \delta x_j \tag{9–145}$$

Then we obtain n differential equations of motion which can be written as a matrix equation of the form

$$[m]\{\ddot{x}\} + [k]\{x\} = \{f(t)\} \tag{9–146}$$

First assume that the applied forces are harmonic functions of time. Using complex notation we can write

$$f_j(t) = F_j e^{i\omega t} \tag{9–147}$$

To find the *steady-state solution*, we note first that the system is linear. Therefore the total steady-state response is the sum of the responses to the individual forces. If we assume solutions of the form

$$x_i = B_i e^{i\omega t} \tag{9–148}$$

and substitute from Eqs. (9–147) and (9–148) into the differential equations of (9–146), we obtain

$$([k] - \omega^2[m])\{B\} = \{F\} \qquad (9\text{–}149)$$

The solution for $\{B\}$ is

$$\{B\} = [T]\{F\} \qquad (9\text{–}150)$$

where $[T]$ is called the *transfer function matrix* and is given by

$$[T] = ([k] - \omega^2[m])^{-1} \qquad (9\text{–}151)$$

A typical element $T_{ij}(i\omega)$ is a *transfer function* which expresses the steady-state output amplitude and phase at x_i due to a unit sinusoidal input force at x_j. For this case of a conservative system $[T]$ is a real symmetric matrix since both $[k]$ and $[m]$ are symmetric and real. We note that $[T]$ exists, except if

$$|([k] - \omega^2[m])| = 0$$

But a comparison with Eq. (9–56) shows that this occurs only when $\omega^2 = \lambda^2$, that is, when the system is driven at one of its resonant frequencies.

In general, the frequency of each forcing function may be different. For this case, the *steady-state solution* at x_i is

$$x_{is} = \sum_{j=1}^{n} T_{ij}(i\omega_j) F_j e^{i\omega_j t} \qquad (9\text{–}152)$$

where the F_j may be complex.

Steady-state solutions can be obtained for more general periodic forcing functions by using a Fourier series representation of the forcing function and superimposing the solutions resulting from individual harmonic terms.

Transfer function methods are also applicable to systems with damping and, in fact, are particularly useful in these cases. On the other hand, we shall find that the use of principal coordinates may be computationally more advantageous for the analysis of many undamped systems.

If the complete solution is desired for a given set of initial conditions, we must also solve for the transient solution. Here we use the methods of Sec. 9–3 and solve for the natural frequencies and amplitude ratios, obtaining a transient solution of the form given by Eq. (9–61), namely,

$$x_{jt} = \sum_{k=1}^{n} A_{jk} C_k \cos(\lambda_k t + \phi_k) \qquad (9\text{–}153)$$

Note that the boundary conditions on the transient solution are

$$x_{jt}(0) = x_j(0) - x_{js}(0) \qquad (9\text{–}154)$$

$$\dot{x}_{jt}(0) = \dot{x}_j(0) - \dot{x}_{js}(0) \qquad (9\text{–}155)$$

from which we solve for the n C's and the n ϕ's.

The Use of Principal Coordinates. Now let us consider the same system in terms of a set of principal coordinates $U_1, U_2, \ldots, U_n$. The equations of motion are similar to those of Eq. (9–85) except that each independent modal system is driven by its corresponding generalized force Q_i. We recall from Eq. (6–51) that

$$Q_i = \sum_{j=1}^{n} f_j \frac{\partial x_j}{\partial U_i} \qquad (9\text{–}156)$$

and from Eq. (9–63), we obtain

$$\frac{\partial x_j}{\partial U_i} = A_{ji} \qquad (9\text{–}157)$$

So we find that

$$Q_i = \sum_{j=1}^{n} f_j A_{ji} \qquad (9\text{–}158)$$

or, in matrix notation,

$$\{Q\} = [A]^T \{f\} \qquad (9\text{–}159)$$

The differential equations of motion are

$$[M]\{\ddot{U}\} + [K]\{U\} = \{Q\} \qquad (9\text{–}160)$$

If we write the individual equations, we obtain

$$\begin{aligned}
M_{11}\ddot{U}_1 + K_{11}U_1 &= Q_1 \\
M_{22}\ddot{U}_2 + K_{22}U_2 &= Q_2 \\
&\;\;\vdots \\
M_{nn}\ddot{U}_n + K_{nn}U_n &= Q_n
\end{aligned} \qquad (9\text{–}161)$$

Now the advantage of using principal coordinates is apparent. Rather than solving n simultaneous differential equations involving coupled coordinates, we can solve for each of the principal coordinates *independently* and superimpose these solutions to obtain the motions of the x's. Thus we can use Eq. (9–63) to obtain solutions of the form

$$x_j = \sum_{i=1}^{n} A_{ji} U_i \qquad (9\text{–}162)$$

The transient portion of the solution for the principal coordinates is obtained by using the methods of Sec. 9–3, where the constants C_i and ϕ_i are evaluated by using Eqs. (9–103) and (9–104). Note that, in this evaluation, we use the initial conditions for the transient solution, not the given initial conditions for the complete solution.

A further advantage of using principal coordinates occurs in obtaining the response to a forcing function having a limited frequency content. One can then use the frequency discrimination of the individual modes to reduce the number of degrees of freedom by omitting those modes whose response is negligible. Another useful approximation is to neglect the inertial forces for those modes whose natural frequency is well above the forcing frequency, and to neglect the elastic forces for those modes whose natural frequency is well below the forcing frequency. In other words, $M_{kk}\omega^2 \ll K_{kk}$ if $\omega^2 \ll \omega_k^2$, and vice versa.

As an illustration of a forced vibration calculation using principal coordinates, consider again the system of Example 9–4. Suppose a force f_2 is applied at x_2, where

$$f_2 = 0.1P \sin \sqrt{\frac{P}{ml}}\, t \tag{9–163}$$

We wish to find the complete solution for x_1, x_2, x_3, assuming that the system is at rest in the equilibrium position at $t = 0$.

First we calculate the generalized forces corresponding to each of the modes. Using Eq. (9–158) we obtain

$$Q_1 = f_2 A_{21} = 0.1P \sin \sqrt{\frac{P}{ml}}\, t$$

$$Q_2 = f_2 A_{22} = 0 \tag{9–164}$$

$$Q_3 = f_2 A_{23} = -0.1P \sin \sqrt{\frac{P}{ml}}\, t$$

These values and those of the $[M]$ and $[K]$ matrices are now substituted into Eq. (9–161) to obtain the following differential equations of motion:

$$\frac{15m}{8} \ddot{U}_1 + \frac{5P}{4l} U_1 = 0.1P \sin \sqrt{\frac{P}{ml}}\, t$$

$$2m\ddot{U}_2 + 4\frac{P}{l} U_2 = 0 \tag{9–165}$$

$$\frac{5m}{4} \ddot{U}_3 + 5\frac{P}{l} U_3 = -0.1P \sin \sqrt{\frac{P}{ml}}\, t$$

To obtain the steady-state solutions, we use trial solutions of the form

$$U_{is} = C_{is} \sin \sqrt{\frac{P}{ml}}\, t \tag{9–166}$$

After substituting into Eq. (9–165) and comparing coefficients, we find that

$$C_{1s} = -0.160l$$

$$C_{2s} = 0 \tag{9–167}$$

$$C_{3s} = -0.0267l$$

Hence the steady-state solutions are

$$U_{1s} = -0.160l \sin \sqrt{\frac{P}{ml}} t$$

$$U_{2s} = 0 \tag{9-168}$$

$$U_{3s} = -0.0267l \sin \sqrt{\frac{P}{ml}} t$$

The transient solutions are of a form similar to that given previously by Eq. (9–62), namely,

$$U_{it} = C_i \sin (\lambda_i t + \phi_i) \tag{9-169}$$

where, for convenience, we have used sines rather than cosines.

To evaluate the constants C_1, C_2, C_3, and ϕ_1, ϕ_2, ϕ_3, we apply the given initial conditions to the complete solution, obtaining

$$U_1(0) = 0 = C_1 \sin \phi_1$$

$$U_2(0) = 0 = C_2 \sin \phi_2 \tag{9-170}$$

$$U_3(0) = 0 = C_3 \sin \phi_3$$

and

$$\dot{U}_1(0) = 0 = C_1\lambda_1 \cos \phi_1 - 0.160\sqrt{\frac{Pl}{m}}$$

$$\dot{U}_2(0) = 0 = C_2\lambda_2 \cos \phi_2 \tag{9-171}$$

$$\dot{U}_3(0) = 0 = C_3\lambda_3 \cos \phi_3 - 0.0267\sqrt{\frac{Pl}{m}}$$

Solving these six equations after substituting the values of the frequencies $\lambda_1, \lambda_2, \lambda_3$, we obtain

$$\phi_1 = \phi_2 = \phi_3 = 0$$

and

$$C_1 = 0.196l$$

$$C_2 = 0$$

$$C_3 = 0.0133l$$

Thus the transient solution in terms of principal coordinates is

$$U_{1t} = 0.196l \sin 0.816\sqrt{\frac{P}{ml}} t$$

$$U_{2t} = 0 \tag{9-172}$$

$$U_{3t} = 0.0133l \sin 2\sqrt{\frac{P}{ml}} t$$

Noting that

$$U_i = U_{it} + U_{is} \tag{9-173}$$

and transforming back to ordinary coordinates using Eq. (9–124), we obtain

$$x_1 = x_3 = 0.147l \sin 0.816\sqrt{\frac{P}{ml}}\, t + 0.0067l \sin 2\sqrt{\frac{P}{ml}}\, t$$

$$- 0.133l \sin \sqrt{\frac{P}{ml}}\, t$$

$$(9\text{--}174)$$

$$x_2 = 0.196l \sin 0.816\sqrt{\frac{P}{ml}}\, t - 0.0133l \sin 2\sqrt{\frac{P}{ml}}\, t$$

$$- 0.133l \sin \sqrt{\frac{P}{ml}}\, t$$

We observe that the second mode is absent in both transient and steady-state solutions. The steady-state solution is missing because the system is being driven at a point of zero amplitude for this mode; hence the generalized force Q_2 is zero. The lack of a transient second-mode solution is a coincidence in the sense that the initial conditions for the given mode must match those for its steady-state solution in order for the transient solution to be zero.

The lack of the second mode in the steady-state solution shows that a mode cannot be driven at a *node* point, that is, at a point of zero amplitude. This result can be used to drive selectively certain modes and not others. More generally, if one can drive the system harmonically and in phase at m points ($m \geq 2$), then the relative amplitudes of the forces can be chosen such that the steady-state motion of ($m - 1$) modes is zero. For example, if a system having no inertial coupling is driven harmonically at all n coordinates by forces proportional to the product of the mass and the kth mode amplitude coefficient at each coordinate, then the kth mode will be the only mode with a non-zero steady-state solution. All other Q's will be zero because of the orthogonality of the other modes with respect to the mass matrix $[m]$, as can be seen by noting that, in this instance,

$$Q_l = \sum_{i=1}^{n} A_i^{(l)} m_{ii} A_i^{(k)} \tag{9--175}$$

By Eq. (9--71), Q_l is zero except when $l = k$, assuming distinct modes.

9-6. VIBRATIONS WITH DAMPING

Free Vibrations. Now let us consider the free vibrations of a system containing *viscous dampers*. An idealized viscous damper is a massless device which transmits a force which is directly proportional to the difference in the values of certain velocity components at its two ends. Normally we think in terms of linear coordinates, in which case ordinary forces are involved. But viscous dampers can also be applied to angular coordinates to yield corresponding moments.

The equations of motion for a damped system having n degrees of freedom can be found by adding terms of the form $c_{ij}\dot{x}_j$ to the equations of (9–52) which were obtained previously for an undamped system. Here we introduce the *damping coefficient* c_j which is the force acting on the damper in the positive direction at x_i due to a unit velocity at x_j. It follows that the total damper force acting *on the system* at x_i is just $-\sum_{j=1}^{n} c_{ij}\dot{x}_j$. We again assume small motions, so c_{ij} is evaluated at the reference configuration and is *constant*. Thus we can write the equations of motion in the form

$$[m]\{\ddot{x}\} + [c]\{\dot{x}\} + [k]\{x\} = 0 \qquad (9\text{--}176)$$

where $[m]$, $[c]$, and $[k]$ are real symmetric matrices.

Another method for obtaining these equations of motion is to use a modified form of Lagrange's equations, namely,

$$\frac{d}{dt}\left(\frac{\partial L}{\partial \dot{x}_i}\right) - \frac{\partial L}{\partial x_i} + \frac{\partial F}{\partial \dot{x}_i} = 0 \qquad (9\text{--}177)$$

where the Lagrangian $L = T - V$ is found in the conventional manner, and where

$$F = \frac{1}{2}\sum_{i=1}^{n}\sum_{j=1}^{n} c_{ij}\dot{x}_i\dot{x}_j \qquad (9\text{--}178)$$

is known as *Rayleigh's dissipation function*. For dissipative systems, F is a positive semidefinite quadratic form in the velocities, and is equal to one-half of the instantaneous rate of mechanical energy dissipation from the system. We note that no individual damper can transmit a force without simultaneously dissipating energy. Nevertheless, undamped motions may be possible if they are such that no damper transmits forces, with the result that no energy is dissipated.

Now let us solve for the free vibrations of the given dissipative system. We assume trial solutions of the form

$$x_j = A_j C e^{\lambda t} \qquad (9\text{--}179)$$

and substitute into Eq. (9–176). After canceling the common factor $Ce^{\lambda t}$, we obtain

$$(\lambda^2[m] + \lambda[c] + [k])\{A\} = 0 \qquad (9\text{--}180)$$

These homogeneous equations possess a nontrivial solution if, and only if, their determinant is zero, that is,

$$|(\lambda^2[m] + \lambda[c] + [k])| = 0 \qquad (9\text{--}181)$$

This is the characteristic equation for the system and is of order $2n$ in λ. Since the coefficients are real, the $2n$ roots must either be real or else must occur in complex conjugate pairs. Furthermore, if we assume there are no energy sources within the system, the real parts of all roots must be negative or zero.

For each root λ_k, we can solve for the corresponding modal column $\{A^{(k)}\}$ by using Eq. (9–180). If the root is real, then the amplitude ratios of the corresponding modal column are entirely real. But if λ_k is complex, then the amplitude ratios are, in general, complex. Also, the modal column $\{A^{(k)} *\}$, corresponding to the complex conjugate root λ_k^*, is just the complex conjugate of the column $\{A^{(k)}\}$. This can be seen by noting that a reflection of each term of Eq. (9–180) about the real axis does not destroy the equality.

So if we assume $A_1 = 1$ for each mode and solve Eq. (9–180) for the modal column, we obtain

$$\{A_j\} = -[(\lambda^2 m_{ij} + \lambda c_{ij} + k_{ij})]^{-1}\{(\lambda^2 m_{i1} + \lambda c_{i1} + k_{i1})\}$$

$$(i, j = 2, 3, \ldots, n) \qquad (9\text{–}182)$$

The general solution for the free vibrations is obtained by superimposing the solutions corresponding to each of the $2n$ roots; hence we have

$$x_j = \sum_{k=1}^{2n} A_{jk} C_k e^{\lambda_k t} \quad (j = 1, 2, \ldots, n) \qquad (9\text{–}183)$$

The $2n$ C's are real, or occur in complex conjugate pairs. Hence they can be determined from the $2n$ initial conditions.

In comparing the solutions of Eq. (9–183) with those which were given previously in Eq. (9–61) for an undamped system, we notice several important differences. First we see that there are $2n$ modal columns rather than just n. Furthermore, those columns corresponding to real roots are not paired as complex conjugates with any other column of amplitude coefficients. So we find that the modal matrix $[A]$ is not square; hence it has no inverse, and one cannot solve for the principal coordinates from a knowledge of the x's alone, as in Eq. (9–65). Usually the $\dot{x}$'s will also be required. We conclude, then, that even though a single mode can exist, independent of the other modes, it is generally not convenient to use principal coordinates in the analysis of heavily damped systems.

Secondly, for the case of complex conjugate roots, we see that the amplitude coefficients are usually complex. This means that the various x's do not move in phase when oscillating in a single mode corresponding to a complex conjugate pair of roots. All coordinates, however, would show the same frequency and damping ratio under these conditions.

Forced Vibrations. The forced vibrations of a similar damped system are described by a set of differential equations of the form

$$[m]\{\ddot{x}\} + [c]\{\dot{x}\} + [k]\{x\} = \{f(t)\} \qquad (9\text{–}184)$$

In considering the methods of solution of these equations, let us first obtain the *steady-state solution* for a *harmonic* forcing function. Using complex notation, we take

$$f_j(t) = F_j e^{i\omega t}$$

as in Eq. (9–147). Also we assume solutions of the form

$$x_i = B_i e^{i\omega t}$$

and substitute into Eq. (9–184). The resulting equations are

$$([k] - \omega^2[m] + i\omega[c])\{B\} = \{F\} \qquad (9\text{–}185)$$

or, using the notation of Eq. (9–150),

$$\{B\} = [T]\{F\}$$

where the *transfer function matrix* $[T]$ is given by

$$[T] = ([k] - \omega^2[m] + i\omega[c])^{-1} \qquad (9\text{–}186)$$

A typical element $T_{ij}(i\omega)$ is a complex number and represents the amplitude and phase of the steady-state response at x_i as a result of a unit sinusoidal force input at x_j, assuming all other input forces are zero. Again we find that $[T]$ is symmetric, that is, $T_{ij} = T_{ji}$.

The steady-state solutions are given by Eq. (9–152):

$$x_{is} = \sum_{j=1}^{n} T_{ij}(i\omega_j) F_j e^{i\omega_j t}$$

Once again it can be seen that Fourier series methods are applicable to the problem of finding the response for any set of periodic input functions. Also, the complete solution may be obtained by adding the transient solution and solving for its $2n$ constants from the initial conditions, in a manner similar to that used for the case of undamped systems.

Now let us find the response at x_i due to an *arbitrary* force $f_j(t)$ applied at x_j. We shall assume for convenience that the system is at rest at its equilibrium position when $t = 0$. Using the method of convolution, we can write an expression for $x_i(t)$ which is similar to Eq. (3–194), namely,

$$x_i(t) = \int_0^t h_{ij}(t - \tau) f_j(\tau) \, d\tau \qquad (9\text{–}187)$$

where $h_{ij}(t)$ is the response at x_i due to a unit impulse applied at x_j when $t = 0$. So one can solve for the complete forced motion of the system if one knows $[h]$, the impulse response matrix. If the initial conditions are not zero, then a transient solution of the form of Eq. (9–183) is added. The constants C_k are evaluated from the actual initial conditions.

It is interesting to note that the impulse response function $h_{ij}(t)$ and the transfer function $T_{ij}(i\omega)$ are related. For example, let us use complex notation with Eq. (9–187) to calculate the response to a unit sinusoidal forcing function $f_j(t) = e^{i\omega t}$. We assume in this instance that $f_j(t)$ was applied infinitely long ago, so that only the steady-state solution remains. Hence we obtain

$$x_i(t) = \int_{-\infty}^{t} h_{ij}(t - \tau_1) e^{i\omega \tau_1} \, d\tau_1 \qquad (9\text{–}188)$$

where we use τ_1 as the dummy variable of integration. But from Eq. (9–152) we see that the response can be written in the form

$$x_i(t) = T_{ij}(i\omega)e^{i\omega t} \qquad (9\text{–}189)$$

Now we equate the right-hand sides of Eqs. (9–188) and (9–189) and also set $t = 0$. Making the substitution $\tau_1 = -\tau$, we obtain

$$T_{ij}(i\omega) = \int_0^\infty h_{ij}(\tau)e^{-i\omega\tau}\,d\tau \qquad (9\text{–}190)$$

Thus $T_{ij}(i\omega)$ is recognized as the *Fourier transform* of the corresponding function $h_{ij}(t)$, where we note that $h_{ij}(t)$ is zero for all negative values of its argument. Conversely, $h_{ij}(t)$ may be obtained as the *inverse Fourier transform* of $T_{ij}(i\omega)$.

$$h_{ij}(t) = \frac{1}{2\pi}\int_{-\infty}^\infty T_{ij}(i\omega)e^{i\omega t}\,d\omega \qquad (9\text{–}191)$$

Convergence difficulties can arise in evaluating the integral of Eq. (9–190) for the case of undamped systems. As a practical matter, however, one can use Laplace transform tables in these evaluations, making the substitution of $i\omega$ for the Laplacian operator s.

REFERENCES

1. Frazer, R. A., W. J. Duncan, and A. R. Collar, *Elementary Matrices*. New York: Cambridge University Press, 1960.

2. Goldstein, H., *Classical Mechanics*. Reading, Mass.: Addison-Wesley Publishing Company, 1950.

3. Halfman, R. L., *Dynamics—Systems, Variational Methods, and Relativity*. Reading, Mass.: Addison-Wesley Publishing Company, 1962.

4. Langhaar, H. L., *Energy Methods in Applied Mechanics*. New York: John Wiley & Sons, Inc., 1962.

5. Rayleigh, Lord, *The Theory of Sound*. New York: Dover Publications, Inc., 1945.

6. Tong, K. N., *Theory of Mechanical Vibration*. New York: John Wiley & Sons, Inc., 1960.

7. Whittaker, E. T., *A Treatise on the Analytical Dynamics of Particles, and Rigid Bodies*, 4th ed. New York: Dover Publications, Inc., 1944.

PROBLEMS

9–1. Two simple pendulums, each of mass m and length l, are coupled by a spring of stiffness k connected between their midpoints. Using the angles θ_1 and θ_2 as coordinates, solve for the frequencies and amplitude ratios of the natural modes. Assume small motions about the equilibrium position at $\theta_1 = \theta_2 = 0$.

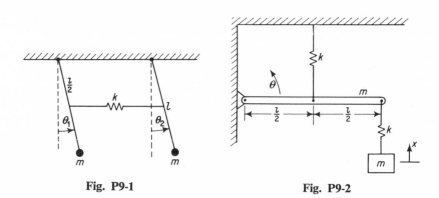

Fig. P9-1   **Fig. P9-2**

9–2. Write the differential equations for small motions of the given system near its equilibrium position at $x = 0$, $\theta = 0$. Solve for the mode frequencies and amplitude ratios.

9–3. For the system of problem 6–8, consider a reference condition in which $\dot\theta = \dot\theta_0$ and $\phi = 0$. Assuming that $R = l$, solve for the non-zero natural frequency and the corresponding amplitude ratio of a small oscillation about the reference condition.

9–4. Two thin uniform rods, each of mass m and length l, are connected by pin joints to form a double pendulum. Assuming small motions that are restricted to a given vertical plane, and letting the coordinates θ_1 and θ_2 designate the angles (measured from the vertical) of the upper and lower bars, respectively, solve for the natural frequencies and amplitude ratios. Use Rayleigh's principle.

9–5. A solid circular cylinder of mass m, radius r, rolls inside a thin cylindrical shell of mass m, radius $2r$, which, in turn, rolls in a cylindrical depression of radius $6r$. Assuming no slipping, find the frequencies and amplitude ratios for small oscillations about the equilibrium position. The absolute rotation angles of the cylinder and the shell are θ and ϕ, respectively.

9–6. A finite-difference representation of a cantilever beam of length $3l$ and mass $3m$ is shown in the figure. The levers are assumed to be rigid and massless with pinned joints. The bending stiffness is lumped into springs of stiffness k which produce a bending moment at the joints that is proportional to the difference in angular orientation of the adjacent levers. Find: (a) the natural frequencies and amplitude ratios; (b) the steady-state solution for y_1 and y_2 when a vertical force

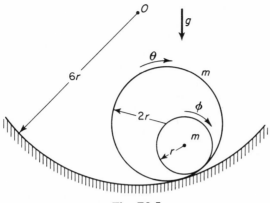

Fig. P9-5

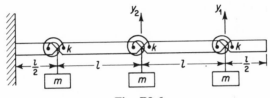

Fig. P9-6

$f_2 = F_2 \cos 2.5\sqrt{(k/ml^2)}t$ is applied at y_2. Assume small motions near the equilibrium position.

9-7. A thin uniform bar of mass m and length $3R$ can roll without slipping on a fixed circular cylinder of radius R. From each end of the bar, a mass m is suspended by a spring of stiffness k. For small motions about the equilibrium position in which the bar is horizontal and the contact point is at its center, solve for the mode fre-

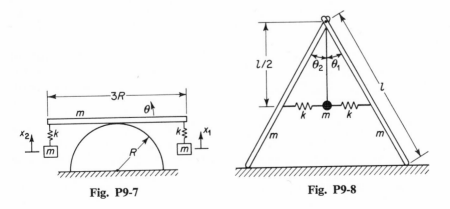

Fig. P9-7   **Fig. P9-8**

quencies and amplitude ratios. Let $mg = 2kR$ and assume that x_1, x_2, and θ are absolute displacements. Check the orthogonality of the calculated modes.

9–8. The upper ends of two thin uniform rods, each of mass m and length l, are connected by a pin joint. The other ends slide on a frictionless floor. A particle of mass m is suspended from the pin joint by a massless string of length $l/2$ and is connected by equal horizontal springs to the rods. Assuming that the system remains in the same vertical plane, find the non-zero natural frequencies and the corresponding amplitude ratios for small oscillations around an equilibrium position at $\theta_1 = \theta_2 = 30°$.

9–9. Solve for the natural frequencies and corresponding amplitude ratios of the plane vibrations of the system shown in the figure. Assume small motions.

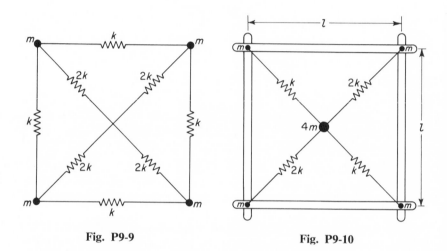

Fig. P9-9   Fig. P9-10

9–10. Find the natural frequencies and amplitude ratios for small motions of the given system in its own plane. The bars are connected by pin joints and are assumed to be massless, but four particles of mass m each are located at the joints.

9–11. Suppose that a spring of stiffness k_0 and a mass m_0 are attached in tandem to the mass m_1 of Example 9–1, forming a system of three masses and three springs connected in a manner similar to Fig. 9–2. Show that the addition of this mass and spring results in lowering the lowest natural frequency and raising the highest natural frequency for the system. Furthermore, if $k_0 \ll k_1$ and $\lambda_0^2 < 2000 \text{ rad}^2/\text{sec}^2$ where $\lambda_0^2 = k_0/m_0$, then the highest natural frequency is approximated by

$$\lambda^2 \cong \lambda_2^2 + \frac{0.0165 k_0}{1 - (\lambda_0/\lambda_2)^2}$$

where $\lambda_2 = 46.18$ rad/sec refers to the original system.

9–12. A thin uniform bar of mass m and length $2l$ is supported by two springs of stiffness k at its ends. A linear damper is connected from its midpoint to the ground, as shown. Using x and θ as coordinates, solve for the motion of the system if it is released with zero velocity from an initial small displacement $x(0) = x_0$, $\theta(0) = \theta_0$. $c < \sqrt{8\ km}$.

9–13. Consider a system which is identical with that of Example 9–1 except that a viscous damper with a damping coefficient $c = 50$ lb sec/ft is connected in parallel with the spring k_1. Solve for the eigenvalues and the modal matrix.

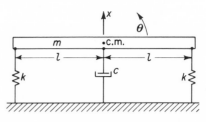

Fig. P9-12

9–14. For a damped system described by Eq. (9–176), show that

$$(\lambda_k + \lambda_l)\{A^{(l)}\}^T[m]\{A^{(k)}\} + \{A^{(l)}\}^T[c]\{A^{(k)}\} = 0, \qquad (\lambda_k \neq \lambda_l)$$

9–15. Consider small angular oscillations of the satellite of problem 8–29. Solve for the natural frequencies and amplitude ratios for the case in which $I_{xx} : I_{yy} : I_{zz} = 3 : 4 : 5$. What are the phase relationships?

APPENDIX A

INERTIAL PROPERTIES OF
HOMOGENEOUS BODIES

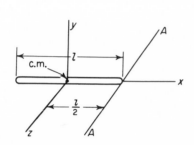

Thin rod

$$I_{xx} = 0$$

$$I_{yy} = I_{zz} = \frac{ml^2}{12}$$

$$I_{AA} = \frac{ml^2}{3}$$

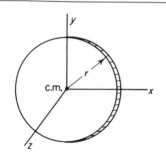

Thin circular disk

$$A = \pi r^2$$

$$I_{xx} = I_{yy} = \frac{mr^2}{4}$$

$$I_{zz} = \frac{mr^2}{2}$$

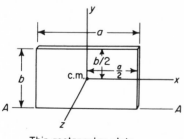

Thin rectangular plate

$$A = ab$$

$$I_{xx} = \frac{mb^2}{12}$$

$$I_{yy} = \frac{ma^2}{12}$$

$$I_{zz} = \frac{m}{12}(a^2 + b^2)$$

$$I_{AA} = \frac{mb^2}{3}$$

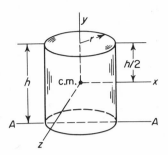

Right circular cylinder

$$V = \pi r^2 h$$

$$I_{xx} = I_{zz} = \frac{m}{12}(3r^2 + h^2)$$

$$I_{yy} = \frac{1}{2} mr^2$$

$$I_{AA} = \frac{m}{12}(3r^2 + 4h^2)$$

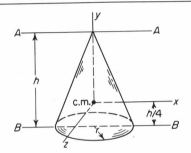

Right circular cone

$$V = \frac{1}{3}\pi r^2 h$$

$$I_{xx} = I_{zz} = \frac{3m}{80}(4r^2 + h^2)$$

$$I_{yy} = \frac{3}{10} mr^2$$

$$I_{AA} = \frac{3m}{20}(r^2 + 4h^2)$$

$$I_{BB} = \frac{m}{20}(3r^2 + 2h^2)$$

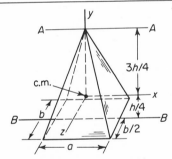

Right rectangular pyramid

$$V = \frac{abh}{3}$$

$$I_{xx} = \frac{m}{80}(4b^2 + 3h^2)$$

$$I_{yy} = \frac{m}{20}(a^2 + b^2)$$

$$I_{zz} = \frac{m}{80}(4a^2 + 3h^2)$$

$$I_{AA} = \frac{m}{20}(b^2 + 12h^2)$$

$$I_{BB} = \frac{m}{20}(b^2 + 2h^2)$$

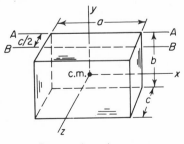

Rectangular prism

$$V = abc$$

$$I_{xx} = \frac{m}{12}(b^2 + c^2)$$

$$I_{yy} = \frac{m}{12}(a^2 + c^2)$$

$$I_{zz} = \frac{m}{12}(a^2 + b^2)$$

$$I_{AA} = \frac{m}{3}(b^2 + c^2)$$

$$I_{BB} = \frac{m}{12}(4b^2 + c^2)$$

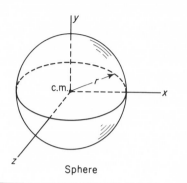

$$V = \frac{4}{3}\pi r^3$$

$$I_{xx} = I_{yy} = I_{zz} = \frac{2}{5}mr^2$$

Sphere

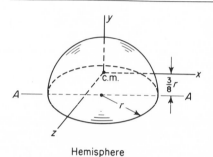

$$V = \frac{2}{3}\pi r^3$$

$$I_{xx} = I_{zz} = \frac{83}{320}mr^2$$

$$I_{yy} = \frac{2}{5}mr^2$$

$$I_{AA} = \frac{2}{5}mr^2$$

Hemisphere

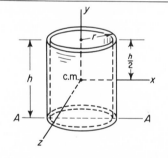

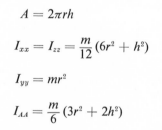

$$A = 2\pi rh$$

$$I_{xx} = I_{zz} = \frac{m}{12}(6r^2 + h^2)$$

$$I_{yy} = mr^2$$

$$I_{AA} = \frac{m}{6}(3r^2 + 2h^2)$$

Thin circular cylindrical shell

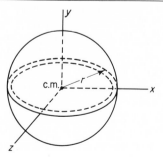

$$A = 4\pi r^2$$

$$I_{xx} = I_{yy} = I_{zz} = \frac{2}{3}mr^2$$

Thin spherical shell

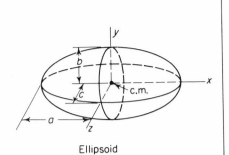

Ellipsoid

$$V = \frac{4}{3}\,\pi abc$$

$$I_{xx} = \frac{m}{5}\,(b^2 + c^2)$$

$$I_{yy} = \frac{m}{5}\,(a^2 + c^2)$$

$$I_{zz} = \frac{m}{5}\,(a^2 + b^2)$$

APPENDIX B

ANSWERS TO SELECTED PROBLEMS

CHAPTER 1

1-4. $A_1 = A_x - A_y - A_z$, $\quad A_2 = \sqrt{2}\, A_y$, $\quad A_3 = \sqrt{2}\, A_z$

1-6. 0.252 ft; 0.0885 sec

1-7. 71.55 ft/sec; 0.0032 rad/sec

1-8. $T_{AB} = T_{CD} = \sqrt{2}\, ma$; $\quad T_{BC} = ma$

CHAPTER 2

2-1. $\mathbf{a}_P = a\alpha[(\sin\theta - \alpha t^2)\mathbf{e}_r + (1 + \cos\theta)\mathbf{e}_\theta]$

2-3. $\mathbf{r}_P = \dfrac{a_A}{(\omega^4 + \dot\omega^2)}(\omega^2\mathbf{i} + \dot\omega\mathbf{j})$

2-5. $\mathbf{a} = 159.4\mathbf{e}_r - 695.5\mathbf{e}_z$ ft/sec^2

2-7. $\mathbf{a} = \left\{\left[\left(1 - \dfrac{r_2}{r_1}\right)\dfrac{r_2^2}{r_1} - \left(1 - \dfrac{r_2}{r_1}\right)^2 b\cos\phi\right]\dot\phi^2 - \left(1 - \dfrac{r_2}{r_1}\right)b\sin\phi\,\ddot\phi\right\}\mathbf{e}_n$
$\quad + \left\{\left(1 - \dfrac{r_2}{r_1}\right)b\cos\phi\,\ddot\phi - \left(1 - \dfrac{r_2}{r_1}\right)^2 b\sin\phi\,\dot\phi^2\right\}\mathbf{e}_t$

2-8. $\mathbf{a} = \dfrac{v_0^2}{R}\left[1 + 2\cos\theta + \dfrac{a}{R}(1 + \cos\theta) + \dfrac{R}{a}\cos\theta - \dfrac{a\dot R}{v_0 R}\sin\theta\right]\mathbf{e}_n$
$\quad - \dfrac{v_0^2}{R}\left[2\sin\theta + \dfrac{R}{a}\sin\theta + \dfrac{a}{R}\sin\theta + \dfrac{a\dot R}{v_0 R}(1 + \cos\theta)\right]\mathbf{e}_t$

2-9. (a) $\dot{\boldsymbol\rho} = \mathbf{v}_P - \mathbf{v}_{O'}$; (b) $(\dot{\boldsymbol\rho})_r = \mathbf{v}_P - \mathbf{v}_{O'} - \boldsymbol\omega \times \boldsymbol\rho$;
(c) $\ddot{\boldsymbol\rho} = 0$; (d) $(\ddot{\boldsymbol\rho})_r = \boldsymbol\omega \times (\boldsymbol\omega \times \boldsymbol\rho) - 2\boldsymbol\omega \times (\mathbf{v}_P - \mathbf{v}_{O'})$;
(e) $\dfrac{d}{dt}[(\dot{\boldsymbol\rho})_r] = -\boldsymbol\omega \times (\mathbf{v}_P - \mathbf{v}_{O'})$

2-10. (a) $\mathbf{v}_{P'P} = (r'\omega' - r\omega)\mathbf{e}_t$, $\quad \mathbf{a}_{P'P} = (r'\omega'^2 - r\omega^2)\mathbf{e}_n$;
(b) $\mathbf{v}_{P'P} = r'(\omega' - \omega)\mathbf{e}_t$, $\quad \mathbf{a}_{P'P} = r'(\omega' - \omega)^2\mathbf{e}_n$

2-12. $\mathbf{a} = (-R\dot\phi^2\sin^2\theta + a\dot\phi^2\sin\alpha\sin\theta\cos\theta + 2a\Omega\dot\phi\sin\alpha\sin\theta)\mathbf{e}_r$
$\quad + (-R\dot\phi^2\sin\theta\cos\theta + a\dot\phi^2\sin\alpha\cos^2\theta + 2a\Omega\dot\phi\sin\alpha\cos\theta$
$\quad + a\Omega^2\sin\alpha)\mathbf{e}_\theta + (-a\dot\phi^2\cos\alpha - 2a\Omega\dot\phi\cos\alpha\cos\theta$
$\quad - a\Omega^2\cos\alpha)\mathbf{e}_\phi$

CHAPTER 3

3-2. $\mathbf{v} = \omega(-1.501\mathbf{i} + 2.179\mathbf{j})$ ft/sec

3-4. $v = \sqrt{(mg/b)} \tan [\sqrt{(bg/m)}\,(t^* - t)]$

where $t^* = \sqrt{(m/bg)} \tan^{-1} (\sqrt{(b/mg)}\, v_0)$;

$$v^2 = \frac{mg}{b}\left[\left(1 + \frac{bv_0^2}{mg}\right) e^{-(2bz/m)} - 1\right]; \quad z_{\max} = \frac{m}{2b} \ln\left(1 + \frac{bv_0^2}{mg}\right)$$

3-5. $\Delta t = \dfrac{1}{\omega_n\sqrt{1 - \zeta^2}} \tan^{-1}\left(\dfrac{2\zeta\sqrt{1 - \zeta^2}}{1 - 2\zeta^2}\right)$

3-7. $x = 0; \quad 0 \le t \le \dfrac{\mu mg}{kv_0};$

$$x = v_0\left[t - \sqrt{\frac{m}{k}} \sin \sqrt{\frac{k}{m}}\left(t - \frac{\mu mg}{kv_0}\right)\right] - \frac{\mu mg}{k}, \quad t \ge \frac{\mu mg}{kv_0}$$

3-12. $r_{\max} = 3.06\,a$

3-13. $l\ddot{l} + \dot{l}^2 = a^2\Omega^2; \quad l = a\Omega t; \quad F = 4ma\Omega^3 t$

3-14. $\theta = 48.2°$

3-15. $v = \sqrt{\dfrac{2ag}{3}}\left(\dfrac{1}{\mu^2} - 3\right)^{1/4}$

3-16. $v_{\max} = \frac{1}{2}\sqrt{5gR}; \quad F = \frac{6}{5}mg$

CHAPTER 4

4-3. $F_r = -m\left[A\beta^2 \sin \beta t + \dfrac{r_0^4\omega_0^2}{(r_0 + A \sin \beta t)^3}\right]$

4-5. $\gamma_0 = \dfrac{1}{2} \sin^{-1}\left[\left(1 + \dfrac{1}{e}\right)\dfrac{gs}{v_0^2}\right]$

4-6. $\mathbf{v}'_A = v_0(0.525\mathbf{i} + 0.475\mathbf{j}); \quad \mathbf{v}'_B = v_0(0.475\mathbf{i} - 0.475\mathbf{j})$

4-7. $\mathbf{v}'_1 = v_0(\frac{3}{4}\mathbf{i} + \frac{1}{4}\mathbf{j}); \quad \mathbf{v}'_2 = v_0(\frac{1}{4}\mathbf{i} - \frac{1}{4}\mathbf{j}); \quad \omega'_1 = -\omega'_2 = -\dfrac{v_0}{2l}$

4-9. $v_0 = 3l_0 \sqrt{\dfrac{3k}{m}}$

4-10. $v = 9 \sqrt{\dfrac{3Fl}{8m}}$

4-11. $v_{\max} = v_0 \left(\sqrt{1 + \dfrac{2k}{mv_0^2 h}} - 1 \right)$

4-12. $v_{\max} = \dfrac{m_0 v_0}{m_0 + m} \left(\sqrt{1 + \dfrac{2k(m_0 + m)}{m_0 mv_0^2 h}} - 1 \right)$

4-17. (a) 1490 lb;   (b) 602 lb;   (c) 166 g, 54 g

4-18. (a) $0.905v$ at 51.3° above horizontal;   (b) Height of bounce is $v^2/4g$ and position is $2v^2/g$ from initial impact point.

4-20. (a) $e = 0.5$;   (b) $e = 1$

4-21. $0 \le \theta \le 2 \tan^{-1}(a/g)$, $T = 4.77 \sqrt{l/g}$

4-22. $\dot{\theta} = \sqrt{2g/l}$

4-23. $(r_1)_{\max} = 1.653\,l$, $P_{\min} = 0.435\,(mv^2/l)$

4-24. $\dot{\phi} = \dfrac{a^2 \Omega}{(2R^2 - a^2)\cos\theta} \left[1 \pm \sqrt{1 - \dfrac{2Rg}{a^4 \Omega^2}(2R^2 - a^2)\cos\theta} \right]$

CHAPTER 5

5-6. $v_r = 2737$ ft/sec

5-7. $r_{\min} = 1.236\,R$, $v_{\max} = 29{,}690$ ft/sec

5-8. $T = \dfrac{24\pi}{5} \sqrt{7R/g_0}$, $P = 0.0367\,mg_0$

5-9. $l = 1.231$ ft

5-10. $\dfrac{\partial \epsilon}{\partial v} = \dfrac{2}{\epsilon v}(1 - \epsilon^2)\left(\dfrac{a}{r} - 1 \right)$

5-14. (a) $\gamma = 56.5°$;   (b) 22.8 min before appearance over horizon.

5-15. $\gamma = 39.3°$, $\epsilon = 2.407$

5-16. $\dfrac{\Delta v}{v} = -8\pi \times 10^{-4}$, $\dfrac{\Delta r}{r} = 16\pi \times 10^{-4}$

5-17. (a) $\mathbf{v}_P = \mathbf{v}_E = 14{,}260\mathbf{i} - 1787\mathbf{j} - 14{,}260\mathbf{k}$ ft/sec
   $\mathbf{a}_P = \mathbf{a}_E = -8.42\mathbf{i} - 2.08\mathbf{j} - 8.29\mathbf{k}$ ft/sec²

 (b) $\mathbf{v}_P = 14{,}260\mathbf{i} - 14{,}260\mathbf{k}$ ft/sec
   $\mathbf{a}_P = -8.29\mathbf{i} - 8.29\mathbf{k}$ ft/sec²
   $\mathbf{v}_E = 14{,}260\mathbf{i} - 1525\mathbf{j} - 14{,}260\mathbf{k}$ ft/sec
   $\mathbf{a}_E = -8.18\mathbf{i} - 8.29\mathbf{k}$ ft/sec²

CHAPTER 6

6-1. $\psi = 71.8°$

6-2. $0.314a$

6-3. $\theta = 67.2°$

6-4. (a) $ml^2[(3 + 2\cos\phi)\ddot{\theta} + (1 + \cos\phi)\ddot{\phi} - \sin\phi(\dot{\phi}^2 + 2\dot{\theta}\dot{\phi})]$
$+ mgl[2\sin\theta + \sin(\theta + \phi)] = 0;$
$ml^2[(1 + \cos\phi)\ddot{\theta} + \ddot{\phi} + \dot{\theta}^2\sin\phi] + mgl\sin(\theta + \phi) = 0$

(b) $ml^2(5\ddot{\theta} + 2\ddot{\phi}) + mgl(3\theta + \phi) = 0;$
$ml^2(2\ddot{\theta} + \ddot{\phi}) + mgl(\theta + \phi) = 0$

6-9. $P = 3mg\cos\theta, \quad T = 7.42\sqrt{\dfrac{l}{g}}$

6-11. $ma(a\ddot{\theta} - a\dot{\phi}^2\sin\theta\cos\theta - g\sin\theta) = 0; \quad \dfrac{d}{dt}[ma^2\dot{\phi}(1 + \sin^2\theta)] = 0;$

$\theta_{\max} = 116.7°$

6-12. $r_{\min} = \tfrac{2}{3}r_0, \quad \dot{r}(0) = -\sqrt{2gr_0/3}$

6-15. $m(\ddot{r} - r\dot{\phi}^2\sin^2\alpha + g\cos\alpha) = 0; \quad \dfrac{d}{dt}(mr^2\dot{\phi}\sin^2\alpha) = 0;$

$\dot{r}_{\max} = 1.185\sqrt{ga}$

6-16. $\delta\ddot{r} + (3g/a)\cos\alpha\,\delta r = 0; \quad \omega = \sqrt{3g\cos\alpha/a}$

6-17. Angular acceleration $\ddot{\phi} = \dfrac{mg\sin 2\gamma}{2r_0(m_0 + m\sin^2\gamma)}$

vertical acceleration $\ddot{z} = -\dfrac{(m_0 + m)g\sin^2\gamma}{m_0 + m\sin^2\gamma}$

6-18. $(1 + \cos^2\theta)\ddot{\theta} - \sin\theta\cos\theta(\dot{\theta}^2 + \dot{\phi}^2) = 0;$

$\dfrac{d}{dt}[(1 + \sin^2\theta)\dot{\phi}] = 0; \quad \theta_{\min} = 35.1°$

6-19. (a) $m(\ddot{r} - r\dot{\theta}^2 - r\dot{\phi}^2\sin^2\theta + g\cos\theta) + k(r - l) + c\dot{r} = 0;$
$r\ddot{\theta} + 2\dot{r}\dot{\theta} - r\dot{\phi}^2\sin\theta\cos\theta - g\sin\theta = 0;$
$\dfrac{d}{dt}(r^2\dot{\phi}\sin^2\theta) = 0$

(b) Steady conical motion with $\dot{\phi} = 1.382\sqrt{g/l}$

6-20. $3m\ddot{x} = \lambda\sin\theta; \quad 3m\ddot{y} = -\lambda\cos\theta; \quad ma^2\ddot{\theta} = -a\lambda/2\sqrt{3};$

$\dot{\theta} = -\dfrac{2\sqrt{3}}{a}v_0\cos(\theta/\sqrt{5}); \quad \lambda = 9.30\dfrac{mv_0^2}{a}\sin(2\theta/\sqrt{5})$

CHAPTER 7

7-3. Particles, each of mass $m/4$, are located at $(\sqrt{3/2}\,\alpha, 0, 0)$,
$(-\alpha/\sqrt{6}, -\beta, -\gamma/\sqrt{3})$, $(-\alpha/\sqrt{6}, 0, 2\gamma/\sqrt{3})$,
$(-\alpha/\sqrt{6}, \beta, -\gamma/\sqrt{3})$, where $\alpha = \sqrt{(-I_{xx} + I_{yy} + I_{zz})/m}$,
$\beta = \sqrt{(I_{xx} - I_{yy} + I_{zz})/m}$, $\gamma = \sqrt{(I_{xx} + I_{yy} - I_{zz})/m}$

7-5.

$$[l] = \begin{bmatrix} \dfrac{1}{2} & \dfrac{1}{2} & \dfrac{-1}{\sqrt{2}} \\[2mm] \dfrac{1}{2} & \dfrac{1}{2} & \dfrac{1}{\sqrt{2}} \\[2mm] \dfrac{1}{\sqrt{2}} & \dfrac{-1}{\sqrt{2}} & 0 \end{bmatrix},$$

$$[I'] = \frac{1}{2} \begin{bmatrix} 7 & -1 & -\sqrt{2} \\ -1 & 7 & -\sqrt{2} \\ -\sqrt{2} & -\sqrt{2} & 6 \end{bmatrix} \text{ lb sec}^2 \text{ ft}$$

7-7.

$$\begin{Bmatrix} x \\ y \\ z \end{Bmatrix} = \frac{1}{\omega_y^2 \alpha_x^2} \begin{bmatrix} \omega_x^2(\omega_x^2 + \omega_y^2) - \alpha_x^2 & \omega_x\omega_y(\omega_x^2 + \omega_y^2) & -\omega_x\omega_y\alpha_x \\ \omega_x\omega_y(\omega_x^2 + \omega_y^2) & \omega_y^2(\omega_x^2 + \omega_y^2) & -\omega_y^2\alpha_x \\ -\omega_x\omega_y\alpha_x & -\omega_y^2\alpha_x & 0 \end{bmatrix} \begin{Bmatrix} a_x \\ a_y \\ a_z \end{Bmatrix}$$

7-9. $I_1 = 365.5$ lb sec^2 ft
$I_2 = 516.5$
$I_3 = 618.0$
$\Phi = 41.6°$

$$[l] = \begin{bmatrix} 0.806 & 0.383 & -0.451 \\ -0.198 & 0.893 & 0.404 \\ 0.557 & -0.236 & 0.796 \end{bmatrix}$$

7-13. $\dot{x} = \left[\dfrac{5m_0 g \sin\alpha \cos\alpha}{7(m_0 + m) - 5m_0 \cos^2\alpha} \right] t$

7-14. $\theta = -\dfrac{2b}{\sqrt{(I/m) + b^2}} \tan^{-1}\left(\dfrac{a}{\sqrt{(I/m) + b^2}} \right)$

7-20. $\ddot{r} - r\dot{\theta}^2 - g\sin\theta = 0$; $\frac{1}{3}(l^2 + 3r^2)\ddot{\theta} + 2r\dot{r}\dot{\theta} - \frac{1}{2}g(2r + l)\cos\theta = 0$;
$F = \frac{2}{7}mg$ upward

7-21. $v = \sqrt{14gR/3}$

7-23. Disk rolls around circle with the speed of the center $v = \frac{2}{5}v_0$.

7-24. $x = 16l/11$

7-25. $\ddot{\theta} = \dfrac{2g\sin\theta}{l(4 - 3\cos^2\theta)^2}(12 - 16\cos\theta + 9\cos^2\theta)$; $F_y = mg/4$

7-27. $\omega = \dfrac{3}{2l}\sqrt{v_0^2 + \frac{4}{3}gl}$

7-29. $r_{\min} = 0.548 r_0$

7-30. (a) $\omega_1' = \dfrac{(I_1 r_2 \omega_1 - I_2 r_1 \omega_2) r_2}{I_1 r_2^2 + I_2 r_1^2}$, $\quad \omega_2' = \dfrac{(I_2 r_1 \omega_2 - I_1 r_2 \omega_1) r_1}{I_1 r_2^2 + I_2 r_1^2}$

$\quad$ (b) $\quad t = \dfrac{I_1 I_2 (r_1 \omega_1 + r_2 \omega_2)}{\mu N (I_1 r_2^2 + I_2 r_1^2)}$

7-31. $\omega_1' = \dfrac{\omega_1}{2} - \dfrac{\omega_2}{4}$, $\quad \omega_2' = -\dfrac{\omega_1}{4} + \dfrac{7\omega_2}{8}$, $\quad t = \dfrac{mr_1}{8\mu N}(2\omega_1 + \omega_2)$

7-33. $\dot{\theta}^2 = \dfrac{72g}{l}\left(\dfrac{1 - \cos \theta}{5 + 27 \sin^2 \theta}\right)$

7-34. $\mu = 0.351$

7-35. Slipping cannot begin for $\cos^{-1}(0.818) < \theta < \cos^{-1}(0.626)$ or for $\theta > \cos^{-1}\left(\frac{1}{3}\right)$.

7-36. $\ddot{x} - \dfrac{1}{2}(l\ddot{\theta} \sin \theta + l\dot{\theta}^2 \cos \theta) = 0$; $\quad \ddot{\theta} - \dfrac{3\ddot{x}}{2l} \sin \theta = 0$; $\quad \dot{\theta}_{\max} = 2\dot{\theta}_0$

7-38. $l = \sqrt{\dfrac{I + 2ma^2}{2m}}$

7-39. $\dot{\theta}_1 = \dfrac{12 m_0 v_0}{(12 m_0 + 7m) l}$, $\quad \dot{\theta}_2 = \dfrac{-18 m_0 v_0}{(12 m_0 + 7m) l}$

7-40. $(l^2 + k^2)\ddot{\theta} + (Rl - l^2 - k^2)\ddot{\phi} + gl\theta = 0$; $(m_0 + m)R^2\ddot{\phi} + m(Rl - l_2 - k^2)(\ddot{\theta} - \ddot{\phi})/- mRl\ddot{\phi} = RF(t)$; $l = k^2/R$;

7-42. $v_A = \frac{6}{5}\sqrt{2gh}$

CHAPTER 8

8-1. $\{\Delta \omega\} = [I]^{-1}\{\mathcal{M}\}$

8-3. $(\alpha - \beta)_{\max} = 19.4°$

8-6. $\mathcal{M}_{\min} = \sqrt{\dfrac{I_{yy}(I_{yy} - I_{xx})\omega_y^2 + I_{zz}(I_{zz} - I_{xx})\omega_z^2}{1 - I_{xx}/I_{zz}}}$

8-9. $M_x = -\dfrac{ml^2}{24}\omega_y^2 \sin 2\omega_x t$, $\quad M_y = \dfrac{ml^2}{12}\omega_x \omega_y \sin 2\omega_x t$

8-14. $\omega = \dfrac{(I + 2ma^2)\omega_0}{I + 2m(a + l \sin \theta)^2}$,

$\quad \dot{\theta}^2 = \dfrac{\omega_0^2(I + 2ma^2)}{ml^2(1 + \cos^2 \theta)}\left[1 - \dfrac{I + 2ma^2}{I + 2m(a + l \sin \theta)^2}\right]$

8-16. $\dot{\psi}(0) = \dfrac{2a^2\Omega}{4l^2 + a^2}$, $\quad |\dot{\theta}| = \dfrac{2a^2\Omega}{4l^2 + a^2}\sqrt{1 + \dfrac{2gl(4l^2 + a^2)}{a^4\Omega^2}}$

8-17. (a) $\dot{\psi} = -0.511\sqrt{g/a}$; (b) $\dot{\psi} = -20\sqrt{g/r}$, $\quad 2\beta = 2.54°$

8-18. (a) $\dot{\psi} = -0.433\sqrt{g/r}$; (b) $\dot{\psi} = -0.360\sqrt{g/r}$

8-20. $\omega = \dfrac{\omega_0}{2\sin^2\theta}$, $\quad \dot\theta^2 = \dfrac{1}{4}\omega_0^2(2 - \csc^2\theta) + \dfrac{3g}{\sqrt{2}\,l}(1 - \sqrt{2}\cos\theta)$,

$(\omega_0^2)_{\min} = 5g/\sqrt{2}\,l$

8-21. (a) $\omega = 2.89\sqrt{g/a}$, $\quad \alpha = 30.0°$;   (b) $2\beta = 41.1°$

8-22. $v_P = 4.04\sqrt{ga}$

8-23. (a) $\omega_t = \dfrac{m_0 v_0 l}{I_t}$;   (b) $2\beta = 2\tan^{-1}\dfrac{m_0 v_0 l}{I_a \Omega}$;

(c) $\dfrac{2\pi I_t}{ml}\left[1 + \left(\dfrac{I_a \Omega}{m_0 v_0 l}\right)^2\right]^{-1/2}$

8-24. $\beta = 25.6°$

8-25. $\theta_{\max} = 1.78°$, $\quad \dot\psi_{\max} = 0.311\sqrt{k/I_t}$

8-26. (a) $\mathscr{M} = -0.50\mathbf{i} + 1000\mathbf{k}$ lb sec² ft;   (b) $2\beta = 0.2$ rad;

(c) $T = 6.25$ sec

8-27. (a) $\mathscr{M} = -0.50\mathbf{i}$ lb sec² ft;   (b) 0.222 rad;   (c) $T = 6.28$ sec

8-29. $I_{xx}\ddot\phi + \omega_0^2(I_{zz} - I_{yy})\phi + (I_{xx} + I_{yy} - I_{zz})\omega_0\dot\theta = 0$

$I_{yy}\ddot\theta + 4\omega_0^2(I_{zz} - I_{xx})\theta - (I_{xx} + I_{yy} - I_{zz})\omega_0\dot\phi = 0$

$I_{zz}\ddot\psi + 3\omega_0^2(I_{yy} - I_{xx})\psi = 0$

CHAPTER 9

9-4. $\lambda_1 = 0.856\sqrt{g/l}$, $\quad \lambda_2 = 2.30\sqrt{g/l}$, $\quad A_2/A_1 = 1.431$, $\quad -2.10$

9-5. $\lambda_1 = 0.360\sqrt{g/r}$, $\quad \lambda_2 = 1.135\sqrt{g/r}$, $\quad A_\phi/A_\theta = 2.48$, $\quad -2.15$

9-6. $\lambda_{1,2} = (\sqrt{2} \mp 1)\sqrt{k/ml^2}$, $\quad A_2/A_1 = 0.414$, $\quad -2.414$;

$y_1 = \dfrac{32l^2}{41k}F_2\cos 2.5\sqrt{\dfrac{k}{ml^2}}\,t$, $\quad y_2 = -\dfrac{84l^2}{41k}F_2\cos 2.5\sqrt{\dfrac{k}{ml^2}}\,t$

9-7. Symmetric mode is $\lambda_2 = \sqrt{k/m}$. Antisymmetric modes are

$\lambda_1 = 0.533\sqrt{k/m}$ and $\lambda_3 = 3.06\sqrt{k/m}$.

9-8. $\lambda_1 = \sqrt{\dfrac{6k}{11m}}$, $\quad \lambda_2 = \sqrt{3\left(\dfrac{k}{m} + \dfrac{g}{l}\right)}$

9-9. $\lambda_1 = \lambda_2 = \lambda_3 = 0$, $\quad \lambda_4 = \lambda_5 = \lambda_6 = \lambda_7 = \sqrt{2k/m}$, $\quad \lambda_8 = 2\sqrt{k/m}$

9-10. $\lambda_1 = \sqrt{k/m}$, $\quad \lambda_2 = \sqrt{3k/2m}$, $\quad \lambda_3 = \sqrt{2k/m}$

9-13. $\lambda_{1,2} = -1.241 \pm i11.18$, $\quad \lambda_{3,4} = -30.0 \pm i32.8$

$A_2/A_1 = 0.559 \pm i0.1443$, $\quad -3.27 \pm i1.727$

INDEX